Lecture Notes in Computer Science 8116

Commenced Publication in 1973
Founding and Former Series Editors:
Gerhard Goos, Juris Hartmanis, and Jan van Leeuwen

T0223451

Thora Tenbrink John Stell
Antony Galton Zena Wood (Eds.)

Spatial Information Theory

11th International Conference, COSIT 2013
Scarborough, UK, September 2-6, 2013
Proceedings

 Springer

Volume Editors

Thora Tenbrink
Bangor University
School of Linguistics and English Language
College Road, Bangor, LL57 2DG, UK
E-mail: t.tenbrink@bangor.ac.uk

John Stell
University of Leeds
School of Computing
Woodhouse Lane, Leeds, LS2 9JT, UK
E-mail: j.g.stell@leeds.ac.uk

Antony Galton
Zena Wood
University of Exeter
College of Engineering, Mathematics and Physical Sciences
Harrison Building, North Park Road, Exeter, EX4 4QF, UK
E-mail: apgalton@ex.ac.uk, z.m.wood@exeter.ac.uk

ISSN 0302-9743　　　　　　　　　　e-ISSN 1611-3349
ISBN 978-3-319-01789-1　　　　　　　e-ISBN 978-3-319-01790-7
DOI 10.1007/978-3-319-01790-7
Springer Cham Heidelberg New York Dordrecht London

Library of Congress Control Number: 2013945092

CR Subject Classification (1998): H.2.8, H.2, I.2, H.3, E.1

LNCS Sublibrary: SL 1 – Theoretical Computer Science and General Issues

Typesetting: Camera-ready by author, data conversion by Scientific Publishing Services, Chennai, India

Printed on acid-free paper

Springer is part of Springer Science+Business Media (www.springer.com)

Preface

This volume contains papers presented at the 11th meeting of COSIT, the Conference on Spatial Information Theory. The range of topics covered highlights the interdisciplinary nature of the conference, which has been one of its distinguishing features ever since the first meeting in 1993. COSIT interprets Spatial Information Theory in a broad sense and has helped to promote interactions between participants from diverse disciplines including geography, linguistics, mathematics, psychology, philosophy, computer science, and anthropology to mention only a few.

For COSIT 2013, we received 62 full-paper submissions and 28 of these were selected for presentation and inclusion in these proceedings. In addition to the selected submissions there are talks by three invited speakers: Karen Emmorey, Director of the Laboratory for Language and Cognitive Neuroscience at San Diego State University; Jens Riegelsberger, Google Geo User Research Team; Trevor Bailey, College of Engineering, Mathematics and Physical Sciences, University of Exeter. The full COSIT program includes a day of workshops and tutorials, and a one-day Doctoral Colloquium.

This year's COSIT takes place in Scarborough, on the east coast of England. The conference is based in the Royal Hotel and takes place from September 2–6. Scarborough is a characterful Victorian seaside resort whose connections with spatial information theory before the advent of COSIT include William Smith (1769–1839), who was for a time resident in the town. Smith pioneered geological map-making in the UK, and his research on correlations between geological strata in different locations based on the occurrence of particular types of fossils exemplifies the kind of interaction between space and time that frequently occurs in COSIT papers today. Scarborough's Rotunda Museum, dating from 1829, originally based the arrangement of its exhibits on the ideas of Smith, who also suggested the design of the building. Besides its historical connections, Scarborough is conveniently located for the tourist. Nearby locations include Whitby, the destination for the conference afternoon excursion, and famous, among other things, as the setting for part of Bram Stoker's novel *Dracula*.

The editors would like to thank all the reviewers for their work, and also all the authors who supported the event by submitting papers. We are also grateful to Brandon Bennett for leading the local organization, and to Max Dupenois for help with the website.

June 2013

Thora Tenbrink
John Stell
Antony Galton
Zena Wood

Organization

General Chairs

Brandon Bennett   University of Leeds, UK
Antony Galton   University of Exeter, UK

Program Chairs

John Stell   University of Leeds, UK
Thora Tenbrink   University of Bangor, UK

Sponsorship Chair

Zena Wood   University of Exeter, UK

Steering Committee

Christophe Claramunt	Naval Academy Research Institute, France
Anthony Cohn	University of Leeds, UK
Michel Denis	LIMSI-CNRS, Paris, France
Matt Duckham	University of Melbourne, Australia
Max Egenhofer	University of Maine, USA
Andrew Frank	Technical University Vienna, Austria
Christian Freksa	University of Bremen, Germany
Nicholas Giudice	University of Maine, USA
Stephen Hirtle	University of Pittsburgh, USA
Werner Kuhn	University of Münster, Germany
Benjamin Kuipers	University of Michigan, USA
David Mark	SUNY Buffalo, USA
Dan Montello	UCSB, USA
Reinhard Moratz	University of Maine, USA
Kathleen Stewart	University of Iowa, USA
Sabine Timpf	University of Augsburg, Germany
Barbara Tversky	Stanford University, USA
Stephan Winter	University of Melbourne, Australia
Michael Worboys	University of Greenwich, UK

Program Committee

Scott Bell	University of Saskatchewan, Canada
Sven Bertel	Bauhaus-Universität Weimar, Germany

Michela Bertolotto	University College Dublin, Ireland
Mehul Bhatt	University of Bremen, Germany
Thomas Bittner	SUNY Buffalo, USA
Gilberto Camara	INPE, Brazil
Christophe Claramunt	Naval Academy Research Institute, Brest, France
Eliseo Clementini	University of L'Aquila, Italy
Helen Couclelis	University of California Santa Barbara, USA
Kenny Coventry	University of East Anglia, UK
Leila De Floriani	University of Genoa, Italy
Maureen Donnelly	SUNY Buffalo, USA
Matt Duckham	University of Melbourne, Australia
Max Egenhofer	University of Maine, USA
Carola Eschenbach	University of Hamburg, Germany
Sara Irina Fabrikant	University of Zurich, Switzerland
Andrew Frank	Vienna University of Technology, Austria
Christian Freksa	University of Bremen, Germany
Mark Gahegan	University of Auckland, New Zealand
Nicholas Giudice	University of Maine, USA
Stephen Hirtle	University of Pittsburgh, USA
Christopher Jones	Cardiff University, UK
Alexander Klippel	Pennsylvania State University, USA
Christian Kray	University of Münster, Germany
Werner Kuhn	University of Münster, Germany
Lars Kulik	University of Melbourne, Australia
Patrick Laube	University of Zurich, Switzerland
Harvey Miller	University of Utah, USA
Daniel R. Montello	University of California, Santa Barbara, USA
Reinhard Moratz	University of Maine, USA
Bernard Moulin	Laval University, Canada
Nora Newcombe	Temple University, USA
James Pustejovsky	Brandeis University, USA
Martin Raubal	ETH Zurich, Switzerland
Jochen Renz	ANU, Australia
Kai-Florian Richter	University of Melbourne, Australia
Andrea Rodriguez	University of Concepcion, Chile
Christoph Schlieder	University of Bamberg, Germany
Angela Schwering	University of Münster, Germany
Kathleen Stewart	University of Iowa, USA
Kristin Stock	University of Nottingham, UK
Holly Taylor	Tufts University, USA
Sabine Timpf	University of Augsburg, Germany
David Uttal	Northwestern University, USA
Jan Oliver Wallgrün	Pennsylvania State University, USA
Nico Van de Weghe	Ghent University, Belgium
Robert Weibel	University of Zurich, Switzerland

Stephan Winter	University of Melbourne, Australia
Thomas Wolbers	University of Edinburgh, UK
Diedrich Wolter	University of Bremen, Germany
Michael Worboys	University of Greenwich, UK
May Yuan	University of Oklahoma, USA

Table of Contents

Spatial Change

Wayfinding and Assistance

Representing Spatial Data

Handling Language Data

Spatial Language and Computation

Spatial Ontology

Spatial Reasoning and Representation

Spatial Primitives from a Cognitive Perspective: Sensitivity to Changes in Various Geometric Properties

Toru Ishikawa

Graduate School of Interdisciplinary Information Studies & Center for Spatial Information
Science, University of Tokyo, 7-3-1 Hongo, Bunkyo-ku, Tokyo 113-0033, Japan
`ishikawa@csis.u-tokyo.ac.jp`

Abstract. This study addressed the issue of spatial concepts by examining the
perception of changes in shape, orientation, size, and cyclic order caused by the
transformations of deformation, rotation, scaling, and reflection. 49 participants
viewed 36 geometric configurations to which different types and degrees of
transformations were applied, and answered how much they thought the
configurations were different from each other. Participants perceived deformed
configurations as more dissimilar as the degree of deformation became larger.
Participants' perception of geometric properties, however, did not conform to
the mathematical classification of transformations. They discriminated between
deformed, rotated, scaled, and reflected configurations when the degree of
deformation was small; but the perceived difference became smaller as the
degree of deformation became larger. Furthermore, mental-rotation ability
affected the sensitivity to geometric properties, with low-spatial people
attending to changes in orientation caused by rotation and reflection.
Implications for spatial learning and education are discussed.

Keywords: Spatial thinking, Spatial concepts, Geometric transformations,
Spatial ability, Bidimensional regression.

1   Introduction

The issue of spatial thinking has recently attracted theoretical and pedagogical
interest, especially since the National Research Council (NRC) published the report
Learning to Think Spatially in 2006. Importantly, spatial thinking plays critical roles
in various disciplines and in everyday life (Newcombe, 2010; Uttal & Cohen, 2012).
For instance, research has shown that spatial ability is related to success in the areas
of science, technology, engineering, and mathematics (Casey, Nuttall, & Pezaris,
1997; Keehner et al., 2004; Kozhevnikov, Motes, & Hegarty, 2007).
   Geospatial science is one of the "spatial" disciplines that involve sophisticated
spatial thinking. Golledge (1992, 2002) argued that geographic knowledge builds
upon various spatial concepts and relations, and Goodchild (2006) pointed to the
importance of spatial thinking in geographic information science by calling it the
"fourth R." Ishikawa and Kastens (2005) discussed that geoscience involves various
spatial tasks and that some students have difficulty with such high-level spatial
thinking (i.e., there exist large individual differences in the ability or skill).

T. Tenbrink et al. (Eds.): COSIT 2013, LNCS 8116, pp. 1–13, 2013.

The NRC report discussed spatial thinking in terms of spatial concepts, representation, and reasoning. Of these elements, spatial concepts have been extensively discussed in the literature of geospatial learning and education. For example, Golledge, Marsh, and Battersby (2008) classified spatial concepts into primitives (identity, location, magnitude, and space-time) and derivatives at higher levels (e.g., arrangement, distribution, distance, adjacency, connectivity, scale, and projection). Similar classifications of spatial concepts were proposed by Gersmehl and Gersmehl (2007), Janelle and Goodchild (2009), and Kuhn (2012).

Although these conceptually derived classifications provide significant insights, it remains to be examined whether these classifications are cognitively viable. In other words, it is of interest to see if people perceive the components of spatial concepts proposed in the literature as being different from each other and whether people perceive some components as more salient than others. It is thus theoretically and pedagogically significant to reveal the components of spatial thinking and the structure and relationships of spatial concepts as conceived by humans, beyond simply equating spatial thinking to spatial ability (e.g., Hegarty, 2010; Ishikawa, 2012). The major objective of the present study is to address these issues empirically.

Although there are some studies that examined spatial concepts from a cognitive perspective (e.g., Golledge, 1992; Golledge et al., 2008; Lee & Bednarz, 2012), more empirical cognitive-based research is needed on the perception and classification of different spatial primitives for a better understanding of human spatial thinking. Traditionally, cognitive studies of an understanding of geometric properties have been conducted in relation to Piaget's theory of intelligence. Piaget and Inhelder (1948/1967) discussed the ontological development of children's conception of space in the sequence of topological, projective, and Euclidean geometries. Notably, this argument was applied to the discussion of spatial microgenesis in terms of landmark, route, and survey knowledge (Siegel & White, 1975). The Piagetian transition of the three types of spaces or geometries has been cognitively scrutinized, particularly in the context of K-12 learning and education (e.g., Kidder, 1976; Mandler, 1983, 2012; Martin, 1976).

Mathematically, geometric properties are defined in terms of properties that are preserved or left invariant through a group of transformations (Gans, 1969). For example, topological transformations preserve openness, interior, order, and connectedness. In addition to these properties, projective transformations preserve collinearity and cross-ratios; similarity transformations (radial transformations or scaling) preserve angle-size; and Euclidean transformations (displacements or rigid motions, i.e., translations, rotations, and reflections) preserve length.

With these issues in mind, this study addresses the question of spatial concepts by examining people's perception of differences in spatial properties caused by various geometric transformations. Specifically, an experiment was conducted in which participants viewed various geometric configurations to which different degrees and types of transformations were applied, and indicated how much they thought the configurations were different from each other. Of the spatial concepts discussed in the geospatial literature, identity, arrangement, distance, direction, shape, orientation, and scale were picked up and assessed in terms of changes in shape (lengths and angles), orientation (rotation and reflection), size (scaling), and cyclic order (reflection).

In this article, the degree to which people perceive pairs of geometric configurations as similar (or discriminate between them) is referred to as "sensitivity" to associated geometric properties, and its characteristics and relationships with spatial aptitudes (spatial ability and sense of direction) are examined. By doing that, this study aims to look into the cognitive classification of spatial properties and compare it to the mathematical classification of geometric transformations.

2   Method

2.1   Participants

A total of 49 students (39 men and 10 women) participated in the experiment. They were undergraduate and graduate students in engineering, computer science, urban planning, or environmental studies. Their ages ranged from 20 to 35, with a mean of 23.8 years.

2.2   Materials

Experimental Stimuli of Geometric Configurations. As experimental stimuli, 36 geometric "configurations" (or arrangements) of three dots were created. The identity of the three dots was varied and represented by three colors (black, white, and red). In the present study, the dots were shown without being linked by lines, so as not to induce participants to regard the task simply as the classification of triangles.

The configurations varied with respect to the degree of deformation and the types of transformations applied. First, a configuration with three dots arranged at vertices of an equilateral triangle was created as an original configuration (Figure 1, leftmost panel #1). Then, the original configuration was deformed (or the shape was changed) into eight configurations, with the lengths and angles among the three dots being changed with the constraint that the scale factor was fixed at 1 (Figure 1, panels #2-9). The degree of deformation was varied on the basis of bidimensional regression coefficients computed between the coordinates for randomly generated three dots and those for the three dots in the original configuration. Since bidimensional regression attempts to maximize the correspondence between two configurations to be compared through translation, rotation, scaling, values for bidimensional regression coefficients do not become too small, for example down to 0, even when coordinates are randomly generated. In the present case of three dots, the values ranged from .73 to 1 (shown in Figure 1). In this article, the transformation that created this set of nine configurations is called *deformation*.

By applying three different types of transformations to these nine configurations, three more sets of nine configurations were created. A second set was created through an application of a *rotation*. Specifically, the nine configurations created above by deformation were rotated 90°. Half the configurations, which were chosen randomly, were rotated 90° to the right, and the other half were rotated 90° to the left.

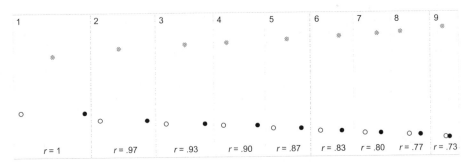

Fig. 1. Nine configurations (#1-9) that were deformed to different degrees. The leftmost panel shows the original configuration. Values for *r* indicate the bidimensional regression coefficients between the original (#1) and deformed (#2-9) configurations. In the experiment, the stimulus configurations were presented to participants paired with the original configuration. (Red dots are shown in grey in Figs. 1 and 2 for black-and-white printing; they were shown in red in the experiment.)

A third set of configurations was created through an application of a radial transformation or *scaling*. The nine configurations created by deformation were scaled by a factor of 2 or 0.5. Half the configurations, which were chosen randomly, were enlarged, and the other half were reduced.

A fourth set of configurations was created through an application of a *reflection*, with the nine configurations created by deformation being reflected or flipped over. A reflection can be conceived as changing the cyclic order of the three dots (i.e., ordered clockwise vs. counterclockwise), and thus the transformation was examined in this study to see if participants perceive the changes caused by it as qualitatively different (i.e., nonmetric rather than metric) from the other transformations.

As examples, the three transformed configurations for the original configuration are shown in Figure 2.

| Original (#1) | Rotated | Scaled | Reflected |

Fig. 2. Three transformations applied to the original configuration (#1, leftmost panel): rotation, scaling, and reflection (the second, third, and fourth panel from the left, respectively)

Card Rotations Test. Participants took the Card Rotations Test, which is a major spatial test assessing people's ability to rotate imagined pictures mentally (Ekstrom et al., 1976). In the test, participants viewed 20 items, each consisting of one card in a standard orientation and eight alternative cards, and answered whether the alternative cards were the same as the standard (i.e., rotated into different orientations) or different from the standard (i.e., flipped over). They received one point for each correctly identified card and lost one point for each wrongly identified card.

Participants were allowed 6 min to complete this test. Mental-rotation ability has been found to correlate with various spatial tasks, such as the understanding and use of maps in the field (e.g., Liben & Downs, 1993), and so it was assessed in this study as a possible correlate with the perception of differences in geometric properties.

Sense-of-Direction Scale. Participants filled out the Santa Barbara Sense-of-Direction (SBSOD) scale, which consists of fifteen 7-point Likert-type questions about navigational abilities or preferences (Hegarty et al., 2002). It is scored so that a higher score indicates a better SOD, ranging in value from 1 to 7. People having higher SOD scores tend to do better on spatial tasks that require survey or configurational understanding in large-scale environments. Participants' scores on this scale were assessed in this study to see whether a skill related to large-scale spatial learning is also related to the perception of differences in geometric properties.

2.3   Design and Procedure

Participants viewed the 36 configurations in random order. The configurations were always shown paired with the original configuration (Figure 1, panel #1). Half the times the original configuration was shown to the left, and the other half to the right. After viewing each pair of configurations, participants answered whether they thought the pair was "spatially the same" or "different" on a 7-point scale (1 = *spatially the same*; 7 = *spatially different*). Namely, the responses indicated a perceived degree of dissimilarity of the transformed configuration to the original configuration. Participants were instructed to interpret what "spatially the same" means as they would like. (A similar method was used by Levinson [1996] to study the use of spatial frames of reference. In that study, participants viewed two stimuli that preserved either an egocentric or allocentric orientation of a previously seen stimulus, and answered which they thought was the "same" as the previously seen stimulus.)

After completing all 36 configurations, participants filled out the SBSOD scale and took the Card Rotations Test (due to time constraints, 20 of the 49 participants took the Card Rotations Test). They finished all these tasks within 20 min on average.

2.4   Hypotheses and Possible Results

Concerning the effects of the degree of deformation, participants would perceive the deformed configurations as more dissimilar to the original configuration as the degree of deformation becomes larger (i.e., the regression line of perceived dissimilarity on bidimensional regression coefficients would have a negative slope). This hypothesis is in line with the findings in the psychophysics literature that perceived stimulus values are given as an increasing function of physical stimulus values (Gescheider, 1997).

Concerning the effects of different types of transformations, some variations could be possible. One possibility is that participants would respond in line with the mathematical classification of geometric transformations. If participants' perception shares characteristics with Euclidean transformations, their responses would not change with a rotation or reflection, because these transformations preserve Euclidean properties. Thus regression lines for their responses for rotated and reflected configurations would coincide that for deformed configurations. Their responses to scaled configurations, however, would differ, because scaling does not preserve the Euclidean property of length. If participants' perception shares characteristics with similarity transformations, regression lines for rotated, reflected, and scaled configurations would coincide that for deformed configurations, as rotation, reflection, and scaling preserve angle-size. By contrast, if participants "live" in the world of topology, they would perceive all configurations as the same, and the regression lines would have a slope of 0; except that they would respond to reflected configurations differently as long as they regard a reflection as breaking cyclic order.

Another possibility is that participants' responses do not conform to the mathematical classification of transformations. Then, rotation, scaling, and reflection would change participants' perception, and so the regression lines for these three transformations would deviate from that for deformation. And in that case, there are two further possibilities. If the effects of rotation, scaling, and reflection are independent of the degree of deformation, the four regression lines would be parallel. Or, if the effects differ depending on the degree of deformation, the slopes for the four regression lines would be different. For the second possibility, the manner in which the slopes for the four regression lines differ is of interest. That is, depending on whether the sensitivity to rotation, scaling, or reflection becomes larger or smaller for more deformed configurations, the slopes for the three transformations would be steeper or less steep negatively than that for deformation.

Therefore the present study examines the size and the equivalence of slopes for regression lines for the four sets of configurations.

3   Results

3.1   Effects of the Types of Transformations and the Degree of Deformation

Participants' responses were examined through a repeated measures analysis of variance (ANOVA), with the degree of deformation (the nine panels in Figure 1) and the types of transformations (deformation, rotation, scaling, and reflection) as within-subject variables. Following the general recommendation (Girden, 1992), univariate and multivariate tests were conducted at the .025 level each (when both tests are significant, statistics for the univariate test are reported).

There were significant main effects of degree of deformation and type of transformation, $F(8, 41) = 48.19$, $p < .001$; $F(3, 46) = 48.62$, $p < .001$, respectively; and a significant interaction between the two variables, $F(24, 25) = 6.49$, $p < .001$. The existence of a significant interaction shows that participants were sensitive to the four types of transformations, but their sensitivity to rotation, scaling, and reflection differed depending on the degree of deformation.

3.2   Regression for Deformation, Rotation, Scaling, and Reflection

The effects of the degree of deformation and the types of transformations were further examined through regression analysis, with participants' responses being regressed on

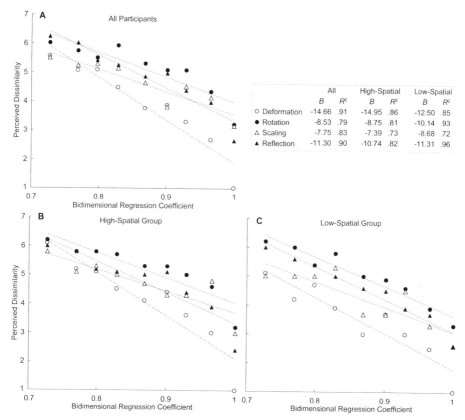

Fig. 3. Relationships between perceived dissimilarity and the degree of deformation for all participants (A), the high-spatial group (B), and the low-spatial group (C). Lines depict linear regression lines. B = unstandardized regression coefficient; R^2 = coefficient of determination.

the degree of deformation for each type of transformation separately (Figure 3A). (When Stevens' power law was applied to the data, the fit of a regression model was equivalent to when the linear regression model was applied, with an increase in determination coefficient values .08 at most.)

The slopes for the regression lines for the four types of transformations are significantly different from each other, except that the slopes for rotation and scaling are not significantly different (Bonferroni, $\alpha = .05/6$). All regression lines have a negative slope, showing that participants perceived the configurations as more dissimilar to the original configuration as the degree of deformation increased.

But because the slopes are different, distance between the regression lines along the vertical axis is different at different values for the bidimensional regression coefficients. When an additional transformation (rotation, scaling, or reflection) was applied to the deformed configurations, their effects on perceived dissimilarity differed depending both on the degree of deformation and on the types of transformations. When the degree of deformation was small, participants perceived rotated, scaled, or reflected configurations as dissimilar to the original, or they were sensitive to the changes in rotation, scaling, and reflection in addition to the changes in shape. As the degree of deformation became large, they became insensitive to rotation, scaling, and reflection. That is, the vertical distance between the four regression lines becomes smaller and eventually almost negligible.

The slope for reflection is steeper than that for rotation or scaling. Namely, the rate of participants' becoming insensitive was slower for reflection than for rotation and scaling (the latter two of which were comparable in the rate), and participants' sensitivity to reflection kept comparatively invariant across different degrees of deformation. These results indicate that participants were most sensitive to the changes in shape, and then to the changes caused by reflection, and least sensitive to the changes caused by rotation and scaling.

For the transformations of scaling and reflection, to confirm that participants took these properties into consideration, their responses were regressed on bidimensional regression coefficients that were computed without scaling or reflection, respectively. That is, in overlaying the original and transformed configurations, only translation and rotation were allowed. Then the regression for scaling became nonsignificant, and the regression for reflection yielded a positive slope indicating a counterintuitive tendency that participants perceive the configurations as more similar as the degree of deformation became larger. These results show that participants took scaling and reflection into consideration in comparing pairs of configurations.

3.3   Effects of Spatial Aptitudes

Mental-Rotation Ability. Effects of spatial ability on participants' responses were examined through an analysis of covariance (ANCOVA), with their scores on the Card Rotations Test being entered as a covariate into the repeated measures ANOVA conducted in section 3.1. A significant effect of mental-rotation ability was observed in terms of a three-way interaction of type of transformation, degree of deformation, and mental rotation, $F(24, 432) = 1.84, p < .01$. It indicates that the interaction of type of transformation and degree of deformation differs depending on the level of mental-ration ability. In light of this result, the difference in the relationships between the degree of deformation and the types of transformations is examined in the next section, by conducting regression analysis for high- and low-spatial groups of participants separately.

Sense of Direction. Similarly, effects of sense of direction were examined through an ANCOVA with participants' scores on the SBSOD scale being entered as a covariate. There were no significant effects observed for the SBSOD scores.

3.4   Regression for the High- and Low-Spatial Groups

Since mental-rotation ability was found to interact with the degree of deformation and the types of transformations, participants' responses were further examined through regression analysis for participants with a high and low mental-rotation ability. To do that, participants were classified into two groups by a median split of their scores on the Card Rotations Test (Mdn = 142.5). A mean score was 151.8 (SD = 6.8) for the high-spatial group and 114.7 (SD = 40.6) for the low-spatial group. The two mean scores were significantly different, $t(18)$ = 2.85, $p < .05$.

For both groups, the slopes for regression lines for the four types of transformations are significantly different from each other (Figures 3B and 3C).

Next, for each transformation, the slopes for regression lines for the high- and low-spatial groups were compared. For deformation, the slope for the high-spatial group is steeper than that for the low-spatial group. By contrast, for rotation and scaling, the slopes for the high-spatial group are less steep than those for the low-spatial group. These results show that the high-spatial group of participants perceived deformed configurations as more dissimilar to each other (or discriminated between them more sensitively), and became insensitive to the difference between rotated, scaled, and deformed configurations at a faster rate, than did the low-spatial group of participants. For the low-spatial group, when the configurations were deformed greatly, they still perceived rotated and reflected configurations as dissimilar to deformed configurations (Figure 3C); that is, the low-spatial group's sensitivity to rotation and reflection was greater than that of the high-spatial group.

Also, Figures 3B and 3C show that the regression lines are closer to being parallel for the low-spatial group than for the high-spatial group. It suggests that the low-spatial group perceived rotated, scaled, and reflected configurations as more dissimilar to the original configuration deformed to various degrees. Stated differently, the high-spatial group's perception was more in line with the mathematical classification of transformations and geometric properties.

3.5   Multidimensional Scaling of Responses by High- and Low-Spatial People

To examine the effects of mental-rotation ability in more detail, the high- and low-spatial groups' responses were analyzed through multidimensional scaling (MDS). Figure 4 shows a two-dimensional solution for each group obtained from ordinal MDS with the PROXSCAL method. A stress value was .09 for the high-spatial group and .10 for the low-spatial group. (Kruskal [1964] discussed that a stress value below .10 shows a fair fit.) The classification was further examined through cluster analysis of the two-dimensional coordinates, with three clusters being identified for each group (see Clusters I-III and the dendrogram shown in Figure 4).

For the high-spatial group, Cluster I consists of the original configuration (panel #1 in Figure 1) and its rotated, scaled, and reflected images, and they are positioned far away from each other. Thus the transformations of rotation, scaling, and reflection are discriminated from each other when applied to the original (or non-deformed) configuration. Cluster II consists mainly of scaled configurations and Cluster III consists of rotated and reflected configurations. These two clusters are clearly separated from the first cluster, and are tightly clustered. In comparison, for the low-spatial group although three clusters were similarly identified, the grouping of and the distinction between the three clusters were less clear. These results show that the high-spatial group discriminated between the four types of transformations and placed those configurations farther apart mentally than the low-spatial group.

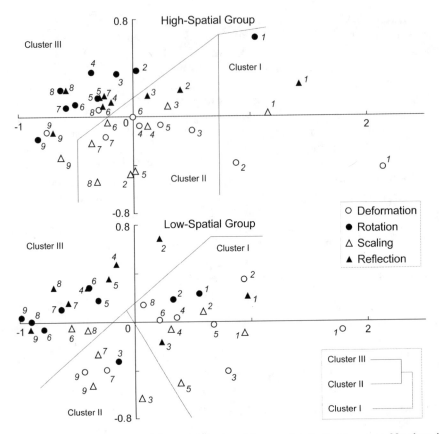

Fig. 4. MDS solutions for the high-spatial (top) and low-spatial (bottom) groups. Numbers in italics correspond to the panel numbers in Figure 1 and indicate the degree of deformation.

4   Discussion

This study addressed the issue of spatial concepts through an empirical examination of the perception of changes in shape (deformation), orientation (rotation or reflection), size (scaling), and cyclic order (reflection).

Results show that people are sensitive to the changes in shape (lengths and angles), and perceive deformed configurations as more dissimilar to the original configuration as the degree of deformation becomes larger, or as the correspondence between the transformed and original configurations in terms of bidimensional regression coefficients becomes smaller. Using this relationship between perceived dissimilarity and the degree of deformation as a baseline, people's sensitivity to, or discrimination between, deformed, rotated, scaled, and reflected configurations was examined. Regression lines for the four types of transformations do not coincide, nor are they parallel, contrary to the inference based on the mathematical classification of transformations and associated invariant geometric properties. Thus cognitive and mathematical classifications of spatial properties are different.

Importantly, participants' sensitivity to rotation, scaling, and reflection differs depending on the degree of deformation, as well as differing from each other. As the degree of deformation becomes larger, participants become less sensitive to the changes in rotation, scaling, and reflection, and thus do not differentiate rotated, scaled, or reflected configurations from the configurations that receive deformation only. People become insensitive to rotation and scaling faster than to reflection; people's sensitivity to reflection is relatively invariant across different degrees of deformation. This observation may stem from the fact that a reflection changes the cyclic order of the three dots. Namely, although a reflection is a rigid motion and preserves Euclidean properties, it can also be interpreted as breaking topology. It may explain why the slope for reflection is steeper than that for rotation or scaling (i.e., the sensitivity to reflection stays relatively invariant). These results indicate that people's sensitivity to the transformations is largest for deformation, and then for reflection, and smallest for rotation and scaling.

Concerning spatial aptitudes, mental-rotation ability, but not sense of direction, has a significant effect on people's perception of spatial properties. This is another indication of the difference between small- and large-scale spatial abilities (Hegarty et al., 2006). The results from regression and MDS analyses show that people with a high mental-rotation ability perceive deformed configurations as more dissimilar to each other (or discriminate between them and place them mentally farther apart from each other) than people with a low mental-rotation ability. When the configurations are deformed to a great degree, people with a low mental-rotation ability perceive the rotated and reflected configurations as more dissimilar to the deformed configurations; that is, their sensitivity to rotation and reflection is greater than that of high-spatial people. It implies that low-spatial people attend to, or are distracted by, mathematically superficial changes in orientation caused by rigid motions (i.e., a rotation and a reflection). This tendency is not observed for high-spatial people, whose responses are more aligned with the mathematical classification of transformations.

This study revealed the characteristics of the perception of spatial concepts and demonstrated the differences between cognitive and mathematical ways of viewing geometric properties. Moreover, the results pointed to the relationship between the conception of spatial properties and the level of spatial ability. These results provide theoretical insights into the discussion of spatial concepts. In particular, the spatial concepts discussed in the geospatial literature thus far—identity, arrangement, distance, direction, shape, orientation, and scale, which were examined in this study in terms of changes in shape, orientation, size, and cyclic order—differ in the degree of cognitive salience, with the changes in shape being most salient. Also, the sensitivity to changes in shape interacts with that to the other changes, caused by rotation, scaling, and reflection. Although the validity and generality of the method used in this study (asking participants to indicate the extent to which they think pairs of configurations are "spatially the same") and the effects of contextual factors including individual or group differences (e.g., Klippel, Weaver, & Robinson, 2011) need to be inspected in further studies, these results could allow researchers to discuss the components of spatial thinking more specifically than equating it to spatial ability (e.g., Hegarty, 2010; Ishikawa, 2012).

The present results also provide pedagogical implications. For example, to help students with a low spatial ability to acquire a better understanding of geometric transformations and properties, instructors may keep in mind that such low-spatial students tend to be distracted by non-essential changes in orientation when studying the concepts of Euclidean transformations. Furthermore, an understanding of these transformations may be related and applied to geospatial tasks; for example, deformation and scaling relate to an understanding of map projections and scale, and rotation is required for spatial orientation with a map. Although one cannot specify a causal direction based on the present data only, the results could offer guidance on possible instructional methods for mathematics and geoscience education, and lead to bringing research findings into the classroom (Manduca, Mogk, & Stillings, 2003).

References

Casey, M.B., Nuttall, R.L., Pezaris, E.: Mediators of gender differences in mathematics college entrance test scores: A comparison of spatial skills with internalized beliefs and anxieties. Developmental Psychology 33, 669–680 (1997)

Ekstrom, R.B., French, J.W., Harman, H.H., Dermen, D.: Kit of Factor-Referenced Cognitive Tests. Educational Testing Service, Princeton (1976)

Gans, D.: Transformations and Geometries. Appleton-Century-Crofts, New York (1969)

Gersmehl, P.J., Gersmehl, C.A.: Spatial thinking by young children: Neurologic evidence for early development and "educability". Journal of Geography 106, 181–191 (2007)

Gescheider, G.A.: Psychophysics: The Fundamentals, 3rd edn. Erlbaum, Mahwah (1997)

Girden, E.R.: ANOVA: Repeated Measures (Sage University Paper Series on Quantitative Applications in the Social Sciences, Series No. 07–084). Sage, Newbury Park (1992)

Golledge, R.G.: Do people understand spatial concepts: The case of first-order primitives. In: Frank, A.U., Formentini, U., Campari, I. (eds.) GIS 1992. LNCS, vol. 639, pp. 1–21. Springer, Heidelberg (1992)

Golledge, R.G.: The nature of geographic knowledge. Annals of the Association of American Geographers 92, 1–14 (2002)

Golledge, R.G., Marsh, M., Battersby, S.: A conceptual framework for facilitating geospatial thinking. Annals of the Association of American Geographers 98, 285–308 (2008)

Goodchild, M.F.: The fourth R? Rethinking GIS education. Arc News 28(3), 11 (2006)

Hegarty, M.: Components of spatial intelligence. Psychology of Learning and Motivation 52, 265–297 (2010)

Hegarty, M., Montello, D.R., Richardson, A.E., Ishikawa, T., Lovelace, K.: Spatial abilities at different scales: Individual differences in aptitude-test performance and spatial-layout learning. Intelligence 34, 151–176 (2006)

Hegarty, M., Richardson, A.E., Montello, D.R., Lovelace, K., Subbiah, I.: Development of a self-report measure of environmental spatial ability. Intelligence 30, 425–447 (2002)

Ishikawa, T.: Geospatial thinking and spatial ability: An empirical examination of knowledge and reasoning in geographical science. The Professional Geographer (2012) (in press; first published online), doi:10.1080/00330124.2012.724350

Ishikawa, T., Kastens, K.A.: Why some students have trouble with maps and other spatial representations. Journal of Geoscience Education 53, 184–197 (2005)

Janelle, D.G., Goodchild, M.F.: Location across disciplines: Reflection on the CSISS experience. In: Scholten, H.J., Velde, R., Manen, N. (eds.) Geospatial Technology and the Role of Location in Science, pp. 15–29. Springer, Dordrecht (2009)

Keehner, M.M., Tendick, F., Meng, M.V., Anwar, H.P., Hegarty, M., Stoller, M.L., Duh, Q.: Spatial ability, experience, and skill in laparoscopic surgery. American Journal of Surgery 188, 71–75 (2004)

Kidder, F.R.: Elementary and middle school children's comprehension of Euclidean transformations. Journal of Research in Mathematics Education 7, 40–52 (1976)

Klippel, A., Weaver, C., Robinson, A.: Analyzing cognitive conceptualizations using interactive visual environments. Cartography and Geographic Information Science 38, 52–68 (2011)

Kozhevnikov, M., Motes, M., Hegarty, M.: Spatial visualization in physics problem solving. Cognitive Science 31, 549–579 (2007)

Kruskal, J.B.: Multidimensional scaling by optimizing goodness of fit to a nonmetric hypothesis. Psychometrika 29, 1–27 (1964)

Kuhn, W.: Core concepts of spatial information for transdisciplinary research. International Journal of Geographical Information Science 26, 2267–2276 (2012)

Lee, J., Bednarz, R.: Components of spatial thinking: Evidence from a spatial thinking ability test. Journal of Geography 111, 15–26 (2012)

Levinson, S.C.: Frames of reference and Molyneux's question: Cross-linguistic evidence. In: Bloom, P., Peterson, M., Nadel, L., Garrett, M. (eds.) Language and Space, pp. 109–169. MIT Press, Cambridge (1996)

Liben, L.S., Downs, R.M.: Understanding person-space-map relations: Cartographic and developmental perspectives. Developmental Psychology 29, 739–752 (1993)

Mandler, J.M.: Representation. In: Mussen, P.H. (ed.) Handbook of Child Psychology, 4th edn., pp. 420–494. Wiley, New York (1983)

Mandler, J.M.: On the spatial foundations of the conceptual system and its enrichment. Cognitive Science 36, 421–451 (2012)

Manduca, C., Mogk, D., Stillings, N.: Bringing Research on Learning to the Geosciences. Science Education Resource Center, Carleton College (2003)

Martin, J.L.: A test with selected topological properties of Piaget's hypothesis concerning the spatial representation of the young child. Journal of Research in Mathematics Education 7, 26–38 (1976)

National Research Council: Learning to Think Spatially. National Academies Press, Washington, DC (2006)

Newcombe, N.S.: Increasing math and science learning by improving spatial thinking. American Educator 34(2), 29–43 (2010)

Piaget, J., Inhelder, B.: The Child's Conception of Space (trans. Langdon, F.J., Lunzer, J.L.). Norton, New York (1967; original work published 1948)

Siegel, A.W., White, S.H.: The development of spatial representations of large-scale environments. Advances in Child Development and Behavior 10, 9–55 (1975)

Uttal, D.H., Cohen, C.A.: Spatial thinking and STEM education: When, why, and how? Psychology of Learning and Motivation 57, 147–181 (2012)

Transitional Spaces: Between Indoor and Outdoor Spaces

Christian Kray, Holger Fritze, Thore Fechner, Angela Schwering, Rui Li,
and Vanessa Joy Anacta

University of Münster, Institute for Geoinformatics
Weseler Straße 253, Germany
{c.kray,h.fritze,t.fechner,schwering,rui.li,v.anacta}@uni-muenster.de
http://ifgi.uni-muenster.de

Abstract. Traditionally, spaces have been classified as being located either indoors or outdoors. On closer inspection, however, this distinction is not as clear cut as usually assumed. For example, when navigating complex urban landscapes, pedestrians frequently traverse tunnels, enclosed footbridges or partially roofed courtyards. In this paper, we investigate this type of spaces between indoor and outdoor areas. We present an initial definition of transitional spaces based on a conceptual analysis, and then report on results from an empirical study with 103 pedestrians, whom we interviewed in an urban area. A statistical and linguistic analysis of the outcomes of the study provides evidence for the existence of transitional spaces and their use. The outcomes also support an initial set of characteristics and properties that further clarify these areas. The results pave the way for the further investigation of transitional spaces, e.g. in terms of providing effective navigation support through them.

Keywords: transitional spaces, navigation, urban areas.

1 Motivation

Urban spaces can be quite complex to navigate, in particular for people who are not familiar with certain areas of a city. The structure and fabric of cities and their use has therefore been an active area of research for many years [20]. While this has led to a better understanding of how people navigate in general, how we can support this and what elements might be relevant, the problem can still not be considered to be solved. Even when equipped with modern context-aware guide applications people still frequently struggle with the task of navigating urban areas. One reason for this is the complex and dynamic nature of urban landscapes, which include indoor areas and outdoor areas, can be very crowded, and are also subject to access restrictions and temporary changes.

Previous work has investigated how people understand and navigate indoor areas or outdoor areas [5,11], and into how they negotiate routes that cross indoor and outdoor spaces [3]. However, the simple distinction between indoor and outdoor spaces may not be sufficient to describe all areas found in complex urban environments. Arcades, transportation hubs, foot tunnels or partially roofed

T. Tenbrink et al. (Eds.): COSIT 2013, LNCS 8116, pp. 14–32, 2013.

open spaces are examples for areas, which illustrate that this dichotomy might not quite suffice to adequately capture all phenomena found in urban areas. Such *transitional spaces* share properties with both indoor and outdoor spaces but so far have not been analyzed more thoroughly. Since indoor and outdoor spaces differ significantly in a number of ways – for example, in terms of how people navigate these spaces and how they describe routes through them – transitional spaces may constitute a source for confusion when navigating urban areas. In this paper we therefore take a closer look at the different types of areas people traverse when moving about in cities, and we focus in particular on those spaces that may not be easily classified as either being indoors or outdoors. Our goal was to characterize transitional spaces, to gather initial evidence whether for human users of urban spaces such areas differ considerably from indoor and outdoor areas, and to gain an initial understanding of how people describe routes traversing transitional spaces.

The following section reviews related work in this area, which informs a conceptual analysis of what characterizes transitional spaces. In section 4, we present an overview of the methodology we used to further investigate transitional spaces. The three types of analysis we carried out are then reported on in the following sections. Section 8 reviews the results we obtained, discusses limitations of our study and points out implications of the outcomes. The paper closes by summarizing our main contributions and highlighting areas for further research.

2   Related Work

A large number of people move about in urban environments every day for a variety of reasons, e.g. when going to work, carrying out touristic activities, or when pursuing leisure activities [9]. These seemingly mundane activities can actually be quite challenging, i. e. when faced with unknown areas, and may require people to negotiate a highly complex environment. Frequently, pedestrians navigating city spaces traverse areas located outside buildings (e. g. pavements, parks and so on) and spaces, which are considered parts of building interiors (e. g. shops or public buildings). In addition, they also pass through multiple levels and zones, which are difficult to classify as being either indoors or outdoors, such as arcades, malls or roofed footbridges.

Indoor and outdoor spaces are commonly considered to considerably differ in several ways. Firstly, indoor environments such as buildings tend to have a smaller size than outdoor environments. Conceptually, size has been an important differentiator between different types of spaces. For example, Downs and Stea categorized space into *small-scale perceptual space* or *large-scale geographic space* [7]; a similar distinction has been proposed by Montello [22]. A second key difference distinguishing indoor and outdoor environments is dimensionality. Frequently, indoor environments (buildings) enable people to move between different floors or levels whereas this is usually not the case for outdoor spaces. Indoor environments thus are usually classified as being three-dimensional (or at least 2.5-dimensional) spaces unlike commonly two-dimensional outdoor environments. This additional vertical dimension can considerably contribute towards greater cognitive loads when negotiating indoor spaces [12]. Staircases, for example, have the potential to not only negatively affect people's sense of orientation but can also cause problems for systems meant to support navigation [19].

A third relevant difference relates to the use of landmarks, e.g. in order to facilitate wayfinding. Typical landmarks inside interior spaces differ from typical landmarks in outdoor environments. Due to the limited field of view in indoor environments, some useful landmarks available in outdoor environments such as the sun, mountains or water bodies, cannot be seen while inside buildings. Consequently, it is predominantly local landmarks that can be used for indoor wayfinding, while wayfinding in outdoor environments has the flexibility to rely on both global (e.g. skylines or sun) and local landmarks conveniently [25]. A fourth difference stems from the built nature and structure of indoor environments. Walls, doors and other obstacles making up the fabric of indoor spaces restrict movement whereas outdoor environments frequently enable people to move freely in all directions. A beneficial side effect of the constructed nature of indoor environments is that wayfinding is not affected by weather conditions, which can have a considerable impact on wayfinding in outdoor environments. Finally, the line of sight inside buildings is usually much more constrained than in outdoor spaces.

In summary, indoor and outdoor spaces are usually considered to be quite different in terms of their structure and properties but also in terms of how people perceive and navigate them. Signage, for example, may play a more prominent role in indoor environments than it does in outdoor spaces [8], whereas directions based on cardinal directions clearly are more useful in outdoor environments. Consequently, these aspects should be taken into account as well when generating directions to support people navigating such areas [16,23]. Urban areas are, however, complex structures [20] comprising not only indoor and outdoor areas but also some locations, for which it may be difficult to decide whether they are indoors or outdoors (e.g. shopping malls, arcades and underpasses). Such spaces might share properties of indoor and outdoor areas [24], and include multiple traversable levels [11], each of which may be classified differently in terms of whether it is indoors or outdoors.

Frequently, urban navigation entails traversing both indoor and outdoor environments, and transitioning between indoor and outdoor areas holds considerable potential for confusion and loss of orientation [8,13]. Early approaches to support wayfinding often focused on outdoor navigation [5] but more recent work in this area frequently investigates navigation assistance for people moving inside buildings [23,2,19]. In comparison, only very few approaches have been proposed that integrate navigation of indoor and outdoor areas [3,15]. Since people perceive indoor and outdoor areas differently, and since their navigation strategies may vary as well [8], planning routes that span both types of spaces can still be considered a significant challenge [4].

On a technical level, this issue also poses challenges in terms of integrating different representations of indoor and outdoor areas. While there are standards and approaches to model urban areas [10] and indoor environments [14,21] as well as ways to infer 3D information from existing representations [6,18], connecting these models, e. g. to enable seamless route planning, is still difficult [4]. A particular challenge relates to those areas that are neither indoors nor outdoors. These *transitional zones* are not well understood, neither in terms of how to represent them, nor in terms of how people navigate through them or how they describe routes traversing them. A related issue is the actual transition between indoor and outdoor spaces, which poses several challenges: it is not always clear when one type of space has been left and another one has been entered, e. g. in case of partially covered areas such as arcades or underpasses [24]. Given the different nature and navigation behavior in both types of spaces, this uncertainty can also contribute to orientation- and navigation-related problems pedestrians face in urban areas. These observations are further reasons to consider the introduction of transitional spaces on the conceptual and technical level.

3 Transitional Spaces

Based on the discussion in the previous section and on some of the criteria mentioned therein, it makes sense to more systematically contrast indoor and outdoor spaces. Such a comparison will also provide a starting point for the characterization of transitional zones as they might share certain properties of both indoor and outdoor areas, or exhibit distinct characteristics different from indoor/outdoor spaces. The previous section already briefly discussed some examples of transitional zones but in order to arrive at a working definition of such spaces and to support the identification of comparison criteria, a more detailed analysis will be useful.

When looking for such examples, several potential candidates quickly emerge. Shopping malls often combine indoor and outdoor areas as well as areas, which are roofed but are not fully enclosed. In addition, they can serve as a frequently used passage or shortcut from one part of the city to another. As shopping centers can have a considerable spatial extent – e. g. a city block – it may be faster and more convenient to traverse them instead of circumventing them, in particular in adverse weather conditions. Foot tunnels and underpasses constitute further examples for such spaces. They share characteristics with outdoor environments such as footpaths or streets while at the same time being completely enclosed and being man-built. They have a linear structure with horizontal extent, but usually little or no vertical extent, which characterizes many indoor environments. They offer partial protection from weather conditions (i. e. precipitation) but not from others (i. e. wind, temperature). Similar properties are shared by roofed footbridges, which are found in many large cities, where motorways have to be crossed by pedestrians.

A further example are arcades, which are common in many cities – in particular where pedestrians require protection from weather conditions such as frequent rain or very hot temperatures. Arcades are usually roofed and surrounded by walls on one or more sides but frequently not separated at all from outdoor environments. They may share the same pavement as adjacent walkways or streets, and pedestrians can easily exit them to indoor or outdoor areas at any point. Similar to tunnels or underpasses, arcades are often publicly owned and everyone has access to them at any time. In this respect, shopping malls differ slightly since they are usually owned by a company and have specific opening hours, though in practical terms the difference is often small as anyone can walk through a mall during most of the day.

Transportation hubs such as train stations, airports or metro stations are subject to similar access restrictions, i. e. they have specific opening times and are not fully public. They also often include areas, which could be classified as either indoor or outdoor areas, as well as some spaces, which may be difficult to place in either category. For example, the roofed areas near the tracks at a train station could be considered part of the train station itself (and thus as being indoors) or could be construed of as being outdoors (since they are not fully enclosed). Signage plays an important role in such hubs [8], which is partially due to the inherent homogeneity or symmetry. Tracks or gates are usually labeled following a predefined logical rule set, and these labels are used throughout a transportation hub to support wayfinding. Explicit landmarks frequently either do not exist or if they do (e. g. a unique feature such as a specific shop or piece of art), are not commonly used for directions.

Other aspects to consider when analyzing transitional spaces include the function of those areas, the typical duration a person remains in those areas, and the number of people 'using' the space. Here, we can observe that a large number of the examples discussed above share some properties: many of them are not intended for people to remain in them for extended periods of time, and they often do not constitute a destination of their own right – tunnels, raised walkways, and transportation hubs fall into that category. Shopping malls and (to a lesser degree) arcades can also be understood in such a way. However, people do tend to spend more time in these places and also specifically go there, e. g. in order to buy things or to meet up with friends. A characteristic shared by almost all the examples discussed so far is the number of people using these transitional spaces: frequently, these areas are very busy and many people pass through them during the day.

Table 1 summarizes the criteria discussed in the previous paragraphs, and contrasts indoor, outdoor and transitional areas using those criteria. The list of criteria and the typical values associated with each category is based on an analysis of several example sites (such as the ones discussed above or the ones used in study). The set of criteria and values were discussed at a series of meetings of the researchers involved in this project and went through a number of iterations until stable. The table lists typical values observed for each type of space for each category. It is worth pointing out that for each category it is also possible to find examples, which differ from the typical values reported in the table. For example, a small courtyard garden as found in some traditional

Japanese homes, could be considered to be indoors despite not having a roof. This example also highlights the potential for the influence of cultural aspects on what is perceived as being a transitional space. Our initial investigation focused on European settings.

We also used the criteria listed in table 1 to select locations for our survey study, which we report on later on in this paper. Based on this discussion, we can also propose a working definition for the concept of transitional spaces.

> **Transitional spaces** can be defined as spaces that can be neither consistently classified as being indoors nor being outdoors and that share properties with either category. Transitional spaces are generally located between indoor and outdoor spaces.

This definition can serve as a starting point to gain a better understanding of transitional spaces, and to investigate their properties and uses. The following sections describe how we approached the topic and what results we obtained.

4 Methodology

In order to take into account the complex nature of transitional spaces and to gain a deeper understanding of them, we followed a three-pronged approach combining a qualitative site analysis, an empirical study and a linguistic analysis. In doing so, we tried to obtain evidence for the existence of transitional zones and to gain a deeper understanding of their properties, nature and use.

The *site analysis* was based on our working definition of transitional spaces and the criteria we identified in the previous section. We reviewed a number of sites using these criteria to see whether they would confirm our initial observations regarding the properties of transitional zones. Additionally, we were interested in seeing whether we would indeed find examples for such zones. Finally, this analysis also provided us with a number of candidate sites, which we could then use in a user study to gather direct feedback from pedestrians.

We then carried out an *exploratory survey study* in a city center in order to get direct input from people who are using these spaces. Since the transitory nature of transitional spaces seemed to be a relevant factor, the study specifically looked into how people navigate and describe complex urban environments, which include transitional spaces. We interviewed 103 pedestrians in the city center of Hanover, Germany. During the interviews, we asked them to describe short routes through the city, which included areas that were good candidates for transitional zones. We also asked them directly to characterize several locations/decision points in terms of whether they would consider them indoors or outdoors. In addition to classification data, the study also resulted in a number of route descriptions that covered transitional spaces. This body of linguistic data provided a further means to analyze the characteristics of transitional zones.

The third step of our methodology was thus a *linguistic analysis* of the route descriptions given by the participants of the survey study. We specifically investigated which spatial expressions (including but not limited to prepositions) they

Table 1. Comparison of indoor, outdoor, and transitional spaces: typical values observed for a number of key properties

	outdoor	indoor	transitional
access	often unrestricted	usually regulated, gated	varying: unrestricted / regulated, often gated
traversal	usually free movement	often along defined paths	mainly along defined paths
landmarks	global / local	mainly local, signage important	mainly local, signage may be important
line of sight	often unobstructed	usually limited	usually limited
function	varying	varying	often traversal, waiting
enclosure	usually none	usually fully enclosed	usually partially enclosed
protection from elements	usually none	often complete	usually partial
length of stay	varying	varying but often long	frequently short
frequentation	varying	varying	often busy
dimensionality	usually 2D	often 2.5 to 3D	often essentially 2D
ownership	often public	usually non-public	mainly non-public

were using when referring to different locations along the routes. Our goal here was to see whether indoor, outdoor and transitional spaces are described in the same way or whether there were systematic differences. Additionally, we were interested in finding out whether the linguistic data confirmed or contradicted the direct classification that we had obtained from participants.

The following three sections report on the details of how these three steps were implemented and what results we obtained. The penultimate sections interrelate the results, discuss implications and point out limitations of our approach.

5   Site Analysis

The definition given in section 3 locates transitional spaces between indoor and outdoor spaces and also characterizes them along several dimensions. Based on these considerations we reviewed urban areas in Northern Germany in order to identify areas that might qualify as transitional spaces. Using photographs, virtual globes and other sources, we looked at locations such as shopping malls, transportation hubs, event locations, or sites combining these functions. We were particularly looking for inner city locations, which contained several transitional spaces within walking distance, so that a study investigating them with actual

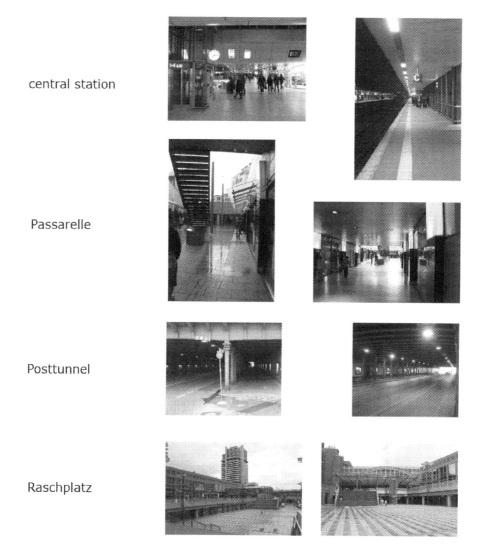

Fig. 1. Photographs of the four sites we analyzed: the central station, the Passarelle, the Posttunnel, and the Raschplatz

users would be feasible. The area around the central station in Hanover proved to be well suited, as it contained multiple candidate sites, which were in close vicinity to each other, and which could also support a study in the area. In the following paragraphs, we review four of these sites, which could be classified as being transitional zones: the central station of Hanover, the Passarelle[1], the Posttunnel, and the Raschplatz.

---

[1] The official name for this shopping area is "Niki-de-Saint-Phalle-Promenade", which it was given ten years ago. Most people still refer to it using its old name "Passarelle".

The *central station* is at the center of the area, around which the other sites are located. With more than 600 trains and 250,000 persons passing through it per day[2] it is a busy and highly frequented site. In addition to its function as a transportation hub, the station itself also serves as multi-story shopping center. On the basement level, it connects to the *Passarelle*. This shopping arcade in a former subway tunnel consists of a single story linear structure below street level. For most of its length, it is upwardly open (i. e. there is no roof). It provides access to the central station, which is located at one end of the Passarelle, and to the ground-level pedestrian precinct through staircases and escalators. As part of a bigger continuous pedestrian area with a total length of about 2500m it sees a lot of foot traffic throughout the day. The *Posttunnel* is a 150m long tunnel-like underpass underneath the railway tracks next to the central station. It is frequented by cars, busses and bikes as well as pedestrians, and it connects major streets on both sides of the central station. The *Raschplatz* plaza is situated behind the central station. This open space is at the same height level as the basement level of the central station. One side connects to the continuation of the Passarelle through the central station, the other sides provide access to ground level via staircases.

Figure 1 shows photographs of the four sites we analyzed. Table 2 summarizes our evaluation of each site with respect to the comparison criteria described in section 3. (Property values might correspond to more than one category.) The analysis followed an iterative approach similar to the one used in the elicitation of the criteria in section 3. Specific values were discussed in a series of group meetings until a consensus was reached. In the table, property values are set in different typefaces: values corresponding to outdoors spaces are in bold, those relating to indoor spaces in italics, and values corresponding to transitional spaces are underlined. Combinations of these typefaces (e.g. bold and underlined) indicate values that can represent different types of spaces (e.g. outdoor and indoor spaces).

The table highlights a number of aspects. Firstly, we can observe that neither site is uniquely classifiable as a particular type of space. Secondly, it is also obvious that the Raschplatz shares many properties with outdoor spaces but only very few with indoor and transitional spaces. Thirdly, central station, Passarelle and Posttunnel each score highly in terms of properties shared with transitional spaces. Finally, we can see that while the latter three sites have a similar distribution of shared properties across the three types of spaces, the actual values in each category vary considerably. In summary, we can thus conclude that except for the Raschplatz all the sites analyzed in this section could be classified as transitional spaces according to our definition and the criteria we specified. In order to confirm these findings, we used these sites to gather direct feedback from users of the identified transitional spaces in the context of an empirical study.

6   Empirical Analysis

In September 2012, we carried out interviews at two different locations in the city center of Hanover. Both locations were in the pedestrian zone near the

---

[2] Source: http://www.bahnhof.de/#station/18705, accessed March 10th, 2013.

Table 2. Classification of the chosen sites in Hanover according to Tab. 1: properties found in **outdoor spaces are bold**, *indoor properties are italic* and transitional properties underlined; multiple categorization are possible. The three bottom rows reflect the number of properties shared per type of space for each site.

	Central Station	Passarelle	Posttunnel	Raschplatz
access	*semi-restricted (valid ticket zones)*	**staircases, escalators, unrestricted**	**unre-stricted**	**staircases, unrestricted**
traversal	*multiple defined paths*	*one path*	*one path*	**free movement**
landmarks	*local, signage*	*local, signage*	local	**global, signage**
line of sight	*limited*	*limited*	*very limited*	**rather un-limited**
function	**shopping, commuting, waiting**	**shopping**	traversal	**traversal, plaza**
enclosure	complete (ground level), partially (tracks)	partially open air, underground	*tunnel*	**almost none**
protection from elements	full (ground level), partially (tracks)	partially	protection from rain but windy	**no protection**
length of stay	*varying*	short	short	**varying**
frequentation	busy	busy	*varied*	*varied*
dimensionality	*3D*	**2D**	**2D**	**2D**
ownership	*non-public*	*non-public*	**public**	**public**
outdoor	2 properties	3 properties	3 properties	10 properties
indoor	7 properties	4 properties	4 properties	2 properties
transitional	8 properties	9 properties	7 properties	3 properties

central station at junctions of highly frequented shopping streets. During the course of one day we interviewed 103 participants, who we approached opportunistically. The demographic characteristics assessed were gender, age group, familiarity with the city of Hanover and the familiarity with the city center. The age distribution spans a broad range from teenagers to elderly people with a high numbers in the twenties and low numbers in the forties as illustrated in Figure 2. The graphics also shows that the number of female participants is slightly larger than the one of male participants (overall 57% female, 43% male).

Familiarity was assessed on a five-level Likert scale ranging from 1 'very familiar' to 5 'not familiar at all'. The familiarity with the city of Hanover in general

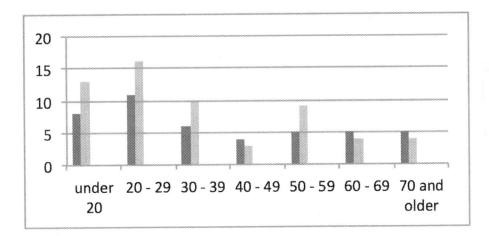

Fig. 2. Age and gender distribution of the participants. Blue bars (left) indicate males, orange bars (right) indicate females

was on average 2.7 (median: 3) and for the city center on average 2.24 (median: 2). After having collected demographic information, we asked participants to provide us with route descriptions for a pedestrian route from the interview location (denoted as B and C, cf. Figure 3) to a destination (denoted as A, cf. Figure 3) behind the central station, which was well known to the citizens. The routes had an approximate length of 700 m and 1100 m respectively. The descriptions were provided orally and we transcribed them manually in parallel. The last part of the interview consisted of a classification task for the four nearby sites presented in the previous section. The participants had to place each location on a five-level Likert scale with the values "indoors", "rather indoors", "neither nor", "rather outdoors", "outdoors". As the four locations were nearby but not directly visible from the interview location we presented participants with a collection of several photographs for each site, showing different aspects and being taken from different perspectives.

Figure 3 (a) shows an overview of the study area, which contains the four sites under investigation. The sites are labeled in the figure as well as color-coded according to the color scheme depicted in Figure 3 (b). In addition, subfigure (a) also shows the routes described by the participants, which are color-coded according to the scheme shown in subfigure (c). The thickness of lines shown in subfigure (c) indicates how often a specific option was chosen by participants: the thicker the line, the more frequently it was chosen. The two main levels are depicted by the grey 3D objects – the ground floor is shown over the satellite image at the top, and the basement level is shown below.

Several key observations emerged when analyzing the information we gathered during the study. Figure 3 (b) highlights a strong disagreement between participants with respect to where to place the four sites under considerations. Whereas for the central station and the Raschplatz there clearly was a preferred category (i.e. outdoor area), this was not true for the Passarelle and the

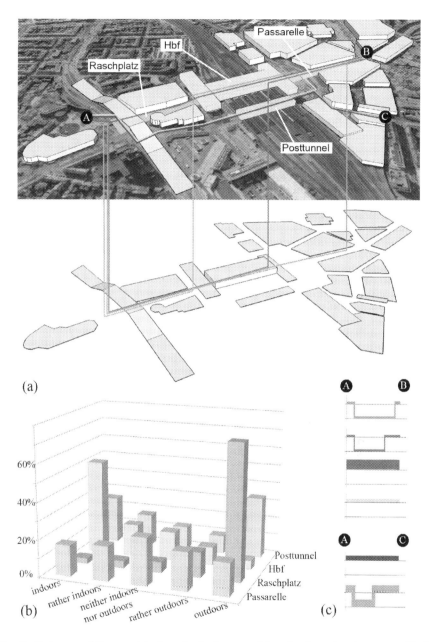

Fig. 3. Outcome of the empirical analysis characterizing the four sites in terms of being indoors or outdoors (b), also visualizing the height profile (c) and the trajectories (a) of the routes based on the directions given by the participants. Locations and routes are color-coded and labeled in subfigure (a) and the same coloring and labeling scheme is used in the diagrams (b) and (c)

Posttunnel. The central station was categorized as being an indoor area by about 50% of the participants; the remaining 50% of the votes were distributed over the other choices. Numbers decreased with the degree of "outdoorsiness". The Raschplatz showed an inverted pattern: 70% considered it being outdoors, while the remaining votes were given to the other choices with number decreasing as the degree of "indoorsiness" grew. In case of the Passarelle, the most frequently selected category (neither indoors nor outdoors) received only a few more votes than the other categories. For the Posttunnel, many people tended to either classify it as indoors or outdoors and fewer people chose an intermediate value. While participants thus did not unanimously put the three transitional sites into one category, the overall distribution of votes differed considerably from site to site.

Route choice (in the direction-giving task) also varied – there was no predominantly chosen route. For example, the most direct route from location B to destination A it is an almost straight line. However, we identified four alternative routes including some that required the traversal of several height levels as well as some routes that included the Posttunnel and thus constituted a considerable detour. All routes described by participants did include indoor, outdoor, and transitional segments (cf. Figure 3 (a) & (c)).

Overall, the results of our empirical study highlight the need to further deepen our understanding of transitional zones and also confirm that urban routes can be quite complex. In the following section, we therefore take a closer look at the route directions participants gave.

7   Linguistic Analysis

In order to validate the classification given by participants on whether they thought a given space was indoor, outdoor, or transitional, we followed an approach proposed by Landau and Jackendoff (1993) [17]. They analyzed how objects and places can be represented with language specifically based on count nouns and spatial expressions. The authors presented a list of linguistic expressions, which they considered important in understanding spatial relations expressed in words. In their study, they discussed comprehensively those linguistic expressions through the asymmetry between figure and reference objects as well as analysis in terms of the geometric aspects. The authors recommended an approach to assess the extent of different spatial relations by counting the frequency of the linguistic expressions used. We hence applied a similar approach in our attempt to explore how people refer to different types of spaces based on the spatial expressions mentioned. However, extensive linguistic analysis of the route instructions produced by participants is beyond the scope of this paper.

Our hypothesis was that different usage of linguistic expressions supports the claim that people distinguish the three types of space in their daily practices. These linguistic expressions are adjectives and adverbs that were used to describe wayfinding instructions with respect to the spaces – including the central station, Passarelle, Posttunnel, and Raschplatz. Therefore, the expressions serve

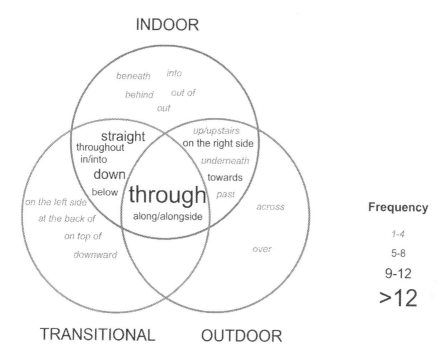

Fig. 4. Spatial expressions relating to indoor, outdoor, and transitional spaces; font size indicates overall frequency of usage (lowest frequency in italics for readability)

as determiners of how spaces could be classified as indoor, outdoor or transitional space. To carry out our linguistic analysis, we used a tool called AntConc [1]. Firstly, in our empirical study we paid particular attention to the spaces (differentiated in the theoretical analysis) that participants indicated in their wayfinding instructions and to the associated linguistic expressions, which they used to refer to these spaces. Secondly, the different spaces and their equivalent names were selected in the tool; phrases containing a selected spatial object and its associated linguistic expressions were then filtered. Thirdly, all linguistics expressions linked to the three different spaces were then counted and summarized. Table 3 gives a complete list of German terms and their English translations. Finally, based on the classification of spaces by participants, we examined if any two or all spaces share similar uses of linguistic expressions. Fig. 4 visualizes the usage of spatial expressions with respect to the different types of spaces and their overall frequency.

The linguistic analysis shows that some expressions are exclusively used with indoor spaces, while others only appear with outdoor spaces or transitional spaces. Some expressions are used to refer to two types of spaces indoor/transitional or indoor/outdoor. We find no expressions that are solely shared by transitional and outdoor space, but only a few which are commonly used to refer to all three types of spaces.

Table 3. Spatial expressions and their frequency of use for the different types of spaces

| spatial expressions | | types of spaces | | | total count |
English	German	indoor	outdoor	transitional	count
through	durch	79	5	14	98
down	runter	6	-	4	10
straight	geradeaus	8	-	1	9
along, alongside	entlang	5	-	3	8
below	unten	4	-	3	7
towards	zum	5	2	-	7
in, into	in	2	-	4	6
on the right side	rechts	3	3	-	6
through	hindurch	3	-	2	5
underneath	unter	3	-	1	4
up, upstairs	hoch	2	2	-	4
over	über	-	3	-	3
on the left side	links	-	-	2	2
past	vorbei	1	1	-	2
across	drüber	-	1	-	1
at the back of	hinten	-	-	1	1
behind	hinter	1	-	-	1
beneath	drunter	1	-	-	1
downwards	hinunter	-	-	1	1
into	hinein	1	-	-	1
on top of	oben auf	-	-	1	1
out	aus	1	-	-	1
out of	heraus	1	-	-	1
	sum	126	17	37	180

Two expressions are used with all three types of space. The participants use 'through' 98 times in the wayfinding instructions; 'through' is thus by far the most commonly used preposition. It refers to all three types of spaces, but it is predominantly used for indoor space. 'Along or alongside' is the only other linguistic expression used to refer to all three spaces. Considering both types and frequency of all expressions, most spatial expressions (70.0%) are used with indoor space, 9.4% are used with outdoor space and 20.6% are used with transitional space. Indoor space also shows the largest variability of spatial expressions used: 17 different expressions out of 23 expressions in total are used to refer to indoor space. Seven different expressions are used with outdoor space and 12 different expressions are used with transitional space.

Considering only the types of expressions, there are a few that are used exclusively within only one type of space: within all types of expressions that are used for indoor areas, 29% are exclusively used for indoor space, 22.2% of outdoor expressions are used exclusively for outdoor space, and 36.4% of transitional expressions are used exclusively for transitional space.

The data analysis revealed that most of the expressions used to describe transitional space are also used in describing indoor space. 'Straight' was predominantly used with indoor space (89%) while 'down', 'below', 'in/into', and 'throughout' were almost equally used with both spaces. The four expressions 'to the left side', 'at the back of', 'down' and 'on top of' were used exclusively for transitional spaces. Outdoor space and transitional space do not exclusively share any spatial expressions. Five expressions are used for referring to outdoor and indoor space: 'past', 'to-wards', 'underneath', 'on the right side', and 'up/upstairs'.

8 Discussion

In this section, we review implications of our findings and discusses limitations our study were subjected to, based on the factual outcomes of the analyses we carried out.

The empirical survey partially confirmed the outcomes of our theoretical analysis, and provided initial evidence that our working definition of transitional spaces might be useful. The participants classified the Raschplatz as being clearly outdoors, which is what we were expecting based on the theoretical analysis. The same can be said about the Passarelle. In the cases of the Posttunnel and the central station, we observed more complex patterns. The participants tended to classify the Posttunnel as either indoors or outdoors but rarely as neither one nor the other despite the fact that the theoretical analysis had identified many similarities with the Passarelle. The central station resulted in the biggest disagreement of theory and practice. Formally classified as transitional space the participants mainly regarded it as an indoor space. However, with the exception of the Raschplatz, people did not produce a consistent classification for the presented spaces. This failure to clearly classify a space can be conceived of as an indicator of transitional spaces while the actual classification patterns of different transitional spaces may vary considerably.

Both the site analysis and the survey study were subject to a number of limitations. The categories we used emerged during the investigation of inner city spaces in Northern Germany. It is possible that further relevant categories exist, and that the set of categories might vary according to country or cultural sphere. The survey study involved more than a hundred participants but they did not constitute a representative selection of the population. Our results are thus not as general as they could have been. In addition, cultural differences are also not accounted for due to the study being carried out in a single city. Finally, we chose to write down the directions we requested from participants rather than to record them. This was done to not intimidate potential participants and to increase the number of participants but at the same time, this practice introduced an additional error source. Obviously, this also constitutes a limitation with respect to the linguistic analysis. In addition, the body of directions used for the study was small (slightly more than a hundred sets of directions).

From the linguistic analysis we can draw a number of conclusions. The data shows that the frequency and the variability of expressions used with indoor space was highest. This might be due to the spatial complexity of indoor spaces, which require complex wayfinding instructions to describe the different interaction possibilities. All three spaces have expressions that are uniquely used with one space. This provides evidence that three spaces can be distinguished and are conceptualized differently. However, they also all share expressions used with other spaces. This is the case for transitional space as well, which supports our claim that there exists a transitional space in between indoor and outdoor space. With respect to overlapping expressions, transitional space shows more similarity with indoor space. On possible explanation for this observation could be a similarity in terms of complexity: transitional space is similarly complex as indoor space, often limiting movement and visual access.

The study demonstrated differences of spaces based on how participants used expressions in giving route instructions. While we cannot use the expressions to infer whether the place it refers to is in indoor, outdoor, or transitional space, but the usage and frequency of expressions provides evidence for the existence of differences between these spaces. The use of spatial expressions may also depend on the physical characteristics of the place wherein one expression may describe various spaces. Expressions such as 'down', 'downwards', 'under', beneath', 'throughout' or 'below' indicate a three-dimensional structure of the environment. Different levels predominately exist in indoor and transitional spaces, and to a lesser degree in outdoor spaces.

A further limitation of our study was that we only used four places to classify indoor, outdoor or transitional space. Due to the small number of places, we were not able to thoroughly investigate factors such as the physical characteristics. Consequently, a logical next step for further research would be to considerably expand the number of places given to participants in a follow-up study. In addition, three of the spaces under investigation were connected to each other (Passarelle, central station, Raschplatz). While this is not necessarily a typical pattern for transitional spaces in urban areas – further studies are needed to establish such patterns – it might potentially have affected the judgement of participants (though the Raschplatz was classified very differently from the other two spaces that were connected).

Summarizing the outcomes of our studies, we can thus observe that there are spaces, which people cannot clearly classify as being either indoors or outdoors. The spaces differ in terms of their properties and in terms of how people describe routes traversing them. Three of the sites we looked at matched our working definition of transitional spaces and exhibited properties we considered indicative of such areas. Given the well-documented differences between indoor and outdoor spaces, our findings have implications for urban navigation and provide a good starting point for future research. Systems providing personalized navigation support, for example, might have to adjust what directions they give (and how) when supporting users navigating transitional zones. While our study was not sufficiently extensive to fully define and investigate transitional zones, it can nevertheless serve as a starting point for future research into those spaces.

9   Conclusion

Previous work commonly only distinguishes between indoor and outdoor space. In this paper, we reported on initial evidence demonstrating that two types of spaces are not enough: there is a third type of space – transitional space – that is neither consistently classified as being indoors nor consistently classified as being outdoors, and it shares properties with both indoor and outdoor space. We theoretically analyzed four example sites from an urban area, and found that three of them can be considered to be transitional spaces. An empirical study, involving 103 participants, largely confirmed the outcomes of the theoretical analysis but also brought to light that classification patterns for transitional spaces may vary considerably but they all share the property that no single category is vastly preferred. In a linguistic analysis, we examined wayfinding instructions given by participants, and the results showed that the types of spatial expressions as well as their variability differ among different types of spaces. Some spatial expressions were solely used for transitional space, whereas only some were used for two or all three spaces. Linguistically, transitional spaces showed more similarity with indoor space than with outdoor space. All three analyses thus provided initial evidence for the existence of transitional spaces as a third type in addition to indoor and outdoor spaces.

While this paper presented first theoretical, empirical and linguistic evidence for the existence of a new type of space besides indoor and outdoor space, further research in this area is clearly needed. In addition to studies on a larger scale and in different cities/countries, a detailed geospatial analysis of transitional spaces would be beneficial. The former would not only verify the findings of this paper but also expand the catalog of properties and help to identify the impact of cultural aspects. The latter may also result in specific spatial configurations that enable us to identify transitional spaces more easily. Both lines of research would also benefit the deeper understanding of the classification patterns we observed, and might help to identify further ones. Finally, the issue of how people navigate transitional space and how technology can support them in this task, is another promising area for further research that also is of high practical relevance.

References

1. Anthony, L.: AntConc 3.2.2.1 (Computer Software). Waseda University, Tokyo (2011), http://www.antlab.sci.waseda.ac.jp
2. Baus, J., Kray, C., Cheverst, K.: A Survey of Map-based Mobile Guides. In: Meng, L., Zipf, A., Reichenbacher, T. (eds.) Map-based Mobile Services, pp. 197–216. Springer, Heidelberg (2005)
3. Butz, A., Baus, J., Krüger, A., Lohse, M.: A Hybrid Indoor Navigation System. In: Proceedings of IUI 2001, pp. 25–32. ACM Press (2001)
4. Cheewinsiriwat, P.: Development of a 3D geospatial data representation and spatial analysis system. PhD dissertation, University of Newcastle Upon Tyne, UK (2009)
5. Cheverst, K., Davies, N., Friday, A., Efstratiou, C.: Developing a Context-aware Electronic Tourist Guide: Some Issues and Experiences. In: Proceedings. of CHI 2000, pp. 17–24. ACM Press (2000)
6. Döllner, J., et al.: Smart Buildings – a concept for ad-hoc creation and refinement of 3D building models. In: Proceedings of the 1st Int. Workshop on Next Generation 3D City Models, Bonn, Germany (2005)

7. Downs, R.M., Stea, D.: Maps in minds.: Reflections on cognitive mapping. Harper & Row (1977)
8. Fontaine, S., Denis, M.: The production of route instructions in underground and urban environments. In: Freksa, C., Mark, D.M. (eds.) COSIT 1999. LNCS, vol. 1661, pp. 83–94. Springer, Heidelberg (1999)
9. Fund, U. N. P. UNFPA state of world population: unleashing the potential of urban growth. (United Nations Population Fund, 2007) (2007)
10. Gröger, G., Kolbe, T.H., Nagel, C., Häftele, K.-H.: OGC City Geography Markup Language (CityGML) Encoding Standard. Discussion Paper, OGC Doc. No. 12-019 (2012)
11. Hölscher, C., Meilinger, T., Vrachliotis, G., Brösamle, M., Knauff, M.: Up the down staircase: Wayfinding strategies in multi-level buildings. Journal of Environmental Psychology 26, 284–299 (2006)
12. Hölscher, C., Büchner, S.J., Meilinger, T., Strube, G.: Adaptivity of wayfinding strategies in a multi-building ensemble: The effects of spatial structure, task requirements, and metric information. Journal of Environmental Psychology 29(2), 208–219 (2009)
13. Hölscher, C., Tenbrink, T., Wiener, J.M.: Would you follow your own route description? Cognitive strategies in urban route planning. Cognition 121(2), 228–247 (2011)
14. Isikdag, U., Aouad, G., Underwood, J., Trodd, G.: Representing 3D building elements using spatial data structures. In: Proceedings of Computers in Urban Planning and Urban Management (CUPUM), London, UK (2005)
15. Kray, C., Kortuem, G., Krüger, A.: Adaptive Navigation Support with Public Displays. In: Proceedings of IUI 2005, pp. 326–328 (2005)
16. Krüger, A., Butz, A., Müller, C., Stahl, C., Wasinger, R., Steinberg, K.-E., Dirschl, A.: The Connected User Interface: Realizing a Personal Situated Navigation Service. In: Proceedings of IUI 2004, pp. 161–168 (2004)
17. Landau, B., Jackendoff, R.: "What" and "Where" in spatial language and spatial cognition. Behavioral and Brain Sciences 16(2), 217–265 (1993)
18. Lewis, R., Séquin, C.: Generation of 3D Building Models from 2D Architectural Plans. Computer Aided Design 30(10), 765–780 (1998)
19. Link, J., Smith, P., Viol, N., Wehrle, K.: Footpath: Accurate map-based indoor navigation using smartphones. In: International Conference on Indoor Positioning and Indoor Navigation (IPIN), pp. 1–8. IEEE (2011)
20. Lynch, K.: The image of the city. MIT Press (1960)
21. Meijers, M., Zlatanova, S., Pfeifer, N.: 3D Geoinformation Indoors: Structuring for Evacuation. In: Proceedings of Next Generation 3D City Models, Bonn, Germany (2005)
22. Montello, D.: Scale and multiple psychologies of space. In: Campari, I., Frank, A.U. (eds.) COSIT 1993. LNCS, vol. 716, pp. 312–321. Springer, Heidelberg (1993)
23. Oppermann, R., Specht, M.: A Nomadic Information System for Adaptive Exhibition Guidance. In: International Cultural Heritage Meeting (ICHIM 1999), Washington D.C, pp. 103–109 (1999)
24. Slingsby, A.D., Longley, P.A.: A Conceptual Framework for Describing Microscale Pedestrian Access in the Built Environment. In: Proceedings of GISRUK, Nottingham, UK (2006)
25. Steck, S.D., Mallot, H.A.: The role of global and local landmarks in virtual environment navigation. Presence: Teleoper. Virtual Environ. 9, 69–83 (2000)

Representing and Reasoning about Changing Spatial Extensions of Geographic Features

Claudio E.C. Campelo and Brandon Bennett

School of Computing, University of Leeds, Leeds, LS2 9JT, UK
{sccec,b.bennett}@leeds.ac.uk

Abstract. This paper presents a *logical framework* for representing and reasoning about *events* and *processes* which are characterised by changes in the spatial extension of geographic features. This work addresses some crucial (and still open) issues on the modelling of geographic dynamics, such as how to represent the relationship between events, processes and geographic features and how to relate event and process *types* to their particular *instances*; how to handle spatial and temporal *vagueness* to associate specific spatio-temporal boundaries with events and process instances; and how to provide a flexible representation which can be applied to distinct geographic phenomena as well as which can accommodate different *standpoints* on these phenomena. Particular focus is placed on phenomena which can be identified from spatio-temporal data of the geographic domain, and therefore a discussion is given on the way the logical framework can be linked to data. Finally, we present a system prototype implemented to experiment the proposed logical framework with real data.

Keywords: geographic events, geographic processes, changing geographic features, spatio-temporal reasoning, vagueness.

1 Introduction

The Geographic Information Science community has been demanding for more conceptualised ways of representing and querying geographic information, in particular information describing both space and time, enabling more comprehensive analysis of dynamic elements of geographic space. Representing geographic phenomena in terms of *events* and *processes* has been suggested by many authors [12,14,18,28], and such conceptual entities appear to be significant in the way humans reason about changes affecting geographic space.

Defining an appropriate representation for geographic events and processes requires dealing with issues regarding the relationship between these concepts and also between them and geographic features. Other crucial issues are how to define the relation between event and process types and their particular instances, and how to handle different kinds of *vagueness* to associate specific spatial and temporal boundaries with process instances[1].

---

[1] For a comprehensive review of issues and challenges for representing geographic processes see [10].

T. Tenbrink et al. (Eds.): COSIT 2013, LNCS 8116, pp. 33–52, 2013.

Foundational ontologies have been proposed to represent events and processes, such as BFO [9,19,21], DOLCE [21] and SWEET [25]. Undoubtedly, upper-level ontologies can be used as valuable guidelines for the development of semantic models and applications; nevertheless, their concepts are not defined in sufficient level of detail to allow their use for reasoning purposes. Formal theories of spatial changes [20,27] and for modelling events and processes [3,17,19] have also been proposed. However, events and processes are often approached in the general sense, and their applicability to the geographic domain would require further extensions and refinement. Moreover, although some works provide important directions, most of them are not yet implemented, and therefore their suitability for handling real-world data is not often discussed.

This paper presents an *logical framework* for representing and reasoning about geographic events and processes, encompassing their relationship with geographic features, which are said to participate in them. We are particularly interested in geographic phenomena which can be described in terms of spatial changes affecting geographic features. Examples are deforestation, urbanisation and desertification. We propose an approach to handling vagueness based on standpoint semantics [6], which enables the proposed reasoning mechanism to define temporal boundaries for geographic processes so that particular instances of a given process type can be determined based on individual viewpoints. We have implemented a system prototype to evaluate the applicability of this theoretical framework to process real data, and thus experiments using this application are also described in the paper.

There has been many disagreements in the literature about the appropriate representation of events and processes. The debate covers issues relating to the classification of these entities either as *endurants* or *perdurants* and tackles questions concerning how these entities are related to each other (e.g., whether one is a subclass of the other). In this paper we do not intend to go into this debate by defending that our representation approach is the most appropriate from the philosophical point of view. Rather, we propose a representational approach based on semantic analysis discussed in previous work, and then experiment an implementation of such a formalism for handling real geographic data, as a call for further research approaching the *application* of such theories to support the development of modern Geographic Information Systems (GIS).

The remainder of this paper is structured as follows. The next section presents the some fundamentals and the basic syntax elements of our logical framework. This is followed by a discussion in Section 3 on our approach to representing event occurrences and their relation to the concept of process activeness. Then Section 4 presents the mechanism to identify process instances by defining their temporal boundaries. Following this, the method of identifying active processes at the data level is described in Section 5. Then Section 6 presents the system prototype we have implemented to evaluate our theoretical model. Experiments carried out using this prototype are then discussed in Section 7. Finally, Section 8 concludes the paper and points to future work.

2 Logical Framework

This section outlines the logical framework we have named *REGEP* (REasoning about Geographic Events and Processes), comprising formal descriptions of space, time, events, processes, geographic features and their related aspects. The formalism is described using definitions and axioms in first order logic, indexed by D and A, respectively. Free variables are implicitly universally quantified with maximal scope.

2.1 Fundamentals

Before presenting our proposed formalism, we now overview some fundamentals underlying this framework.

Space and Time. The formalism incorporates the Region Connection Calculus (RCC) [24] as the theory of space. Here spatial regions are used to represent portions of the earth's surface under some specified 2-dimensional projection.

The Allen's Interval Algebra [2,1] is incorporated as the theory of time. Here an interval may be *punctual*, if its beginning is the same as its end. Such intervals correspond to single time instants. Thus we use explicit time variables t and i to represent time instants and proper time intervals, respectively.

Spatial Region Coverages. Since spatial regions are used here to represent portions of the earth's surface, we associate a region with an attribute which can describe a certain geographic characteristic, which is referred to a spatial region *coverage*. The concept of 'coverage' employed here is not restricted to land coverages. It can also denote properties which can be measured (by sensors or human observation), such as 'urbanised', 'arid', 'temperature > 10 °C' 'water covered', 'heavily populated'.

Vagueness. A characteristic of geographic information is that it may be affected by vagueness, leading to additional representational difficulties [5]. Different methods have been proposed to handle vagueness in geography, such as Fuzzy Logic [29] and Supervaluation Theory [15]. Particularly relevant to geographic information is the *sorites vagueness*. This type of vagueness arises when qualitative concepts (e.g., tallness) are employed to distinguish among objects (e.g., people) which in fact exhibit continuous variation in the observables (e.g., person's height) relevant to the ascription of the concept [4]. *Supervaluation semantics* [15] has been shown as appropriate for the formalisation of sorites vagueness.

Formal approaches have been developed based on the supervaluationist account of vagueness. One such an approach is the *standpoint semantics* [6], whose main idea is to define a finite number of parameters related to observable properties in order to describe each possible precisification of a vague predicate, and then assign different threshold values to these parameters. Thus an assignment of threshold values is regarded as a *standpoint*. In standpoint semantics, the syntax for defining a predicate allows additional arguments to be attached to it corresponding to semantic variation parameters. Specifically, where a vague n-ary predicate V depends on m parameters we write it in the form:

$$\mathsf{V}[p_1, ..., p_m](x_1, ..., x_n)$$

The following example illustrates the use of this syntax, where the threshold $height_{thresh}$ is employed to specify whether a given height is positively relevant to classify m as a mountain.

$$\text{Mountain}[height_{thresh}](r) \equiv_{def} \text{height}(m) \geq height_{thresh}$$

A vague reasoning approach based on standpoint semantics has been presented in [7,8] for identifying and classifying vague geographic features, by specifying thresholds based on spatial attributes such as area and elongatedness. In our work, different standpoints are defined according to judgements about the temporal characteristics of a process type, so that particular instances (process tokens) can be determined based on different viewpoints. This is described in Section 4.

Geographic Features. We are particularly interested in geographic features which can be modelled as the maximal well-connected region[2] of some particular coverage. Examples are forests (which can be regarded as the maximum extension of a certain type of vegetation), deserts (which can be defined based on the level of precipitation) and sea (represented as the maximum extension of water body over a specified level of salinity). However, in order to allow analysis to be carried out at different levels of detail, the notion of 'well-connected' is relaxed. This means that the extension of a feature can be determined by the aggregation of regions[3] which are in fact disconnected, but whose distance in between is less or equal a given threshold, which is referred to here as the *aggregation factor*. Thus this factor is a standpoint semantics parameter used for the feature predicate, which takes the form Feature[aggr](f).

Geographic features will be regarded as a particular kind of *endurant entity*. Although they differ in some way from artefacts or organisms, they share many properties with other kinds of endurants. For instance, geographic features are able to undergo change (e.g., changes in shape or area); and they can change some of their parts while keeping their identity (e.g., a forest can be partially deforested while being still the same forest). Geographic features are regarded here as discrete individuals, and can be referred to by a proper noun (e.g., Amazon Forest, Atlantic Ocean, Antartic Desert), a count noun (e.g., a glacier or even an oil slick on the sea), or by more complex sentences, such as 'mountains over 1,500m in height'. That is, any spatial region with explicit and well-defined spatial extension, which can be individuated based on a certain aspatial and atemporal characteristic (i.e., a region coverage) can be regarded as a feature.

Events and Processes. We conceive events as *perdurant entities*, that is, entities whose properties are possessed timelessly and therefore are not subject to change over time. This is perhaps the most accepted view amongst different authors (e.g., [9,16,17,19,21]). On the other hand, we regard a process as an entity which is subject to change over time (e.g., a process may be said to be accelerating or slowing

---

[2] The term 'well-connected region' is used here in agreement with the discussion and definitions given in [13].

[3] Actually, the feature extension is determined by the concave-hull which includes the aggregated regions.

down), and therefore a process is *not* regarded as perdurant entity as defended by some authors (e.g., [19,21]). Our view of processes is in agreement with the concept of *time-dependent entities* described by Galton [16,17].

We also agree with Galton [16] in the view that "event is not something that can be said to exist from moment to moment in this way, rather it is something that, once it has happened, we can retrospectively ascribe to the time interval over which it occurred" (p. 04). Thus 'the forest is shrinking' does not describe an event, but rather a *process active* at a certain time instant. An event is usually associated with *precise temporal boundaries*, which may be denoted by the *culmination of a process* (i.e., when the goal in initiating it is realised). Hence, once after determined this instant of culmination, we can retrospectively ascribe the shrinkage event to the time interval over which it occurred.

Some authors support the view that processes are entities which are regarded as self-connected wholes, and therefore cannot contain temporal gaps. For instance, Grenon and Smith [19] state that 'processes have beginnings and endings corresponding to real discontinuities, which are their bona fide boundaries' (p. 153). Moreover, the authors defend that 'a given process may not be occurring at two distinct times without occurring also at every time in the interval between them' (p. 153). Nevertheless, this assumption is still the subject of controversy for the modelling of geographic processes, since there are many examples of geographic processes where the existence of gaps seems to be acceptable.

To illustrate, suppose one intends to monitor deforestation in a given forest based on spatial data collected once a day, every day. Then it was observed that the forest shrank every day during 300 days, except between the 84° and the 86° days, and between the 145° and the 165° days. Deciding whether the same instance of a process proceeded over such 10-month period or distinct instances were separated by those periods of inactivity might depend on many factors. Judgement variables include the sort of geographic phenomena which is being analysed (e.g. deforestation), the agents involved (e.g. human action or wildfire originated from spontaneous combustion), the purpose (e.g. deforestation caused by human actions with purpose of wood trading), amongst others.

Hence, in our framework, a process is regarded as an entity which may be affected by temporal gaps. Our approach to determining processes' boundaries aims to provide a flexible mode of representing and reasoning about geographic processes. This approach is based on standpoint semantics and therefore threshold parameters are given to determine the range of variation over which the predicate is judged to be applicable.

2.2 Syntax

This section describes the basic syntax of the logical language named $\Re$ employed in the framework. Relevant predicates and logical relations employed in this framework shall be defined in Sections 3, 4 and 5.

The vocabulary of the logical language $\Re$ used in the framework comprises variables of 12 *nominal types* which can be quantified over. This can be specified by a tuple $\mathcal{V} = \langle \mathcal{V}_t, \mathcal{V}_i, \mathcal{V}_r, \mathcal{V}_f, \mathcal{V}_c, \mathcal{V}_u, \mathcal{V}_v, \mathcal{V}_e, \mathcal{V}_\varepsilon, \mathcal{V}_b, \mathcal{V}_p, \mathcal{V}_\pi \rangle$, representing, respectively, Time Instants, Time Intervals, Spatial Regions, Geographic Features, Coverage Types,

Feature Types, Event-classifiers, Event-types, Event-tokens, Process-classifiers, Process-types, Process-tokens (the denotation of some of these variables shall be discussed later in this section).

The logical language $\Re$ contains the following *functions* to transfer information between distinct semantic types. Function ext: $\mathcal{V}_f \rightarrow \mathcal{V}_r$ returns a spatial region representing the spatial extension of the specified geographic feature. Functions b: $\mathcal{V}_i \rightarrow \mathcal{V}_t$ and e:$\mathcal{V}_i \rightarrow \mathcal{V}_t$. return, respectively, the time instant corresponding to the beginning and the end of a given time interval.

The following *auxiliary functions* are employed to perform calculations over elements of the domain. length: $\mathcal{V}_i \rightarrow \mathbb{Z}$ gives an integer value representing the length of a given interval i^4. area: $\mathcal{V}_i \rightarrow \mathbb{R}$ returns a real number representing the area of a region r.

Atomic propositions of $\Re$ include the 13 Allen's [1,2] relations between time intervals and the RCC-8 relations between spatial regions. In addition, we employ the operators $\prec$, $=$, and $\preceq$ to represent temporal precedence and equality between time instants. The following temporal relations are also defined:

$\ln(i_1, i_2) \equiv_{def}$ Starts(i_1, i_2) $\vee$ During(i_1, i_2) $\vee$ Finishes(i_1, i_2) $\vee$ Equals(i_1, i_2).
$\ln(t, i) \equiv_{def} b(i) \preceq t \preceq e(i)$.

Finally, if φ and ψ are *propositions* of $\Re$, then so are the following: $\neg\varphi$, $(\varphi \wedge \psi)$, $(\varphi \vee \psi)$, $(\varphi \rightarrow \psi)$, $\forall v[\varphi]$, where v is a variable of one of the nominal types described earlier.

2.3   Representing Events

This section describes the elements of our logical framework employed to represent events (i.e., event classifiers, types and tokens) and presents elementary logical relations which hold between events and geographic features. The semantic categorisation used in our formalism is based on that used in Versatile Event Logic (VEL) [3]. The representation of event occurrences shall be discussed in Section 3 and in Section 4.1.

Event classifiers identify general categories of events, independently of particular occurrences or participants. That is, it describes something that might happen in space and time without specifying any temporal information or relating any type of geographic feature. We tend to apply *natural language verbs* (in the third person singular conjugation) to name these classifiers. Examples of such verbs are 'falls', 'expands' and 'shrinks'. Verbs whose denotation is attached to a particular sort of spatial object are avoided, such as 'rains'. In this case, we would prefer to represent 'the raining event' in terms of the 'fall of raindrops', for instance.

This sort of abstraction is desired in the representation of events so that a wider range of geographic phenomena can be represented by associating distinct geographic features with events classifiers. For example, 'desertification' and 'urbanisation' could be represented in terms of *expansion* of 'arid' and 'built-up' regions, respectively. This is particularly applicable to model the type of phenomena we are interested, which

---

[4] The value returned by this function is an integer number which may vary according to the temporal granularity (e.g., days, microseconds) adopted to represent elements of domain, however this change does not affect the semantics.

are those which can be represented in terms of spatial transformations of geographic features. Nonetheless, defining event classifiers at this level of abstraction is not a requirement for the applicability of this representational approach. That is, less abstract events classifiers of can be adopted in cases where the knowledge engineer finds more appropriate their use for modelling a particular domain or situation.

Events are also structured in terms of *types* and *tokens*. An *event type* involves a particular instance of a *geographic feature* as its participant. On the other hand, an *event token* denotes a *particular occurrence* of an event type, and is therefore associated with a *time interval* on which it occurs. For example, 'Amazon rainforest shrinks' describes an event type, since this might occur different times, corresponding to different instances of this event. Whereas 'Amazon forest shrank from May/2006 to July/2006' describes a particular occurrence of this type, that is, an event token. As seen in this last example, an event token can be referred to by using the past simple tense of the verb corresponding to its classifier.

We adopt a reified representation for event classifiers. Therefore an event type e denoting an expansion, for example, is related to an event classifier by Event-Classifier($expands, e$). On the other hand, an event type is treated as complex nominals (i.e. functional terms). Thus an event type e is represented by $e = $ event(v, f), where v is an event classifier and f a geographic feature which participates in this event.

The relation between an event classifier v and an event type e is defined as follows:

D 1   Event-Classifier$(v, e) \equiv_{def} \exists f [e = $ event$(v, f)]$

For convenience we also define:

D 2   Participant-In-Event$(f, e) \equiv_{def} \exists v [e = $ event$(v, f)]$

An event token is represented by a pair $\varepsilon = \langle e, i \rangle$, where e is an event type and i is the interval of occurrence of this event token. However, not all possible pairs $\langle e, i \rangle$ denote an existing event token. Thus the subset of existing event tokens is given by those pairs for which the proposition Occurs-On(e, i) is true.

2.4 Representing Processes

This section describes the elements of our logical framework employed to represent processes (i.e., process classifiers, types and tokens) and presents essential logical relations which hold between processes and geographic features. A more extensive discussion on the representation of geographic processes shall be given in Sections 3 and 4.

Similarly to as discussed for events, *process classifiers* are used to describe processes without any association with temporal information or participants. Processes are also structured in terms of *types* and *tokens*. Whereas a process type denotes a series of changes involving a particular feature, a process token denotes an instance of a certain type of process. Hence, tokens are said to *proceed* on a specific time interval. The structure employed to represent process classifiers, process types, process tokens and the relation between processes and their participants is analogous to the one employed for events. Thus the following relations are also defined.

D 3   Process-Classifier$(b, p) \equiv_{def} \exists f [p = \mathsf{process}(b, f)]$

D 4   Participant-In-Process$(f, p) \equiv_{def} \exists b [p = \mathsf{process}(b, f)]$

A process token is represented by a pair $\pi = \langle p, i \rangle$, where p is a process type and i is the interval over which this process token is said to proceed. Therefore the subset of valid process tokens is composed by those pairs for which the proposition Proceeds$[a_{th}](p, i)$ is true ($[a_{th}]$ is a standpoint semantics parameter applied to handle temporal vagueness in this predicate and is described in detail in Section 4).

2.5   Relating Events and Processes

There has been disagreements in the literature about how *processes* and *events* are related to each other. While Sowa [26] defines event as a subclass of process, Pustejovsky [23] defines process as a subclass of event. They are also described as non-overlapping categories [2] or as subclasses of the same class (occurrence) [22]. Galton [17] suggests that events and processes can be related in many ways (events can be described in terms of other events or in terms of processes, while processes can be described in terms of events or in terms of other processes).

Following Galton [17], in our logical framework, a process can be described in terms of their constituent events, whilst an event can be described as a 'chunk of a process'. The relation Constituted-Of(b, v) associates an event classifier with a process classifier. Asserting a fact using this relation means that occurrences of an event classified by v over a certain interval denote that a process classified by b proceeds in that interval. Conversely, the relation Is-Chunk-Of(v, b) associates a process classifier with an event classifier. Asserting a fact using this relation means that the occurrence of an event (classified by v) on given time interval i, is determined by the fact that a process (classified by b) proceeds on i. The meaning of these relationships between events and processes shall be further clarified in Section 3 and in Section 4.1.

3   Process Activeness and Event Occurrences

Existing spatio-temporal datasets do not usually consist of explicit assertions of events occurrences or process activity. Rather, they often contain elements describing changes of objects' properties over time, from which events and processes can be inferred. For example, movements of objects can be identified from data on the position of such objects at different time instants. Similarly, the expansion of a built-up region can be inferred from data describing its spatial extensions at different times.

Deciding on whether data best reproduce the intended denotation of events or processes is a crucial issue for the development of a theoretical framework which is supposed to be implemented to operate on real datasets. On the one hand, a piece of data representing a single change affecting a feature (e.g., an increase in area between two distinct time instants t_1 and t_2) is not sufficient to infer the occurrence of an event, since information on the feature's area after t_2 would be required to determine the precise temporal boundary which is expected for an event. On the other hand, such a piece of data would not reproduce the density and homogeneity characteristics associated with a process. That is, whereas it can be inferred that an expansion process was active for some time between t_1 and t_2, it is not possible to ensure that the process was active during the whole interval.

In our proposed logical framework, we stick to the approach of abstracting away from the lack of information for representing the activeness of process. This is to say that, if it is known that a geographic feature changes from instant t_1 to t_2 and nothing is known about the period between them, we assume that the process which characterises this change is active at all time instants from t_1 to t_2. The *activeness* of a process is represented by the predicates Active-At(p, t) and Active-On(p, i). Whilst the former determines that a process of type p is active at a time instant t, the latter specifies that a process is active on a time interval i, meaning it is going on at every time instant within that interval. Since the former is defined in low level manner, that is, closer to the way data is represented, we leave the discussion on its definition to be given in Section 5.3, after we overview our approach to representing spatio-temporal data and ground geographic features in Section 5.1. Whereas the latter is defined as follows.

D 5   Active-On$(p, i) \equiv_{def} \forall t [\, \mathsf{In}(t, i) \rightarrow \mathsf{Active\text{-}At}(p, t)\,]$

Beyond the predicates to represent the activeness of a process, it is also convenient to define a predicate which verifies whether a process is inactive at a certain time instant or on a given time interval. The latter is particularly useful for defining the the predicate which determine event occurrences (presented later in this section) and the predicate which specifies whether a process proceeds (presented in Section 4). However, one should notice that there is a difference between a process being inactive during an interval and it not being active, that is, Inactive-On$(p, i) \not\equiv \neg$Active-On(p, i). For instance, if an interval includes some parts where a process is active and others where it is not active, then $\neg$Active-On(p, i) will hold. will hold. Hence, the predicates denoting a process inactivity are defined as follows.

D 6   Inactive-At$(p, t) \equiv_{def} \neg$Active-At(p, t)

D 7   Inactive-On$(p, i) \equiv_{def} \forall t [\, \mathsf{In}(t, i) \rightarrow \neg \mathsf{Active\text{-}At}(p, t)\,]$

We discussed that an event can be said as 'made of' a process, and that the culmination of the process (i.e., when the goal in initiating it is realised) denotes the occurrence of an event. Therefore, in our logical framework, an event token is modelled as a chunk of a process bounded by temporal discontinuities, meaning that the process is inactive on both time intervals which meets and is met by the interval on which the event occurs.

Furthermore, since we accept that events can also be regarded as constituent of processes, this might lead to a continuous cycle. Thus we realised the need for distinguishing event tokens which cannot contain other events of the same type. These are called *primitive* event tokens. Expressly, if a primitive event e occurs on an interval i, there is no sub-interval of i on which an event of the same type of e occurs.

It can be seen that allowing the existence of nested event tokens of the same type implies that events can also affected by gaps, since an event token is necessarily bounded by temporal discontinuities, as discussed above. Therefore another property of primitive event tokens is that they are not affected by temporal gaps, meaning that the process of which it is made must be active throughout the whole interval on which the event is said to occur. The predicate to represent a primitive event occurrence is defined in D8, whilst the representation of non-primitive ones is discussed in Section 4.

D 8   Occurs-On-Prim$(e, i) \equiv_{def} \exists v p b i' i''[e = \text{event}(v, f) \land p = \text{process}(b, f) \land$
$\quad \text{Is-Chunk-Of}(v, b) \land (\forall t[\text{In}(t, i) \rightarrow \text{Active-At}(p, t)]) \land$
$\quad \text{Meets}(i', i) \land \text{Met-by}(i'', i) \land$
$\quad \text{Inactive-On}(p, i') \land \text{Inactive-On}(p, i'')]$

Once defined predicates for modelling process activeness and event occurrences, the relation Is-Chunk-Of(v, b) (introduced in Section 2.5) can defined in terms of them. This is as follows.

D 9   Is-Chunk-Of$(b, v) \equiv_{def} \forall p f e i t[p = \text{process}(b, f) \land e = \text{event}(v, f) \land$
$\quad \text{Occurs-On-Prim}(e, i) \land \text{In}(t, i) \leftrightarrow \text{Active-At}(p, t)]$

Figure 1 exhibits a stretch of the timeline to illustrate possible primitive event tokens for an event type e. In this figure, 3 occurrences of the same event type e are shown (on the intervals i_1, i_2 and i_3). On the other hand, no event of type e occur (as a primitive token) on intervals i_4, i_5 or i_6, as such occurrences would not be in accordance with the properties of primitive event tokens discussed above.

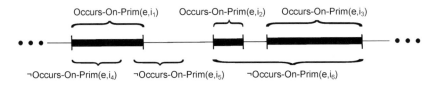

Fig. 1. Stretch of the timeline exhibiting primitive event tokens of type e. 3 distinct occurrences of the same event type (e) are shown (on intervals i_1, i_2 and i_3, respectively). No primitive event token of type e are identified on intervals i_4, i_5 or i_6.

4   Determining Processes' Boundaries

We have described our approach to representing an event occurrence based on the fact that a process is active throughout the whole interval on which the event occurs. However, we have discussed in Section 2.1 that a process may be affected by *temporal gaps*. We describe such temporal gaps as *periods of inactivity* of a process. Thus a given process may be regarded as active and inactive at different times within an interval on which it is said to proceed. Such interval determines the explicit boundaries for a process and is applied to identify distinct instances of a certain process type.

Our approach to determining such boundaries of geographic processes is based on standpoint semantics, so that particular instances of a given process type can be determined based on individual viewpoints. Therefore we say that a process of type p

proceeds on a time interval i if it is active on every subinterval i' of i, unless i' is shorter than a given *activeness threshold* a_{th} (i.e., a_{th} is a standpoint parameter for that predicate).

This definition, however, leads to another important representational issue. As discussed so far, it seems acceptable to say that a process may proceed over relatively short time intervals between its constituent events, representing periods of inactivity of that process. However, it seems controversial to say that a process proceeds on a time interval i which is started or finished by periods of inactivity, without taking into account the past or future to look for occurrences of constituent events on intervals outside i. To illustrate, let us suppose a process whose inactivity threshold is 2 weeks. Then assume the process has been inactive for 5 days until the present time. We could say the process is still active, since 5 days $<$ 14 days. However, how could we guess the process has not reached its culmination point?

We address this issue by defining a predicate Proceeds-On-Max$[a_{th}](p, i)$, which captures the maximum interval i over which a process proceeds. Hence the interval i must be bounded by temporal discontinuities which are *longer* than the threshold a_{th}. In other words, i is the maximum interval on which there exists no significant temporal gaps (longer than the specified threshold) in the process. This predicate is defined as follows.

D 10   Proceeds-On-Max$[a_{th}](p, i) \equiv_{def}$
$\exists i'i''[\, \text{Starts}(i', i) \,\wedge\, \text{Active-On}(p, i') \,\wedge$
$\text{Finishes}(i'', i) \,\wedge\, \text{Active-On}(p, i'') \,\wedge\, \text{Before}(i', i'')\,] \,\wedge$
$\exists i'i''[\, \text{Meets}(i', i) \,\wedge\, \text{length}(i') > a_{th} \,\wedge\, \text{Inactive-On}(p, i') \,\wedge$
$\text{Met-by}(i'', i) \,\wedge\, \text{length}(i'') > a_{th} \,\wedge\, \text{Inactive-On}(p, i'')\,] \,\wedge$
$\forall i'[\, \text{In}(i', i) \,\wedge\, \text{Inactive-On}(p, i') \,\rightarrow\, \text{length}(i') \leq a_{th}\,]$

According to the predicate described above, there may be many different intervals i on which a process of the same type (and consequently affecting the same individual geographic feature) proceeds. Therefore we say that each of these intervals determines the temporal boundaries of an *individual* process, whose spatial boundaries are established by the spatial extension of affected features. Once processes are individuated, we can determine different (and possibly non-overlapping) intervals on which the same individual process is said to proceed. This is specified by the the predicate Proceeds-On$[a_{th}](p, i)$, which is defined as follows.

D 11   Proceeds-On$[a_{th}](p, i) \equiv_{def} \exists i'[\text{Proceeds-On-Max}[a_{th}](p, i') \,\wedge\, \text{In}(i, i')]$

Figures 2 and 3 illustrate different situations in which a process of type p is said to proceed. Figure 2 shows a process token which proceeds on interval i. Although two periods of inactivity are identified within the process token, they are ignored as their duration are shorter than the specified activeness threshold a_{th}. The example of Figure 2 also exhibits how the logical constructs that appear in the Definition 10 match the illustrative situation. On the other hand, in the example of Figure 3, a smaller threshold is specified, meaning that shorter periods of inactivity are permitted to regard that the process token exists. Figure 2 illustrates 2 distinct Proceeds-On(p, i)

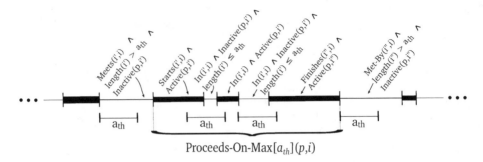

Fig. 2. Stretch of the timeline exhibiting a process of type p which proceeds (maximally) on an interval i, where a_{th} represents a given activeness threshold. There are 3 intervals on which the process is active and 2 intervals on which the process is inactive but whose lengths are shorter than the threshold. The process token is bounded by intervals over which the process is inactive and whose lengths are longer than a_{th}.

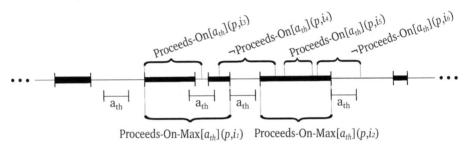

Fig. 3. The same stretch of the timeline shown in Figure 2. Here a shorter activeness threshold a_{th} has been specified and then 2 process tokens of type p are identified for i_1 and i_2. These tokens are separated by an interval over which the process is inactive and whose length is longer than a_{th}.

4.1 Deriving Events from Processes

It has been shown that primitive events are identified directly from data and therefore are not inferred from processes. However, we have discussed that any interval over which a process proceeds, bounded by intervals over which it does not proceed, can be regarded as a chunk of this process and therefore can be used to represent an event occurrence. Hence we employ a predicate to represent event occurrences in general (including non-primitive ones), which is defined in terms of processes they are made of. From Definition D12, it can be seen that the interval on which an event occurs is equivalent to the the maximum interval on which a process is said to proceed.

D 12 $\text{Occurs-On}[a_{th}](e, i) \equiv_{def} \exists v f p b i' i''[e = \text{event}(v, f) \land p = \text{process}(b, f) \land$
$\text{Is-Chunk-Of}(b, v) \land \text{Proceeds-On-Max}[a_{th}](p, i)]$

Finally, we specify Axiom A1 to determine that a primitive token is also regarded as a event token in general where the activeness threshold is zero, meaning that the event is made of a process which is not affected by temporal gaps.

A 1 $\mathsf{Occurs\text{-}On\text{-}Prim}(e, i) \rightarrow \mathsf{Occurs}[0](e, i)$

5   Identifying Processes in Spatio-temporal Data

We have discussed that the process activeness can be identified from primitive elements grounded upon the data. This section we present the logical apparatus which allow us to define the predicate $\mathsf{Active\text{-}At}(p, t)$ in such a way that processes can identified in a temporal series of topographic data.

5.1   Data Representation and Grounding

Implementing a system to verify such a logical framework with real data requires establishing a method of grounding the symbols upon elements of data. This requires work at multiple levels, both to select the appropriate set of predicates to be grounded and to formulate a suitable representation for the data. The approach to representing the spatio-temporal data and to grounding the logical framework upon the data is described in previous work [11]. This modelling approach is called STAR (Spatio-temporal Attributed Regions), and allows the prototype to operate on distinct spatio-temporal topographic datasets.

In the STAR approach, a spatial region is represented by a 2-dimensional *polygon g* (which specifies a portion of the earth's surface under some specified projection). This region is associated with an *attribute a* describing a property of that portion of geographic space which can be obtained by observation or measurements (e.g. 'vegetated', 'arid', or 'heavily populated'), and with a *timestamp s* denoting a time instant at which the attribute holds for the polygon. Facts of the form $\mathsf{Star}(a, g, s)$ are explicitly asserted in the Knowledge Base (KB), which is connected to a deductive mechanism that allows implicit facts to be derived from other existing ones.

Implicit facts derived by the system include, for example, spatial regions whose coverage have changed over time. For instance, a different extension of a 'forested' region can be inferred from the appearance of overlapping built-up regions. Furthermore, spatial extensions of geographic features can be derived from facts representing spatial regions associated with certain properties. For example, the extension of a 'desert' can be inferred from facts representing regions whose annual precipitation rate are below a particular threshold.

By implementing the STAR approach, our system becomes equipped with a grounding mechanism which assigns to variables c, f, r, t, i the appropriate values denoted by data elements (i.e., attributes, geometries and timestamps), so that primitive facts of the KB are represented in the form $\mathsf{Star}(c, r, t)$. In addition, this mechanism also includes definitions of the concepts of *feature-life*, *life-part* and *minimum-life-part (MLP)* in terms of facts of the form $\mathsf{Star}(c, r, t)$, so that higher level concepts (e.g., events, processes) can be defined only in terms of them and therefore without concerns about data structure. A feature life can be described as a spatio-temporal volume representing the portion of space which the feature occupies throughout the time interval over which it is said to exist. A life-part is any slice of such a spatio-temporal volume, whilst an MLP is a life-part for which there are data describing its beginning and end, but there are no data describing the period between them. An MLP is represented by the predicate $MLP(f, r_b, t_b, r_e, t_e)$, where r_b and r_e denote the spatial regions occupied by a feature f at time instants t_b and t_e, respectively.

5.2   Spatial Changes

Since we aim to represent processes in terms of spatial changes affecting geographic features, we employ the the relation *Spatial Change* $SC(b, r_b, r_e)$ to capture the intended denotation of a given process classifier. As for geographic features, this relation is defined in terms of primitive elements of our framework. However, a distinct definition of this relation should be provided for each process classifier p, where r_1 and r_2 represent spatial regions of a certain feature's MLP. Hence, these definitions determine how geometric computations should be computed at data level so that a spatial change can be identified within a feature's MLP.

To illustrate how a spatial change can be defined, we present definitions of the SC relation for 4 different process classifiers: *expanding*, *shrinking*, *extending* and *contracting* (Definitions D13 to D16). These changes are illustrated in Figure 4.

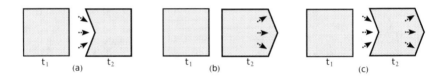

Fig. 4. Examples of spatial changes. In (a) a feature contracts and shrinks; in (b) a feature extends and expands; in (c) a feature extends and contracts, but neither shrinks or expands.

As shown in Definitions D13 and D14, *expansion* and *shrinkage* can be determined by comparing the area occupied by two given spatial regions.

D 13 $SC(expanding, r_1, r_2) \equiv_{def} \text{area}(r_2) > \text{area}(r_1)$

D 14 $SC(shrinking, r_1, r_2) \equiv_{def} \text{area}(r_2) < \text{area}(r_1)$

Whereas *expands* and *shrinks* are defined quantitatively, *extension* and *contraction* can be defined qualitatively, in terms of mereological relationships between two given spatial regions. These spatial changes are specified formally in Definitions 15 and 16 respectively. In these definitions, PP, PO, EC are the RCC relations *proper-part*, *partially overlaps* and *externally connected*, respectively.

D 15 $SC(extending, r_1, r_2) \equiv_{def} PP(r_1, r_2) \vee PO(r_1, r_2) \vee EC(r_1, r_2)$

D 16 $SC(contracting, r_1, r_2) \equiv_{def} PP(r_2, r_1) \vee PO(r_1, r_2) \vee EC(r_1, r_2)$

5.3   Defining Process Activeness

In Section 3, we discussed the assumption held for representing the activeness of a process: if there is a piece of data that describes a change affecting a geographic feature between time instants t_1 and t_2 and nothing else is known between them, then a process characterised by this change is said to be active from t_1 to t_2. Here such a piece of data corresponds to a feature's MLP, which represents the most detailed information held about changes affecting a geographic feature.

Thus we can now present our approach to defining process activeness. The predicate Active-At(p, t) is defined in a general manner so that it can be applied to any process classifier associated with a spatial change affecting geographic features. In this definition, a process-type associated with a process classifier b is said to be active at a time instant t having a geographic feature f as its participant if t is between the time instants t_b and t_e representing the beginning and end of a certain f's MLP, in which f changes as described by the process classifier.

D 17   Active-At(p, t) $\equiv_{def}$ $\exists b r_b t_b r_e t_e [$
$$p = \text{process}(b, f) \wedge \text{MLP}(f, r_b, t_b, r_e, t_e) \wedge$$
$$\text{SC}(b, r_b, r_e) \wedge (t_b \leq t < t_e)]$$

6   Application

To validate the theoretical framework proposed in this paper, we have implemented a *system prototype* which can process real geographic data. The system takes temporal series of topographic data as an input and allows logical queries to be formulated about the data, returning textual and graphical information on events, processes, and the geographic features which participate in them.

The system prototype's main screen is shown in Figure 5, where the smaller window on the front is a command-line terminal used to formulate user queries and to visualise textual results; and the window on the back contains an interactive map for visualising spatio-temporal results in a graphical way. In addition, the system provides basic GIS functionalities which can improve the analysis of results in many different forms. For example, query results might be overlaid on other thematic map layers.

The *system architecture* is shown in Figure 6. The system is structured in three main layers, named *data*, *processing* and *visualisation* layers. The data layer comprises the KB and the deductive mechanism which complies the aforementioned STAR approach. Facts are stored in a spatially enabled Data Base Management System (DBMS)[5], whilst the deductive mechanism is implemented using SWI-Prolog[6] programming language.

The processing layer comprises the *Interpretation* and *GIS Engines*. The former, also implemented in SWI-Prolog, includes the REGEP logical framework, which processes the user queries, and a mechanism to ground it upon the data. The latter is implemented

---

[5] We have been using the PostgreSQL DBMS (www.postgresql.org) and PostGIS (www.postgis.net), which adds to the former the capability of handling spatial information.

[6] SWI-Prolog, www.swi-prolog.org

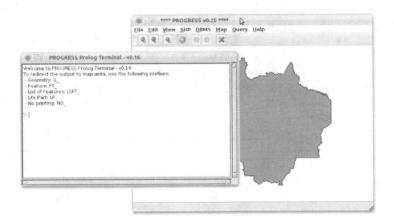

Fig. 5. System prototype's main screen

in Java[7] programming language and uses built-in spatial functions provided by the DBMS. This module provides the GIS functionalities mentioned above, processes geometric computations required by the grounding module and renders maps (which may contain spatial information representing the result of a user query). The visualisation layer provides mechanisms for formulating user queries and visualising their results. The user interaction can be performed both textually (via a Prolog-like command-line terminal) and graphically (via an interactive map). The next section gives an overview on the experiments carried out using this prototype.

7 Results of Using the System with Topographic Data

Experiments using this prototype have been conducted in the form of a case study, investigating the phenomenon of deforestation in Amazon rainforest in a 7-year period (between June 2004 and May 2011). The dataset used in this case study consists of a set of polygons representing regions deforested in Brazilian Amazon for each calendar month[8].

The initial KB contains 47,459 facts of the form $Star(c, r, t)$, where c is a type of region coverage which denotes the spatial region r is 'deforested' at time instant t. From these facts, for each distinct time instant, the maximal extensions of well-connected regions of this coverage are identified, representing spatial extensions of geographic features which are regarded as existing at each particular time instant. That is, implicit facts of the form $Star(u, r, t)$ are derived by the system, where u denotes the type of

---

[7] The implementation of the GIS Module make use of the GeoTools (www.geotools.org), a Java code library which provides standards compliant methods for the manipulation of geospatial data.

[8] These data is available online at http://www.obt.inpe.br/deter/. These are an output of DETER project at INPE, which uses remote sensing techniques to detect land cover changes within the Brazilian Amazon area.

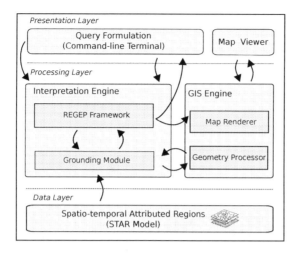

Fig. 6. Architecture of the system prototype

geographic feature and r is the spatial region whose extension corresponds to the extension of that feature at time instant t (i.e., $\text{ext}(f, t) = r$). Geographic features analysed in these experiments are denoted by 'deforested regions', whereas deforestation is analysed in terms of expansion of these features. The features described here have been generated by specifying an aggregation factor of 50Km, which resulted in an average of 38.92 features per time instant, ranging from 15 to 46.

To illustrate the way queries can be formulated and how the prototype can be interacted with, we now describe an example of a logical query and the results returned by the system.

Query 1: *Where was Amazon being deforested between 15/09/2005 and 30/04/2006?*

As deforestation is characterised here in terms of the expansion of features of type 'deforested', this query can also be described as *"show the geographic features of type 'deforested' that were expanding between 15/09/2005 and 30/04/2006?"*. Hence, for a given activeness threshold of 3 months, this query's equivalent representation in first-order logic is as follows:

$\exists f p [\text{Feature}[50\text{Km}](f) \wedge (\text{f-type}(f) = \text{deforested}) \wedge (p = \text{process}(\text{extending}, f))$
$\wedge \text{Proceeds-On}[3\text{months}](p, [15/09/2005, 30/04/2006])]$?

Logical queries submitted to PROGRESS are specified using Prolog syntax, so that they can be processed by the interpretation engine. Figure 7 shows one way in which Query 1 could be written to be input to the system, as well as the results it returns. The predicate `time_threshold`(0,3,0,0,0,0) represents the activeness threshold, and its parameters denote, respectively, the number of years, months, days, hours, minutes and seconds; `int_in`(...) corresponds to the temporal relation $\text{In}(i_1, i_2)$ between time intervals. Variables of a query can named using special prefixes handled by PROGRESS's Terminal to help the system control the result output. In the query of this illustration, `LFT_` is used to inform to the system that the variable is a list of features and that its value must be shown on the map (rather than on the Terminal), whilst `NO_` commands the system to hide the variable's value from the output.

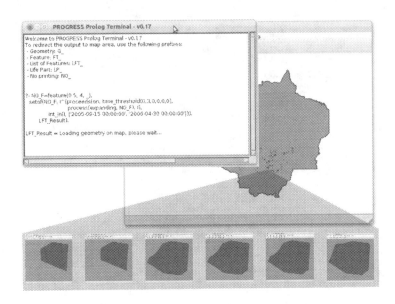

Fig. 7. Example of query formulation and system interaction

Figure 7 also exhibits the results displayed on the map area. The system provides a navigation mechanism which allows the user to verify the extension of the features at different time instants. On the bottom of Figure 7, it is shown the extension of a particular feature for 6 consecutive months within the specified interval. From these illustrations, it can be seen that the feature's extension remained unchanged for a certain period of time, however this period is shorter than the 03 months activeness threshold and therefore the feature was still regarded as a process participant.

The query shown in Figure 7 uses a built-in predicate *setof* to collect solutions together by repeatedly backtracking and generating alternative values for a *result set*, corresponding to different instantiations of the free variables of the goal. As the only values of interest were those of F variable, the *Interval* variable is existentially quantified *I*. The following query illustrates a different scenario where time intervals are also of interest.

Query 2: *Where and when was Amazon deforested before 2011?*

This query could be rewritten as *"show the geographic features of type 'deforested' which expanded before 2011 and the respective time intervals on which these expansions occur"*. This query could be specified in our prototype as follows.

```
?- FT_F=feature(0.5,4,_), NO_E=event(expands, FT_F),
   occurs(on, NO_E, I), int_before(I,
   ['2011-01-01 00:00:00', '2011-01-01 00:00:00']).
```

In the last example query, it can be seen that FT_F and I are both free variables, and therefore their values are gradually displayed on the map and on the terminal (respectively), corresponding to different solutions for the query.

8 Conclusions and Further Work

We presented a logical framework for representing and reasoning about events and processes that are conceptualised in terms of spatial changes affecting the spatial extensions of geographic features. The field of events and processes representation is still the subject of considerable controversy in the literature, and therefore previous work have avoided providing precise definitions for certain concepts. Conversely, our formalism includes a number of precise definitions, with the aim of applying this for processing real topographic data. Furthermore, we presented a system prototype that implements the proposed theory. This helps demonstrate the framework is effective for identifying dynamic geographic entities in spatio-temporal data, by processing logical queries specified in a high level of abstraction. Apart from the use of this framework for supporting the development of modern GIS, this can also be employed as a basis for other formalisms, such as those for modelling causal relations between events and processes.

This work can be extended in many directions. For instance, the modelling of more complex relationships between events and processes would allow the framework to be used to reason about a broader range of geographic phenomena. Moreover, since we regard a process as an entity that is subject to change over time, the modelling of certain properties of processes (e.g. a process may be said to accelerate) is also a desired feature to be incorporated to our framework.

References

1. Allen, J.: Maintaining knowledge about temporal intervals. Communications of the ACM 26(11), 832–843 (1983)
2. Allen, J.: Towards a general theory of action and time. Artificial Intelligence 23(2), 123–154 (1984)
3. Bennett, B., Galton, A.: A unifying semantics for time and events. Artificial Intelligence 153(1-2), 13–48 (2004)
4. Bennett, B.: Modes of concept definition and varieties of vagueness. Applied Ontology 1(1), 17–26 (2005)
5. Bennett, B.: Spatial vagueness. In: Jeansoulin, R., Papini, O., Prade, H., Schockaert, S. (eds.) Methods for Handling Imperfect Spatial Information. STUDFUZZ, vol. 256, pp. 15–47. Springer, Heidelberg (2010)
6. Bennett, B.: Standpoint semantics: A framework for formalising the variable meaning of vague terms. In: Cintula, P., Fermuller, C., Godo, L., Hajek, P. (eds.) Understanding Vagueness — Logical, Philosophical and Linguistic Perspectives. College Publications (2011)
7. Bennett, B., Mallenby, D., Third, A.: Automatic grounding of vague geographic ontology in data. In: Proceedings of the Second International Conference on GeoComputaton (2007)
8. Bennett, B., Mallenby, D., Third, A.: An ontology for grounding vague geographic terms. In: FOIS 2008, pp. 280–293. IOS Press (2008)
9. Bittner, T., Smith, B.: Formal ontologies of space and time. IFOMIS, Department of Philosophy. University of Leipzig, University at Buffalo and NCGIA 17, Leipzig, Buffalo (2003)
10. Campelo, C.E.C., Bennett, B.: Geographical processes representation: Issues and challenges. In: Podobnikar, T., Ceh, M. (eds.) Universal Ontology of Geographic Space: Semantic Enrichment for Spatial Data. IGI Global, USA (2012)

11. Campelo, C.E.C., Bennett, B., Dimitrova, V.: From polygons and timestamps to dynamic geographic features: Grounding a spatio-temporal geo-ontology. In: Castano, S., Vassiliadis, P., Lakshmanan, L.V.S., Lee, M.L. (eds.) ER 2012 Workshops 2012. LNCS, vol. 7518, pp. 251–260. Springer, Heidelberg (2012)

12. Claramunt, C., Theriault, M.: Toward semantics for modelling spatio-temporal processes within GIS. Advances in GIs Research I, 27–43 (1996)

13. Cohn, A.G., Bennett, B., Gooday, J., Gotts, N.: RCC: a calculus for region-based qualitative spatial reasoning. GeoInformatica 1, 275–316 (1997)

14. Devaraju, A., Kuhn, W.: A Process-Centric ontological approach for integrating Geo-Sensor data. In: FOIS 2010 (2010)

15. Fine, K.: Vagueness, truth and logic. Synthese 30(3), 265–300 (1975)

16. Galton, A.: On what goes on: The ontology of processes and events. In: Bennett, B., Fellbaum, C. (eds.) Proceedings of FOIS Conference 2006, vol. 150. IOS Press (2006)

17. Galton, A.: Experience and history: Processes and their relation to events. Journal of Logic and Computation 18(3), 323–340 (2007)

18. Galton, A., Mizoguchi, R.: The water falls but the waterfall does not fall: New perspectives on objects, processes and events. Applied Ontology 4(2), 71–107 (2009)

19. Grenon, P., Smith, B.: SNAP and SPAN: towards dynamic spatial ontology. Spatial Cognition & Computation, 69–104 (2004)

20. Hornsby, K., Egenhofer, M.J.: Identity-based change: A foundation for spatio-temporal knowledge representation. International Journal of Geographical Information Science 14, 207–224 (2000)

21. Masolo, C., Gangemi, A., Guarino, N., Oltramari, A., Schneider, L.: Wonderweb EU project deliverable d18: The wonderweb library of foundational ontologies (2003)

22. Mourelatos, A.: Events, processes, and states. Linguistics and Philosophy 2(3), 415–434 (1978)

23. Pustejovsky, J.: The syntax of event structure. Cognition 41(1-3), 47–81 (1991)

24. Randell, D., Cui, Z., Cohn, A.: A spatial logic based on regions and connection. KR 92, 165–176 (1992)

25. Raskin, R., Pan, M.: Knowledge representation in the semantic web for earth and environmental terminology (sweet). Computers & Geosciences 31(9), 1119–1125 (2005)

26. Sowa, J., et al.: Knowledge representation: logical, philosophical, and computational foundations, vol. 511. MIT Press (2000)

27. Stell, J., Worboys, M.: A theory of change for attributed spatial entities. Geographic Information Science, 308–319 (2008)

28. Worboys, M., Hornsby, K.: From objects to events: GEM, the geospatial event model. Geographic Information Science, 327–343 (2004)

29. Zadeh, L.A.: Fuzzy sets. Information and Control 8(3), 338–353 (1965)

Trust and Reputation Models
for Quality Assessment
of Human Sensor Observations

Mohamed Bishr[1] and Werner Kuhn[2]

[1] Paluno_The Ruhr Institute for Software Technology, Duisburg-Essen University,
45127 Essen, Germany
mohamed.bishr@paluno.uni-due.de
http://paluno.uni-due.de/
[2] Institute for Geoinformatics, University of Muenster, Muenster, Germany
kuhn@uni-meunster.de

Abstract. With the rise of human sensor observation as a major source of geospatial information, the traditional assessment of information quality based on parameters like accuracy, consistency and completeness is shifting to new measures. In volunteered geographic information (VGI) these conventional parameters are either lacking or not explicit. Regarding human observation quality as fitness for purpose, we propose to use trust and reputation as proxy measures of it. Trustworthy observations then take precedence over less trustworthy observations. Further, we propose that trust and reputation have spatial and temporal dimensions and we build computational models of trust for quality assessment including these dimensions. We present the case study of the $H2.0$ VGI project for water quality management. Through agent based modeling, the study has established the validity of a spatio-temporal trust model for assessing the trustworthiness and hence the quality of human observations. We first introduce a temporally sensitive trust model and then discuss the extension of the temporal model with spatial dimensions and their effects on the computational trust model[1].

1 Introduction

The term Web 2.0 was coined to highlight web applications where users become information contributors. Today even search engines are tapping into the social web by enabling user feedback on search results and corporations are looking into social search solutions enabled by social networks to get relevant results by tapping into the users' social circles. On the social web, reputation and trust become essential, much like they are in modern societies in general [1]. Both concepts, trust and reputation, can now be used to help people make judgments about the quality and relevance of information produced by other users.

---

[1] Research leading to this work was done while the first author was at the Muenster Semantic Interoperability Lab (MUSIL), University of Muenster, Germany.

T. Tenbrink et al. (Eds.): COSIT 2013, LNCS 8116, pp. 53–73, 2013.
© Springer International Publishing Switzerland 2013

While trust has always been at the top of the semantic web layer cake, its role remains open for interpretation by research. In [2] trust is discussed from the perspective of society. The author argues that the role of government agencies as the sole trustworthy providers of geospatial information (GI) has receded in much the same manner as that of traditional encyclopedias. In the world of GI, VGI [3] is arguably playing a similar role to wikipedia-like information sources. Yet, traditional information quality parameters like accuracy, consistency, completeness, and provenance are either lacking or not explicit in VGI [1,2] as much as in all Web 2.0 content. Since trust and reputation models have been used to filter collaborative content, they can also be used as proxy measures of the quality of VGI. Information quality is then viewed as fitness for purpose, which is an adequately subjective and context-dependent measure.

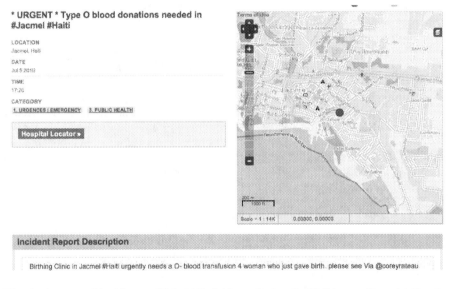

Fig. 1. A reported incident on Ushahidi platform during the Haitian earthquake disaster

The massive community effort following the Haitian earthquake produced invaluable user generated content that shaped the relief effort. The Ushahidi[2] project platform allows users to submit reports through SMS, MMS, email, and an online web interface. Figure 1 shows a geo-tagged report during the Haitian earthquake crisis. Ushahidi received alerts from people on the ground using SMS and producing a massive influx of reports during the early days of the crisis. Rescue messages were also sent by volunteers on the ground through Twitter, where other volunteers in various parts of the world translated the messages from Creole to English, geo-tagged them and placed them on Ushahidi. Finally, the U.S. state department intervened to assist in the geo-location and verification of the flow of messages to assist in delivering trustworthy information to the

---

[2] http://www.ushahidi.com/

Red Cross and the U.S. Coast Guard [4]. Ushahidi is centered around real-time geo-tagged reports that contain actionable messages and has been widely used for purposes in and beyond disaster response[3]

Ushahidi was hailed by army officials involved in the relief effort as having saved lives every day [5]. However, the relief efforts could have been improved if the community was able to curate and classify reports in a more efficient manner to ensure that higher quality reporting takes precedence. Similar arguments have been made about wild fires occurring between 2007-2009 in Santa Barbara, California [6].

Until today, there are no reliable methods to control the quality of user generated content in general and specifically in real time environments where ad-hoc networks of users form around situations like a natural disaster or a rapidly changing and important phenomenon like water quality, as our use case below illustrates. We address this lack of controlling and filtering methods for VGI by developing spatio-temporal trust and reputation models in two stages. The first model incorporates the temporal dimension and the second the spatial dimensions along with time. This way we isolate the effects of time and space on the models to prove their effectiveness in filtering observations. The models enable trustworthy observations to take precedence over untrustworthy observations, based on spatial and temporal dependence.

2   Related Work

In [2], it is argued that VGI is characterized by the scaling up of closed loops, where closed loops are adaptive systems incorporating feedback. The author uses the analogy of adjusting the water temperature in a shower to represent a closed loop where VGI is the *hot water* which needs to be managed. The tremendous increase in the flow of GI from volunteers means there is a growing need to manage, filter and integrate data reflecting different world views and different levels of quality. Approaching this need through computational models of trust and reputation offers a potent addition to current data management methods in GIScience to manage and filter collaborative content.

This section covers the fundamentals necessary to understand the paper. We start with a cross-disciplinary discussion of trust to establish the context of our research when dealing with such an overloaded term. Throughout this discussion we also lay the foundations for considering the spatio-temporal aspects of trust and their role in this research. We then briefly discuss previous work on the quality of VGI in comparison to the approach taken in this paper.

2.1   Trust and Reputation: A Sociological, Economic and Computer Science Perspective

Traditional strands of sociology have held two complementary views on society. The first is concerned with the societal whole, complex structures and social

---

[3] e.g. monitoring in elections Egypt. http://crisismapper.wordpress.com/2010/11/20/ushahidiegyptwhenopendataisnotsoopenorwhenpeoplejusdon'tgetit/

systems; the second with societal members and individual actions. This second strand highlighted the importance of trust as an element in individual interactions and based on individual actions [7]. Individuals rely on those involved in *representative activities* or those who act on their behalf in matters of economy, politics, government, and science, implying high degrees of trust by the individuals. In the online world similar behaviors occur in collaborative activities like Wikipedia, where the millions of users rely on about 1% of the others users to create 90% of the content for them, indicating the implicit role of trust in lubricating the wheels of the system.

In [7, p.27] trust is defined as *a bet about the future contingent actions of others.* This is the definition we adopt in our work and it has two components, belief and commitment. A belief that a certain person will act in a favorable way and a commitment to a certain action based on that belief. Such definition is consistent with other definitions like that of Deutsch [8] and both definitions have a game theoretic view which is further consistent with the economic view on trust [9]. A natural implication of such definitions is that trust acts as a mechanism for the reduction of complexity in society as suggested by Luhmann [10], which can also mean that trusting information coming from individuals implies something positive about the quality of this information.

Gambetta [11] was among the first to use values for trust. While trust is subjective by nature – what is trusted by one person can be distrusted by another – the exercise of placing values representing trust is an essential first step in formalizing the concept. This approach has shown value in the computer science literature, by placing trust values on individuals in social networks on a continuous binary scale, and then using such values to detect important email messages from spam, for example [12]. Gambetta argues that *Trust (or, symmetrically, distrust) is a particular level of subjective probability with which an individual assesses that another will perform a particular action, both before he can monitor such action and in a context in which it affects his or her own action [11, p.217]..* In an online context things are different from this view in that a low value on the trust scale implies, in most practical situations, that one does not have enough information to assess trust computationally in an efficient way. This lack of knowledge does not necessarily mean a lack of trust, but rather implies that one simply cannot determine if a person is trustworthy, which is different from distrusting. A low computed value of trust implies weak trust and not distrust.

Other relevant work in economics studies the relation between trust and geographic proximity. In Buskens [9] distance is used as a proxy measure for social network density of buyer and supplier networks. It is asserted that partners in proximity are always preferred partners [13], implying a higher degree of trust. This is also due to the fact that social network embeddings resulting from firms located closer to each other mean that they have more common ties among themselves and with third parties nearby, further strengthening the effects of geographic distance on trust.

Buskens also asserts that the cost to the buyer as a result of abusing trust by the trustee increases with distance. This means that the larger the distance, the more difficult it is to resolve problems, leading to lower trust. A similar conclusion is made in [14], where distance is found to have a positive effect on the probability of a subcontractor and a customer governing their relation by a formal contract (i.e., having less trust). The explanation is provided in [15] where the subjects indicated that personal contact is important for establishing trust, and reducing geographic distance was necessary for easing this personal contact. Whether such effects also exist in modern online social networks is not entirely clear, but there is strong evidence of similar effects of geographic proximity on how online friendships emerge in online social networks [16].

From an information science perspective, studies on credibility of information on the web have dealt with web site appearance and design, as well as other signs of authority such as the trustworthiness and reputation of the entity behind the website, be it a person or a company [17–21]. To avoid unnecessary complexities in this paper, we look at trust assessment of some information entity a user encounters online as a single unit of decision or a black-box, without going deeper into the underlying cognitive aspects of the decision. We are simply interested in the decision of a user about whether or not to trust an information entity and not in the process by which the decision is derived in the mind of the individual. We also hypothesize that, if the information entity has spatial and temporal dimensions, these can be expected to have some effect on the trustworthiness of the information [1, 22]. One might trust another in providing information about Berlin because he lives there and not about London where he has never lived.

Finally, of interest to this work are the properties of online trust between individuals as identified in [12], namely:

- *Composability and transitivity:* composability pertains to the ability to combine different trust values, which in turn entails transitivity. Computationally, we can make trust transitive along chains of connections (see e.g. [12, 23, 24]).
- *Personalization and asymmetry:* this pertains to two intuitive facts about trust. Trust is a personal opinion, so that two people will typically have different trust values for the same person. This also means that trust is asymmetrical; if A trusts B with some value x, it can happen that B trusts A with a lower or higher value y.
- *Contextualization:* trust is highly contextualized. Trusting you to review my paper does not entail trusting you to fix my car. However, in this paper we assume that the context of trust is always fixed as a simplifying assumption.

2.2   Reputation and Trustworthiness

Trustworthiness of a certain person is not an intrinsic quality of that person. This is to say that trusting someone does not actually mean the person is trustworthy, it simply means the trustor decided to place trust in that person; whether or not the trustee will honor this trust or defect is another question.

When trustworthiness is viewed this way, reputation is then said to be the perception of trustworthiness of a person by the community [25]. Reputation of a person is not an act of that person, but a quality bestowed upon that person by the community, and it depends on many factors including previous behavior, community perception of the person, the capacity of the community to sanction bad behavior and propagation through word of mouth.

In this paper we propose that reputation is the collective trust vested in a person by the community. If Bob trusts Alice with a value x, then the collective of x values by all other members of the community is said to be Alice's reputation. Reputation is also a contextualized notion, although slightly differently from trust. One can see that acquiring a bad reputation as a student could imply a bad reputation in other contexts such as work, or family. However, for our work we consider that reputation in one context has no influence on other contexts. Thus reputation of person x in context C can be represented as:

$$R_x = (\widehat{T_x}, C) \tag{1}$$

In other words reputation of person x is a function of the collective trust in person x by the community, where this collective trust is denoted $\widehat{T_x}$ in context C. Following traditional economic thought, rational agents would naturally like to maintain good reputations and have an incentive to make quality observation contributions in order to maintain their reputation. It could be said that quality of information contributed by a person is related to their reputation and people try to maintain a good reputation by providing high quality information. Finally, the way in which we define reputation in relation to trust implies that reputation will be subject to the same spatio-temporal influences as trust.

2.3   Quality of VGI

Quality is one of the first issues that arise when one wants to use VGI for any purpose. Early work on quality of GI focused on accuracy with the assumption that there is a universal truth to which GI can be compared with respect to its accuracy [26]. This view, however, is unrealistic given the near impossibility of measuring a universal truth. Thus the focus later shifted to modeling of uncertainty in GI [6] implying a more subjective view on the topic.

Quality of VGI has in several cases been studied by comparing authoritative data sources with their VGI counter parts. This approach means that in many cases we are comparing recent VGI like OpenStreetMap that is dependent on technologies like GPS with high positional accuracy against older mapping data that might have lower positional accuracy [6]. We believe this approach to quality modeling is limited in most cases because of these positional accuracy issues, and also because of the fast changing nature of VGI to reflect changing world views and truths.

In [27] an approach to analyzing spatial semantic interactions of points of interest (POI) in OpenStreetMap combines point pattern analysis with semantic similarity. The spatial-semantic interaction derived can measure the likelihood of some feature types co-occurring within a specific semantic and spatial range. Consequently, a recommender system can use these results to suggest POI types to the mapper adding new POIs based on the spatial-semantic interaction within its proximity. This in turn can increase the overall quality of the VGI product. The authors argue that the same recommendations can mean that some POIs reported in an area might be unlikely, and hence the map editors can be alerted to examine them. This approach to quality requires the existence of a dataset like Open-StreetMap to run the analysis, but it does not deal with the real-time nature of many VGI applications where on the fly decisions on data quality are essential, such as the use case discussed later in this paper or disaster and crisis management situations. Also, this approach does not utilize the simple fact that VGI is after all a social phenomenon, being about people interacting in space and time. However, we believe recommender systems are an essential building block in the research agenda on VGI quality. Another approach using workflows to control positional and syntactical errors in VGI is presented in [28], which can be combined with the recommender system approach to improve quality of VGI by assisting the volunteers in making more informed and consistent contributions.

To our knowledge, our approach is novel in its introduction of trust as a proxy measure of VGI quality and its formalization of the social construct of trust to build spatio-temporally sensitive computational models for real-time quality assessment of VGI.

3 Trust as a Proxy Measure of Observation Quality

With respect to VGI, we make two observations. Firstly, human observations have obvious spatial and temporal aspects. Secondly, computational trust and reputation models aimed at VGI have, to the best of our knowledge, not been attempted earlier. Current trust models developed in the context of social networks [12] are static in nature - they, in many cases, rely on user ratings and do not take into account space and time. To introduce such models, we need to establish some foundations on the spatial and temporal aspects of trust, and on how such models can deal with VGI.

3.1 Spatial Aspects of Trust

Based on [9] we recognized earlier that social network density has a positive effect on trust relationships in real-world social networks. In this research, we are concerned with online communities. We note that online communities and real-world communities could both, to a limited degree, be used as proxies for one another. From the fact that network density is affected by geography such that it enhances trust relationships, we hypothesize that geographic proximity positively affects trust relationships also in online communities.

In [16] friend-formation patterns in a large scale spatially situated social network harvested from online sources are studied. The probability of befriending a certain person turns out to be proportional to the number of people in geographic proximity. This does not, of course, say much about trust in particular, but it does imply some relation since we postulate that friendship rests on an inherent trust component at some level. The question here is to what level geographic proximity impacts trust between people, and consequently recommendations about observations. Our claim is that representing the spatial dimension explicitly in trust and reputation models for spatial information like VGI allows for building effective quality models for VGI.

3.2   Temporal Aspects of Trust

An attempt to formalize the notion of trust in distributed artificial intelligence is introduced in [29]. Marsh produces a theoretical model of trust and incorporates sociological and psychological aspects, including the temporal nature of trust. A problem with Marsh's model is that, in the open environment of the web, much of the information needed for his model is not available [12]. Users tend to give each other trust ratings in a limited context that mostly has to do with trusting the user's recommendations about some product. However, we find Marsh's model highly relevant in that it formalizes how humans build and maintain trust. Particularly interesting is the time aspect dealt with in the model. Also, an implication of the work reported in [30] is that trust is developed over time as a result of continuous interactions and does not arise spontaneously. Hence, we assert that trust is built slowly over time, but it is also tarnished immediately when abused. Some trust models have tried to incorporate this intuitive assumption (see [31]). Thus, trust relationships develop and decay with the temporal dimension.

3.3   Informational Trust

Several researchers argue that trust holds only between people, which would require an explanation for the semantics of statements like *I trust this information*. Trusting a company like Lufthansa to take you to your destination is, in fact, trusting the people behind the company and their acts, in a way personifying the company as an institution. How can we then argue for using trust in information as a measure of information quality in this paper? We do this by developing the notion of *informational trust*.

We adopt the definition of trust in [7], as *a bet about the future contingent actions of others* for *interpersonal trust*, establishing a social tie between a trustor and a trustee [32]. One can argue that trust in objects is based on trust in the person(s) responsible for or producing these objects. Following this argumentation, we propose the notion of *people-object transitivity of trust*, which relates to trust transitivity discussed earlier. In our view, *interpersonal trust* implies the transition of trust from the trustee to information objects conveyed by the trustee. The trustor can then assert trust in the information conveyed by the trustee. We call the result *informational trust*; where a trusting tie between a trustor and an information object such as VGI is mediated by interpersonal trust between the VGI originator and the VGI consumer.

Given our earlier discussion on the spatial and temporal aspects of trust, we further propose to model *informational trust* by spatial and temporal characteristics. It is plausible to assert that trust decays and develops over time, and that trust develops slowly, but can easily be tarnished if abused. It is also intuitive that people's proximity to observed phenomena impacts how others trust their observations. Informational trust and its spatio-temporal dependencies are at the core of the computational model presented and evaluated in this paper.

4   A Computational Model of Trust and Reputation

In this section we start by presenting the use case for this paper based on the $H2.0$ [4] initiative, which is based on the Human Sensor Web project [5] for water availability monitoring. Then, we move to the design of a temporally sensitive computational trust and reputation model and show its application to the presented use case.

4.1   Case Study

The system on which this use case is based has been implemented in Africa [6] and is described in [33]. The project was co-funded by google.org and UN-Habitat. The project develops a Water Supply Monitoring System (WSMS), based on the vision of humans as sensors [3]. The project utilizes sensor web technology to implement a VGI system by which the local population can report on the quality of water wells in the area as well as inquire about the available water wells that have drinkable water within the same area.

Figure 2 illustrates the WSMS system showing the major user groups and interactions of the users with the system. The users are the general public, the subscribers, and the reporters. Reporters act as sensors and make observations, reporting them to the WSMS. The subscribers to the WSMS receive notifications upon request on the water quality of the water wells in the area. The general public users are neither subscribers nor reporters, but use a web portal to access the information from the WSMS.

The WSMS system in general assumes an ideal environment where all users act altruistically for the common good of all. While this assumption is reasonable in many cases, it falls short of its promise in this scenario, where the motivation for the abuse of the system by some users is high. We postulate various scenarios where the system is misused or abused, among which:

- inexperienced users make faulty reporting; this results in low quality information in the system despite the good intention of the users;
- malicious users from competing tribes attempt to manipulate the system to keep the high quality water wells for themselves by misguiding others;

---

[4] http://www.h20initiative.org/

[5] http://www.h20initiative.org/article/17001/Human_Sensor_Web

[6] http://geonetwork.itc.nl/zanzibar/

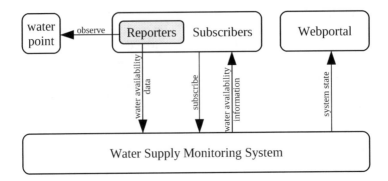

Fig. 2. Overview of the Water Supply Monitoring System [33]

– as in many web 2.0 applications, the flow of VGI becomes so large as to make effective information triage hard.

The result of these factors is low or at least uncertain information quality or an information overload that makes the WSMS an unreliable source of information for all its users. To mitigate such risks the project implemented a simplified version of the computational model discussed in this paper. Here, we will go beyond that and use agent based modeling of the system to evaluate the computational model presented in the next section.

4.2   Computational Model Design

We focus here on the computational artifacts of the model and touch on the dynamic aspects without much detail, except when it is necessary to explain initialization and post-initialization phases of the model.

To help understand the design of the computational model, we indicate here that it has an initialization phase and a post-initialization phase. In figure 3 we illustrate the main artifacts of the computational model at both phases. Water wells in our model are denoted w_z where the set of all wells is $W = \{w_1, ..., w_z\}$. The volunteers in the same figure $i \in volunteer$ where each volunteer makes a report r_i about a water well. Thus the set of all drinkability reports for a specific water well w_z is $D_z = \{r_1, ..., r_i\}$. Volunteer i makes a drinkability report r_i about water well, where $r_i \in \{d, u\}$. Here $d = drinkable$ and $u = undrinkable$.

During the initialization phase we have no information other than that about new volunteers i and their reports r_i where $r_i \in \{d, u\}$. To initialize the system we use the majority rule by taking the toll of drinkable $r_i = d$ versus undrinkable $r_i = u$ reports for any water well w_z and making a prediction on the water well status from the volunteer reports. This rule is illustrated in equation 2. In other words if the number of all drinkable reports $r_i = d$ for a water well w_z is higher than that of the undrinkable reports $r_i = u$ for the same water well then the water

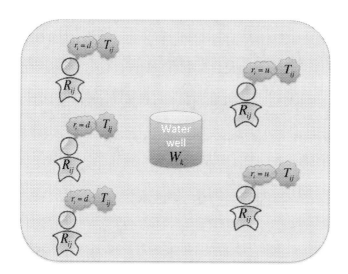

Fig. 3. An Illustration of the main elements of the computational model interactions. w_k is a water well. R_{ij} is the reputation of the user i in the system at history moment j, and T_{ij} is the informational trust value of a report r_i by person i at a history moment j.

well status will be predicted to be drinkable and vice versa. Initial reputation values are assigned based on compliance of a volunteer with the majority rule.

$$w_z = \begin{cases} drinkable, & \text{if } \#\{r_i \in D_z | r_i = d\} \geq \#\{r_i \in D_z | r_i = u\} \\ undrinkable, & \text{if } \#\{r_i \in D_z | r_i = d\} < \#\{r_i \in D_z | r_i = u\} \end{cases} \quad (2)$$

We gather information about the correctness of a volunteer's reporting during initialization based on compliance with the majority, gradually building user reputations through an approach similar to [31]. Equation 3 is used to reward users who do honest reporting with increasing reputation while equation 4 is used to punish those who make fraudulent or false reporting. We use two equations so that we can adjust equation parameters to control the rate of increase or decrease in reputation to cause a slow build up in user reputations and a fast tarnish when abuse occurs.

$$\widetilde{R}_{ij} = \alpha R_{ij} + (1 - \alpha)P_{pos} \quad (3)$$

where $P_{pos} \geq 0$

$$\widetilde{R}_{ij} = (1 - \alpha)R_{ij} + \alpha P_{neg} \quad (4)$$

where $P_{neg} < 0$

In equations 3 and 4 R_{ij} is the reputation of a volunteer i at a specific moment j in the model's history where $R_{ij} \in [-1, 1]$, and $\widetilde{R}_{ij}$ is the newly computed value of the reputation for the same person i at moment j, based on the previous reputation value and the other model parameters. The parameters $P_{pos} \in [0 - 1]$

and $P_{neg} \in [-1, < 0]$ control the reward rate and punishment rate for increasing or tarnishing reputations of volunteers and α is a control parameter to be determined experimentally to further optimize the reward and tarnish rates.

Since users in the model now have reputation values R_{ij} we introduce trust values for each individual report from each volunteer. Each report r_i created by a volunteer i at a specific moment in the model's history j is assigned an *informational trust* value T_{ij}.

We now proceed to integrate the temporal dimension of *informational trust* into the model. The water wells are by their nature not stable with respect to water quality such that a well that is drinkable at some time might become undrinkable as time passes. The *informational trust* of any report is, thus, subject to decay in its truth value, given the possible change in water well status, no matter whether the volunteer making the report was honest. Such a decay rate for the *informational trust* of the reports has to be determined for each model implementation depending on the phenomena being observed. We introduce an exponential time decay function 5. In this equation informational trust T_{ij} assigned to a report from a volunteer i at a specific moment in the model's history j decays in time t_j where t_j is the time passing from the moment j when the report is made. The parameter k controls the decay rate of *informational trust*. Thus the status of the water well will change with time as the truth value is impacted by the gradually decaying trustworthiness of volunteer reports.

$$\widetilde{T}_{ij} = T_{ij} e^{-k \times t_j} \tag{5}$$

Equation 5 integrates the temporal dimension of *informational trust* into the model. In the post-initialization phase, we have volunteer reputations R_{ij} and informational trust values T_{ij} for each report $r_i \in \{d, u\}$ of all the reports on a water well w_k. At this point, we can predict the final status of w_k. Let us denote the indices of the drinkable reports of all volunteers in all reporting histories j as $I^d := \{(i, j) \in volunteer \times history | r_{ij} = d\}$ and the indices of the undrinkable reports of all volunteers in all reporting histories as $I^u := \{(i, j) \in volunteer \times history | r_{ij} = u\}$. In both cases the reports are sub-indexed with j to denote the moment in the model history when the report is made. The final prediction of the water well status is given by equation 6.

$$w_z = \begin{cases} drinkable, \text{ if } \sum_{(i,j) \in I^d} R_{ij} \cdot T_{ij} \geq \sum_{(i,j) \in I^u} R_{ij} \cdot T_{ij} \\ undrinkable, \text{ if } \sum_{(i,j) \subset I^d} R_{ij} \cdot T_{ij} < \sum_{(i,j) \in I^u} R_{ij} \cdot T_{ij} \end{cases} \tag{6}$$

Equation 6 is somewhat similar to equation 2 in that it produces the binary values of drinkable or undrinkable. However, instead of using majority rules and absolute counts of reports, it uses sums of reputations and informational trust values to arrive at a prediction for the status of a water well.

4.3   Case Study Scope and Limitations

When modeling our case study using agent models, we make simplifying assumptions. For example, the water quality changes according to a probability distribution. Also, there is no feedback from the system users. In the following section we utilize minimal data input from the environment to do trust and reputation computations in our agent based model. Thus, one is tempted to ask questions, such as: will the subscribers not want to give feedback when they receive erroneous information from the system? is there no spatial correlation in the water quality of neighboring wells? are the events and processes behind water quality deterioration known? Generally there is a significant amount of information available from the environment beyond the reputation of individual reporters that helps determine the reliability of VGI. Yet, our agent-based model does not model any of the phenomena and focuses on trust and reputations and their temporal and spatial dimensions. The rationale behind this are as follows:

- *Focus of this research:* looking at the hypothesis behind this research, namely, "Spatially and Temporally Sensitive Trust and Reputation Models are Suitable Proxy Measures for VGI Quality" it becomes clear that we study spatiotemporal trust and reputation models as quality measures for VGI. Research questions concern the effects of trust and reputation themselves and the effects of spatial and temporal aspects on the efficacy of such models. For this purpose, it is not helpful to model correlations between water quality in adjacent wells or the events and processes behind water quality deterioration or even user feedback. Assuming fixed values or modeling such variables by random distributions enables us to isolate their effects from the variables of interest.
- *Inherent complexity of ABM's:* ABM's can become complex and their resulting data can become difficult to interpret. This complexity of ABM's rapidly expands as the number of variables in the simulation grows [34]. Trying to model too many variables in the environment in a single simulation impacts the ability to isolate the effects of a single variable. Consequently, modeling the above questions, while interesting, will result in increased complexity so as to negatively impact the understanding of the phenomenon and the variables in which we are interested.
- *Simple, but not simpler, argument:* The distinguished economist and pioneer of economic geography, Paul Krugman, stated that "When working in a very new area, it is entirely forgivable to make outrageous simplifications in pursuit of insights, with the faith that the model can be brought closer to facts on later passes" [35, p.39]. There is no doubt that understanding trust and reputation models for VGI quality in this research is a novel angle on the topic. A further novelty is understanding the spatial and temporal aspects of trust and reputation and their role in VGI quality assessment. It is not yet possible to further study this topic without first building the fundamental understandings gained through our models.

With the above points in mind, the resulting ABM's discussed later are in fact relatively complex. They have eight and nine different parameters respectively, resulting in a tedious number of permutations when studying the parameter space. For example, when studying the spatial aspects discussed in section 6, the experimental design resulted in a requirement of 3240 runs at 6000 ticks each. Due to the computationally intensive nature of this particular ABM with the social network and a bipartite graph it was estimated, while our simplifying assumptions hold, that on multiple computing cores (approximately eight), this experiment would run for over a week.

5   Implementation and Results

Due to practical problems with the WSMS system in Africa, such as power failures, lack of awareness and expiry of the project's funding, we opted for another approach to test our model with this use case. Using agent based modeling, we developed a simulation of the WSMS scenario, which has many benefits, including the possibility of testing the model's resilience under different conditions, such as low reporting volumes, malicious attacks, and changing volunteer behaviors. First we define performance measures for our experiments based on the developed ABM and then we show the results of our experiments. Details on the developed ABM can be found in [36].

5.1   Defining Experimental Performance Measures

To help understand the ABM of the WSMS and our performance measures we point to one aspect of the simulation, the types of volunteers. Three types exist,

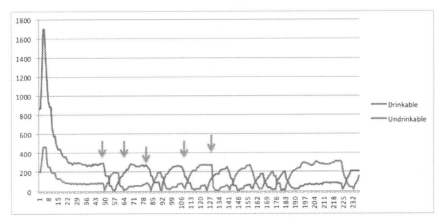

Fig. 4. *The behavior of the trust and reputation computational model in the base run. Days are shown on the $x - axis$ and values of $\sum_{(i,j)\in I^d} R_{ij} \cdot T_{ij}$ and $\sum_{(i,j)\in I^u} R_{ij} \cdot T_{ij}$ for a single well on the $y - axis$. The vertical arrows are the moment in time when the water well status switched between drinkable or undrinkable. Our computational model adjusts to reflects the true status of the water well (Precision) as it changes after sometime (latency).*

namely cooperative volunteers, which are consistently honest by reporting the true status of the wells, non-cooperative volunteers, consistently dishonest, and adaptive volunteers, acting as honest or dishonest modeled through a Bernoulli probability distribution.

With these types of volunteers we can simulate different combinations of volunteer behavior to test our model. We ran the ABM for several experimental settings and searched the parameter space of the computational model using genetic algorithms to identify optimal parameter values for an objective. This objective is for the computational model to correctly predict the true status of the water well from volunteer reports.

A main base run is illustrated in figure 4. We proceeded to analyze characteristic behaviors of the computational model and identified three performance measures, two for precision and one for latency.

There are two different ways we calculate the precision of the computational model. The first tells us how often the predicted status matched the actual status of the water well at any moment in time. Let s_t^i denote the state of well i at tick t, and p_t^i denote the prediction of the model for well i at tick t. For every time-point T, the average *hit rate* $\bar{p}(T)$ of the model is:

$$\bar{p}(T) = \frac{1}{T|wells|} \sum_{i \in wells} \sum_{t=1..T} I(s_t^i = p_t^i), \tag{7}$$

Where $I(x)$ equals 1, if x is true, and zero otherwise. A hit is a single instance of our model correctly predicting the current status of the water well. This measure is basically an average hit rate of the predictions taken over all wells and time.

Secondly, whenever the status of a well changes, the model should adapt to the new situation and stay in the correct state until the real status changes again. This is because when the model is unstable due to e.g. attacks by uncooperative volunteers the hit rate can be reasonable, but the model is not stable in making predictions over time. Between two real status changes of the water wells, the model is called *adapted* if after an initial incorrect prediction (latency), it switches to the correct prediction and keeps it until the real state changes again. We denote the number of times the model adapted to the state of well i until time t as a_t^i. Furthermore, let c_t^i denote the number of times well i changed state until time t. Then the *adaption rate* at time T $\bar{a}(T)$ is:

$$\bar{a}(T) = \frac{1}{|wells|} \sum_{i \in wells} \frac{a_T^i}{c_T^i} \tag{8}$$

Finally, the latency measure quantifies the phenomenon we observed in base runs that, after each status change, it takes some time for the computational model until it receives enough volunteer reports to reflect the change. On each occasion, we count the number of simulation ticks it takes for the model to adapt. The latency measure is simply an average of the latencies for all the adaptions where the model succeeded.

Let l_t^i denote the number of ticks the model required to adapt to the change of well i that happened in time t. Then the latency of the model in time T, $l(T)$ is:

$$l(T) = \frac{1}{\sum_{t=1..T} \sum_{i \in wells} a_T^i} \sum_{t=1..T} \sum_{i \in wells} l_t^i \qquad (9)$$

5.2   Analysis and Results

We now focus on the discrete temporal dimension, denoted by parameter k and its effects on the performance measures of the model. We also show the combined effects of different types of volunteer populations and parameter k on the same performance measures. Our aim is to show how the temporal dimension affects performance and also how this is related to numbers and types of volunteers. Our analysis shows that when all parameters are fixed, increasing the number of volunteers raises the precision of the model up to a point, after which more volunteers do not contribute to improving precision. We also show that k has a direct favorable effect on latency.

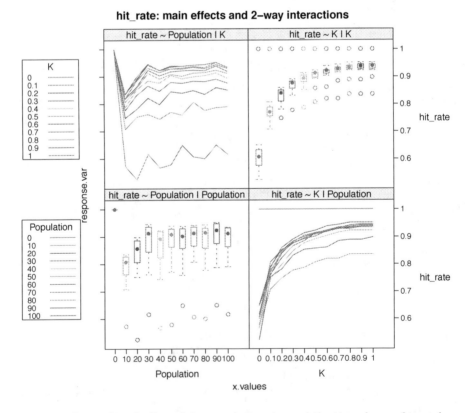

Fig. 5. The combined effect of the population size and the time decay of trust k

It is interesting to see the combined interaction of volunteer populations and time on the model's precision. The results of the analysis on these effects from our experiments are summarized in figure 5. The figure shows the interaction plot of the population size and time affecting model precision. In each sub-plot, the y-axis represents the precision of the model, particularly the hit rate (only for when adaptation is optimal), the x-axis represents the input of the corresponding column, and the line shades represent the corresponding row. For example, in the top-left corner, the x-axis is the population size, the y-axis is the precision, and the line shades are the different k values. In these plots, we used results from parameter combinations with equal numbers of adaptive, cooperative, and uncooperative volunteers, and the population sizes on the x-axis reflect the number of one type only, so that they have to be multiplied by three to get the total number of volunteers.

The plots in the diagonal in figure 5 have in the lower-left corner the population size and in the top-right corner the k value. These results informed us that increasing population increases precision at lower numbers of volunteers, but when more than 30 volunteers per type are present, a further increase does not help precision anymore.

Both the top-left and bottom-right plot show the same interaction between population size and time, from different viewpoints. On the top-left the bottom most line (strongly zigzag line) representing the $k = 0$ case stands out from the rest. Here precision is hardly better than 50%, and an increase of population size (the x-axis) does not help much. Additionally, one can observe an instability in precision when the population size changes, from the zig-zagging of this same line. For larger values of ks this irregularity disappears from the output (observe the remaining lines), and a larger k clearly results in better precision. The same conclusions can be drawn from the bottom-right plot. For $k = 0$ at the x-axis of this bottom-right plot, the precision is low, and the different lines are very close. As k increases the precision increases and at higher population levels converges closer to 1 (each line represents a higher population from bottom to top). This is the main reason for the model's transient behavior, namely for low k values the model performance is impacted negatively and as the value increases the effect is reversed and the time decay of trust improves model performance dramatically. Also, precision tends to converge to 1 because k improves latency and because latency is correlated with hit rate, as deeper analysis indicated. Thus it is to be expected that the time decay of trust also improves the model's hit rate as it improves its latency.

To summarize the interaction of the population size and time, we can say that at a low value of k ($k < 0.1$), controlling the decay rate, the population size does not matter much, the precision being equally low at any number of volunteers. For increasing k, the effect of population on model precision is positive and the effect of time on precision is strongly positive through improving both latency and hit rate. These effects are solidified once we have at least 30 honest volunteers in the system. An optimal situation is one where population numbers and time result in high precision and low latency.

6   Discussion on the Spatial Aspects of Trust

The model presented so far incorporates the temporal dimension of trust. In an extension of this model, we took into consideration the proximity of the reporter to the reported phenomenon (the water wells). We used distances ranging from 10 to 2000 units. The results of our experiments are summarized in the interaction diagrams in figure 6.

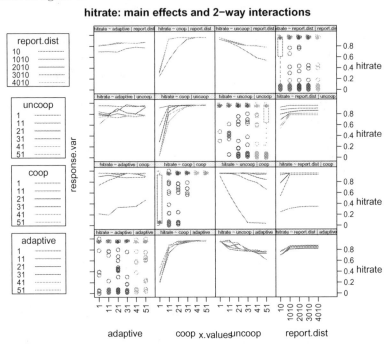

Fig. 6. The 2-way interactions of the different volunteer populations and reporting distance and the combined effects on the model's hit rate performance measure.

Looking at the main effects on hit rate (diagonal charts from top left most chart to bottom right most in figure 6), we observe that, regardless of reporting distance, the number of adaptive volunteers has a very limited effect on the hit rate. The only effect of adaptive volunteers is that, for higher reporting distances, the overall hit rate increases, but remains stable for all variations of adaptive populations. For the cooperative volunteers, several things can be observed. Once we have at least eleven cooperative volunteers, the effect on the hit rate stabilizes at all reporting distances. Essentially, reporting distance has a higher influence on performance when the number of cooperative volunteers is lower. In the third chart from the left at the top we observe that the increasing number of uncooperative volunteers reduces the overall hit rate of the model, yet this effect is offset by the increase in reporting distance. In other words, reporting distance has a direct positive impact on the model's hit rate when populations of malicious volunteers increase.

Finally, in the second column of charts, we observe that the higher the number of cooperative agents, the higher the hit rate will be on all three interaction charts. The top one shows that if the reporting distance is 10, hit rate grows more slowly with the number of cooperative volunteers, than if the reporting distance is higher. This means that higher reporting distances improve the prediction in favor of the cooperative/honest volunteers. The same analysis conducted for the adaptation performance measure revealed similar results.

7   Conclusions

We offer a novel take on modeling the quality of collaborative content, focusing on VGI as an example. Assessing the quality of VGI has often been treated as a traditional data quality problem. However, on one side of the VGI coin there is the data delivered, but on the other there are the reporting volunteers and their social, spatial, and temporal settings. This is also true for other forms of collaborative content. We argued that an alternative quality paradigm for VGI can rely on the social dimension using the notions of trust and reputation to identify trustworthy contributions. Trust is then viewed as a proxy measure of VGI quality.

We looked at collaborative content and asserted that in all cases and particularly for VGI, the quality of such content has strong spatial and temporal dependencies. Extending the social notion of trust with spatial and temporal dimensions allows for a new class of computational spatio-temporal trust and reputation models. In our use case, they have been shown to be effective tools for information triage and filtering. Trustworthy or high quality observations in that sense are given precedence over untrustworthy or low quality observations. Our analysis shows that the computational models developed were able to resist faulty VGI reporting as well as malicious attacks trying to eschew the system. We have also shown that the temporal and spatial dimensions of trust both had positive effects on the performance of the computational model and were in fact behind its strong resilience to false reports and malicious attacks. The temporal dimension contributed strongly to model precision and reduced latency. While taking the spatial dimension into account improved model precision under different volunteer behaviors.

We must point out, however, that the spatial dimension was represented in a rudimentary fashion through the distance between volunteers and the phenomenon on which they are reporting information at the time of reporting it. Such a representation is a useful simplification for research, but often not adequate in practice. For example, we might still trust somebody to give us information about Berlin, even if the person now lives in London, after having lived in Berlin for ten years. Humans have familiarity spaces [37,38], which are areas on which they are experts but might not at the moment be present there. For example, one may have three urban activity spaces with which one is familiar in a city: the area around work, the area around home, and the area around the city center. It follows that one can be trusted about these areas more than about others, as well as potentially having a far reaching network of activity spaces in different cities across different continents. Still, when reporting on a traffic jam, being at the jam might imply being more trustworthy than a person with this particular familiarity space. In the future, we want to experiment with more complex and socially richer representations of space, studying their effects on trust and reputation.

Acknowledgments. The research leading to these results has received funding from the European Union's Seventh Framework Programme FP7/2007-2013 under grant agreement 317631 (OPTET).

References

[1] Bishr, M., Kuhn, W.: Geospatial information bottom-up: A matter of trust and semantics. In: 10th Agile Conference, pp. 365–387. Springer (2007)

[2] Kuhn, W.: Volunteered geographic information and giscience, pp. 13–14. NCGIA, UC Santa Barbara (2007)

[3] Goodchild, M.F.: Citizens as sensors: the world of volunteered geography. GeoJournal 69(4), 211–221 (2007)

[4] Knowles, D.: Twitter message sparks big rescue effort, http://www.aolnews.com/tech/article/twitter-message-sparks-big-rescue-effort-in-haiti/19337648 (February 20, 2012)

[5] Zook, M., Graham, M., Shelton, T., Gorman, S.: Volunteered geographic information and crowdsourcing disaster relief: a case study of the haitian earthquake. World Medical & Health Policy 2(2), 2 (2010)

[6] Goodchild, M.F., Glennon, J.A.: Crowdsourcing geographic information for disaster response: a research frontier. International Journal of Digital Earth 3(3), 231–241 (2010)

[7] Sztompka, P.: Trust: A Sociological Theory. Cambridge University Press (1999)

[8] Deutsch, M.: Cooperation and trust: Some theoretical notes. In: Nebraska Symposium on Motivation, vol. 10, pp. 275–319. University of Nebraska Press, Lincoln (1962)

[9] Buskens, V.W.: Social networks and trust. Springer (2002)

[10] Luhmann, N.: Trust and Power. John Wiley and Sons (1979)

[11] Gambetta, D.: Can We Trust Trust? In: Trust: Making and Breaking Cooperative Relations. Basil Blackwell, New York (1990)

[12] Golbeck, J.: Computing and Applying Trust in Web-based Social Networks. PhD thesis (2005)

[13] Nohria, N., Eccles, R.G.: Networks and organizations: structure, form, and action. Harvard Business School Press (1992)

[14] Lyons, B.: Contracts and specific investment: An empirical test of transaction cost theory. Journal of Economics and Management Strategy 3(2), 257–278 (1994)

[15] Lorenz, E.: Neither Friends nor Strangers: Informal Networks of Subcontracting in French Industry. In: Trust: Making and Breaking Cooperative Relations. Basil Blackwell, New York (1988)

[16] Liben-Nowell, D., Novak, J., Kumar, R., Raghavan, P., Tomkins, A.: Geographic routing in social networks. Proceedings of the National Academy of Sciences 102(33), 11623 (2005)

[17] Fogg, B., Tseng, H.: The elements of computer credibility. In: Proceedings of the SIGCHI Conference on Human Factors in Computing Systems: the CHI is the Limit, pp. 80–87. ACM Press, New York (1999)

[18] Hawkins, D.T.: What is credible information? Online (Weston, CT) 23(5), 86–89 (1999)

[19] Liu, Z.: Perceptions of credibility of scholarly information on the web. Information Processing and Management 40(6), 1027–1038 (2004)

[20] Metzger, M.J.: Making sense of credibility on the web: Models for evaluating online information and recommendations for future research. Journal of the American Society for Information Science and Technology 58(13), 2078 (2007)

[21] Rieh, S.Y., Danielson, D.: Credibility: A multidisciplinary framework. Annual Review of Information Science and Technology 41, 307–364 (2007)

[22] Bishr, M., Mantelas, L.: A trust and reputation model for filtering and classifying knowledge about urban growth. GeoJournal 72(3), 229–237 (2008)

[23] Gray, E., Seigneur, J.-M., Chen, Y., Jensen, C.: Trust propagation in small worlds. In: Nixon, P., Terzis, S. (eds.) iTrust 2003. LNCS, vol. 2692, pp. 239–254. Springer, Heidelberg (2003)

[24] Guha, R., Kumar, R., Raghavan, P., Tomkins, A.: Propagation of trust and distrust. In: Proceedings of the 13th International Conference on World Wide Web, pp. 403–412. ACM Press, New York (2004)

[25] Mezzetti, N.: A socially inspired reputation model. In: Katsikas, S.K., Gritzalis, S., López, J. (eds.) EuroPKI 2004. LNCS, vol. 3093, pp. 191–204. Springer, Heidelberg (2004)

[26] Goodchild, M.F., Gopal, S.: The accuracy of spatial databases. Taylor and Francis (1989)

[27] Mülligann, C., Janowicz, K., Ye, M., Lee, W.-C.: Analyzing the spatial-semantic interaction of points of interest in volunteered geographic information. In: 10th Conference on Spatial Information Theory, pp. 350–370 (2011)

[28] Ostermann, F.O., Spinsanti, L.: A conceptual workflow for automatically assessing the quality of volunteered geographic information for crisis management. In: Proceedings of AGILE (2011)

[29] Marsh, S.: Formalising Trust as a Computational Concept. PhD thesis (1994)

[30] Cheskin, S.: ecommerece trust: Building trust in digital environments. Archetype/Sapient (1999)

[31] Walter, F., Battiston, S., Schweitzer, F.: A model of a trust-based recommendation system on a social network. Autonomous Agents and Multi-Agent Systems 16(1), 57–74 (2007)

[32] Mayer, R., Davis, J., Schoorman, F.: An integrative model of organizational trust. Academy of Management Review 20(3), 709–734 (1995)

[33] Jürrens, E.H.: Swe-connection to a gsm-interface to establish water availability monitoring in africa. Master's thesis, Institute for Geoinformatics, University of Muenster, Muenster, Germany (2009)

[34] Axelrod, R., Hamilton, W.D.: The evolution of cooperation. Science 211(4489), 1390–1396 (1981)

[35] Krugman, P.: Development, geography, and economic theory. MIT Press (2002)

[36] Bishr, M.: Trust and reputation models for human sensor observations. PhD thesis (2011)

[37] Schönfelder, S., Axhausen, K.: On the variability of human activity spaces. Place Zurich (2002)

[38] Newsome, T.H., Walcott, W.A., Smith, P.D.: Urban activity spaces: Illustrations and application of a conceptual model for integrating the time and space dimensions. Transportation 25(4), 357–377 (1998)

Using Maptrees
to Characterize Topological Change

Michael Worboys

School of Computing and Mathematical Sciences
University of Greenwich, London, England
http://www.worboys.org

Abstract. This paper further develops the theory of maptrees, introduced in [13]. There exist well-known methods, based upon combinatorial maps, for topologically complete representations of embeddings of connected graphs in closed surfaces. Maptrees extend these methods to provide topologically complete representations of embeddings of possibly disconnected graphs. The focus of this paper is the use of maptrees to admit fine-grained representations of topological change. The ability of maptrees to represent complex spatial processes is demonstrated through case studies involving conceptual neighborhoods and cellular processes.

Keywords: maptree, topology, topological change, spatial information theory, conceptual neighborhood, cellular process.

1   Introduction

The maptree was introduced in [13] as a structure that could represent fine topological details of spatial scenes. In this paper we exploit the maptree structure to provide an analysis of some types of spatial change.

By a spatial scene we mean an embedding of a graph in a surface. The general theory of maptrees applies to any orientable, closed surface embedded in Euclidean 3-space. These surfaces, according to the 1863 Möbius classification theorem for orientable, closed surfaces, are homeomorphic to a sphere with g handles (g-holed torus), for $g \geq 0$. The non-negative integer g is referred to as the *genus* of the surface. From henceforth we assume all surfaces are orientable. For the purposes of this paper, the only closed surface we consider is the sphere, and we show that the theory can also be applied to the Euclidean plane.

As discussed in [13], the maptree structure allows a finer level of detail of topological structure and relationships to be captured than some other approaches (for example, [10,5]). Similarly, in terms of change, the earlier theory developed by the author and colleagues in [9,8,11] allows, for example, a formal distinction between a hole emerging from a point in the center of a region and the same region merging with itself to create a hole, but cannot distinguish between a merge of two regions at a point or at a linear boundary. In this paper, we use the finer structural detail provided by the maptree to formally characterize some of the intricacies that can occur when a spatial scene undergoes topological change.

T. Tenbrink et al. (Eds.): COSIT 2013, LNCS 8116, pp. 74–90, 2013.

It appears that we have not gained anything by transforming one spatial configuration (a graph embedding) into another (a maptree). However, a maptree is a labelled tree, and hence unlike a general embedding it is a computationally tractable structure.

Section 2 of the paper reviews the salient definitions and properties of maptrees, both in the planar and spherical cases. This section draws on the corresponding material in [13], and is included for completeness, and for readers who may not be familiar with these constructions. However, we do include a new method of constructing maptrees from stars. In section 3, we discuss topological change, focussing on merge and split operations and demonstrating the higher level of detail that can be captured. Section 4 presents two case studies on transitions in a well-known conceptual neighborhood and on a fundamental cellular process as demonstrators of the representational power of our approach. We conclude with a discussion and consideration of future work needed.

2   Background

The maptree construction brings together two separate formal descriptions of topological configurations, namely combinatorial maps [4,12] and adjacency trees [2,11]. We begin by developing the background theory of combinatorial maps.

2.1   Combinatorial Maps

A *graph* is defined in the usual way as a set of vertices and edges between vertices, except that we allow edges that connect vertices to themselves, and also multiple edges between two vertices. A graph is *connected* if any pair of its vertices may be linked by a chain of adjacent edges. Informally, an *embedding* of a graph in a surface is a drawing of the graph on the surface in such a way that its edges may intersect only at their endpoints. Graph embeddings in closed surfaces have the property that the complement in the surface of an embedding of a connected graph is a collection of regions or *faces*, and each of these faces is a 2-manifold. If, furthermore, each of the faces is homeomorphic to a disc, the embedding is called a *2-cell embedding*. When the graph is embedded in the Euclidean plane, then one of the faces will be of infinite extent, and called the *external face*.

Let $A = \{a, b, \ldots, k\}$ be a finite collection of elements. We call any bijective function $\phi : A \to A$ a *permutation* of A. Essentially, we can think of ϕ as rearranging the elements of A. Now, any permutation can always be written as a collection of cycles $(a_1 a_2 \ldots a_n)$, where $a_2 = \phi a_1$, $a_3 = \phi a_2$, and so on, and $a_1 = \phi a_n$. So, for example, suppose $A = \{a, b, c, d, e\}$, and $b = \phi a$, $c = \phi b$, $a = \phi c$, $e = \phi d$, and $d = \phi e$. Then ϕ may be written in cycle notation as $\phi = (abc)(de)$.

Suppose now that we have a collection of permutations of A, $\Phi = \{\phi_1, \ldots, \phi_m\}$. Then Φ is *transitive* if, given any elements $x, y \in A$, we can transform x to y by a sequence of permutations from Φ. That is,

$$x \xrightarrow{\phi_{i_1}} x_1 \xrightarrow{\phi_{i_1}} \ldots x_p \xrightarrow{\phi_{i_p}} y$$

With these preliminaries taken care of, we can now define a combinatorial map.

Definition 1. *A combinatorial map* $M\langle S, \alpha, \tau \rangle$ *consists of:*

1. *A finite set S of elements, called* semi-edges, *where the number of semi-edges is even. We can write S as $S = \{a, \overline{a}, b, \overline{b}, \ldots, k, \overline{k}\}$*
2. *A permutation α of S*
3. *A permutation τ of S, which in cyclic form is $\tau = (a\overline{a})(b\overline{b}) \ldots (k\overline{k})$*

subject to the constraint that the collection of permutations $\{\tau, \alpha\}$ is transitive.

An example of a combinatorial map is given by $M_1\langle S, \alpha, \tau \rangle$, where $S = \{a, \overline{a}, b, \overline{b}, c, \overline{c}\}$, $\alpha = (\overline{a}c\overline{b}\overline{c})(a)(b)$, and $\tau = (a\overline{a})(b\overline{b})(c\overline{c})$. It is easy to check that the collection $\{\tau, \alpha\}$ is transitive.

Combinatorial maps provide formal representations of 2-cell graph embeddings, as is seen from the following theorem due to Edmonds [4] and Tutte [12].

Theorem 1. *(Edmonds, Tutte) Each combinatorial map provides a topologically unique (up to homeomorphism of the surficial embeddings) representation of a 2-cell graph embedding in a closed surface. Conversely, every 2-cell graph embedding in a closed surface can be uniquely (up to permutation group isomorphism) be represented by a combinatorial map.*

Given a combinatorial map $M\langle S, \alpha, \tau \rangle$, the 2-cell embedding is constructed as follows. Each edge of the embedded graph is represented by a pair of semi-edges, called a *facing pair*, transposed by τ. Each cycle of α defines the ordering of semi-edges around each face of the embedding. Each face is defined as the region on the left while traversing the semi-edges of a cycle of α. We say that combinatorial map $M\langle S, \alpha, \tau \rangle$ *represents* the 2-cell embedding so constructed. Because of the transitivity property, any graph represented by a combinatorial map must be connected. We may also note that the constituent cycles of α are sufficient to uniquely reconstruct the embedding. The constituent cycles of α are termed the α-*cycles* of M. Figure 1 illustrates this construction using combinatorial map, $M_1\langle S, \alpha, \tau \rangle$.

We now focus on the nature of the closed surface in which the configuration represented by a combinatorial map is embedded (unique by Theorem 1). To determine the surface from the combinatorial map, we need to determine the number of vertices of the embedded graph. To do this, we calculate a further permutation β by the formula $\beta = \tau\alpha^{-1}$ where the product is a composition of functions, and α^{-1} denotes the inverse function of α. In our example $\beta = (a\overline{a}c)(b\overline{b}\overline{c})$. We may note that each cycle of β represents a vertex of the embedded graph, and the cycle itself represents the ordering of semi-edges around the vertex.

Now we invoke the famous result of Euler and Poincaré:

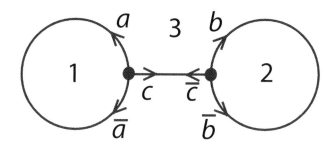

Fig. 1. Embedding of $\mathbf{M}_1$ in the sphere

Theorem 2. *(Euler-Poincaré) Given a 2-cell embedding of a graph in a surface of genus g, where V, E, and F are the numbers of vertices, edges, and faces, respectively. We have $V - E + F = 2 - 2g$.*

Given a combinatorial map, $\mathbf{M}\langle S, \alpha, \tau \rangle$, the genus of its embedding surface can then be calculated by observing that V, E and F are the number of constituent cycles of β, τ and α, respectively. We sometimes refer to the genus of the combinatorial map, meaning the genus of its embedding surface. For the above map $\mathbf{M}_1\langle S, \alpha, \tau \rangle$, $V = 2, E = 3, F = 3$, and so $2 - 2g = 2$, and $g = 0$ which accords with our knowledge that the embedding surface is a sphere.

2.2 Star Representation of a Combinatorial Map

As already observed, a combinatorial map is completely specified by its collection of α-cycles. This leads to the following visualization tool that will be helpful later.

Definition 2. *Given a combinatorial map $\mathbf{M}$ with α-cycles $\alpha_1, \ldots, \alpha_n$, then the star associated with $\mathbf{M}$ is an edge-labeled tree with a central black node from which edges connect to n white nodes, the ith edge being labeled with α_i, $(1 \leq i \leq n)$.*

White nodes represent faces of the embedding and the black node represents the connected network of edges. The star of example $\mathbf{M}_1$ is shown in figure 2.

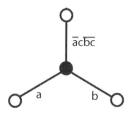

Fig. 2. Star representation of $\mathbf{M}_1$

2.3   Planar Embeddings

The above results on combinatorial maps apply to closed surfaces, and therefore do not apply directly to the Euclidean plane. However, any graph that is embeddable in the sphere is embeddable in the plane, and conversely. The plane places an extra piece of structure on the embedding in that we have the notion of the infinite face. An embedding of a graph in the sphere may correspond to many topologically distinct embeddings of it in the plane, depending which spherical face becomes the infinite face. Figure 3 shows two non-homeomorphic planar embeddings representing $\mathbf{M}_1$. Of course, when viewed as embeddings on the sphere, they are homeomorphic. Note also, for reason of clarity, we show only one of each pair of semi-edges.

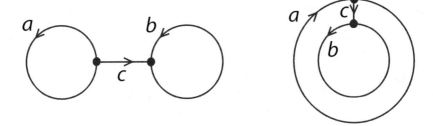

Fig. 3. Two homeomorphic embeddings of $\mathbf{M}_1$ in the sphere that are non-homeomorphic when viewed as planar embeddings

In order that a combinatorial map can uniquely specify a planar 2-cell embedding, all we need to do is specify which cycle represents the boundary of the external face. We can then invoke a slightly extended version of Theorem 1 (Edmunds, Tutte) to guarantee topological uniqueness of the representation.

With respect to the star representation, our approach is to make the white node representing the external face the root of a rooted tree. More formally, we have the following definition.

Definition 3. *Given a combinatorial map $\mathbf{M}$ with α-cycles $\alpha_1, \ldots, \alpha_n$, then a p-star associated with $\mathbf{M}$ is an edge-labeled rooted tree with a central black node from which edges connect to n white nodes, the ith edge being labeled with α_i, $(1 \leq i \leq n)$. The root of the tree is the white node whose incident edge is labeled by the infinite face.*

Viewed as a p-star, figure 2, with the root node being by convention at the top, represents the planar embedding on the lefthand side of figure 3.

2.4   Maptrees

Combinatorial maps have the limitation that they can only represent embeddings of connected graphs. (in fact, as we have said, every 2-cell embedding must be the embedding of a connected graph). We now extend the algebraic representation of graph embeddings as combinatorial maps to also represent embeddings of disconnected graphs. Assume we have a collection of maximal connected components (henceforth called components) of a graph embedding, such that each of the components is a two-cell embedding. To specify the topology of the embedding, it is not sufficient to give the collection of stars (or p-stars) representing the components. To see this, consider the two graphs, each embedded in the surface of a sphere, shown on the lefthand and righthand sides of figure 4. Both embeddings would have the same representation as a collection of three stars. However, it is easy to see that these embeddings are not topologically equivalent, in the sense that there does not exist a homeomorphism of the sphere that maps one embedding to the other. How graph embeddings stand with respect to one another becomes an issue, and we now develop the extra structure to represent this. The essential idea is to join the component stars together, merging them at appropriate white nodes.

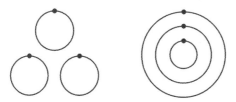

Fig. 4. Two non-homeomorphic embeddings in the sphere

The process is described in detail in [13], but here, we adopt a slightly different approach to the construction, illustrating with the embedding shown in figure 5. In this figure we have labeled the faces $1, 2, \ldots$ (some no longer 2-cell) created by the embedding. We begin by considering stars of the components, as shown in figure 6, where are also labelled the white nodes with the names of the regions that they represent. We then join the stars together, merging white nodes when they have the same face name. The resulting maptree is shown in figure 7.

We are now ready to give the formal definition of a maptree. As a preliminary, we introduce the construction of a bw-tree, that is able to uniquely represent how the components and regions stand in the relation to each other, but does not provide details about the topology of the components themselves. This is a construction closely related to that of the adjacency tree, discussed in the next section.

Definition 4. *A* bw-tree *X is a colored tree with the nodes colored black or white, respectively, subject to the condition that no two adjacent nodes have the same color.*

We now combine the bw-tree construct with the combinatorial maps of the components.

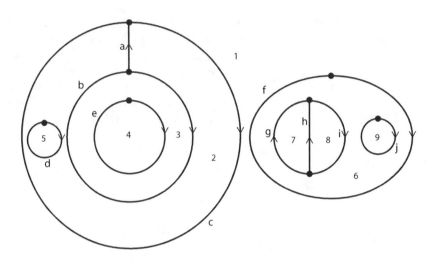

Fig. 5. Embedding of a disconnected graph

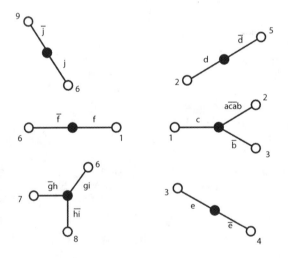

Fig. 6. Stars of the components of the embedding in figure 5

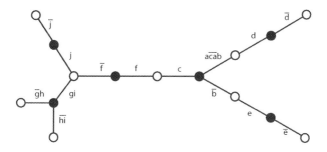

Fig. 7. Maptree of the embedding in figure 5

Definition 5. *Let* $\mathcal{M}$ *be a finite collection of combinatorial maps, represented by stars. A* maptree *is an edge-labelled bw-tree* $\mathcal{T}_{\mathcal{M}}$ *formed by the merge of the component stars at white nodes.*

As above, the interpretation of maptree $\mathcal{T}_{\mathcal{M}}$ is that each black node of $\mathcal{T}_{\mathcal{M}}$ represents the 2-cell embedding of the map associated with that node. If two black nodes are connected via a white node, then the two cycles labeling the two edges joining the nodes represent the regions that "face up to each other" in the embedding.

We have the following proposition, discussed in more detail in [13].

Proposition 1. *Let* $\mathcal{M}$ *be a finite collection of combinatorial maps of genus zero, and* $\mathcal{T}_{\mathcal{M}}$ *be a maptree. Then* $\mathcal{T}_{\mathcal{M}}$ *provides a unique representation (up to homeomorphism of the sphere) of the disconnected graph embedding of the connected components represented by* $\mathcal{M}$.

2.5   Maptrees for Planar Embeddings

We can modify the map tree construction to account for planar embeddings by distinguishing the infinite exterior region, represented by one of the white nodes of one of the p-stars. We make this node the root of a rooted tree. The other p-stars are arranged under the root in such a way as their local roots connect with leaves of upper p-stars. As an example, the planar maptree for the planar embedding shown in figure 5 is shown in figure 8, where the root is at the top of the figure.

3   Adjacency Trees, Maptrees, and Topological Change

In this section we relate topological change to structural changes to the maptree. We begin by throwing away some of the detail that maptree representations

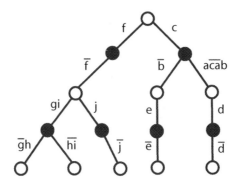

Fig. 8. Maptree of the planar embedding in figure 5

provide, and focusing on the nesting of regions inside each other. Insights gained are then applied to the more refined case. Adjacency trees were introduced by Buneman [2] to represent the nesting structures of disk-like structures in the plane. As with all the work in this paper, their natural embeddings are in closed surfaces, but can be applied to the case of the plane by distinguishing the infinite face.

Definition 6. *Given an embedding of a collection of simple closed curves in a closed surface, its associated* adjacency tree *is a tree whose nodes are the faces of the embedding and where two nodes are joined by an edge if and only if the corresponding faces share a common bounding curve.*

Clearly, adjacency trees provide some useful characterization of surficial embeddings, but have no details about boundary intricacies. However they can be used to represent basic types of topological change (see [9,8,11]). For example, the upper part of figure 9 shows a region merging with itself and in the process engulfing some of its exterior space. (This is a planar embedding interpretation). The lower part shows the corresponding changes in the form of an unfolding operation on the adjacency tree. We may note that all these changes are reversible, as indicated by the double-headed arrows.

Figure 10 shows the merging of two regions and the corresponding folding operation on the adjacency tree. In this case the change is irreversible if we wish to recapture the qualities of the original regions from the merged region.

The third example, shown in figure 11 has similar start and end regions as the first, but this time the process is not self-merge but region insertion in the form of a hole in an existing region. The corresponding adjacency tree operation is the insertion of a new node and edge. This change is reversible, and results in a region deletion.

The first and third of these examples show that adjacency trees can capture the distinction between a hole forming from a self-merge and the insertion of a region.

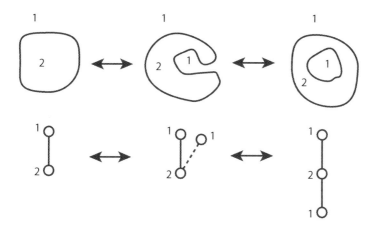

Fig. 9. Adjacency tree unfolding corresponding to a region self-merge

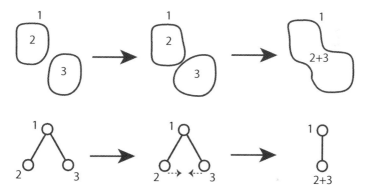

Fig. 10. Adjacency tree folding corresponding to the merge of two regions

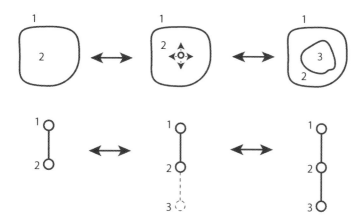

Fig. 11. Adjacency tree insertion corresponding to creation of a hole

3.1   Maptrees and Topological Change

We have indicated by means of examples how using folding/unfolding and in-
sertion/deletion operations on adjacency trees can be used to represent basic
kinds of changes to regions. We go on to demonstrate how maptree operations
serve to represent in greater detail varieties of topological change. We focus on
the operations of region merge and self merge, but the other cases are similar. These cases, which cover a large part of the atomic topological changes
available to graph embeddings, give a flavor of the more find-grained kinds of
analyses that maptrees facilitate. Throughout, we will use rooted maptrees, and
hence be considering action in the plane. A very similar analysis follows in the
spherical case.

Figures 12 and 13 are refinements of figure 9 in that in the self-merge of the
region with itself we can distinguish merge at a point from merge at a line.

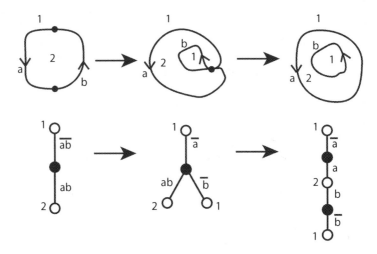

Fig. 12. Self-merge of a region at a point

In the first stage of the self-merge in figure 12, shown in the middle section
of the figure, the self-merge still leaves a single connected component, and the
maptree star unfolds an edge and node from the top, using the central black
node as fulcrum. The edge label $\overline{ab}$ splits into two labels $\overline{a}$ and $\overline{b}$. In the second
stage of the self-merge, resulting in the formation of a hole, the single component
splits into two. This is represented in the transformation of the maptree with an
unfolding about lower black node labeled 2, where the edge label ab splits into
two labels a and b .

The first stage of the line self-merge in figure 13 is a non-topological change
bringing the edges c and d that will merge into closer proximity. In the next
stage these edges merge, resulting in the single edge e. Formally, $d \mapsto e, c \mapsto \overline{e}$.
The unfolding is similar to the previous example, except for the elimination of

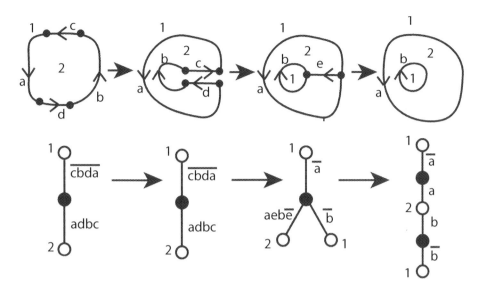

Fig. 13. Self-merge of a region at a line

$\bar{c}$ and $\bar{d}$ in the unfolding, and of e and $\bar{e}$ in the second. Again, these operations are reversible, with a self-split being represented by two foldings.

It is worth pointing out that if we work from right to left with either of the operations given in figures 12 and 13 but with focus on the two regions labeled 1, then the process of merge of the two regions, either at a point or a line, is represented by the folding operations given. Thus, region merges at a point or a line are represented similarly by folding/unfolding and edge label operations on maptrees.

4   Two Case Studies

In this section we develop two fairly substantial case studies that illustrate the utility of maptrees for representing topological change. The first might be called classical in that it provides a fine-grained representation of transitions in a well-known conceptual neighborhood diagram. The second provides an example of an fundamental process in cellular biology.

4.1   Conceptual Neighborhoods Revisited

Conceptual neighborhoods have been an important tool for spatial reasoning since their development by Freksa in [7] and use by many others (e.g., [6]). In this subsection we discuss the conceptual neighborhood diagram for the case of two regions We will see that maptrees provide a more fine-grained analysis of these planar topological changes than were obtained by earlier approaches. Figure 14 shows the diagram in question, and it provides both the region-region

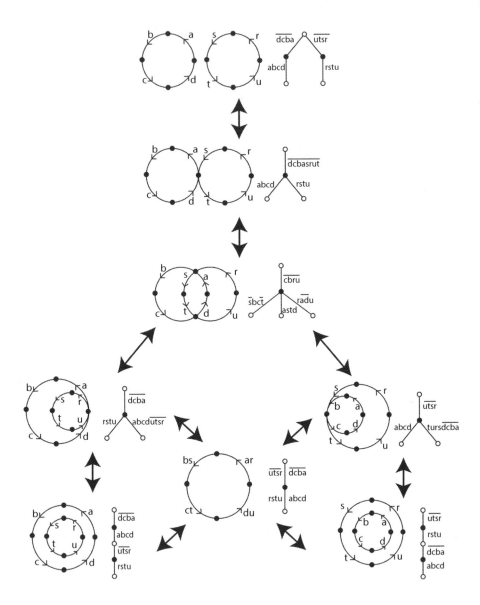

Fig. 14. Conceptual neighborhood with maptree representation

relationships and their maptree representation as disjoint regions meet at a point, overlap, are in a tangential proper part relationship at a point, are in a non-tangential proper part relationship, and are equal. Although we have not the space to show it, the maptree configurations would be different if the regions interacted at more than one point, at a line, or in a more complex manner. Thus the maptree representation can indeed capture more detail than RCC and other formulations.

As in the previous example, the transformations are represented in the maptree by various foldings and unfoldings along with edge label symbolic manipulations. The transition from disjoint to tangent regions is covered in the previous subsection, and is represented in the maptree by a folding around a black node and appropriate concatenation of cycles. The transition from tangent to overlapping regions is represented in the maptree by an unfolding of a new white node and some redistribution of edge labels. (This transition is actually a composition of rather simpler transpositions, but space precludes a full analysis here.) Transition to tangential proper part has some qualities of the inverse of the previous move, and the move to non-tangential proper part is similar to the inverse of the first transition from disjoint to tangent. We have shown the representation of equality of regions as having alternative labelings, based either on the edge labels a, b, c, d or r, s, t, u. We should note that the similarities to maptree operations in the previous section are no accident, as on the sphere, the transition from disjoint to tangent is identical to the transition from non-tangential to tangential proper part, as is the transition from overlapping to tangential proper part. In the case of the sphere, we lose the notion of a distinguished root node in the maptree.

4.2 Cellular Topology

Our second illustration of the power of the maptree structure is in the modeling of part of the cellular process, phagocytosis, where solid particles are engulfed by the cell membrane to form an internal phagosome. Cellular processes have been the subject of much formal analysis (see, for example, [1,3]), however topological structure has not always been well-represented. A simplified form of the cellular transitions is shown in figure 15. As we follow the process round, we see foreign matter, bounded by edge (a) is disjoint and then tangent with the cell bounded by (bcd). The cell then makes geometrical changes in preparation to engulfing the foreign matter (no topological change) and then engulfs the foreign matter by firstly self-merging at a point and then thickening out its membrane. The foreign matter then detaches itself from the cell membrane and is enclosed in a vacuole (cd), called a phagosome, within the cell.

This process can be represented through all its stages by maptrees and their structural transformations, as shown in figure 16. We have seen most of these processes already. The tree unfolds about its root, representing the tangential attachment of the foreign body to the cell. We have omitted the geometrical change to the cell as this results in no change to the tree. The self merge is represented by unfolding about the central black node, and thickening of the

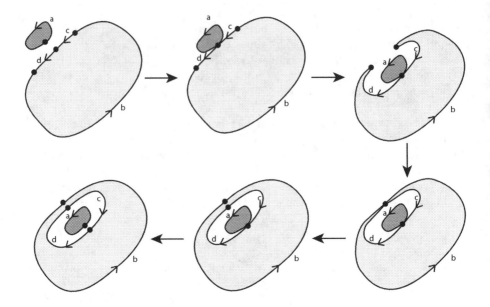

Fig. 15. Schematic of phagocytosis

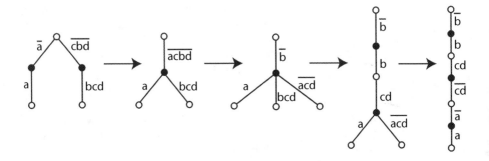

Fig. 16. Maptree transitions representing aspects of phagocytosis

membrane by unfolding about a white node. The final stage of detachment is represented by a further unfolding and results in a linear maptree structure.

5   Conclusions

This paper has further developed a topological representation of closed surficial embeddings that we call a maptree. We have demonstrated that the representation does indeed contain sufficient information to generate unique embeddings, up to homeomorphism of the embedding surface. We have also shown how a simple modification provides a representation for planar embeddings. This work continues the line of work on formal description of spatial relationships found for example in [10,5]). However the map tree allows representation of multiple spatial regions, and allows a much greater level of complexity in the topological configurations to be represented.

We have also continued to develop a theory of topological change. We have illustrated how simple transitions between maptrees, involving foldings and unfoldings as well as an algebra of edge labels, can allow useful insights into the structure of basic topological changes. We have also illustrated its applications with two larger examples. The full theory, with its interplay between structural changes to the tree and its edge labels has yet to be fully elaborated, and this is the subject of future work. We also see great potential for maptrees in the formal descriptions of cellular processes, and this is the subject of current research.

Much more needs to be done in mining this rich vein of formal representation. Structures that can represent three dimensional configurations exist in the connected case, but add considerable complexity. The extension to disconnected embeddings as 3-manifolds is also a subject of ongoing research.

Acknowledgments. This material is partly based upon work supported by the US National Science Foundation under grant number IIS-0916219 and by the University of Greenwich.

References

1. Barbuti, R., Maggiolo-Schettini, A., Milazzo, P., Pardini, G., Tesei, L.: Spatial P systems. Natural Computing (2010)
2. Buneman, O.P.: A grammar for the topological analysis of plane figures. In: Meltzer, B., Michie, D. (eds.) Machine Intelligence, vol. 5, pp. 383–393. Elselvier (1970)
3. Cardelli, L.: Brane calculi. In: Danos, V., Schachter, V. (eds.) CMSB 2004. LNCS (LNBI), vol. 3082, pp. 257–278. Springer, Heidelberg (2005)
4. Edmonds, J.R.: A combinatorial representation for polyhedral surfaces. Notices Amer. Math. Soc. 7, 646 (1960)
5. Egenhofer, M.J., Franzosa, R.D.: Point-set topological spatial relations. International Journal of Geographical Information Systems 5(2), 161–174 (1991)
6. Egenhofer, M.J., Mark, D.M.: Modeling conceptual neighborhoods of topological line-region relations. International Journal of Geographical Information Systems 9(5), 555–565 (1995)
7. Freksa, C.: Temporal reasoning based on semi-intervals. Artificial Intelligence 54, 199–227 (1992)
8. Jiang, J., Nittel, M., Worboys, S.: Qualitative change detection using sensor networks based on connectivity information. GeoInformatica 15(2), 305–328 (2011)
9. Jiang, J., Worboys, M.: Event-based topology for dynamic planar areal objects. International Journal of Geographical Information Science 23(1), 33–60 (2009)

10. Randell, D.A., Cui, Z., Cohn, A.: A spatial logic based on regions and connection. In: Nebel, B., Rich, C., Swartout, W. (eds.) KR 1992. Principles of Knowledge Representation and Reasoning: Proceedings of the Third International Conference, pp. 165–176. Morgan Kaufmann, San Mateo (1992)
11. Stell, J., Worboys, M.: Relations between adjacency trees. Theoretical Computer Science 412, 4452–4468 (2011)
12. Tutte, W.T.: What is a map? In: New Directions in the Theory of Graphs, pp. 309–325. Academic Press, New York (1973)
13. Worboys, M.: The maptree: A fine-grained formal representation of space. In: Xiao, N., Kwan, M.-P., Goodchild, M.F., Shekhar, S. (eds.) GIScience 2012. LNCS, vol. 7478, pp. 298–310. Springer, Heidelberg (2012)

Resolving Conceptual Mode Confusion with Qualitative Spatial Knowledge in Human-Robot Interaction

Cui Jian and Hui Shi

SFB/TR 8 Spatial Cognition, University of Bremen, Germany
{ken,shi}@informatik.uni-bremen.de

Abstract. This paper presents our work on using qualitative spatial knowledge to resolve conceptual mode confusion occurring frequently during the communication process between human operators and mobile robots. In order to bridge the gap between human's mental representation about space and that of a mobile robot, a qualitative spatial beliefs model is applied. Then with a computational framework based on qualitative spatial reasoning offered by this model, a set of high level strategies are developed and used to support the interpretation of natural language route instructions to mobile robots for navigation tasks.

Keywords: qualitative spatial representation and reasoning, communication of spatial information, activity-based models of spatial knowledge, human robot interaction, mode confusion.

1   Motivation and Introduction

Over the last few decades, much research has been done on intelligent robots for effectively and sensibly acting and interacting with humans in different domains. Typically, these robots are collaboratively controlled by an intelligent system and a human operator, who have to share a set of common resources such as the environment, the ongoing system's behavior and state, the remaining action plan, etc. Thus, problems may occur, when the operator's mental state about the shared common resources is different from the current system observed state, especially for the intended users of intelligent service robot, who are usually uninformed persons without specialized knowledge. These problems are called mode confusion (cf. [1]), referring to situations in which a system behaves differently from an operator's expectation. Due to its undesired consequence, mode confusion has been intensively studied, e.g. in [2], [3] and [4]. Meanwhile, with the rapid development of language technology, interaction with intelligent robots via natural language is gaining significantly increasing interest (cf. [5] and [6]), which leads to a subtype of mode confusions, called conceptual mode confusion.

Our work is focusing on resolving conceptual mode confusion occurring in human-robot joint spatially-embedded navigation tasks, where a mobile robot is instructed by a human operator via natural language to navigate in a partially known environment. Conceptual mode confusion occurs, e.g., when the human operator instructs the robot

T, Tenbrink et al. (Eds.): COSIT 2013, LNCS 8116, pp. 91–108, 2013.

to go straight ahead, take a left turn and pass a landmark on the right, but in that situation the referred landmark is only allowed to be passed after taking a right turn instead of a left turn. How this kind of spatially-related mode confusions can be detected and resolved becomes a very interesting question. [7] and [8], e.g., conducted corpus-based studies on giving route instructions to mobile robots; [9] tried to map sequences of natural language route instructions into machine readable representations; [10] reported on an approach to represent indoor environments as conceptual spatial representations with layers of map abstractions, etc.

Diverging from that literature, some research has been focusing on the perspective of human operators. [11] gave a general overview about recent reports on the human behavior and human cognition, as well as their essential relation with various spatial activities. Especially, [12] and [13] described how human thoughts and language can affect the structure of mental space for different navigation tasks. Similarly, in the scenarios of human-robot collaborative navigation, modern intelligent mobile robots have to rely on quantitative information from either pre-installed map or real-time scanners/sensors to navigate in an environment, and therefore can only accept driving requests consisting of metrical data, such as '89,45 meters ahead, then at that point make a 42,5 degree turning'. On the other hand, while interacting with mobile robots, the human operators' instructions usually contain qualitative references other than precise quantitative terms, such as "straight ahead, and then make a left turn", as well as utilizing conceptual landmarks as reorientation points in cognitive mapping (cf. [14]). There is apparently an interaction gap between a human operator and a mobile robot if either the human operator wants to send a qualitative driving command to the robot, or the robot wants to negotiate with the human about unresolved situation using its internal quantitative information. Therefore, there have been many research efforts on applying qualitative spatial calculi and models to represent spatial environments and reason about situations within those spatial settings (cf. [15], [16], [17], [18]). Adding to this body of literature, we tried to bridge the interaction gap between humans and mobile robots by introducing a qualitative spatial knowledge based intermediate level to support human-robot interaction.

In general, this paper reports our work on resolving conceptual mode confusion during the interaction process between human operators and intelligent mobile robots for spatially-embedded navigation tasks. Based on our previous work on representing and reasoning about the shared spatial related resources using a Qualitative Spatial Beliefs Model (QSBM) (cf. [19]), a computational framework is then implemented with a set of high-level strategies and used to assist human operators as well as mobile robots to detect and resolve conceptual mode confusion to interact with each other more effectively. Specifically, the first two reasoning strategies have been compared and reported in [20], the positive empirical results provided evidence on our theoretical foundation, models and frameworks. However, during the further integration within an interactive system, an additional type of conceptual mode confusions was identified and accordingly, a new high-level strategy is developed and presented in this paper.

The rest of the paper is organized as follows. We first introduce the qualitative spatial beliefs model in section 2 and a model based computational framework in section 3. Then in section 4 we present the high-level strategies, which are developed based on the QSBM model, integrated within the computational framework and used to support the resolving of conceptual mode confusion in human-robot joint navigation tasks. Then we give a conclusion and an outline to the future work in section 5.

2   Using Qualitative Spatial Knowledge to Model a Mobile Robot's Beliefs

While interacting with a mobile robot for navigation tasks, human operators often use qualitative references with vague and uncertain information to communicate with the robot. From the perspective of human operators, the navigation environment is not represented as quantitative data based map fragment as mobile robots usually do, but as a conceptual world with objects, places and the qualitative spatial relations between them. Accordingly, a qualitative spatial beliefs model is developed for representing and reasoning about spatial environments and is used to model a mobile robot's Beliefs. Then with the definition of the qualitative spatial beliefs model, we are able to define a set of update rules to support the interpretation of natural language route instructions for mobile robots in navigation tasks.

2.1   Qualitative Spatial Beliefs Model

There has been a substantial effort to develop approaches and models to represent a mobile robot's beliefs and therefore support corresponding navigation tasks. One of the widely accepted model is called Route Graph (cf. [21], [22]), which models human's topological knowledge on the cognitive level in navigation space. In conventional route graph (cf. Fig. 1 a)), all geographical places are denoted as route graph nodes with particular positions regarding a quantitative reference system, all the routes between places are then abstracted as sequences of route segments with certain lengths, directed from source nodes to target nodes. Route graphs can be used as metrical maps of mobile robots to control the navigation, because they reflect the conceptual topological structure of humans' perspective about space and therefore also ease the interaction with human to a certain extent. However, due to lack of qualitative spatial relations between places and routes, conventional quantitative route graphs are not suitable for supporting interpretation of human route instructions containing qualitative references.

Meanwhile, Double-Cross Calculus (DCC) (cf. [23]) was proposed for qualitative spatial representation and reasoning using the concept orientation grids. In this calculus, as illustrated in Fig. 1 b), a directed segment AB divides the 2-dimensional space into disjoint grids, together with the edges between the grids, 15 meaningful qualitative spatial relations (DCC relations) can be defined, such as "Front", "RightFront", "Right", etc. Thus, DCC model can describe the relative relations between objects in the local navigation map with the directed line from an egocentric

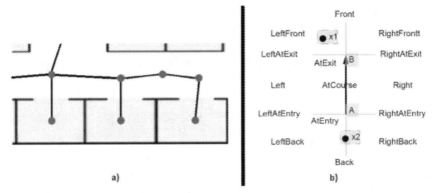

Fig. 1. A conventional route graph a) and the orientation frame of Double-Cross calculus with 15 named qualitative relations b)

perspective, e.g., the relations of x_1 and x_2 with AB can be denoted as "LeftFront" and "Back", accordingly. However, a conventional DCC model does not consider the relations between objects and places within global navigation maps, where they are connected conceptually.

Therefore, in order to benefit from both of these two models, we developed Conceptual Route Graph (abbreviated as CRG, cf. [19]), which combines the structure of conventional route graphs and the Double-Cross Calculus. On the basis of the conventional route graph and its topological representation of the places and routes in a geographical environment, qualitative spatial relations between route graph nodes and directed route segments are integrated regarding DCC relations. Thus, a conceptual route graph can be viewed as a route graph with additional qualitative DCC relations, which shows the following advantages:

- It can serve as a semantic framework for inter-process communication between different components on different levels.
- It can assist in the intuitive interpretation of human route instructions as well as appropriate presentation of internal feedback from the mobile robot system with the integrated qualitative spatial relations.
- It can be used as a direct interface with the mobile robot system for performing navigation tasks via the structure of the conventional route graph.

Given the definition of the conceptual route graph, a Qualitative Spatial Beliefs Model (abbreviated as QSBM) can be defined as a pair of a conceptual route graph and the hypothesis of the current position of a mobile robot in the given conceptual route graph, formally denoted as $<crg, pos>$, where

- crg is a conceptual route graph, formally represented by a tuple of four elements (M, P, V, R), where
  - M is a set of landmarks in a spatial environment, each of which is located at a place in P,

- — *P* is a set of topological places on the conceptual level of the abstracted environment,
- — *V* is a set of vectors from a source place to a different target place, all of which belong to *P*,
- — *R* is a set of DCC relations between places and vectors.
- and *pos* is a directed route segment of *V*, indicating that the robot is currently located at the source place of the route segment *pos* and has the segment pointing to the target place as its orientation.

As a simple example (cf. Fig. 1 b)), let us suppose that x_1 and x_2 are two places representing a copy room and a laboratory accordingly, AB a vector from place A to place B and a mobile robot is now at the position of A looking at the place B. The qualitative spatial relations of the places x_1 and x_2 with the vector AB can be written as <*AB, LeftFront, x_1*> and <*AB, Back, x_2*>, indicating that x_1 is on the left front of AB and x_2 is at the back of AB. Therefore, a simple instance of a QSBM model of Fig. 1 b) can be specified as:

```
< crg =
        (M = {copy room:x₁, laboratory:x₂},
        P = {A, B, x₁, x₂},
        V = {AB, BA},
        R = {<AB, LeftFront, x₁>, <AB, Back, x₂>}),
  pos = BA >
```

2.2   Interpreting Route Instructions with Update Rules

In order to interpret natural language route instructions for navigation tasks, a mobile robot's QSBM instance should be updated according to each interpreted route instructions and provides possible feedback to the human operator. Based on the empirical studies (cf. [7], [8]) and the previous research effort concentrating on the connection between natural language, cognitive models and route instructions (cf. [9], [24], [25]), we developed a set of update rules with respect to the most common used route instructions. Formally, each update rule is defined with the following three elements:

- a name (denoted as *RULE*), which identifies a class of human route instructions
- a set of pre-conditions (denoted as *PRE*), under which this rule can be applied, and
- an effect (denoted as *EFF*), describing how the QSBM instance should be updated after applying the update rule.

With the formal definition, the QSBM update rules are presented regarding the example classes of the common route instructions as follows:

Reorientation
defines the class of the simplest route instructions, which may change the orientation of a robot regarding the current position. "Turn left", "Turn right" and "Turn around"

are the typical expressions of such route instructions. The pre-condition for Reorientation is whether the robot can find a CRG node in its QSBM instance with the following two conditions: 1. it is connected with the current position, and 2. it has the desired spatial relation with the current position; the effect is that the robot faces that found CRG node after the reorientation. Formally it is described as:

```
RULE: Reorientation
PRE: pos = P₀P₁,
     ∃P₀P₂∈V. <P₀P₁, dir, P₂>
EFF: pos = P₀P₂
```

This specifies that the robot is currently at the place P_0 and faces the place P_1 (P_0P_1 is a vector in a QSBM instance), if there exists a vector P_0P_2 in the CRG vector set V with a targeting place P_2, such that the spatial relation of P_2 with respect to the route segment P_0P_1 (or the current position) is the given direction dir to turn, i.e., $<P_0P_1, dir, P_2>$, then the current position will be updated with P_0P_2 after applying this update rule.

Moving Through Motion

contains a landmark, through which the robot should go, e.g., "go through the door" or "go through the lobby". These route instructions require that the given landmark is located at the extension of the current directed path.

```
RULE: MoveThrough
PRE: pos = P₀P₁,
     ∃P₂P₃∈V. (l:P₁ₘ) ∧ <P₁P₂, AtCourse, P₁ₘ> ∧
              <P₀P₁, Front, P₂> ∧ <P₁P₂, Front, P₃>
EFF: pos = P₂P₃
```

In this rule, l is the landmark given in the instruction, (l : P_{1m}) indicates that l is located at the place P_{1m} in the QSBM instance. The pre-condition is to find a route segment P_2P_3 in front of the current robot position P_0P_1, such that the place P_{1m} of the given landmark is located on the path of P_1P_2 (denoted as $<P_0P_1, AtCourse, P_{1m}>$, where AtCourse is a predefined relation in Double-Cross Calculus [23]). After applying the update rule, P_2P_3 is the new robot position.

Directed Motion

refers to the route instructions which usually contain a motion action and a turning action changing the direction of the continuing motion, such as "take the next junction on the right". These instructions usually involve with a landmark (e.g. the "junction"), until which the robot should go, and a direction (e.g. on the "right"), towards which the robot should turn. For example, to deal with the route instruction "take the next corridor on the right", the most important step is to find the first corridor on the right from the robot's current position. Thus, the update rule for directed motions with the first landmark and a turning direction is specified as:

```
RULE: DirectedMotionWithFstLandmarkAndDirection
PRE: pos = P₀P₁,
     ∃P₂P₃∈V. ((l:P₂) ∧ <P₁P₂, dir, P₃> ∧ <P₀P₁, Front, P₂>)
   ∧ ∀P₄P₅∈V. ((l:P₄) ∧ <P₁P₄, dir, P₅> ∧ <P₀P₁, Front, P₄>
              ∧ (P₂≠P₄)) → <P₁P₂, Front, P₄>
EFF: pos = P₂P₃
```

In this rule, l is the targeted landmark and dir is the turning direction; The first pre-condition specifies that the robot should find a route segment P_2P_3, such that the targeted landmark is located at P_2, the spatial relation between P_3 and the segment P_1P_2 before turning is the desired direction dir and P_2 is in front of the robot's current position; The second pre-condition specifies that P_2 is the first place encountered referring to the given landmark at the given direction, instead of an arbitrary one; this condition is satisfied by if there exists a place P_4 with the same feature as P_2, P_4 must be ahead of P_2 from the current perspective. Then the effect is that, the robot position is updated as P_2P_3 after applying the rule. With the definition of this rule, other variants of directed motions, such as "go straight ahead", "go right" or "take the second left" can be specified with similar update rules accordingly.

Passing Motion
refers to the route instructions containing an external landmark to be passed by, typical examples are "pass the laboratory" or "pass the laboratory on the left" with direction information. For these route instructions, the robot should first identify the landmark given in the instruction, and then check whether the landmark can be passed by along the current directed path. Furthermore, the desired direction should be considered as well, if the direction for passing the landmark is given. Accordingly the update rule PassLeft for passing a landmark on the left is specified as:

```
RULE: PassLeft
PRE: pos = P₀P₁,
     ∃P₂P₃∈V. (l:P₁ₘ)
                ∧ <P₀P₁, LeftFront, P₁ₘ> ∧ <P₂P₃, LeftBack, P₁ₘ>
                ∧ <P₀P₁, Front, P₂> ∧ <P₀P₁, Front, P₃>
EFF: pos = P₂P₃
```

The pre-condition checks whether the current directed path satisfies the spatial requirement of the route instruction, i.e., the landmark l is located at the place P_{1m}, which is on the left front of the robot regarding the current position P_0P_1, and left behind the robot with the updated route segment P_2P_3 after executing the update rule.

Besides the above introduced update rules, there are other rules which are accordingly defined to interpret further route instructions, such as **Straight Motion**, which requests the robot to follow the current directed path; or **Moving Until Motion**, which is similar to passing motion but the robot should stop at the position parallel to the referred landmark, etc. All the QSBM update rules are implemented and integrated into a QSBM based computational framework, which is introduced in the next section.

3   A QSBM Based Computational Framework

According to the introduced formal definitions of the Qualitative Spatial Beliefs Model and the QSBM update rules, we developed SimSpace, a QSBM based Computational Framework for supporting and testing the QSBM based interpretation of natural language route instructions. This section starts by introducing the architecture of the SimSpace system, and then describes how a QSBM instance can be generated from a spatial environment, and finally presents how the interpretation of the individual route instructions is supported by the SimSpace system.

3.1   System Architecture

The architecture of the SimSpace system is illustrated in Fig. 2. It consists of two major components and one optional component, together with the external resources presented as follows:

- **The external resources** include
    - the quantitative map data containing quantitative and conceptual information about a certain spatial environment,
    - the conceptual route graph file (.crg), which is a XML-based specification of a the conceptual route graph of a QSBM instance, and
    - the qualitative spatial knowledge based toolkit SparQ (cf. [26]), connected with the SimSpace system to support the qualitative spatial representation and reasoning about the QSBM instance of the spatial environment.

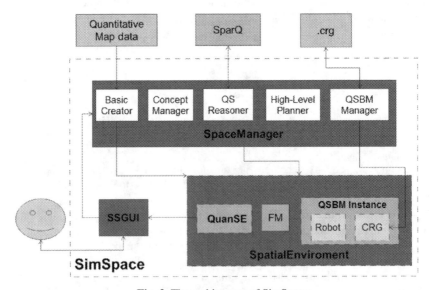

Fig. 2. The architecture of SimSpace

- **The component Spatial Environment** maintains the QSBM instance with the conceptual route graph and the hypothesis of the robot position in the CRG as defined before, as well as the optional quantitative spatial environment (QuanSE) for quantitative data and the optional feature map (FM) component containing the conceptual information of the environment.
- **The processing component Space Manager** is the central processing unit of the SimSpace system with the following functional components:
  - Basic Creator creates a spatial environment instance with quantitative and conceptual data according to the quantitative map data, if given.
  - Concept Manager manages an ontology-based database of the none-spatial conceptual knowledge, such as names of locations or persons, how they are conceptually related, etc. It is used to interpret the conceptual terms in the natural language route instructions.
  - QS Reasoner is responsible for the direct communication with the SparQ toolbox and handles the most basic operations of qualitative spatial representation and reasoning in QSBM.
  - High-Level Planner integrates all the QSBM update rules and implements a set of high-level strategies to choose and apply appropriate update rules to interpret route instructions and resolve conceptual mode confusion. The planning process is detailed in the next section.
  - QSBM Manager generates a QSBM instance according to a quantitative environment if given, manipulates and updates an existing QSBM instance, and saves it into a XML-based specification, if needed.
- **The interaction component SSGUI** is the graphical user interface of SimSpace. It is an optional component and is only used if the SimSpace system is started as a stand-alone application. The current SSGUI visualizes the spatial environment with quantitative and conceptual descriptions, interacts with a human user who is giving the natural language route instructions, and communicates with the Space Manager for the interpretation of incoming route instructions as well as outgoing system responses.

3.2   Construction of a QSBM Instance

In order to support the QSBM based interpretation of route instructions, a conceptual route graph in QSBM regarding a specific spatial environment needs to be constructed. One possible way is to use existing quantitative data. SimSpace takes a conventional quantitative route graph as input and constructs a corresponding DCC-based conceptual route graph in two steps:

- The quantitative data is qualified into DCC relations with the qualify module of the SparQ toolkit.
- The DCC relations cannot be used directly, instead they need to be generalized, i.e., the relations regarding angles near 0, 90 and 180 are accordingly assigned to those matching exactly 0, 90 and 180.

Despite the fact that a slight loss of precision may occur, the generalization is necessary. First and foremost, a conceptual route graph serves as an interface with human operators for navigation tasks, where they usually generalize the perceived qualitative relations mentally; therefore human operators usually use the generalized relations in the route instructions (cf. [27]). Instead of saying 'take a LeftFront turn', e.g., they usually say 'take a left turn'. Moreover, the qualitative spatial reasoning with ungeneralized relations provides too many possible results even after one step calculation, which is unlikely to handle.

A conceptual route graph of a QSBM instance can also be generated dynamically, if it is used for a mobile robot navigating in an unknown or partially known environment. The conceptual route graph is empty or with the partially known graph connections at the beginning, and keeps being updated by the QSBM manager while performing the navigation tasks, either via the collaborative interaction with the human operator, or with the real-time sensory data of the mobile robot.

3.3   Interpreting Route Instructions with SimSpace

With a QSBM instance generated, SimSpace can interpret a human route instruction in the following steps:

- The given natural language route instruction is firstly parsed into a pre-defined semantic representation.
- According to the category of the semantic representation, an applicable QSBM update rule is chosen and its pre-conditions are instantiated. Taking the sample instruction "pass the laboratory on the left" in the previous section, the update rule PassLeft is applied and by assuming AB to be the current robot position, the second pre-condition is instantiated to:

$$\exists P_2 P_3 \epsilon V. \ (\texttt{Laboratory}: P_{\texttt{lab}})$$
$$\wedge \ \texttt{<AB, LeftFront, } P_{\texttt{lab}}\texttt{>} \wedge \ \texttt{<}P_2 P_3\texttt{, LeftBack, } P_{\texttt{lab}}\texttt{>}$$
$$\wedge \ \texttt{<AB, Front, } P_2\texttt{>} \wedge \ \texttt{<AB, Front, } P_3\texttt{>}$$

- Then with the SparQ toolkit, the instantiated pre-conditions are checked against the current state of the QSBM instance. If the current state is matched with the instantiated pre-condition, the current robot position is updated to $P_2 P_3$; If however the current state provides e.g. the following relations:

$$(\texttt{Laboratory}: P_{\texttt{lab}}) \wedge \texttt{<AB, RightFront, } P_{\texttt{lab}}\texttt{>}$$

i.e., the laboratory is located on the right side regarding the perspective of the current robot position, which means that <AB, LeftFront, $P_{\texttt{lab}}$> in the pre-condition cannot be satisfied. In this case, SimSpace interprets these results into a corresponding representation to allow the generation of a clarification to the human operator.

Generally, the SimSpace system combines the implementation of the QSBM and the update rules. As a well encapsulated module, it can be integrated into an interactive mobile robot system to assist in the interaction with human operators via its intuitive qualitative spatial representation and reasoning about the spatial environment and set up a direct communication with the mobile robots via the inherited features from conventional route graphs. It can also be used as a stand-alone evaluation platform for visualizing spatial environments, generating corresponding QSBM instances and testing the interpretation of natural language route instructions. However, in order to interpret a sequence of natural language route instructions and resolve the possible consequent conceptual mode confusions, a set of high-level strategies are needed, which are developed and introduced in the following section.

4   Resolving Conceptual Mode Confusion with High-Level Strategies

For human operators, giving a sequence of route instructions to mobile robots may cause conceptual mode confusions, because before they can organize the appropriate terms for the route instructions, they first need to correctly locate the robot's current position and the desired goal location, and then take the imagined journey in mind to go along the expected route, while working in possible mental rotation during the travelling. Due to this complicated process, a wrongly located place or taken turn, which happens quite often (cf. [28]), would cause the failure of the interpretation of the entire route instructions and consequently lead to conceptual mode confusion situations. In order to cope with these problems, a set of high-level strategies based on QSBM and the QSBM update rules are developed and presented in this section.

4.1   Deep Reasoning

One of the most typical conceptual mode confusions is spatial relation or orientation mismatches (cf. [29]). This type of conceptual mode confusions occurs, if a spatial object is incorrectly oriented in the operator's mental representation, such as the previous example "pass the laboratory on the left", where the laboratory can only be passed on the right; or "take the next junction on the right", where the next junction is only leading to the left.

Given a QSBM instance and the appropriate QSBM update rule, the expected state of route instructions will be checked against the actual QSBM observed state using qualitative spatial reasoning. If an unsatisfied situation is identified, it can be presented to the human operator appropriately. If possible, while checking the instantiation of the route instruction leading to the unsatisfied situation, a corrected spatial or orientation relation is inferred for resolving the confusion. Therefore, the deep reasoning strategy tries to resolve the problematic situation by either giving a reason regarding the current situation to support the human operator to reorganize the route instructions, such as "you can't pass the laboratory on the left, because it is now behind you". More intuitively, it can make a suitable suggestion if existed, such as "you can't take a right turn here, but maybe you mean to take a left turn?"

4.2   Deep Reasoning with Backtracking

Besides the straightforward conceptual mode confusions concerning with the mentally wrongly oriented objects, there are situations where route instructions involve mental travelling or rotation that could easily be made incorrectly. If one instruction is wrongly given, the rest of the route instructions cannot be interpreted because the actual state caused by the wrong instruction does not match the mental state of the operator. Fig. 3 illustrates a typical example of this type of conceptual mode confusions, where the position of the mobile robot is shown by the arrow and the sequence of route instructions is "go straight, go left, left again, pass by B1 on the right". Using the deep reasoning strategy in the previous subsection, the robot will go along the path following the first three instructions, and then have a problem to interpret the last one "pass by B1 on the right". Finally it can only provide a reason why the last instruction cannot be taken, like "you can't pass by B1, because it's now behind you". However, by taking one step backwards, instead of "turn left", if the instruction is "turn right", then the last instruction can be interpreted appropriately.

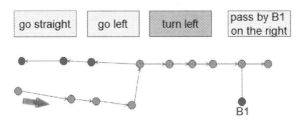

Fig. 3. An example for the conceptual mode confusions caused by an incorrect instruction

Therefore, the strategy "Deep Reasoning with Backtracking" manages the application of QSBM update rules with respect to the matching route instructions as "Deep Reasoning" usually does. Moreover, after applying one update rule for each route instruction, the state of the updated QSBM instance is kept in a transition history. If the checking of the current system state against a route instruction fails, the previous state is reloaded from the transition history. Based on it, possible correction/suggestion can be made, such as "turn right" substituting "turn left" in the example of Fig. 3, so that the interpretation of the remaining route instructions can proceed. In this case, instead of giving a reason regarding the first encountered uninterpretable route instruction, deep reasoning with backtracking tries to locate the potentially wrongly made route instructions in the interpretation history, then resume the checking of the current route instruction with a possible correction/suggestion, and finally finds a successful interpretation if one exists.

4.3   Searching with QSR-Weighted Value Tuples

Although deep reasoning with backtracking interprets more route instructions and better supports resolving conceptual mode confusion compared to the deep reasoning

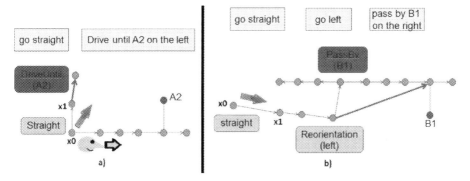

Fig. 4. Two examples of conceptual location mismatch

strategy (cf. [19] and [20]), an additional type of conceptual mode confusions regarding incorrectly located starting or turning position (called "conceptual location mismatch") is found. Fig. 4 illustrates two examples of conceptual location mismatches.

In example of Fig. 4 a), the position of the mobile robot is shown by the thick arrow, slightly nearer to the path leading upwards, and therefore the hypothesized robot position in the QSBM instance is represented by the route segment x0x1 leading upwards. However, the human operator, who is looking at the right direction in the illustrated map, thinks the robot is looking at the same direction and gives the instruction: "straight ahead and drive until A2 on the left". In the example of Fig. 4 b) with the robot position x0x1 and the route instructions "go straight and go left and pass by B1 on the right", after taking left, the operator thinks B1 is now located directly on the right from her/his perspective, and therefore simply ignores a turning point in the conceptual representation of the QSBM instance.

Both deep reasoning and deep reasoning with backtracking cannot provide an appropriate solution for resolving this type of conceptual mode confusions, because suggestions can only be made according to the given route instructions, while in such cases one instruction is missing and no possible suggestion can be made. Thus, the additional strategy of searching with Qualitative-Spatial-Relation-weighted (abbreviated as QSR-weighted) value tuples is developed.

First, we defined a set of QSR weighted value tuples for each starting/turning/decision point with every outgoing direction as:

```
{(route, instructions, qsr-v)*}
```

Here "route" represents the currently taken route, "instructions" includes all the along this route interpreted instructions, and "qsr v" is the cumulative value calculated by

$$\sum (mr_i * sr_i)$$

where mr_i is the matching rate by comparing the taken qualitative spatial direction with the current route direction at the i-th decision point, and sr_i is the success rate of interpreting a route instruction at that point.

With the definition of QSR-weighted value tuples, finding an appropriate interpretation (namely a route) to correspond to a sequence of natural language route instructions is illustrated in Fig. 5. The QSBM manager first initializes an empty set of QSR-weighted value tuples at the current position of the robot (the black point in the middle of the network in Fig. 5). This value-tuple-set is then automatically updated by the QSBM manager (as {(r1, i1, v1), (r2, i2, v2), (r3, i3, v3)} in Fig. 5, where (rx, ix, vx) indicates the tuple of the covered route rx, the interpreted instructions ix and the relating QSR-weighted value vx). Searching agents of the QSBM manager are then travelling along all paths (according to the branching of the current point, e.g. three paths in Fig. 5) on the current QSBM. The value-tuple-set gets updated and expanded by the QSBM based update rules when new branches are encountered or new instructions are interpreted. Finally, a full set of value-tuples is generated. The value tuple with the highest QSR-weighted value is either the best possible solution for interpreting the given route instructions or contains the most relevant information to provide the possible suggestion/correction to resolve the conceptual mode confusions.

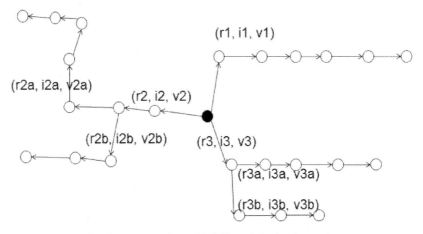

Fig. 5. The searching with QSR-weighted value tuples

With this strategy, the two example situations caused by the conceptual location mismatches in Fig. 4 can be solved as illustrated in Fig. 6.

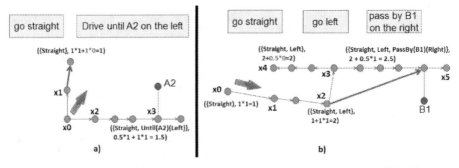

Fig. 6. Illustrated processes of solving conceptual location mismatch using QSR-Weighted value tuples (the route element in each QSR-weighted value tuple is ignored for simplicity)

In the first example of Fig.6 a) with the starting place x0, the set of the QSR-weighted value tuples is:

```
{(x0->x1, {Straight}, 1  * 1 = 1),
 (x0->x2, {Straight}, 0.5 * 1 = 0.5)}
```

Here for the route x0->x1, the matching rate is 1 because the route matches exactly the starting segment x0x1, the success rate of interpreting "go straight" is 1, and therefore the QSR-weighted value is 1*1 = 1. For the route x0->x2, the qualitative relation of the place x2 with the starting segment x0x1 is RightAtExit, so the matching rate is assigned as 0.5, however the success rate is again 1, so the QSR-weighted value is 0.5*1 = 0.5. Similarly, as the searching goes on, the route starting from x0x1 keeps going straight with the matching rate 1, but fails to interpret the instruction "Drive until A2 on the left" and gets the success rate 0; while the other route from x0x2 is also leading straight as well as being able to interpret the second instruction. Therefore, the instructions are interpreted with the route starting from x0x2 and a higher QSR-weighted value 1.5.

In the second example of Fig. 6 b), the searching goes along one route while successfully interpreting the first two instructions "go straight, and then left" with the following QSR-weighted value tuple:

```
(x0->x1-x2->x3, {Straight, Left}, 1*1 + 1*1 = 2)
```

When the searching comes to the turning point x3 with the instruction "pass by B1 on the right", there are two routes going into the left and right directions with the matching rates 0.5, but the route leading to the left cannot interpret the last instruction while the other one can. Therefore, the instructions are interpreted with the route x0->x1->x2->x3->x5, since it has the higher QSR-weighted value 2.5.

On the one hand, the strategy of searching with QSR-weighted value tuples resolves conceptual mode confusion from the perspective of a mapping problem in a directed graph with QSR weighted values. On the other hand, it preserves the functionality of deep reasoning with the QSBM update rules and the QSBM instance on each outgoing path from each decision point. Therefore, it can be viewed as a searching algorithm with multiple deep reasoning agents supporting the interpretation of more route instructions while clarifying more conceptual mode confusions.

5   Conclusion and Future Work

This paper has presented our research work on three important aspects of Cognitive Science: representing and managing, and reasoning with, qualitative spatial knowledge. Specifically, we focus on resolving conceptual mode confusion during the natural language based interaction between a human operator and a mobile robot for spatially-embedded navigation tasks. In order to support effective and intuitive human robot interaction, a qualitative spatial beliefs model has been developed for representing the shared conceptual spatial environment and a model-based computational framework has been accordingly implemented. Together with a set of high-level QSBM-based strategies, especially the strategy of searching with qualitative spatial relation weighted value tuples, mobile robots can be assisted to detect and resolve different types of conceptual mode confusions for more effective and intuitive interaction with human operators.

The reported work continued the pursuit of our goal towards building effective, intuitive and robust interactive frameworks and systems in spatially-related settings. Currently, we are integrating the qualitative spatial beliefs model and the model-based computational framework into a natural language interactive dialogue system for navigating a mobile robot in indoor environments. Regarding the high-level strategies, especially how to combine and apply the deep reasoning with backtracking and searching with QSR-weighted value tuples to better support natural language based spatially-related interaction is being investigated with empirical studies. Further qualitative spatial calculi and models are also being considered to extend the current qualitative spatial beliefs model for supporting more application domains. Human-robot collaborative exploration and navigation in unknown or partially known spatial environment is also another work package to be covered in the next stage.

Acknowledgement. We gratefully acknowledge the support of the German Research Foundation (DFG) through the Collaborative Research Center SFB/TR 8 Spatial Cognition - Subproject I5- [DiaSpace].

References

1. Sarter, N., Woods, D.: How in The World Did We Ever Get into That Mode? Mode error and Awareness in Supervisory Control. Human Factors 37, 5–19 (1995)
2. Bredereke, J., Lankenau, A.: A Rigorous View of Mode Confusion. In: Anderson, S., Bologna, S., Felici, M. (eds.) SAFECOMP 2002. LNCS, vol. 2434, pp. 19–31. Springer, Heidelberg (2002)
3. Lüttgen, G., Carreno, V.: Analyzing Mode Confusion Via Model Checking. Technical Report, Institute for Computer Applications in Science and Engineering, ICASE (1999)
4. Heymann, M., Degani, A.: Constructing Human-Automation Interfaces: A Formal Approach. In: Proceedings HCI-Aero 2002, Cambridge, MA, pp. 119–125 (2002)
5. Augusto, J.C., McCullagh, P.: Ambient intelligence: Concepts and Applications. Computer Science and Information Systems 4(1), 228–250 (2007)
6. Montoro, G., Haya, P.A., Alamán, X.: A Dynamic Spoken Dialogue Interface for Ambient Intelligence Interaction. International Journal of Ambient Computing and Intelligence 2(1), 24–51 (2010)
7. Bugmann, G., Klein, E., Lauria, S., Kyriacou, T.: Corpus-Based Robotics: A Route Instruction Example. In: Proceedings of the 8th International Conference on Intelligent Autonomous Systems (IAS-8), Amsterdam, Netherlands, pp. 96–103 (2004)
8. Shi, H., Tenbrink, T.: Telling Rolland Where to Go: HRI Dialogues on Route Navigation. In: Conventry, K., Tenbrink, T., Bateman, J. (eds.) Spatial Language and Dialogue, pp. 177–190. Cambridge University Press (2009)

9. Lauria, S., Kyriacou, S., Bugmann, G., Bos, J., Klein, E.: Converting Natural Language Route Instructions into Robot Executable Procedures. In: Proceedings of the 2002 IEEE International Workshop on Human and Robot Interactive Communication, pp. 223–228 (2002)

10. Zender, H., Mozos, O.M., Jensfelt, P., Kruijff, G.-J.M., Burgard, W.: Conceptual Spatial Representations for Indoor Mobile Robots. International Journal of Robotics and Autonomous Systems 56(6), 493–502 (2008)

11. Denis, M., Loomis, J.M.: Perspectives on Human Spatial Cognition: Memory, Navigation, and Environmental learning. Psychological Research 71(3), 235–239 (2007)

12. Tversky, B., Kim, J.: Mental Models of Spatial Relations and Transformations from Language. In: Habel, C., Rickheit, G. (eds.) Mental Models in Discourse Processing and Reasoning, pp. 239–258 (1999)

13. Tversky, B.: Structures of Mental Spaces: How People Think About Space. In: Environment & Behavior, pp. 66–80 (2003)

14. Michon, P.-E., Denis, M.: When and Why Are Visual Landmarks Used in Giving Directions. In: Montello, D.R. (ed.) COSIT 2001. LNCS, vol. 2205, pp. 292–305. Springer, Heidelberg (2001)

15. Skubic, M.: Qualitative Spatial Referencing for Natural Human-Robot Interfaces. In: Interactions, vol. 12(2), pp. 27–30. ACM, New York (2005)

16. Wallgrün, J.O., Wolter, D., Richter, K.-F.: Qualitative Matching of Spatial Information. In: Proceedings of the 18th SIGSPATIAL International Conference on Advances in Geographic Information Systems (GIS 2010), pp. 300–309. ACM, New York (2010)

17. Kurata, Y.: 9+-Intersection Calculi for Spatial Reasoning on the Topological Relations between Heterogeneous Objects. In: Proceedings of the 18th SIGSPATIAL International Conference on Advances in Geographic Information Systems (GIS 2010), pp. 390–393. ACM, New York (2010)

18. Liu, W., Wang, S., Li, S., Liu, D.: Solving Qualitative Constraints Involving Landmarks. In: Lee, J. (ed.) CP 2011. LNCS, vol. 6876, pp. 523–537. Springer, Heidelberg (2011)

19. Shi, H., Jian, C., Krieg-Brückner, B.: Qualitative Spatial Modelling of Human Route Instructions to Mobile Robots. In: Proceedings of the 2010 Third International Conference on Advances in Computer-Human Interactions (ACHI 2010), pp. 1–6. IEEE Computer Society, Washington, DC (2010)

20. Jian, C., Zhekova, D., Shi, H., Bateman, J.: Deep Reasoning in Clarification Dialogues with Mobile Robots. In: Coelho, H., Studer, R., Wooldridge, M. (eds.) Proceedings of the 2010 conference on 19th European Conference on Artificial Intelligence (ECAI), pp. 177–182. IOS Press, Amsterdam (2010)

21. Werner, S., Krieg-Brückner, B., Herrmann, T.: Modelling Navigational Knowledge by Route Graphs. In: Freksa, C., Brauer, W., Habel, C., Wender, K.F. (eds.) Spatial Cognition II. LNCS (LNAI), vol. 1849, pp. 295–316. Springer, Heidelberg (2000)

22. Krieg Brückner, B., Frese, U., Lüttich, K., Mandel, C., Mossakowski, T., Ross, R.J.: Specification of an Ontology for Route Graphs. In: Freksa, C., Knauff, M., Krieg-Brückner, B., Nebel, B., Barkowsky, T. (eds.) Spatial Cognition IV. LNCS (LNAI), vol. 3343, pp. 390–412. Springer, Heidelberg (2005)

23. Freksa, C.: Using Orientation Information for Qualitative Spatial Reasoning. In: Frank, A.U., Formentini, U., Campari, I. (eds.) GIS 1992. LNCS, vol. 639, pp. 162–178. Springer, Heidelberg (1992)

24. Denis, M.: The Description of Routes: A Cognitive Approach to the Production of Spatial Discourse. Cahiers de Psychologie Cognitive 16, 409–458 (1997)

25. Tversky, B., Lee, P.U.: How Space Structures Language. In: Freksa, C., Habel, C., Wender, K.F. (eds.) Spatial Cognition 1998. LNCS (LNAI), vol. 1404, pp. 157–175. Springer, Heidelberg (1998)
26. Wallgrün, J.O., Frommberger, L., Wolter, D., Dylla, F., Freksa, C.: Qualitative Spatial Representation and Reasoning in the SparQ-Toolbox. In: Barkowsky, T., Knauff, M., Ligozat, G., Montello, D.R. (eds.) Spatial Cognition 2007. LNCS (LNAI), vol. 4387, pp. 39–58. Springer, Heidelberg (2007)
27. Montello, D.R.: Spatial Orientation and the Angularity of Urban Routes — A Field Study. Environment and Behavior 23(1), 47–69 (1991)
28. Reason, J.: Human Error. Cambridge University Press (1990)
29. Shi, H., Krieg-Brückner, B.: Modelling Route Instructions for Robust Human-Robot Interaction on Navigation Tasks. International Journal of Software and Informatics 2(1), 33–60 (2008)

Event Recognition during the Exploration
of Line-Based Graphics in Virtual Haptic Environments

Matthias Kerzel and Christopher Habel

Department of Informatics
University of Hamburg
D-22527 Hamburg, Germany
{kerzel,habel}@informatik.uni-hamburg.de

Abstract. Pictorial representations are widely used in human problem solving. For blind and visually impaired people, haptic interfaces can provide perceptual access to graphical representations. We propose line-based graphics as a type of graphics, which are suitable to be explored by blind and visually impaired people, and which can be successfully augmented with auditory assistance by speech or non-verbal sounds. The central prerequisite for realizing powerful assistive interaction is monitoring the users' haptic exploration and in particular the recognition of exploratory events. The representational layers of line-based graphics as well as of exploration-event descriptions are specified by qualitative spatial propositions. Based on these representations, event recognition is performed by rule-based processes.

Keywords: line-based graphics, virtual environment haptic representation, geometric diagrams, event recognition.

1 Line-Based Graphics

1.1 Pictorial Representations for Blind and Visually Impaired People

Pictorial representations play a major role in human cognition and especially in human problem solving [1]. We use maps for way-finding. We visualize data, i.e., we (re-)present data in a pictorial format more suitable for thinking, problem solving, and communication, namely in the representational modality of graphs. And we use diagrams of devices or machines for teaching how these artifacts work. For blind and visually impaired people, access to pictorial representations is limited. Therefore, haptic interfaces to graphical representations—named 'haptic representations' as well as 'tactile representations'—have been proposed for partially substituting vision in comprehending different types of pictorial representations, such as graphs, mathematical diagrams, maps, floor plans or line drawings of animals [2, 3, 4, 5, 6]. Whereas visual perception supports comprehension processes that switch quickly between global and local aspects of a pictorial representation, haptic perception is more local and in particular more sequential [7]. Thus, compared to visual representations, in exploring haptic representations users have to integrate distributed,

T. Tenbrink et al. (Eds.): COSIT 2013, LNCS 8116, pp. 109–128, 2013.
© Springer International Publishing Switzerland 2013

sequentially grasped local information over the course of exploration, i.e., over time, into a global 'picture' of pictorial representation.

Realizing haptic representations in virtual environments offers the opportunity of assistive interactions with the user. As haptic exploration without visual feedback is a difficult task, the user should not also be burdened with active control of the interaction, e.g., pressing keys or performing command gestures. Instead, the computer should act like an intelligent and proactive partner; see Beaudouin-Lafon [8] for the idea of computer-as-partner interaction. In agreement with this view, we propose an interaction situation in which concurrently (a) the user explores a pictorial representation via the haptic interface, and (b) the system observes, i.e., monitors the exploration process, in particular the user's movements. The observations are related to an internal model of the system and, additionally, a propositional representation of the current exploration situation id build up. Based on these representations, the system can interact with the user: Additional information about entities that are relevant but not in the present focus of the user, as well as relations and properties of the depicted entities, are provided using speech or non-linguistic sound [5, 9, 10].

In order to foster the user's active control over the interaction, future research might aim at integrating keyboard- or dialog-based information requests into the architecture. However, a certain degree of autonomy of the system is necessary to compensate the user's missing overview, e.g., in some situations the user might not be aware that useful information could be requested.

1.2   Virtual Haptic Environments and Line-Based Graphics

To increase the effectiveness of knowledge acquisition from pictorial representations provided by virtual haptic environments, we have developed prototypes of systems which provide access to different types of line-based graphics—such as maps, floor plans, graphs and diagrams—augmented with auditory assistance provided by speech or sonification [5, 9, 10, 11, 12]. In our approach virtual haptic perception is realized by using a Phantom Omni force feedback device (http://www.sensable.com), Fig. 1(a) depicts this device, as well as an exemplifying geometric diagram. To explore a virtual object haptically, a pen-like handle that is attached to a robot arm is used. The user's pen movements are received by sensors in the robot arm and are evaluated with respect to the position of the tip of the pen. This position information is used to simulate a virtual pen tip, called haptic interface point (HIP), in a virtual environment. The force interaction between the HIP and the 3D model in the virtual environment is calculated as if the HIP and the 3D model were real objects, i.e., when the HIP collides with the 3D model, it is pushed back as if one would try to penetrate a solid surface with a real pen. This computed force feedback is made perceivable to the user through the motors in the robot arm that the pen is attached to. In short, the user feels as if probing a real 3D object with the tip of a real pen.

Virtual haptic exploration with a Phantom device can be characterized as *one-point interaction*, see Fig. 1(b). Thus, the exploration of an object is completely local and sequential. The haptic features of the object have to be explored sequentially and integrated, over time, into an internal model of the pictorial representation that should be consistent [13]. In order to design the interaction as user-friendly as possible, the haptic representation should consist of a small inventory of graphic atoms (see, for visual diagrams [14]) that are easy to explore haptically with the Phantom, and that additionally are cognitively adequate for constructing internal models.

As an alternative approach for haptic access to spatial information vibrotactile interfaces have been investigated. Giudice and colleagues [15] employ vibration to convey spatial information during exploring a touch screen using one fingertip. Like the force feedback realized by the Phantom, these vibrotactile approaches are based on one-point interaction. Thus, the formalism for event recognition, which is presented in sect. 3 and 4, could be extended to these vibrotactile interfaces.

Spatial information for several application domains of virtual haptics like street maps, public transit maps, geometric diagrams, graphs and charts can be based on *lines*. In this view, regions can be seen as second-order entities that are specified by their boundary. For example, a triangle (closed plane figure) can be characterized as well as it is depicted by three straight lines meeting at three vertices, see Fig. 1(a). Therefore, we propose to use line-based graphics as the conceptual basis for representing spatial information in virtual haptic environments. A line-based graphic consists of lines in a plane. These lines may be combined to form complex arrangements and shapes.

Lines are realized in a virtual haptic environment by using a planar 3D model with u-shaped grooves. Fig. 1(b) shows a cross section of such a model with a virtual HIP. On a conceptual level line-based graphics are used as planar spatial representations. The u-shaped grooves guide the user's exploration along a linear trajectory: Haptic groove-following leads to line-following on the conceptual layer and results in line recognition.

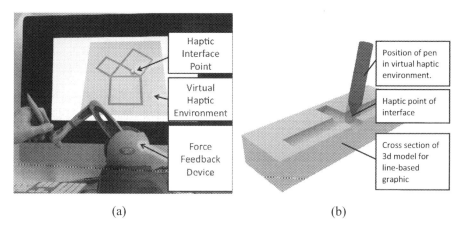

(a)   (b)

Fig. 1. (a) Sensable Phantom Omni and visualization in the domain of geometry (b) Cross section of 3D model used to represent a line-based graphic

1.3   Event Recognition for Exploration Monitoring

As described in sect. 1.2, a crucial deficit of haptic representations compared with visual representations is determined by the sequential and local nature of haptic exploration. For example, when looking at an arrangement of geometric figures as depicted in Fig. 1(a), an observer will immediately be able to identify, distinguish and locate the individual figures. In haptic explorations each figure, and even each part of a figure, has to be explored and identified separately. Thus, at the start of a haptic exploration a line is perceived as a line, and not, for example, as side of a triangle.

To help users overcome this type of problems, the system can provide additional information. Habel, Kerzel and Lohmann [12] suggest event recognition as an interaction paradigm for the generation of verbal assistance during the exploration of virtual environment tactile maps. Based on principles of division of labor, haptic and audio modality can be combined [5]. Verbal assistance provided during haptic exploration can offer non-spatial information that would be given as textual annotation in a visual representation as well as spatial information that is difficult or time consuming to extract from the haptic representation.

Parkas [16], Moustakas et al. [17] and Wang et al. [18] describe touch-based approaches that realize verbal assistance during exploration of tactile presentations. In their systems, touching (using a stylus/pen device or a finger on a touch screen) certain parts of a haptic representation immediately triggers a predefined assistance. This, however, limits the set of possible assistances by not taking into account the history of exploration, thus limiting the interactive capabilities of these systems.

Especially in haptic representation, where several elements overlap or have direct contact to each other, identifying which element is currently explored and should be in the focus of assistance, requires the recognition of *exploration events* (pen movements during exploration). For example, a triangle can be explored haptically by following the three triangle-constituting lines. Without going into details here (but see sect. 3.2), exploration events include both ongoing processes (called extended events) as well as short momentary transitions (called non-extended events). Recognition of extended exploration events takes the history of the exploration into account. Extended events take place over a time interval and often have a complex structure, being composed of several smaller events (see sect. 3).

Thus, event recognition provides the possibility to assist the user in different kinds of situations. The event recognition component has access to the full spatial representation and can interpret exploration movements in this context. The user, on the other hand, only has a local perception and lacks an overview of the explored representation. In haptic perception of space, distances and sizes are perceived differently depending on position and direction of the haptic exploratory movements [19]. This complicates identifying and understanding structures during the exploration, but assistance based on event recognition can help the user to overcome these difficulties.

In the following, we exemplify this conception of event recognition with some exploration events, both extended as well as non-extended ones, during the exploration of a geometric figure illustrating the Pythagorean theorem; see Fig. 2(a).

The Pythagorean theorem is chosen for exemplifying haptic exploration of line-based graphics in an abstract domain. Moreover, we conducted empirical studies of haptic exploration of line based-graphics depicting more complex spatial arrangements that were novel to the participants, see Lohmann, Kerzel and Habel [10] for an example of haptic exploration of line-based graphics from the domain of maps.

The pictorial representation of the Pythagorean theorem, consisting of a triangle and three squares, illustrates the fact that the area of the square over the hypotenuse of a right-angled triangle is equal to the sum of the area of the two squares over the other two sides. In a haptic environment based on line-based graphics, which could for example be used in teaching geometry to blind students, the lines of the figures are realized as grooves. The user explores the outlines of the figures by following these grooves in order to judge shape and size of the explored geometric entities. Fig. 2(b) shows an exemplifying exploration: (1) a side is explored, which is the hypotenuse of the right-angled triangle, but is also a side of a square, namely, the square over this hypotenuse. Thus, this exploration movement can—in future exploration—lead to two extended exploration events. (2) The next exploration movement follows another side of the triangle, which is again also a side of a different square. So the process of exploring square A has ended, while the exploration of the triangle is still going on. (3) Finally, the third exploration movement ends where the exploration started, the third side of the triangle is explored, thus completing the process of exploring the triangle. This fact could be communicated to the user, who might not be aware that the exploration ended in a closed loop.

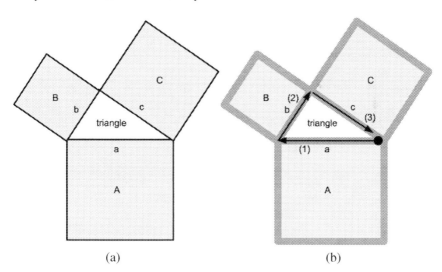

(a)   (b)

Fig. 2. (a) Visual representation for Pythagorean theorem, as used for teaching Geometry. (b) Visualization of haptic realization for Pythagorean theorem, lines are grooves (darker grey), the black arrows indicate exploration trajectories mentioned in the text

This example shows both extended events, like the process of exploring the side of a square or the triangle, and non-extended events, like completing the exploration of the triangle by arriving at the start point of the exploration. Exploration events are not unique with regard to their descriptions. The first exploration event depicted in Fig. 2(b) can be described both as 'exploration of the hypotenuse of a triangle' and as 'exploration of the upper side of a square'. The exploration can also be described in terms of complex events. The exploration of the triangular shape, i.e., the sequence of exploring the three sides, coincides with the exploration of the sides of three different squares. In such cases, the time intervals during which the extended events take place overlap. This exemplifies that not only the internal structure of exploration events may be complex, but also the temporal relations among recognized events may form a variety of temporal relations, see [20] for a representation of qualitative temporal relations. When reacting to the user's exploration, these temporal relations need to be captured during the process of event recognition.

2   Propositional Representations for Spatial Information in Virtual Haptic Environments

Line-based graphics can be based on an inventory of geometric concepts that can be used for qualitative spatial descriptions and reasoning. These propositional qualitative representations are the basis for both representing exploration movements in virtual haptic environments (sect. 3) and a rule-based approach to recognize events (sect. 4). In this section we give a rough overview on the basic concepts and on their use for specifying line-based graphics. (The underlying geometrical concepts belong to incidence geometry and ordering geometry, see [21, 22, 23].)

Lines and Configuration Points. *Lines* in line-based graphics are specified as maximal straight linear entities, i.e., lines are not part of other lines, but lines may have linear parts, namely *segments* (see below). Points on different lines may coincide; these points are called *configuration points,* as they are determined by configurations of lines. The *endpoints* of lines also are configuration points: They are defined as points on the line, which do not lie between two other points on the line [22]. Configuration points can be differentiated as a *junction*, a *tee* or a *crossing* of two lines; these subtypes depend on cardinality- and shape properties of the determining line configurations (see [24, 25] for kindred approaches).

C-Segments and Planar Embedding. Each line may be decomposed by a set of configuration segments. A *configuration segment* (c-segment) is a part of a line that lies between two adjacent configuration points on that line. The segmentation predicate describes which c-segments belong to a line. A line that contains no tee or crossing configuration point consists of only one c-segment.

Sets of c-segments may be considered to be sets of edges of a planar graph. Consequently, the nodes of the planar graph may be seen as counterparts of configuration points. Configuration points are embedded in the plane by defining their position in absolute coordinates. Thus, a line-based graphic equals an embedded planar graph.

Other Segmentation Criteria. The approach to insert segmentations into lines can be extended to other criteria depending on the application domain. A line segment may for example represent the part of a line determined by the projection of another entity onto the segmented line. In the application domain of street maps this is used to represent the part of a street that is next to a landmark. Segments defined by projections are called *p-segments*.

Line Complexes. Line complexes can be formed by a set of transitively connected c-segments. A c-segment may belong to multiple line complexes. All configuration points of c-segments also belong to the line complex(es) that the c-segment belongs to. This model follows an intuitive understanding of typical application domains of virtual haptic representations: When two streets intersect, each street can be said to have a crossing with another street, the crossing does not belong exclusively to just one of the streets. The same applies to two lines in a graph chart.

Depending on the configuration of c-segments, a line complex may have several properties: A line complex is branching if there exists a configuration point that is shared by more than two c-segments. A line complex is closed if the c-segments in the line complex form a closed path (according to graph theory). A non-branching closed line complex defines a region that is bounded by the c-segments. Thus, line complexes may form second order elements of line-based graphics.

Domain-Specific Properties. Each structure in a line-based graphic may also have properties relevant to the application domain. For example, the proper name in the context of the application domain is used as a predicate. This approach can easily be extended to provide other relevant information about the element in the domain.

Table 1. Subset of propositions defining elements of line-based graphics

Propositional description	Verbal description
line(l)	Element l is a straight line on a plane.
endPoint(l, p)	Point p is an endpoint of line l.
configurationPoint(c)	Configuration point c is a point on a line, c can either be an endpoint of a line or a point where two lines meet.
located(c, x, y)	Configuration point c is embedded in a plane at the coordinates (x, y).
c-segmentation(seg, l, $\{s_1, .., s_n\}$)	Line l is segmented by segmentation seg, dividing l into the set of c-segments $\{s_1, .., s_n\}$.
c-segment(s, seg, c_1, c_2)	C-segment s is created by segmentation seg due to configuration points. C-segment s is the part of a line that lies between configuration point c_1 and configuration point c_2.
lineComplex(lc, $\{s_1, .., s_n\}$)	Line complex lc is formed by the set of transitively connected c-segments $\{s_1, .., s_n\}$.
domainName(e, *name*)	Element e has is called *name* in the domain.

3 Exploration Events Based on Propositional Spatial Knowledge

3.1 Exploration Events in Line-Based Graphics

According to Lederman and Klatzky, manual exploration plays the core role for haptic perception [19]. In the case of line-based graphics in virtual haptic environments the manual exploration is mediated through a haptic interface point (HIP) that the user moves within grooves provided by a virtual haptic model. Thus, exploration movements are mainly restricted to movements along linear structures similar to movements of vehicles within a network of roads; see Pfoser and Jensen [26]. (Since the users of our prototypical line-based-graphics systems are instructed to explore a graphic based on lines, only in very few cases a user left the grooves and started exploratory movements in regions outside the grooves; compare Yu and Habel [5] for a similar process of border-following during the haptic exploration of regions depicting rooms.)

Thus, the exploration events to be recognized by the system are movements along the lines of line-based graphics. Storing of and reasoning about observed events presupposes a systematic notation for representations of events. These event representations are the basis for user interaction. They describe what is currently happening in the exploration or what has just happened. The event representation also serves as a history of the exploration, which is utilized to recognize more complex exploration patterns.

According to Shipley and Maguire [27] an observed sequence of movements may have an event representation on different levels of granularity. During a haptic exploration, a recognized exploration event may be the tracing of a single geometric figure. But this exploration also consists of tracing each single edge of the figure. Thus, the observed exploration can also be represented as a sequence of finer grained exploration events. The presented approach does not try to solve the problem of granularity by defining an optimal level of granularity. Instead, a *hierarchical event representation* is used where events are parts of more complex events. Which events are used to generate assistance, depends on the importance of the explored elements in the line-based graphic in the context of the application domain and will not be discussed in this article.

As an example, completely exploring a line complex having the shape of a triangle requires exploring all edges of the triangle, i.e., completely exploring all c-segments that are part of the line complex forming this triangle. In turn, completely exploring a c-segment requires an exploration movement from one configuration point of the c-segment to the configuration point at the other end of the c-segment. This movement is then in turn broken down into non-extended events that describe position changes during the exploration, like leaving or arriving at a configuration point. Each recognized event in this hierarchy offers an opportunity to assist the user during the exploration.

3.2   Types of Exploration Events

Exploration events are distinguished by temporal duration: *Non-extended events* answer the question "What has just happened?" while *extended events* answer the question "What is currently happening?" Non-extended events take place in a single moment. Extended events on the other hand describe exploration processes. Fig. 3 shows a taxonomy of the different event types developed for the representation of haptic exploration of line-based graphics. This taxonomy of exploration events is not exhaustive, i.e., a user may perform exploration movements that have not been classified. Yet the proposed set of exploration events enables the assisting system to react adequately to the user's current exploration movements by focusing assistance on currently explored elements of the line-based graphic.

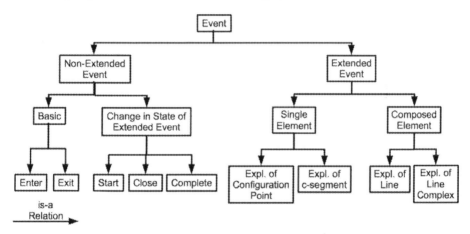

Fig. 3. Taxonomy of exploration events concerning line-based haptic graphics (Expl. stands for exploration)

Basic Non-extended Events. *Basic non-extended events* can be directly observed from the sensor data provided by the haptic device. The detection of basic non-extended events represents the transition from continuous perception to discrete events related to the geometry of a line-based graphic. For each c-segment and configuration point, a space in the virtual haptic environment is defined. Although a point has no area on the conceptual level, configuration points are perceived through active movement in the haptic environment, i.e., repeated movements against the end of a line or repeated tracing of the corner between two lines. Fig. 4(a) shows typical exploration movements for lines and configuration points. Basic non-extended events describe *entering* or *exiting* non-composed elements of the line-based graphics, i.e., c-segments (or other types of segments) and configuration points.

Extended Events: C-Segment and Configuration Point Exploration. *C-segment* and *configuration point exploration* events describe the process of exploring a single element in a line-based graphic. Exploration movements in c-segments often go back and forth. These repeated exploration movements allow the explorer to perceive the angle at which the c-segment is oriented. Likewise, configuration points are explored through movements in its vicinity. The event representation abstracts these movements into a single exploration event.

The spaces associated with configuration points and c-segments overlap. Thus, while exploring a c-segment, the configuration points at its ends can be explored without leaving the c-segment, and vice versa the adjoining part of c-segments can be explored while not leaving a configuration point. Fig. 4(b) shows a 2D sketch of the spaces of c-segments and a configuration point.

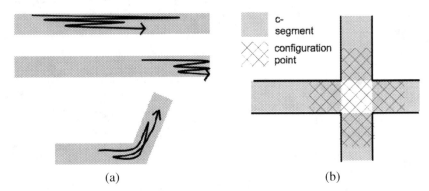

(a)	(b)

Fig. 4. (a) Typical exploration movements (depicted by black arrows) within a line segment, at the end of a line segment, and at a configuration point (2-way junction) (b) Overlapping areas associated with c-segments and a configuration point at a crossing

Extended Events: Line and Line-Complex Exploration. *Line* and *line complex explorations* describe the exploration of more complex elements in line-based graphics that consist of c-segments. The user might not sequentially explore the c-segments but move back and forth between c-segments to gain a better understanding of the spatial layout. As long as the explored element is not left during the exploration, repetitions are abstracted in the event representation.

Non-basic Non-extended Events: Start, Closure, Completion and Subordination. *Non-basic non-extended events* describe changes to ongoing, extended events. In contrast to basic non-extended exploration events, they are not directly observable. Arriving at a configuration point can be directly observed, but realizing that arriving at this configuration point completes exploring a side of a triangle cannot. Recognizing this event relies on knowledge about the history of the exploration.

The recognition of an exploration process causes the *start* of a corresponding exploration event. Likewise, the end of an extended event represents the moment when the extended event is irrevocably over. For instance, the user might start to explore a triangle, but before finishing it, he moves on to explore a different part of the line-based graphic. This is called *closure* as the exploration event is kept in the exploration history, but can no longer be influenced by the current exploration.

If the user later starts to explore the triangle again, this will be modeled as a new exploration event, unrelated to the former exploration event.

Extended exploration events are recognized before they are *completed*. This is useful for the purpose of assistance generation. The user might be interested in knowing that a triangle is explored before the complete contour of the triangle is traced. The completion of an element is recognized separately; this can be used to inform the user that a complete element has been explored and nothing was missed.

Events can be integrated into the hierarchical structure of an open extended event. This is called *subordination*. When an event e_1 is subordinated under an event e_2, this can be interpreted as e_2 growing not only in a temporal, but also in a conceptual way. For instance, the user explored the first side of a triangle. The extended event of exploring the triangle is started. When the user moves to the next side of the triangle, the exploration of this side is subordinated under the exploration of the triangle, thus the exploration event of the triangle now consist of two exploration events of its sides.

3.3   Propositional Representation for Exploration Events

Events are represented using propositions. For each exploration event, propositions are used to specify the type (see, for the taxonomy of exploration events, Fig. 3), the unique name and other properties.

Hierarchical Properties: List of Subordinated Events. Events are recognized by hierarchically aggregating events into more complex events. Basic non-extended events are used to conceptualize extended events describing the exploration of single elements in a line-based graphic; these extended events are in turn used to conceptualize events describing explorations of composed elements. When an event is aggregated from simpler events, these simpler events become subordinated to the more complex event. In the propositional representation each extended event features a list of events subordinated to it.

Temporal Properties: Time Stamp, Qualitative Temporal Relations and Closure. Basic non-extended events have a pseudo timestamp, representing the time point they were observed. By recursively accessing the information about the hierarchical aggregation down to the level of basic non-extended events, temporal information about any event can be computed and compared in terms of qualitative temporal relations. An extended event that is ongoing has the property *open*.

Progress Property: Completion. An extended event is completed when its subordinated events form a defined pattern. For instance, the exploration of a c-segment is completed when both configuration points at its ends have been explored without leaving the c-segment. In that case, the exploration events of both configuration points would be subordinated to the exploration of the c-segment. A completed event has the property *completed*.

Interaction of Closure and Completion. *Open* and *completed* are independent properties of an extended event. An extended event might be both *open* and not *completed*, i.e., the user might be exploring a triangle, but might not yet have finished exploring all parts of the triangle. If the user quits exploring the triangle before all parts have been explored, the exploration event becomes closed, while not being *completed*. The user might also have completed exploring all parts of the triangle, but still continue to trace its contour to get a better understanding of its spatial layout, the exploration of the triangle thus being complete and open. Finally, after the complete exploration of the triangle, the user might continue the exploration at a different figure; the exploration of the triangle thus being still *complete* but no longer *open*.

Event Representations. For each event the criteria for start, closure and completion are listed. The term e is used as a unique name to identify the event. Using the properties *open* and *complete* the state of an extended event is specified. The list $[e_1, .., e_n]$ contains names of all events that are subordinated to event e.

1) The event e is the basic non-extended event of entering or exiting the zone associated with c-segment or configuration point c. Event e takes place at time point t.

enter(e, c, t), exit(e, c, t)

2) The extended event e describes the exploration of a c-segment s. Event e starts when an enter event for s is detected. Event e is completed once both configuration points at the ends of the c-segment have been explored, that is, exploration events for both configuration points are contained in $[e_1, .., e_n]$. During e, no element of the line-based graphic that is not s or one of its configuration points may be entered, otherwise e is closed.

cSegmentEx($e, s, open, completed, [e_1, .., e_n]$)

3) The extended event e describes the exploration of a configuration point c. The event starts when an enter event for c is detected. The event is completed once $[e_1, .., e_n]$ contains explorations of all c-segments meeting in c. During e the exploration may not exit c, otherwise e is closed.

cpEx($e, c, open, completed, [e_1, .., e_n]$)

4) The extended event e describes the exploration of a line l. The event starts, when a c-segment belonging to l is explored. The event is completed once $[e_1, .., e_n]$ contains completed explorations of all c-segments in l. During e no element that is not part of l may be explored, otherwise e is closed.

lineEx($e, l, open, completed, [e_1, .., e_n]$)

5) The extended event e describes the exploration of a line complex lc. The event is completed once $[e_1, .., e_n]$ contains completed exploration events of all c-segments in line complex lc. During e no element that is not part of lc may be explored, otherwise

e is closed. The list of events $[e_1, .., e_n]$ lists all line- and c-segment exploration events that are part of e.

ComplexEx(e, *lc*, *open*, *completed*, $[e_1, .., e_n]$)

3.4   Example: Exploration Events during Exploration

Fig. 5(a) shows a line-based graphic illustrating the Pythagorean theorem; the black arrows indicate the exploration trajectory of the user. The following 11 steps explain how the event representation is updated, each step corresponding to an enter event or to an exit event with regard to a c-segment or a configuration point.

(Step 1) The exploration movement starts in the lower right corner of the triangle. The start position of the exploration movement is both within the space of configuration point c_1 (the lower left corner of the triangle) and c-segment s_1, the hypotenuse of the triangle. In this step, s_1 and c_1 are entered. Therefore, corresponding exploration events for c_1 and s_1 are ongoing. As c-segment c_1 is a common part of line l_1, line complex *triangle* and line complex *square$_1$*, the exploration of c-segment s_1 contributes to all three exploration events of these elements which begin with this step. (Step 2) The black arrow in the hypotenuse shows the first exploration movement. The hypotenuse is traversed from right to left. First the vicinity of the lower left corner (configuration point c_1) is exited. The corresponding exploration event of c_1 is closed. It has not been completed, as not all c-segments that meet in c_1 have already been explored. From the incomplete exploration of c_1 it can be deduced that the user is not aware of how this configuration point is connected to other lines and therefore he should be given suitable assistance. (Step 3) At the end of the first exploration movement, the lower left corner of the triangle is reached, configuration point c_2 is entered. Reaching this point completes the exploration of c-segment s_1. As s_1 is the only c-segment belonging to line l_1, the hypotenuse of the triangle, the exploration of this line is also completed. (Steps 4 to 6) In the second exploration movement, the user traces the left side of the triangle, which is an exploration of c-segment s_2. C-segment s_1 is exited and c-segment s_2 is entered thereafter. Up to this point, the explorations of line complexes *triangle* and *square$_1$* were ongoing. As c-segment s_2 is part of line complex *triangle* and *square$_2$*, but not of *square$_1$*, the exploration of line complex *triangle* continues, exploration of *square$_1$* ends, and a new exploration event for *square$_2$* begins. C-segment s_2 is also part of line l_2, therefore a corresponding exploration event begins. When the exploration movement exits the vicinity of configuration point c_2, the exploration of c_2 ends without having been completed. (Step 7) Configuration point c_3, the top of the triangle, is entered. This completes the exploration of c-segment s_2, the left side of the triangle. (Steps 8 to 10) The third exploration movement continues in a straight line along one side of *square$_3$*. C-segment s_2 is exited and s_3 is entered. C-segment s_3 is part of line complex *square$_3$*, but not of *triangle*. The exploration of *triangle* ends and a new exploration event of *square$_3$* begins. The user is still moving along line l_2, the corresponding exploration event still goes on. As the user continues the exploration in a straight line, it can be deduced that the user might not be aware that the exploration of *triangle* has stopped. This fact could be the basis for assistance to the user. (Step 11) Finally, the exploration movement reaches the topmost corner of $square_3$, configuration point c_4. This completes the exploration of c-segment s_3. As line l_2 consists of the c-segments s_2 and s_3 and both have been completely explored without leaving line l_2, the exploration of line l_2 is also completed. Fig. 5(b) shows selected events created and modified during the exploration process.

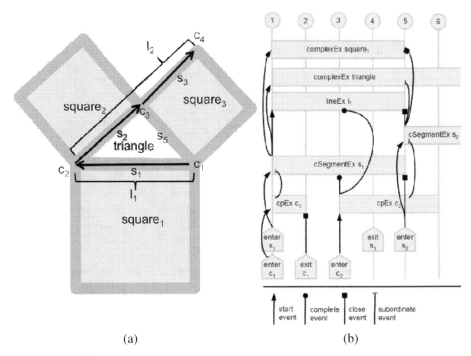

(a)   (b)

Fig. 5. (a) Visual depiction of a haptic exploration trajectory (depicted by black arrows) of the Pythagorean theorem diagram (b) Selected events that are created and modified during the first steps of the exploration

4   Rule-Based Event Recognition

Section 2 and 3 dealt with the representation of spatial information and exploration events. This section will discuss the procedural aspects of event recognition using a rule-based formalism.

4.1   Complex Event Processing

Luckham [28] coined the term *complex event processing* (CEP), to describe the process of recognizing meaningful events from a continuous stream of less complex events in a timely manner. Event recognition during the exploration of virtual haptic line-based graphics is a case of complex event processing. While the stream of position information from the haptic device is monitored for basic non-extended events, these non-extended events are processed in order to recognize more complex, extended exploration events. The computational challenge is to identify complex event patterns in a fast changing knowledge base of events that is subject to unpredictable input.

Because of the ability to perform well in fast changing environments, *rule-based systems* have been established as a paradigm for complex event processing; see Eckert and Bry [29]. Rule-based systems are used in artificial intelligence to codify human problem solving skills as a set of situation-action rules, see Hayes-Roth [30]. In [31], Anderson shows how complex cognitive abilities can be achieved by rule-based systems that not only encompass solving predefined problems, but also reaction to unpredictable input. In event recognition, during the exploration of virtual haptic environments, this unpredictable input is the user's exploration movement, to which the system that monitors the exploration must adapt quickly.

Rule-based systems realize their abilities through interaction of *declarative knowledge* (propositions in a knowledge base) and *procedural knowledge* (a set of rules). The knowledge base contains both static and dynamic information. In the case of event recognition during the exploration of line-based graphics, the spatial information is static. The dynamic part of the knowledge base consists of the representation of events. The procedural knowledge defines how the dynamic part of the knowledge base is altered. These rules consist of preconditions and actions that modify the declarative knowledge, once the preconditions of a rule are fulfilled.

The *inference engine* of a rule-based system constantly monitors the knowledge base, evaluates the preconditions of rules and executes the actions specified by the rules. This process can in return modify the knowledge base, triggering recursive execution of other rules. Forgy [32] developed the Rete algorithm that is based on arranging rule preconditions in a network to handle matching of preconditions against a fast changing knowledge base with a manageable time complexity. Walzer at al. [33] show how the Rete algorithm can be extended to allow for qualitative temporal relations in rule conditions, which are often used in event recognition.

4.2   Architecture for Event Recognition during the Exploration of Line-Based Graphics

The rule-based event recognition is embedded into an architecture for event recognition and assistance generation, see Fig. 6. The architecture of this system is based on an architecture suggested by Habel, Kerzel and Lohmann [12]. While the user interacts with the haptic interface, the virtual haptic environment provides force feedback based on 3D spatial information (i.e., on a 3D model of the line-based graphic). Exploration movements of the user are sent to a component for *basic non-extended event detection*, where the current position of the exploration is related to spatial information of the line-based graphic. Basic non-extended events are added to the knowledge base upon detection. The *inference engine* of the rule-based system constantly monitors the knowledge base taking into account both static spatial information and dynamic information about exploration events. If the precondition of a rule is fulfilled, the rule is executed, creating new exploration events, or modifying existing ones. Based on recognized exploration events, (verbal) assistance is generated and passed to the user.

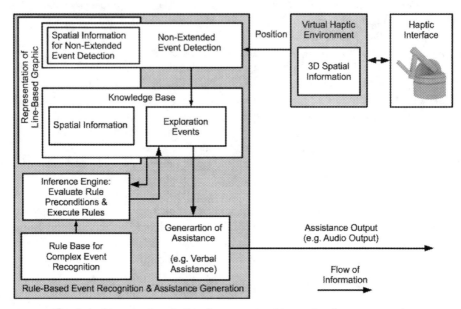

Fig. 6. Architecture for rule-based event recognition and assistance generation

4.3   Rules for Event Recognition

The complex events introduced in sect. 3.3 can be described by formal patterns using first order predicate logic as preconditions of rules. For example, completing the exploration of a line complex requires two main conditions: For all c-segments of the line complex an exploration event has to be completed; and all these exploration events have to happen without the exploration being interrupted by leaving the line complex. However, the user is allowed to explore the line complex in any order and perform repetitions during the exploration as long as the first two criteria are fulfilled.

For each type of extended event, a separate rule specifies when an event starts, ends, is completed (if appropriate) and when another event is subordinated to it. In figure 7 and 8, two rules out of a set of about 20 rules are shown [34].

The conditions of rules are given as a conjunction of propositions. In order for the rule to be executed, all conditions of the rule have to evaluate to *true* according to the knowledge base. Several functions are used in these rules: id() creates a new and unique identifier for an event; in(l, e) evaluates to *true*, if event e is contained in the list of events l and append(l, e) adds event e to the end of list l. Temporal relations are expressed in qualitative terms. The underscore (_) is used to denote that any value is acceptable or that no change is made to a value.

IF
    enter(e_{enter}, s, *time*)
    $\neg\exists\, e_{cp}$: cSegmentEx(e_{cp}, cp, _, _, *list*)
    in(e_{enter}, *list*)
THEN
    Create:
    cSegmentEx($id()$, s, t, f, [e])

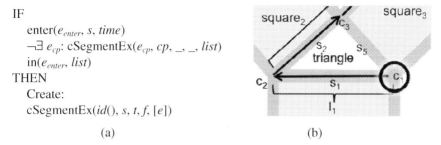

(a)   (b)

Fig. 7. (a) Rule for recognizing the start of an exploration event of a c-segment (b) Starting situation of exemplifying exploration (sect. 3.4)

Fig. 7(b) depicts the starting situation of the exploration from sect. 3.4. C-segment s_1, the hypotenuse, is entered. This creates a corresponding entering event. There is no c-segment exploration event in the knowledge base that is based on this entering event. Therefore, an exploration event of s_1, which is currently ongoing and not completed, is added to the knowledge base by the rule in Fig. 7(a). This falsifies the second condition of the rule, thus preventing the rule to be executed repeatedly. At the same time, configuration point c_1, the lower right corner of the triangle, is entered. A different rule creates a corresponding exploration event of c_1 in the knowledge base.

The rule for subordinating the ongoing exploration of a configuration point, in this case the exploration of c_1, causes this exploration event to be subordinated to the ongoing exploration of c-segment s_1. Spatial knowledge from the propositional representation of the line-based graphic is used to check whether the configuration point is actually one of the endpoints of the c-segment, which holds true in this case.

IF
    cSegmentEx(e_{seg}, s, t, f, *list*)
    c-segment(s, seg, cp_1, _)
    cpEx(e_{cp1}, cp_1, _, _, _)
    in(e_{cp1}, *list*)
    c-segment(s, seg, _, cp_2)
    cpEx(e_{cp2}, cp_2, _, _, _)
    in(e_{cp2}, *list*)
THEN
    Modify:
    cSegmentEx(e_{seg}, _, _, t, _)

(a)   (b)

Fig. 8. (a) Rule for recognizing the completion of an exploration event of a c-segment (b) Completion of the first exploration movement along c-segment s_1 (see sect. 3.4)

Fig. 8 (b) depicts the situation after the first exploration movement from the example in sect. 3.4. has been completed. The exploration movement has just reached configuration point c_2 in the lower left corner of the triangle. An exploration event for the configuration point is added to the knowledge base and subordinated to the ongoing exploration of the traversed c-segment, i.e., it is added to the list of events being subordinated to the exploration of c-segment s_1.

Three exploration events have already been subordinated to the exploration event of s_1, the first is entering s_1, the second is the exploration event of configuration point c_1 where the exploration started; the third is an exploration event for configuration point c_2 which has just been reached. Thus, the preconditions stated in the rule in Fig. 8(a) are fulfilled; the configuration points on both ends of s_1 have been explored now. The state of the exploration event of s_1 is changes to *completed*.

5   Conclusion

Line-based graphics have been introduced as a formal basis for conveying spatial information from different application domains, such as maps, graphs, charts or mathematical diagrams, to blind or visually impaired people through virtual haptic environments. The specification of line-based graphics follows the criteria outlined by Mackworth [35]. This offers the advantages that, firstly, design principles for haptic line-based graphics can be applied to a variety of application domains and that, secondly, users familiar with haptic line-based graphics in one application domain can successfully use their exploration experience in a new domain.

To assist blind or visually impaired people in their exploration of pictorial representations, assistive communication using speech or non-verbal sound has been proved to be helpful. Powerful assistance not only reacts locally to the current point of interaction, but also considers the context of exploration as well as the history of the exploration process. As a core component for such assistance system, we have proposed an event monitor, which recognizes exploration events using rule-based mechanisms. A system prototype, which uses a prior version of the event-recognition component for *Verbally Assisted Virtual-Environment Tactile Maps*, has been successfully evaluated by Lohmann, Kerzel and Habel [10] in the domain of street maps.

Acknowledgement. We like to thank the anonymous reviewers for their thoughtful comments and valuable suggestions. The presented research has been partially supported by DFG (German Science Foundation) in the international research training group 'Cross-modal Interaction in Natural and Artificial Cognitive Systems' (CINACS, IRTG 1247).

References

1. Tversky, B.: Visualizing Thought. Topics in Cognitive Science 3, 499–535 (2011)
2. De Felice, F., Renna, F., Attolico, G., Distante, A.: A Haptic/Acoustic Application to Allow Blind the Access to Spatial Information. In: World Haptics Conference, pp. 310–315 (2007)
3. Shimomura, Y., Hvannberg, E., Hafsteinsson, H.: Haptic cues as a utility to perceive and recognise geometry. Universal Access in the Information Society Online First, doi: 10.1007/s10209-012-0271-2

4. Sjöström, C., Danielsson, H., Magnusson, C., Rassmus-Gröhn, K.: Phantom-based Haptic Line Graphics for Blind Persons. Visual Impairment Research 5, 13–32 (2003)
5. Yu, J., Habel, C.: A Haptic-Audio Interface for Acquiring Spatial Knowledge about Apartments. In: Magnusson, C., Szymczak, D., Brewster, S. (eds.) HAID 2012. LNCS, vol. 7468, pp. 21–30. Springer, Heidelberg (2012)
6. Yu, W., Brewster, S.A.: Evaluation of Multimodal Graphs for Blind People. Journal of Universal Access in the Information Society 2, 105–124 (2003)
7. Loomis, J., Klatzky, R., Lederman, S.: Similarity of Tactual and Visual Picture Recognition with Limited Field of View. Perception 20, 167–177 (1991)
8. Beaudouin-Lafon, M.: Designing Interaction, not Interfaces. In: Proceedings of the Working Conference on Advanced Visual Interfaces, pp. 15–22. ACM (2004)
9. Lohmann, K., Habel, C.: Extended Verbal Assistance Facilitates Knowledge Acquisition of Virtual Tactile Maps. In: Stachniss, C., Schill, K., Uttal, D. (eds.) Spatial Cognition 2012. LNCS, vol. 7463, pp. 299–318. Springer, Heidelberg (2012)
10. Lohmann, K., Kerzel, M., Habel, C.: Verbally Assisted Virtual-Environment Tactile Maps: A Prototype System. In: Proceedings of the Workshop on Spatial Knowledge Acquisition with Limited Information Displays, pp. 25–30 (2012)
11. Alaçam, Ö., Habel, C., Acartürk, C.: Towards Designing Audio Assistance for Comprehending Haptic Graphs: A Multimodal Perspective. In: Stephanidis, C. (ed.) UAHCI/HCII 2013, Part I. LNCS, vol. 8009, pp. 409–418. Springer, Heidelberg (2013)
12. Habel, C., Kerzel, M., Lohmann, K.: Verbal Assistance in Tactile-Map Explorations: A Case for Visual Representations and Reasoning. In: Proceedings of AAAI Workshop on Visual Representations and Reasoning 2010, pp. 34–41. AAAI, Menlo Park (2010)
13. Tversky, B.: Cognitive Maps, Cognitive Collages, and Spatial Mental Models. In: Campari, I., Frank, A.U. (eds.) COSIT 1993. LNCS, vol. 716, pp. 14–24. Springer, Heidelberg (1993)
14. Tversky, B., Zacks, J., Lee, P., Heiser, J.: Lines, Blobs, Crosses and Arrows: Diagrammatic Communication with Schematic Figures. In: Anderson, M., Cheng, P., Haarslev, V. (eds.) Diagrams 2000. LNCS (LNAI), vol. 1889, pp. 221–230. Springer, Heidelberg (2000)
15. Giudice, N.A., Palani, H., Brenner, E., Kramer, K.M.: Learning Non-Visual Graphic Information using a Touch-Based Vibro-Audio Interface. In: Proceedings 14th International ACM SIGACCESS Conference on Computers and Accessibility, pp. 103–110. ACM (2012)
16. Parkes, D.: NOMAD - An Audio-Tactile Tool for the Acquisition, Use and Management of Spatially Distributed Information by Partially Sighted and Blind People. In: Tatham, A., Dodds, A. (eds.) Proceedings of the 2nd International Conference on Maps and Graphics for Visually Disabled People, pp. 24–29. King's College, London (1988)
17. Moustakas, K., Nikolakis, G., Kostopoulos, K., Tzovaras, D., Strintzis, M.G.: Haptic Rendering of Visual Data for the Visually Impaired. IEEE Multimedia 14, 62–72 (2007)
18. Wang, Z., Li, B., Hedgpeth, T., Haven, T.: Instant Tactile-Audio Map: Enabling Access to Digital Maps for People with Visual Impairment. In: Proceedings of the 11th International ACM SIGACCESS Conference on Computers and Accessibility, pp. 43–50. ACM, Pittsburg (2009)
19. Lederman, S.J., Klatzky, R.L.: Haptic Perception: A Tutorial. Attention, Perception, and Psychophysics 71, 1439–1459 (2009)
20. Allen, J.: Maintaining Knowledge about Temporal Intervals. Communication of the ACM 26, 832–843 (1983)

21. Eschenbach, C., Kulik, L.: An Axiomatic Approach to the Spatial Relations Underlying 'left'–'right' and 'in front of'–'behind'. In: Brewka, G., Habel, C., Nebel, B. (eds.) KI 1997. LNCS (LNAI), vol. 1303, pp. 207–218. Springer, Heidelberg (1997)
22. Eschenbach, C., Habel, C., Kulik, L.: Representing Simple Trajectories as Oriented Curves. In: Kumar, A.N., Russell, I. (eds.) FLAIRS 1999, Proceedings 12th International Florida AI Research Society Conference, pp. 431–436. AAAI, Menlo Park (1999)
23. Habel, C.: Representational Commitment in Maps. In: Duckham, M., Goodchild, M., Worboys, M. (eds.) Foundations of Geographic Information Science, pp. 69–93. Taylor & Francis, London (2003)
24. Klippel, A.: Wayfinding Choremes. In: Kuhn, W., Worboys, M.F., Timpf, S. (eds.) COSIT 2003. LNCS, vol. 2825, pp. 301–315. Springer, Heidelberg (2003)
25. Reiter, R., Mackworth, A.: A Logical Framework for Depiction and Image Interpretation. Artificial Intelligence 41, 125–155 (1989)
26. Pfoser, D., Jensen, C.: Indexing of Network Constrained Moving Objects. In: Proceedings of the 11th ACM International Symposium on Advances in Geographic Information Systems, pp. 25–32. ACM (2003)
27. Shipley, T., Maguire, M.: Geometric Information for Event Segmentation. Understanding Events: How Humans See, Represent, and Act on Events. In: Shipley, T.F., Zacks, J.M. (eds.) Understanding Events: How Humans See, Represent, and Act on Events, pp. 415–435. Oxford University Press, New York (2008)
28. Luckham, D.C.: The Power of Events: An Introduction to Complex Event Processing in Distributed Enterprise Systems. Addison-Wesley, Reading (2002)
29. Eckert, M., Bry, F.: Complex Event Processing (cep). In: Informatik-Spektrum 32, pp. 163–167 (2009)
30. Hayes-Roth, F.: Rule-Based Systems. Communications of the ACM 28, 921–932 (1985)
31. Anderson, J.R.: ACT: A Simple Theory of Complex Cognition. American Psychologist 51, 355–365 (1996)
32. Forgy, C.L.: Rete: A Fast Algorithm for the Many Pattern/Many Object Pattern Match Problem. Artificial Intelligence 19, 17–37 (1982)
33. Walzer, K., Breddin, T., Groch, M.: Relative Temporal Constraints in the Rete Algorithm for Complex Event Detection. In: Proceedings of the Second International Conference on Distributed Event-Based Systems, pp. 147–155. ACM (2008)
34. Kerzel, M.: Rule Patterns for Event Recognition During Exploration of Hapic Virtual Environment Line-based Graphics. Technical Report, Department of Informatics, University of Hamburg, Germany (2013)
35. Mackworth, A.K.: Adequacy Criteria for Visual Knowledge Representation. In: Pylyshyn, Z. (ed.) Computational Processes in Human Vision, pp. 464–476. Ablex Publishers, Norwood (1988)

Cognitive Transactions – A Communication Model

Paul Weiser and Andrew U. Frank

Vienna University of Technology, Austria
Department for Geodesy and Geoinformation
Research Group Geoinformation
Gusshausstr. 27-29, 1040 Vienna
{weiser,frank}@geoinfo.tuwien.ac.at
http://www.geoinfo.tuwien.ac.at

Abstract. Whenever a person gets lost and there is no way to access stored spatial information, e.g. in the form of maps, they need to rely on the knowledge of other humans instead. This situation can be modelled as a communication setting where a person lacking spatial knowledge requests information from a knowledgeable source. The result are cognitive transactions in which information over various levels of detail (LoD) is negotiated. The overall goal is to agree on a shared spatial representation with equal semantics, i.e., common ground. We present a communication model that accounts for establishing common ground between two agents. The agents use a modified "wayfinding choreme" language and special signals to negotiate the LoD. Findings of a case study were used to verify and refine our work.

Keywords: Cognitive Transactions, Spatial Negotiation, Level of Detail, Route Descriptions.

1 Introduction

Compared to the navigational skills of some animals (c.f. ants, bees, or birds) humans perform rather poorly at wayfinding. As a result, unless specially trained, either formally (e.g., ship navigators) or experientially (e.g., Inuits), we get lost on a quite regular basis [23]. Tools to store and share spatial information, e.g. maps, help us to accumulate expert knowledge with the goal to overcome our inborn cognitive limitations. Also, the manipulation of physical tools reduces the need to carry out computations by the "manipulation of the mind" [17].

There are situations, however, where we cannot access permanently stored information, but have to rely on the knowledge of other humans instead. A prototypical example is that of route descriptions given by another person orally. They differ from automatically generated or written route descriptions in a number of ways. For example, the information needs to be memorized. Written instructions can be consulted several times along the way; asking for directions is a single event, bound to the location the communication takes place. Also, oral

T. Tenbrink et al. (Eds.): COSIT 2013, LNCS 8116, pp. 129–148, 2013.

instructions frequently combine different levels of detail (LoD), e.g., (1) Walk down the stairs then (2) take train S1; it leaves at 10 a.m from platform A. Note that (1) offers spatial information while (2) offers additional thematic and temporal information.

Allen [3] coined the term cognitive transaction, i.e., the process where one individual (target) seeks information from a knowledgeable source. The LoD at which the source presents information depends, among other things, on the target's knowledge and the source's expectation of the target's knowledge. The target in turn attempts to construct a mental model based on the presented information. Because the source's and target's mental models are not necessarily identical, the communication seeks to eliminate the differences. The result is a negotiation over the correct amount of LoD for each part of the route description. The communication ends, if both individuals believe they have established a shared spatial representation with equal semantics, i.e., common ground (See Section 5).

While there already exists some work on adaptable route descriptions, e.g., in a dialog between humans and machines [31], as well as an analysis of the structure of giving directions ([41], [29], [5], [6]), a comprehensive study of the cognitive processes taking place during a spatial communication setting between two humans is still missing. In this paper, we provide a formal cognitive model of the communication process between an information seeking target and a knowledgeable source. Our model integrates the notion of language advocated by Clark[4], who sees human communication as joint actions between agents, as well as a modified version of "wayfinding choremes" [18]. To achieve this, we introduce special signals used by the agents to indicate granularity changes. Finally, our model makes the differences between the participating agent's mental models explicit and discusses issues on how to establish common ground.

We expect our findings to contribute to the design of better navigation systems, capable of adapting the LoD at which information is presented to a user's knowledge and possible special needs. A formal understanding of how humans abstract from information and convey it through language allows us to build more intuitive computational models. Language as a tool of study seems particularly promising because it allows us to look into otherwise "hidden highly abstract cognitive constructions" ([9], p. 34).

The remainder of the paper is structured as follows. Section 2 discusses relevant previous work. In Section 3 we briefly review several existing approaches to LoD in route descriptions and show how wayfinding choremes can be expanded to take granularity changes into account. In Section 4, we introduce and discuss various signals an agent can use to indicate a granularity changes. In Section 5, we elaborate on the individual phases during a spatial negotiation setting, as well as the issue of establishing common ground from a global perspective. In section 6, we present the results of our case study used to verify and refine our model. The final section concludes with future research directions.

2   Relevant Previous Work

This section discusses the core concepts relevant to our work. We should note, however, that our approach and choice of terminologies are motivated by the goal to build models that resemble the way the world is perceived and acted upon by ordinary humans. This connects our work to the fields of "Naive Geography" [7] and "Geographic Cognitive Engineering" [30].

2.1   Cognitive Transactions, Wayfinding, and Actions

Navigation can be conceptualized as the combination of wayfinding and locomotion [26]. Lynch [24] defines wayfinding as the "consistent use and organization of sensory cues from the external environment". In terms of Norman [28], sensory cues are "knowledge in the world", i.e., information that can be derived from the world adding to an agent's existing "knowledge in the head". Whether or not wayfinding is successful depends on the two types of knowledge available, i.e., information in the head (mental representation) and information in the world. In case a person has a spatial mental model [38] of the geographic space in question, wayfinding is simply the matching of the actions possible based on the representation in the head with possible actions perceived from the environment. If the information in the head alone is not sufficient, information from the world can be an additional aid. For example, a street sign could allow me to carry out an action my mental representation would not have allowed. If neither type of information is sufficiently available, a person is considered to be lost.

In this paper we use the term action to refer to processes performed by an agent. There exists extensive literature concerned with the semantics of process and action. For a recent discussion and comprehensive review of the different approaches see Galton and Mizoguchi [12], who stress the interdependence of objects and processes and advocate to "model reality as it appears to humans engaged in ordinary human activities (p. 3)". From an informal point of view, a route description is a sequence of actions that, if carried out accordingly, allow an agent to navigate from A to B. Generally, we can view the world as having a state at any time. From an agent's perspective, the state of the world can be changed through actions. Actions unfold in space and time and are usually goal-oriented. Therefore, we can conceptualize navigation as a sequence of actions performed by an agent to move from A (source) to B (goal). Furthermore, Allen [3] mentioned the importance of choice points along the route, "affording options with regard to pathways, with intersections being the most typical example".

Klippel [18] provides a formalism of aforementioned principles in which he states that a route can be seen as a sequence of tuples that consist of a route segment (RS) and a decision point (DP). Here, DPs take the role of Allen's choice points [3], thus they require an agent to decide on a further action, e.g., which direction to take. Klippel [18] also showed that humans have seven different conceptualizations of turning actions at DPs. With these action primitives, or "wayfinding choremes (wc)", we can express most turn-by-turn route descriptions. The following formal definition was adapted from Klippel [18].

<Route> ::= <DecisionPoint> <Segment> [<RoutePart>] <DecisionPoint>
<RoutePart> ::= <DecisionPoint> <Segment>
<DecisionPoint> :: = wc_l | wc_r | wc_s | wc_l^{RM} | wc_r^{RM} | wc_s^{RM}

Note that we use only a subset the original seven actions. We use "turn left", "turn right", and "go straight" which we denote by "wc_l", "wc_r", and "wc_s", respectively. Our definition also includes routemarks (RM), i.e., landmarks along the route [18]. RM are DPs with a special status and we indicate turning actions at RMs by wc_l^{RM}, wc_r^{RM}, and wc_s^{RM}, meaning "turn left at routemark", "turn right at routemark", and "pass routemark", respectively.

2.2   Granularity and Navigation

Our work adds to the growing body of literature on level of detail (LoD) in the context of the communication of route descriptions. Hirtle et al. [14] stressed that the LoD at which information should be presented needs to take the context, more specifically, an agent's activity at hand, into account. Unfortunately, commercial route descriptions only fit for prototypical instances of humans or prototypical activities.

The choice of how much information needs to be communicated can be guided by Grice's [13] conversational maxims. The maxims state that the information communicated should be (1) relevant, (2) not overly redundant but (3) sufficiently detailed for an agent to carry out her task. Sperber and Wilson [34] argue that their concept of relevance can account for all of Grice's maxims. If the sender tries to be as relevant as possible, the contextual implications that follow reduce the cognitive load for the receiver. It is important to note that two route descriptions, although different in granularity, could both lead to the same outcome. Likewise, two messages of equal granularity may lead to different outcomes, depending on the needs of the user receiving the message. Frank [10] coined the term "pragmatic information content" to account for that fact.

Commercial applications are far from considering aforementioned requirements. For example, in Figure 1 we see a very detailed route descriptions similar to "Follow x-street for 52 secs, then turn right at intersection y and follow the road for another 23 m". Who would be able to count 52 seconds (clock) or measure 23 meters (odometer) while driving a vehicle?

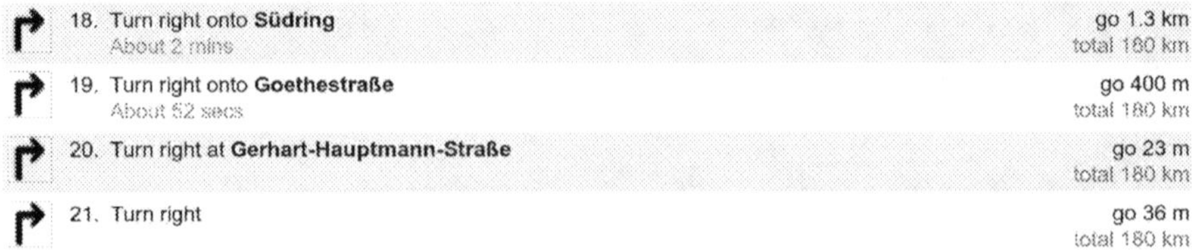

↱	18. Turn right onto **Südring**	go 1.3 km
	About 2 mins	total 180 km
↱	19. Turn right onto **Goethestraße**	go 400 m
	About 52 secs	total 180 km
↱	20. Turn right at **Gerhart-Hauptmann-Straße**	go 23 m
		total 180 km
↱	21. Turn right	go 36 m
		total 180 km

Fig. 1. Excerpt of a route suggested by Google Maps

An example for a less detailed route description (See Figure 2) is the journey planer offered by the German Railway Company (DB AG). It only lists start- and endpoints of a trip with corresponding times as a function of the transport mode. While information regarding other modes of transport can occasionally be accessed via hyperlinks, the actual process of combining those pieces to allow for navigation, i.e., their semantics, is entirely left to the user. Note that we used the same start and end points for both searches (Hamburg Rothenhauschaussee - Rostock Central Station). The route offered by DB AG assumes that we can acquire most of the missing information "in the world", while Google seemingly wants us memorize the route beforehand, so we can access information "in the head"[28]. One could argue that the information offered by DB AG hardly qualifies as a route description in the classical sense. The problem, however, is that people use this information as if it were a route description, mostly because of the lack of better alternatives in this (indoor) context. In general, we can conclude that both descriptions, as well as most automatically generated route descriptions, violate Grice's maxims and do not take the activity at hand into account. Tenbrink and Winter [35], for example, criticize automatically generated descriptions and their "(potentially disruptive) redundancy" as well as their lack of taking prior knowledge into account.

Station/Stop	Date	Time / prognosis		Platform	Products
Klgv Rothenhauschaussee, Hamburg	We, 21.11.12	dep	14:28		Bus 228
Bf. Bergedorf, Hamburg	We, 21.11.12	arr	14:40		
walk 5 min.					
Hamburg-Bergedorf	We, 21.11.12	dep	14:45 +0	1	RE 4313
Rostock Hbf	We, 21.11.12	arr	16:51	6	

Fig. 2. Route suggested by Deutsche Bahn AG

While oral route descriptions have the potential to be erroneous or too vague to be useful for the receiver, they often make use of an adaptive LoD. For example, Hirtle et al. [15] showed that this is the case for situations that are perceived to be cognitively demanding ("tricky"), e.g., the "absence of appropriate signage or landmarks", as well as complex geometric situations. Tenbrink and Winter [35] provide empirical evidence that humans adapt the LoD in route descriptions to an addressee's individual information need. Schwering et al. [33] showed that humans tend to use hierarchically structured route descriptions, using detailed descriptions at decision points and more abstract descriptions for other parts of the route.

2.3   The Structure of Spatial Knowledge

The hierarchical conceptualization of our world (containment relations) becomes evident if one looks at both its spatial and temporal aspects. We often perceive objects as nested within other objects, or as a collection of objects. For example, a tree is an object possibly contained by a collection of trees, called forest. Also, processes can be conceptualized as contained by other processes. For example "withdrawing money from an ATM" can be seen as part of the process "applying for a new passport" [1]. Furthermore, empirical evidence suggests that people organize spatial knowledge hierarchically [25] [35], albeit with inherent systematic errors[38]. Timpf et al. [36] proposed a model that accounts for three levels of abstraction of spatial containment relations in the context of planning a trip.

The commonly used metaphor for the storage of spatial knowledge is the cognitive map. As an alternative, Tversky [38] suggested the terms cognitive collages and spatial mental models. Cognitive collages account for the fact that our knowledge is often incomplete and erroneous. As a result not all of its parts can necessarily be integrated. For geographic areas that are well known or simple, Tversky suggests the term spatial mental model. This type of model allows for spatial perspective-taking and inferences; both are crucial for the communication of spatial knowledge. We adapt the notion of spatial mental model for our communication model and discuss its role in Sections 4 and 5.

3   Approaches to LoD in Route Descriptions

In this section we briefly review common approaches to LoD in route descriptions and specify how the language our agents use to communicate ("wayfinding choremes") can be modified to take granularity changes into account. For a comprehensive overview on the issues of LoD in route descriptions see [35].

3.1   Segments

Klippel et al. [20] propose a data structure based on "spatial chunking", i.e., the combination of elementary information into higher order elements with the goal to reduce the cognitive load on the wayfinder. This essentially reduces redundancy but also ensures that the pragmatic information content [10] stays equal. Klippel et al. [19] propose rewriting rules (granularity changes) of elementary "wayfinding choremes" to allow for higher-order elements. For example, the following sequence $wc_s + wc_s + wc_r$ ("go straight, go straight, then turn right") can be rewritten as dwc^{3r} ("turn right at the third intersection"). Accordingly, we can group the following sequence $wc_s + wc_s + wc_r^{RM}$ "straight, straight, turn right at landmark" to dwc^{3rRM} ("turn right at routemark"). Note that we deviate from Klippel's original notation to allow for a processing of expressions in both ways (the number of preceding straight segments is saved explicitly in the string). The modified rules can be defined as:

(D1) $(n*wc_s) + wc_D \longleftrightarrow dwc^{nD}$

(D2) $(n*wc_s) + wc_D^{RM} \longleftrightarrow dwc^{nDRM}$, where n $\in$ **N**, $D \in \{l,r\}$

3.2   Decision Points

The chunking method to combine similar route elements into higher-order segments works well with turn-by-turn descriptions. Humans, however, make also use of destination descriptions [37], e.g., "take the train to Leuven, Belgium". Such descriptions are basically discrete 2D containment relations, i.e., "Leuven is contained in Belgium". Destination descriptions emphasize "What is there" rather than "How to get there".

Since "wayfinding choremes" are limited to turn-by-turn instructions we can not account for 2D destination descriptions per se. In our modified language we can, however, indicate granularity changes at decision points with routemarks. Let us assume that we store 3 discrete granularity levels of a routemark in our data model. For example, if we wish to indicate that we refer to a routemark at the most specified level we could say "pass the routemark RM with name N along street S with address A". We codify the most detailed level by $"wc_s^{RM}"$. Accordingly, "pass the routemark RM with name N at street S" is codified by $"wc_s^{+RM}"$. Consequently, the most abstract level is indicated by $"wc_s^{++RM}"$. The following definition shows the granularity change of routemarks (from most specific to most abstract).

$$(D3) \quad "wc_D^{RM}" \longleftrightarrow "wc_D^{+RM}" \longleftrightarrow "wc_D^{++RM}" \text{ , where } D \in \{l, r, s\}$$

3.3   Context Specifics

The following subsections briefly discuss context specific approaches to LoD. Their integration into our model, however, is left for further research.

Elaborating on Situations. During communication, the amount of knowledge shared between speaker and addressee determines the correct LoD of each utterance. If I come home after work, open the door and say "Hi! I'm home. Where are you?", my girlfriend (if she is there) could answer "I'm in the bathroom" because we share the knowledge that we are both in the same flat. If, on the other hand, my girlfriend calls me on my phone and I ask her "Where are you?" she will likely be less specific and say "I'm at home" rather than "I'm in the bathroom" because she could be in another bathroom at another flat.

Similarly, if A tells B to "go straight on, until you come to a bank" B could respond "What bank do you mean?" triggering A's possible response "Bank of America". This approach is called elaboration [35] or situational approach to LoD and can be formalized by using .partial function application ([39], [40]). On a conceptual level this approach is similar to changing the granularity at a routemark but could refer to any entity.

Metonymies. In real life we often understand patterns from one domain of experience by projecting it onto another domain of a different kind [22]. We call these mappings metaphors and make use of them very frequently. For example, love is often understood as a journey, e.g., "Our relationship has gotten off the tracks" [22]. While metaphors are mappings between two different domains we also make use of cognitive abstraction processes that map within the same domain. For example, the "UK" are often (falsely) referred to by its part "England". The part, i.e., England, stands as a representative concept for the whole UK. A mapping of such kind is called metonymy.

Humans conceptualize and reason with the help of metonymies [21]. Naturally, this also applies to the the communication of spatial information. For example, a single mode trip can be divided into phases, each of which is required to successfully complete the trip. According to Lakoff [21], the subprocesses of such a trip can be modeled as follows: In the beginning you are at origin of the trip (Location A). You need access to a vehicle (precondition), then you get into the vehicle (embarkation phase), then you drive (center phase), then at your destination you get off (disembarkation phase), and finally you are at your goal (Location B). Metonymic effects show, for example, when we communicate such spatial descriptions. Imagine you meet a friend of yours at a conference and she asks you "How did you get here?". Consider the following two replies:

1. "First I walked to my car, [...] , I started it up, [...] took a right turn at X, [...] , I parked my car, [...], and here I am"
2. "I drove"

Option 1 includes various descriptions that refer to scripts [32] and are implicitly known to most people as well as a detailed turn-by-turn description of the trip. Violating Grice's conversational maxims and having no relevancy in this context it would bore your friend to death. Option 2 instead, uses the center phase of the model mentioned above to stand metonymically for the entire trip. Other conceptualizations are possible, for example "I hopped on the train" (embarkation) or "I borrowed my father's car (precondition).

Apart from allowing to abstract from redundant information, metonymies are a possible source for misinterpretations. If I tell somebody "then you take the 11.15", they will have a hard time unless we have established shared common ground [4] that we both talk about a specific train to take ("time of motion [stands] for an entity involved in the motion"[8]).

4   Signalling between Agents

In this section, we define the notion of signals in the context of communicating route descriptions. Signals are a "method by which one person means something for another"[4]. It is important to mention that signals are not limited to speech alone, but also include gestures, facial expressions, and body language. Thus, the signals presented in this section should not be understood as being bound to a particular modality.

4.1   Principles of Language Use

Our model uses principles developed by Clark [4], who emphasizes the collaborative nature of communication. In his view, the use of language is the exchange of joint actions between a speaker and an addressee. Joint actions are used to achieve a "mutually desired goal", i.e., to solve a particular problem. Clearly, joint actions can only be successful if sufficient information is available to solve the problem. As a consequence, we assume that our source (the agent who provides information) is both willing to share information and has total knowledge of the situation at hand. We can assume this form of knowledge, if we take a spatial mental model [38] for granted.

Joint actions are always taken in respect to each other and try to establish common ground between the communication partners. Thus, both the speaker's meaning and the addressee's understanding are created from common ground. Clark [4] defines common ground as "the part of information we think we share with somebody else". The attempt to establish common ground leads us to the notion of agents who negotiate the LoD at which information is communicated.

In this paper, we distinguish between local and global common ground. We define local common ground as the information p (e.g. a DP or a sequence of DPs) currently negotiated during a joint action. Since "wayfinding choremes" are actions an agent can carry out at decision points, we can conceptualize local common ground as the actions needed to navigate *one* part of the route, which both participants agree upon. In contrast, we define global common ground as the set of *all* actions an agent can carry out to navigate the entire route, on which both participants agree upon. Therefore, global common ground can be quantified as the overall similarities in the belief systems between the communicating agents (See Section 5). Once agreement on local common ground has been established it becomes global common ground.

4.2   The Source's and Target's Signals

In Section 2 and 3 we established a simple language, i.e., an extension to "wayfinding choremes", our agents can use to communicate route descriptions over multiple LoD. We can now define the signals that allow both source (S) and target (T) to negotiate the LoD. The signals were extracted from the interviews we conducted during this research (See Section 6). The open-ended records were coded based on the approach described in Montello [27].

Present. A basic signal for S to present information (a particular DP or a sequence of DPs) at a given LoD (the LoD need not to be constant and can vary).

Accept. A basic signal for T to accept the piece of information (a particular DP or a sequence of DPs) at a given LoD presented by S.

Probe. A basic signal for S to probe information from T. Probing can refer to a particular DP ("Do you know xy-routemark?") or to a sequence of DPs ("Do you know how to get from DP_1 to DP_2?")

Reject. A basic signal for both S and T. If used by T as a response to S probing information it means that T does not know the DP or a sequence of

DPs suggested by S. If used by S as a response to T requesting an LoD change it means that S cannot provide the requested information.

Secure. A basic signal for T to make sure information from S was understood correctly. T's secure signal can refer to a DP or a sequence of DPs.

RequestLoDChange. A basic signal for T to explicitly request a change of LOD (of a particular DP or a sequence of DPs).

LoDChange. A basic signal for S to adjust the LoD of a DP or a sequence of DPs. This can happen explicitly, e.g., as a response to T's request for a change of LoD, but also implicitly, e.g., if T wants to secure information and S reacts to it (See signal combinations in next subsection).

OfferChoice. A basic signal, S can use to offer T a choice on alternative paths T could take.

4.3   Signal Combinations

In this section we elaborate on some typical signal combination we observed during the interviews, and apply them to example utterances made by source (S) and target (T) in our modified "wayfinding choreme" language.

S:Present and T:Accept. This is the simplest signal combination. S presents information and T accepts it. The information becomes global common ground for both participants and thus minimizes differences between both agent's belief systems (See Section 5). More formally, S knows that T knows that p and T knows that p, where p is the currently negotiated part of the route.

Note that all other signal combinations need an implicit accept signal to indicate the end of the negotiation sequence (p can then be added to global common ground). Present does not imply that the information presented is at a constant LoD (See EX 1). In fact, an agent may use a varying degree of LoD in the same sequence.

(EX 1) S: Present $[\mathrm{dwc}^{2\mathrm{l}}, wc_s, wc_s] \longrightarrow T : Accept$

S:Present and T:RequestLoDChange. We can distinguish LoD changes from specific sequences to abstract chunks (EX 2), LoD changes from abstract chunks to specific sequences (EX 3), and LoD adjustments at a decision point with a routemark present (EX 4). Other combinations are possible but not listed here explicitly.

(EX 2) S: Present $[wc_s, wc_s, wc_r] \longrightarrow T : RequestLoDChange \longrightarrow$
S: LoDChange + Present $[\mathrm{dwc}^{3\mathrm{r}}]$

(EX 3) S: Present $[\mathrm{dwc}^{2l}] \longrightarrow T : RequestLoDChange \longrightarrow S : LoDChange$
+ Present $[wc_s, wc_l]$

(EX 4) S: Present $[wc_r^{RM+}] \longrightarrow T : RequestLoDChange \longrightarrow S : LoDChange$
+ Present $[wc_r^{RM}]$

T: Secure and S: (Accept+LoDChange) / Reject. If T tries to secure an information presented by S the usual reason for this is to make sure whether the information was understood correctly and can be added to global common ground. S can reject the secure signal (EX 6), i.e., the utterance was not understood correctly. Alternatively, S can accept the securing attempts by T. S can also accept the secure signal and repeat the same information at a higher LoD to remove potential ambiguities (EX 5).

(EX 5) T: Secure $[\text{dwc}^{2l}] \longrightarrow S : Accept + LoDChange + Present[wc_s, wc_l]$

(EX 6) T: Secure $[\text{wc}_s, wc_s] \longrightarrow S : Reject + Present[wc_s, wc_l]$

S:Probe and T:Accept / Reject. In case S probes information p and T accepts the probing, S has successfully determined that p is part of their common ground. For example, if S establishes a sequence of DPs to be common ground for both S and T, the negotiation could skip this part of the route and continue at the first DP after the probed sequence. In case a probing is rejected S can not assume that p is part of common ground for both S and T (EX 7).

(EX 7) S: probe $[wc_s^{RM}] \longrightarrow T : reject[wc_s^{RM}]$

Other combinations are possible, e.g., S:Probe and T:requestLoDChange, or T: RequestLoDChange and S:Reject, but not discussed here in detail.

5   The Negotiation Phases

In Section 3, we established a language agents can use to communicate route descriptions. In Section 4, we introduced various signals that can be used to adjust the LoD of such descriptions. In this Section, we discuss possible phases during the negotiation process and their effect on the spatial mental representations, i.e., global common ground.

5.1   Motivating Example

Imagine Alice, who is a tourist, visiting Vienna for the first time. After a day of sightseeing she suddenly realizes that she is lost. She is under time pressure because the last train with destination to her home town leaves in one hour. She stops Bob on the street, who turns out to be a local, and asks for the fastest way to the central station.

5.2   Initial Situation

In the following we use the abbreviations N and M to refer to Alice's and Bob's spatial mental model, respectively. Alice's initial situation is as follows: She can neither utilize her "knowledge in the head" (N(Alice)) nor can she acquire the necessary information available in the world to carry out the actions (actions(N)) necessary to find her way to the central station (DP goal). We denote this fact by equation 1, indicating that the actions Alice can carry out leave her stuck at a decision point (DP) that is not her intended goal.

$$actions(N) = DP, DP \neq Goal \tag{1}$$

Bob's situation looks different: He has the necessary knowledge in the head (M(Bob)) and can carry out actions (actions(M)) to get to the central station (DP goal).

$$actions(M) = DP = Goal \tag{2}$$

The initial situation is characterized by the fact the Bob and Alice share none or little common ground. To quantify the notion of common ground we conceptualize spatial mental models as the set of all actions that be carried out on the structure of the environment. The fact that Bob's and Alice's mental models differ in respect to their semantics can be described by the following equation:

$$|\Delta(M, N)| > 0 \tag{3}$$

5.3   Theory of Mind

How does Bob communicate the correct amount of information to Alice? In order to follow the conversational maxims, i.e., to maximize the relevancy of his message, Bob needs to put himself into the shoes of Alice. Knowing (rather the belief to know) what the other person knows can help to choose an initial LoD.

The ability to attribute mental states to oneself and others has been termed "theory of mind". This includes the notion that my own mental states can be different to somebody else's. Grice (1989) makes use of this concept in his example of how pragmatic inferences during a conversation (implicature) are made:

"He has said that p; [...]; he could not be doing this unless he thought that q; he knows (and knows that I know that he knows) that I can see that the supposition that he thinks that q is required; he has done nothing to stop me thinking that q; he intends me to think, or is at least willing to allow me to think, that q; and so he has implicated that q"(p. 31, emphasis by the authors)

Empirical evidence presented by Fussel and Krauss [11] suggests that one takes other people's knowledge into account when communicating a message. As a result, people design the communication of their knowledge depending on the audience ("audience design hypothesis"). More recently in the context of route descriptions, Hoelscher et al. [16] concluded that written route descriptions, as well as the actual traveled route are different when presented to somebody else, as compared to one's own conceptualization of the same route.

While the actual process of "extracting" somebody's knowledge is poorly understood, we assume that the image of the mental states of another person is influenced by many factors. Some include (1) The language spoken: Can the person read signs and therefore extract knowledge from the world?, or, (2) Special needs: The elderly, disabled persons, or children may require additional information based on accessibility and/or safety. See Section 6 for responses of what the participants in our case study thought they had designed explicitly for the target.

We can conclude that Bob's expectation of Alice's knowledge results in an image of Alice's mental model in Bob's mind (N_M, see Figure 3). Bob has to deal with two different belief systems, his own, and one he thinks resembles Alice's. Note that Alice's actual mental model (N) is not necessarily identical to Bob's image of Alice's mental model (N_M). This fact is expressed in Equation 4.

$$|\Delta(M, N_M)| \neq |\Delta(M, N)| \tag{4}$$

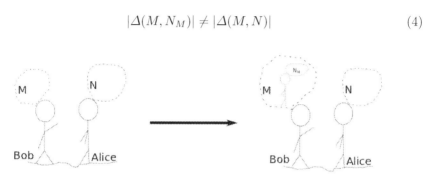

Fig. 3. Applying the Theory of Mind

5.4   Negotiation

This phase makes use of the signals mentioned in Section 4.2. Because we assume Alice's knowledge and Bob's image of Alice's knowledge (global common ground) not to be equal, the negotiation seeks to eliminate or at least minimize the differences. Bob needs to recall his knowledge and describe it to Alice orally at a given LoD (signal present). The LoD chosen can vary for each part of the description and depends on Bob's own mental representation of geographic space, his expectation of Alice's knowledge, and his own perceived difficulty of a certain section of the route resulting in an emphasized treatment of that particular "tricky part" [15].

Also, Bob may repeat certain sections of the route description as well as increase or decrease its LoD (present signal), depending on Alice's reactions to his elaborations (accept, or, secure signals). Alice needs to process the description and memorize it (build her mental model) and signal Bob whether the information he presents is either too little or too much detailed (requestLoDChange signal). In any case, the negotiation of a route part with the help a joint action has the goal to agree on local common ground.

The Effect of Probing on Common Ground. During probing (signal probe), Bob attempts to update his image of Alice's model through inquiring on knowledge on a particular DP or a sequence of DPs, similar to "Do you know how to get to Karlsplatz?". The goal of probing is to minimize the differences between Bob's mental model and his image of Alice's model (See Equation 5), i.e., to increase global common ground. Figure 4 illustrates (successful) probing of knowledge by Bob (probe(k)) and Alice's answer (accept (k)) as well as its effect on Bob's mental image of Alice's model (N'_M). In case probing is rejected the mental models remain unchanged.

$$|\Delta(M, N'_M)| < |\Delta(M, N_M)| \tag{5}$$

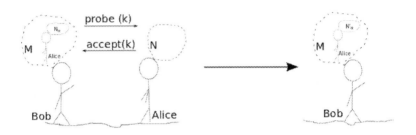

Fig. 4. Probing Information

The Effect of Knowledge Presentation on Common Ground. Instead of probing, Bob may present knowledge (signal present) to Alice at a given LoD for each section of the route description (See Figure 5). The effect of Bob presenting his knowledge is an updated mental model for Alice (N'). Knowledge presentation reduces the differences (See Equation 6) that existed between Bob's and Alice's initial representations (M,N) and the updated representations (M,N'). This results in adding the route part both agree upon (local common ground) to be added to global common ground.

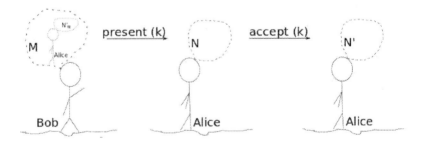

Fig. 5. Bob's knowledge presentation and its effect on Alice's model

$$|\Delta(M, N')| < |\Delta(M, N)| \tag{6}$$

5.5   Final Situation

The communication ends once Bob has finished his description of the route, and both participants think Alice has the relevant knowledge to successfully complete her wayfinding task. In other words, this is the case if the predicted actions of Alice p(N') match the predicted actions of Bob's image of Alice's model $(p(N_M))$.

$$p(N_M) = p(N') = DP = Goal \qquad (7)$$

Note that we cannot say that Alice's actions lead to the goal state (actions(N') = goalState) because we cannot be certain that Alice will find the way previously described to her. Indeed, the model rather describes both Bob's and Alice's beliefs, i.e., what they predict will happen. Both participants believe to have established equal global common ground. In terms of both mental models the (optimal) final state can be described as follows:

$$|\Delta(M, N')| \simeq 0 \qquad (8)$$

6   Case Study

In this section, we discuss the results of our case study, analyzing the interactive communication of route descriptions between participants and experimenter.

6.1   Design

We asked 10 participants (5 female, 5 male) with a mean age of 31.2 to take part in our case study. The participants were asked to describe a route they travel frequently and know well (e.g. from work to home). This particular set-up ensured that participants were likely to have a spatial mental model which allows for perspective taking, inferences about spatial locations, and "total" knowledge of the situation [38]. The route description was meant to be interactive, i.e., the experimenter took the part of the target lacking the knowledge.

6.2   Findings

The Use of Different Approaches to LoD. Most participants provided detailed turn-by-turn instructions throughout the entire route description. 5 of 10 participants made use of destination descriptions, mostly once the description was about to reach the end of the route. Spatial Chunking was used by all but one participants. Participant I (See Table 1) used spatial chunking in 5 of the 17 joint actions. Elaboration on specific DPs happened as a response to requests on LoD Change or as a response to securing signals (See next subsection for a more detailed discussion).

Metonymies were only used by four participants, mostly to indicate line numbers of buses, or subways, or to conceptualize entire route parts with the help of one particular DP, as the following example shows. The route description starts by mentioning to use a subway line (present signal). A means to say "my route description starts at Karlsplatz" using the subway to conceptualize Karlsplatz. We should note, however, that the route description should start at Gusshausstrasse (the place the experiment took place), which is about 5 minutes on foot from Karlsplatz. B's attempt to verify this fact (secure signal) triggers an increase of level of detail (loDChange + present signals) in A's repetition of the first section of the route.

> **A: I only need one public means of transport, (line) U4** (Ich brauch nur ein öffentliches Verkehrsmittel, das ist die U4)
>
> **B: So, you walk from here to Karlsplatz?** (Also, du gehst hier zum Karlsplatz?)
>
> **A: I walk to Karlsplatz, through Resselpark, then I get on U4** (Ich geh zum Karlsplatz, durch den Resselpark, dann steig ich in die U4)

Another observation we made is the explicit referral to knowledge in the world. Instead of giving detailed descriptions for three potentially difficult decision points, one participant referred to knowledge that can be extracted from the world. He recommended to ask the bus driver for directions, as well as to check the time tables of trains because he was not entirely sure what train one should take. Another participant explicitly referred to signs visible in the train station, i.e., "then follow the signs...".

The Use of Signals and Their Combinations. The following table (Table 1) shows the results of our analysis on the total number of joints actions per participant (the number of signal combinations until local common ground was achieved) and the most prominent signal combinations. Signals could also be seen from a more detailed point of view, e.g., we differentiate between "secure DP" and "secure sequence of DP". The table, however, does not show this differentiation and groups all sub-signals together.

We can see that an explicit request for increase of LOD happened relatively seldom, except during the interviews with participants C, E, and, H, in which a request was stated in every fourth joint action. On a more detailed level, however, we noticed that the target's attempts to secure a particular DP or a sequence of DPs almost always resulted in attempts by the source to present the information again, but now at a higher LoD. Apparently, the target's attempts to secure information is perceived by the source as an implicit request for LoD change. We also noticed that a securing signal by T is most of the time accepted by S, indicating that T's understanding of S's meaning was correct. Choices on alternative paths were offered by 5 participants. Only 3 participants used some form of probing.

If we look at how often the present-accept signal combination (T was sure to have received enough information) was used in comparison to the other signals (T was not sure and had to request more information) we could deduce the following: The routes offered by A and I were perceived as the most difficult ones (both about 4x more "insecurity" than present-accept signals), whereas

Table 1. Total no. of joint actions and signal combinations per participant

Participant	Joint Actions	Present-Accept	Secure-Accept	Secure-Reject	Request LoDChange	Probing	Offer Choice
A	12	2	6	2	2	0	0
B	16	7	6	2	0	1	0
C	18	6	4	3	4	0	1
D	25	15	5	2	2	0	1
E	15	5	5	0	4	0	1
F	32	17	9	1	1	3	1
G	18	5	7	1	3	1	1
H	19	6	8	0	5	0	0
I	17	4	8	3	2	0	0
J	35	20	8	3	4	0	0

routes described by D and, F were perceived as the easier ones (both about 1.5x more present-accept than "insecurity" signals).

Theory of Mind. We expected the participants to adjust the message to the needs of the receiver as suggested in Section 5. Because this process happens subconsciously, we asked the participants after they had delivered their description what they assumed the other person could have known and whether this had influenced their description.

As expected, participants referred to the use of routemarks (easily recognizable) and said they tried to keep the description very general. Some mentioned they would have adjusted the description accordingly, if they had communicated the route to somebody who was not familiar with Vienna at all (e.g. foreigners). Parts of the routes participants perceived as difficult themselves were given special treatment leading to more elaborated descriptions. Although participants thought they used very general descriptions, they often used semantics that were not entirely clear to the experimenter (e.g., what does "towards the city limits" mean? Possible interpretations: Right, Left, North, South, etc.). This suggests a gap between one owns perception of space and the audience design hypothesis mentioned in Section 5, a fact already recognized by Hoelscher et al. [16].

7   Conclusions and Future Research

We presented a model that accounts for cognitive transactions in a communication setting between a knowledgeable source and an individual lacking spatial knowledge. Agents use different LoD for the presentation of information and this fact is reflected in the language they use. We modified "wayfinding choremes" to account for granularity changes and developed signals the agents can use to indicate such adjustments. The overall goal of negotiating the LoD is the agreement on common ground. We discussed the differences between local and global common ground, elaborated on the various phases during a spatial negotiation setting, and made their effect on common ground explicit.

We argue that our investigation can be a fruitful approach to design better navigation systems, capable of taking the user's knowledge and special needs into account. Commercial applications have, so far, failed to achieve this goal. Some directions for future work include:

Prototype Application. A short-term goal of our research is the design, of a prototype application capable of simulating the interactive communication process between a computer and a human agent. The model presented in this work serves as the conceptual foundation for its development. To improve the model presented here, it will be necessary to repeat the experiment described in Section 6 on a larger scale. We expect to identify more signals and a more fine-grained view on the various signal combinations.

Language Extension. Klippel [18] noted that wayfinding is a goal-oriented task. We speculate on a simple language that is capable to communicate spatio-temporal tasks in general. This could contribute to the design of better (spatial) personal information management tools [2], e.g., calendars with a sense of space.

References

1. Abdalla, A., Frank, A.U.: Combining trip and task planning: How to get from A to passport. In: Xiao, N., Kwan, M.-P., Goodchild, M.F., Shekhar, S. (eds.) GIScience 2012. LNCS, vol. 7478, pp. 1–14. Springer, Heidelberg (2012)
2. Abdalla, A., Weiser, P., Frank, A.U.: Design Principles for Spatio-Temporally Enabled PIM Tools: A Qualitative Analysis of Trip Planning. In: Vandenbroucke, D., Bucher, B., Crompvoets, J. (eds.) AGILE 2013. LNGC, pp. 323–336. Springer (2013)
3. Allen, G.: From knowledge to words to wayfinding: Issues in the production and comprehension of route directions. In: Hirtle, S.C., Frank, A.U. (eds.) COSIT 1997. LNCS, vol. 1329, pp. 363–372. Springer, Heidelberg (1997)
4. Clark, H.H.: Using Language. Cambridge University Press (1996)
5. Denis, M.: The description of routes: A cognitive approach to the production of spatial discourse. Current Psychology of Cognition 16(4), 409–458 (1997)
6. Denis, M., Pazzaglia, F., Cornoldi, C., Bertolo, L.: Spatial discourse and navigation: An analysis of route directions in the city of Venice. Applied Cognitive Psychology (13), 145–174 (1999)
7. Egenhofer, M.J., Mark, D.M.: Naive geography. In: Kuhn, W., Frank, A.U. (eds.) COSIT 1995. LNCS, vol. 988, pp. 1–15. Springer, Heidelberg (1995)
8. Evans, V., Green, M.: Cognitive Linguistics - An Introduction. Edinburgh University Press (2006)
9. Fauconnier, G.: Mappings in Thought and Language. Cambridge University Press (1997)
10. Frank, A.U.: Pragmatic Information Content: How to measure the information in a route description. In: Perspectives on Geographic Information Science, pp. 47–68. Taylor & Francis (2003)

11. Fussell, S.R., Krauss, R.M.: Coordination of knowledge in communication: Effects of speakers' assumptions about what others know. Journal of Personality and Social Psychology 62(3), 378–391 (1992)
12. Galton, A., Mizoguchi, R.: The water falls but the waterfall does not fall: New perspectives on objects, processes and events. Applied Ontology 4(2), 71–107 (2009)
13. Grice, P.: Studies in the Way of Words. Harvard University Press (1989)
14. Hirtle, S.C., Timpf, S., Tenbrink, T.: The effect of activity on relevance and granularity for navigation. In: Egenhofer, M., Giudice, N., Moratz, R., Worboys, M. (eds.) COSIT 2011. LNCS, vol. 6899, pp. 73–89. Springer, Heidelberg (2011)
15. Hirtle, S.C., Richter, K.-F., Srinivas, S., Firth, R.: This is the tricky part: When directions become difficult. Journal of Spatial Information Science 1, 53–73 (2010)
16. Hoelscher, C., Tenbrink, T., Wiener, J.M.: Would you follow your own route description? Cognitive strategies in urban route planning. Cognition 121(2), 228–247 (2011)
17. Hutchins, E.: Cognition in the Wild. The MIT Press (1995)
18. Klippel, A.: Wayfinding choremes: Conceptualizing wayfinding and route direction elements. PhD Thesis, SFB/TR 8 Spatial Cognition (2003)
19. Klippel, A., Tappe, H., Kulik, L., Lee, P.: Wayfinding Choremes - A language for modeling conceptual route knowledge. Journal of Visual Languages and Computing 16(4), 311–329 (2005)
20. Klippel, A., Hansen, S., Richter, K.-F., Winter, S.: Urban granularities. A data structure for cognitively ergonomic route directions. Geo Informatica 13, 223–247 (2009)
21. Lakoff, G.: Women, Fire, and Dangerous Things: What Categories Reveal About the Mind. University of Chicago Press (1987)
22. Lakoff, G., Johnson, M.: Metaphors We Live By. University of Chicago Press (1980)
23. Levinson, S.: Space in Language and Cognition: Explorations in Cognitive Diversity. Cambridge University Press (2003)
24. Lynch, K.: The Image of the City. The MIT Press (1960)
25. McNamara, T.P., Hardy, J.K., Hirtle, S.C.: Subjective Hierarchies in Spatial Memory. Journal of Environmental Psychology: Learning, Memory, and Cognition 15(2), 211–227 (1989)
26. Montello, D.R.: The Cambridge Handbook of Visuospatial Thinking. Cambridge University Press (2005)
27. Montello, D.R., Sutton, P.C.: An Introduction to Scientific Research Methods in Geography. Sage Publications (2006)
28. Norman, D.A.: The Design of Everyday Things. Doubleday Books (1988)
29. Psathas, G.: Direction-Giving in Interaction. Reseaux (8), 183–198 (1990)
30. Raubal, M.: Cognitive engineering for geographic information science. Geography Compass 3(3), 1087–1104 (2009)
31. Richter, K.-F., Tomko, M., Winter, S.: A dialog-driven process of generating route directions. Computers, Environment and Urban Systems (32), 233–245 (2008)
32. Schank, R.C., Abelson, R.P.: Scripts, Plans, Goals, and Understanding (1977)
33. Schwering, A., Li, R., Anacta, J.A.: Orientation Information in Different Forms of Route Instructions. In: Short Paper Proceedings of the 16th AGILE Conference on Geographic Information Science, Leuven, Belgium (2013)
34. Sperber, D., Wilson, D.: Relevance: Communication and Cognition. Basil Blackwell Oxford (1986)
35. Tenbrink, T., Winter, S.: Variable Granularity in Route Directions. Spatial Cognition and Computation 9, 64–93 (2009)

36. Timpf, S., Volta, G.S., Pollock, D.W., Egenhofer, M.J.: A conceptual model of wayfinding using multiple levels of abstraction. In: Frank, A.U., Formentini, U., Campari, I. (eds.) GIS 1992. LNCS, vol. 639, pp. 348–367. Springer, Heidelberg (1992)

37. Tomko, M., Winter, S.: Pragmatic construction of destination descriptions for urban environments. Spatial Cognition and Computation 9(1), 1–29 (2009)

38. Tversky, B.: Cognitive Maps, Cognitive Collages, and Spatial Mental Models. In: Best, E. (ed.) CONCUR 1993. LNCS, vol. 715, Springer, Heidelberg (1993)

39. Weiser, P., Frank, A.U.: Process Composition And Process Reasoning Over Multiple Levels Of Detail. In: Online Proceedings of the 7th International Conference GIScience, Columbus, Ohio (2012)

40. Weiser, P., Frank, A.U.: Modeling discrete processes over multiple levels of detail using partial function application. In: Proceedings of GI Zeitgeist, Muenster, Germany (2012)

41. Wunderlich, D., Reinelt, R.: How to get there from here. In: Jarvella, R.J., Klein, W. (eds.) Speech, Place, and Action. John Wiley & Sons (1982)

Strategy-Based Dynamic Real-Time Route Prediction

Makoto Takemiya[1] and Toru Ishikawa[2]

[1] Graduate School of Interdisciplinary Information Studies, The University of Tokyo
[2] Center for Spatial Information Science, The University of Tokyo
qq107405@iii.u-tokyo.ac.jp, ishikawa@csis.u-tokyo.ac.jp

Abstract. People often experience difficulty traversing novel environments. Predicting where wayfinders will go is desirable for navigational aids to prevent mistakes and influence inefficient traversals. Wayfinders are thought to use criteria, such as minimizing distance, that comprise wayfinding strategies for choosing routes through environments. In this contribution, we computationally generated routes for five different wayfinding strategies and used the routes to predict subsequent decision points that wayfinders in an empirical study traversed. It was found that no single strategy was consistently more accurate than all the others across the two environments in our study. We next performed real-time classification to infer the most probable strategy to be in use by a wayfinder, and used the classified strategy to predict subsequent decision points. The results demonstrate the efficacy of using multiple wayfinding strategies to dynamically predict subsequently traversed decision points, which has implications for navigational aids, among other real-world applications.

Keywords: navigation, route prediction, individual differences, spatial cognition, spatial abilities.

1 Introduction

Imagine stepping off a train into a new city for the first time. Using the map on your phone, you exit the station and head towards your intended destination, but after a few intersections you make a wrong turn and start taking a much longer route to your goal. Although you eventually reach your destination, you arrive late and cause much trouble. If only you had been warned before starting off on the wrong path in the first place, you might have had a better outcome.

The objective of the present work is to help with situations such as these. We aim not to detect mistakes in wayfinding, but rather to prevent them from occurring, through probabilistic prediction of suboptimal traversals of decision points and proactive influencing of wayfinders via electronic navigational aids. The main objective of this work is to predict decision points that wayfinders will traverse based on assumptions about their potential wayfinding strategies. Moreover, we do not only consider the case where wayfinders use a single strategy throughout the duration of their traversal, but also explore the case where

T. Tenbrink et al. (Eds.): COSIT 2013, LNCS 8116, pp. 149–168, 2013.

wayfinders dynamically change strategies to adapt to the environment that they are traversing.

1.1 Wayfinding Strategies

As opposed to aimless perambulation, wayfinding is a motivated activity to reach a destination, thus wayfinding tasks require both locomotive ability and cognitive reasoning skills [31]. In navigating spatial environments, people are theorized to use diverse criteria for route selection that comprise various wayfinding strategies [14]. These criteria were developed to explain the behavior of wayfinders, so they are not necessarily related to the actual cognitive processes that wayfinders employ, but they are useful for studying route choice through environments.

Among many possible route selection criteria that wayfinders can use, Golledge [14] explored the criteria in List 1, and empirically observed that the use of route selection criteria (strategies) seemed to change with the configuration of the environment and wayfinders' perceptions thereof.

List 1: route selection criteria in Golledge [14]
 – shortest distance
 – least time
 – fewest turns
 – most scenic
 – first noticed
 – longest leg first (also cf. [10])
 – many curves
 – many turns
 – different from previous
 – shortest leg first

Hölscher, Tenbrink, and Wiener [20] grouped wayfinding strategies into two broad classes: those based on the graphical structure of the street network, with decision points as nodes connected by streets, and directional strategies, where wayfinders choose routes based on their orientation to the perceived destination. Of the strategies considered by Golledge [14], shortest distance, fewest turns, many curves, and many turns can be considered to be based on the graphical structure of an environment, although other strategies could also be thought of as graph-based. Other wayfinding strategies, such as first noticed, longest leg first, different from previous, and shortest leg first, are arguably directional because wayfinders can use the perceived direction to the goal to choose routes with each strategy, regardless of the macroscopic graph structure.

Golledge [14] found that strategies to minimize distance, time, and turns were frequently used by wayfinders. Dee and Hog [8] applied the shortest-distance and fewest-turns strategies to behavioral modeling for surveillance camera footage and found that these two strategies enabled detecting surveillance events of interest. In addition to the strategies in List 1, Kato and Takeuchi [24] reported that wayfinders used memorization-based strategies such as memorizing turn sequences or landmarks in the environment (e.g., numbers on buildings) to keep track of where they were.

Similar to the fewest-turns strategy, a strategy to minimize the structural complexity of a route is also possible. Minimizing the complexity of route traversals has been studied by Duckham and Kulik [9]. As implemented in Bojduj et al. [2] and Richter et al. [35], the complexity of a decision point was a function of the number of branches, number of competing branches (separated by less than 45°), and distance. Distance in and of itself does not necessarily equate with complexity, but if the distance is too long, then the physical strain required to traverse a route could burden a wayfinder.

Previous work studying route choice on maps has suggested a possible preference for traversing decision points in the cardinal southern part of an environment, pointing to a possible semantic association between northern travel and going uphill [4,5]. Actual use of this strategy by situated wayfinders has, to our knowledge, not been demonstrated and may be related to culture.

A preference for wider streets (major roads) can also be considered as a strategy, since major roads form the top layer of a hierarchy of streets in a road network [23].

Since graph-based strategies require detailed knowledge and comprehension of the road network, wayfinders may rely more on directional heuristics and local information when knowledge of the environment is uncertain [32]. For instance, Bailenson et al. [1] put forth the initial-segment strategy, where wayfinders showed a preference for initially straighter routes. Minimizing the angular deviation from the perceived goal direction [6,18,19] can be considered a directional strategy that looks at the routes immediately available and chooses the one that seems to go in the direction of the destination. Minimizing the angular deviation to the goal may increase the overall complexity of a route, because for some structural configurations, wayfinders will have to make many turns in order to go straight to the goal.

1.2   Route Prediction

Although much previous work has focused on route prediction for car drivers [25,26,27,36], relatively few authors have studied route prediction for pedestrians. One exception is Laasonen [28], who studied route prediction using pedestrian cellular GPS data to determine where people would go, but this method requires empirically observed behavior and hence may not be applicable to novel environments.

Previous work has calculated probabilities of where car drivers will likely go based on a given partial trajectory, a history of where the driver has previously been, and knowledge about driving behavior of the individual [27]. While approaches like this can be used to predict where drivers will go in a principled way, such approaches may not extend to new environments for which training data do not exist. Krumm [25] used a discrete Markov chain to predict where drivers would turn with 90% accuracy, but their tested environment was fairly simple and thus chance was approximately 50%. Although probability-based route predictions derived from the structure of roads and empirically observed training data can yield useful results, the cognitive plausibility of such methods has not been adequately demonstrated. Thus in the present work we aim to combine wayfinding strategies and probabilistic inference for dynamic prediction of wayfinders in a real-time context.

1.3   Strategies Considered in the Present Work

In this present work, we used wayfinding strategies to generate routes to probabilistically predict decision points traversed by observed wayfinders. Landmarks and scenery, as well as travel time, were not considered in the wayfinding strategies we used, although minimizing distance travelled and minimizing time can result in the same traversals for some environments. Both of the routes in our empirical study (Section 2) started in areas surrounded by short streets, so a strategy of choosing the longest leg first was not applicable, although empirical work has suggested that in some situations wayfinders may prefer longer lines of sight and larger numbers of visible branching paths [10]. Rational wayfinders who desire to reach their destination efficiently were assumed, so the many-curves and many-turns strategies of maximizing changes in orientation were not analyzed. Our present work considered the following five strategies:

List 2: strategies considered in the present work
  − shortest distance
  − least complexity
  − minimum deviation
  − preference for southern routes
  − preference for wider streets

The shortest distance, least complexity, and preference for wider streets (major roads) are included as graph-based strategies. The minimum deviation strategy is a directional strategy for minimizing the angular deviation from the perceived destination. The method presented in this paper allows probabilistic inspection of strategy use, so the preference for southern routes strategy is included to demonstrate how our method can be used to investigate strategy use on an exploratory basis.

Of the many possible strategies available, we only consider five strategies as a starting point for studying wayfinding strategy-based route prediction. Future work should consider more strategies, especially strategies using landmarks and scenic routes.

2   Empirical Study

We performed an empirical wayfinding study in Nara, Japan [37,38]. Two routes with two different pairs of start and goal locations were specified, and 30 participants (15 female) were given maps and asked to navigate to the goal for each route, as quickly as possible. Participants were not informed as to how their data would be analyzed and possible wayfinding strategies were not discussed, so as to avoid a priming effect. After traversing the routes, wayfinders were classified into "good" and "poor" groups based on the distance travelled and successful arrival at the goal. Good and poor labelled route traversals from all participants are shown in Figure 1. Specifically, poor labels were determined using the following criteria, and all participants who did not match the criteria were given good labels.

1. Participant failed to reach the goal.
2. Participant's route traversal was 15% longer than the optimally shortest route from start to goal.

Using these criteria to classify good and poor wayfinders, we observed that poor wayfinders' route traversals were an average of 21% longer for route 1 and 22% longer for route 2.

In addition to traversing two routes, participants also completed the Santa Barbara Sense-of-Direction (SBSOD) scale [17] and the mental rotation test (often used in studies on map use, see [29]), in order to assess their spatial abilities.

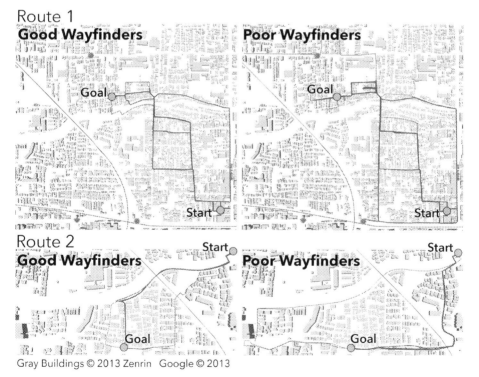

Fig. 1. Good and poor wayfinders' traversals for routes 1 and 2. Darker lines denote more participants traversing a route.

3   Predicting Routes of Wayfinders in Real-Time

To test whether wayfinders' routes could be reliably predicted by wayfinding strategies, for each strategy we generated 50 routes using the A* heuristic search algorithm between every decision point in our study area and the goal decision point, for each of the two routes. For example, to generate routes in accordance with the shortest-distance strategy, the A* algorithm found routes from the start to the goal while trying to minimize overall distance, whereas to generate routes for the minimum-deviation strategy, the sum of the angular deviation from the start to the goal was minimized. When considering points for traversal in the A* search, if two points had a heuristic score within 10% of each other, then one point was randomly chosen over the other. This had the effect of creating variability in the routes generated from the wayfinding strategies, while still adhering to the strategy's overall route selection (in our case, decision-point selection) criteria.

For computational efficiency, we discretized the representation of route traversals into ordered lists of decision points; spaces between decision points were not considered in the present work. To predict subsequent decision points that a wayfinder would visit after the current point, the routes for a selected strategy were used to calculate the transition probabilities from the current decision point to the neighboring points. This allowed for real-time prediction, in the sense that only incomplete information about a wayfinder's route traversal from the starting point up to and including the current decision point was required to make a prediction. Using only information available during traversal is a requirement for wayfinding aids to be practically developed, so this constraint motivated our methodology. Prediction results for each of the five strategies considered in our study are in Section 5.

The efficacy of using generated routes instead of empirically observed routes as training data for classification was shown by Dee and Hog [8], where generated shortest and simplest paths were able to outperform paths from a training set of observed humans for event detection in surveillance video. Also, Takemiya and Ishikawa [38] used computationally generated routes to classify wayfinding performance in real-time and Takemiya, Richter, and Ishikawa [39] showed that the probability of decision points being used in generated routes correlated with use of the decision points in cognitively ergonomic route directions.

4   Dynamic Real-Time Route Prediction

If wayfinders were to use many different strategies when choosing routes through environments, always using the same strategy to predict individual wayfinders, as in the previous section, may lead to sub-optimal prediction results. Significant individual differences are known to exist in the configurational understanding of routes [22], which are also likely related to differences in individual spatial abilities [16]. Differences in abilities, in turn, can lead to the use of different wayfinding strategies [24]. Therefore, classifying which strategy a wayfinder is using and then using the classified strategy to predict where a wayfinder will go should yield good prediction results across a variety of environments and wayfinders with varying spatial abilities. This is the approach we used for dynamic real-time route prediction.

4.1 Classification

To classify the strategies used by wayfinders in our empirical study, we used the sets of routes generated by each strategy to calculate the probability of each strategy being used, given information about which decision points had been traversed, up to and including the current point. This was accomplished as shown in the following equations, adapted from [7,30], where a score (*score*) in the range [-1, 1] was calculated for each of j decision points (d_j) with respect to the probability of the point being associated with a strategy (S, $score > 0$) or with being associated with the complement of the strategy (i.e., all other strategies; S', $score \leq 0$). The scores for each point traversed were summed to get an overall score (*eval*) for a route traversal (r_i) for each strategy.

$$eval(r_i) = \sum_j score(d_j) \tag{1}$$

$$score(d_j) = \frac{P(d_j|S) - P(d_j|S')}{P(d_j|S) + P(d_j|S')} \tag{2}$$

We previously [37,38] used Equations 1 and 2 to discriminate between good and poor wayfinders based on information available in real-time during route traversal. The present work uses the same equations, but rather to classify the strategies used, instead of route performance.

We used an Extra-Trees classifier[1] to select the decision points used for classifying each route. An Extra-Trees classifier is an ensemble classifier that uses a predetermined number (we used 50) of randomized decision trees created using subsamples of features in a training data set [13]. By training an Extra-Trees classifier on our generated routes, the classifier learned which features of the data, in our case decision points, were important for discriminating between strategies. This information was then used to reject decision points below a certain threshold of importance for classification. The decision-point selection was performed using the routes generated for each strategy and was thus independent of the human traversals in our empirical study, which were used for validation; this guaranteed a clean separation between training and validation data sets. Out of 171 total decision points in the graph of the section of Nara used in our study, 54 points were selected for route 1 and 36 for route 2. Only these selected points were used when evaluating a route traversal (Equation 1).

In addition to selecting decision points for our strategy classifier, we also assigned a temporal weighting to the score of decision points in Equation 1 by giving a higher weight to scores of recently traversed points than to points

---

[1] Part of scikit-learn [34], http://scikit-learn.org/

that were traversed earlier. This was done with the assumption that wayfinders may dynamically change wayfinding strategies based on the environment, during navigation, so more recently traversed points are more likely to be associated with the current strategy employed. Albeit rare, if multiple strategies were tied for the highest probability, a strategy was chosen in the order of precedence in List 2. This was decided upon to make the algorithm deterministic, although randomly choosing among tied strategies would also be a reasonable approach.

4.2   Prediction

Once a probable strategy was classified, we then proceeded to predicting subsequent decision points traversed. This was carried out in exactly the same way as for the individual strategy predictions, described in Section 3.

Since we assumed rational, goal-driven wayfinding, we inferred that wayfinders would not want to revisit previously traversed decision points. Therefore, if the next point predicted was already previously traversed by a wayfinder, the currently classified strategy was ignored and the closest point to the goal from the current decision point was used as the prediction. This is a reasonable method because a real-time navigational aid could feasibly keep a record of where a person has already traversed and then use that information when predicting subsequent points.

5   Results

Since our primary goal was to create a practical, strategy-based method for predicting wayfinders in real-time, we first tested the accuracy of prediction for all of our wayfinders (Section 5.1). To investigate a possible relationship between individual differences in wayfinding performance and prediction accuracy, we then calculated our prediction results for the good and poor wayfinders. Next, to study the relationship between spatial abilities and prediction accuracy, we predicted groups of wayfinders who did well and poorly on the SBSOD scale and the mental rotation test.

Beyond creating a method for predicting wayfinders, we also wanted to explore the probable use of wayfinding strategies by participants. To do this, we classified the likely wayfinding strategy used at each point during a traversal, and then correlated the classified strategy uses with the good and poor labels, SBSOD scores, and mental-rotation scores (Section 5.2).

Finally, to quantify differences between the environmental structures of routes 1 and 2, we calculated graph-theoretic metrics for the decision points in our environments (Section 5.3).

5.1   Prediction

For each of the two route traversals for the 30 participants in the empirical study (Section 2), we predicted the subsequent decision points that each person traversed using each wayfinding strategy in List 2, as well as our dynamic route predictor. We used accuracy, defined below, to quantify the efficacy of prediction.

$$accuracy = \frac{correct\ predictions}{(correct + incorrect\ predictions)} \tag{3}$$

For each decision point that a participant traversed, we predicted the next point that they visited. We then took the predicted point and predicted the next decision point after that. This point, in turn, was then used to predict the next point after that, and this chaining of predictions was carried out for the length of each participant's route traversal.

For all results, the random baseline was calculated by randomly choosing a next decision point out of the set of untraversed points connected to the current point. Random results were calculated 100 times and then averaged.

Prediction accuracies for each of the five independent strategies, our novel dynamic real-time predictor, and the random baseline, are shown in Table 1. To investigate the relationship between individual differences and prediction accuracy, Table 2 shows differences in next-point prediction accuracies between good and poor participants. Table 3 shows the result of dividing the 30 participants based on SBSOD scores, while Table 4 shows the result of segregating participants by their scores on the mental rotation test. Participants whose SBSOD score was 4.0 (inclusive; the midpoint of the 7-point scale) or under were classified as having a poor ability, while scores of over 4.0 were classified as good. Similarly, participants with scores less than 22 on the mental rotation test were classified as poor, while 22 or over were good.

Table 1. Next 3-decision-points prediction results for routes 1 and 2. All predictions are significantly different from the random baseline (binomial test; $p < .01$)

	Route 1			Route 2		
shortest distance	0.74	0.64	0.55	0.74	0.60	0.52
minimum deviation	0.68	0.57	0.48	0.78	0.67	0.60
least complexity	0.69	0.55	0.47	0.72	0.70	0.63
preference for wider streets	0.68	0.49	0.37	0.78	0.67	0.59
preference for southern routes	0.56	0.08	0.08	0.76	0.72	0.69
dynamic real-time predictor	0.75	0.63	0.55	0.90	0.83	0.78
random baseline	**0.32**	**0.12**	**0.05**	**0.33**	**0.12**	**0.04**

Figure 2 shows the prediction accuracies for good, poor, and all participants, as well as the random baseline predictions, for our dynamic real-time predictor for routes 1 and 2. Predictions at each decision point are significantly above chance (binomial test; $p < .001$). Since different participants traversed different numbers of decision points on their way to the goal, we only show analysis for decision points up to the shortest number of points traversed by all participants, which was 16 decision points for route 1, and 7 for route 2. These are prediction chains, so decision point 1 means predictions for the next decision point, along the length of each participants' traversal, whereas 2 means predictions for the next 2 decision points, from each point in a traversal. As predictions get further separated from the current point, accuracy decreases.

Table 2. Good/poor next-point prediction results (for route 1, 21 good and 9 poor participants; for route 2, 16 good and 14 poor participants)

	Route 1		Route 2	
	Good	Poor	Good	Poor
shortest distance	0.79	0.63	0.75	0.73
minimum deviation	0.74	0.57	0.79	0.76
least complexity	0.71	0.65	0.75	0.70
preference for wider streets	0.71	0.64	0.79	0.76
preference for southern routes	0.59	0.48	0.80	0.71
dynamic real-time predictor	0.78	0.67	0.94	0.85
random baseline	**0.32**	**0.33**	**0.33**	**0.33**

Table 3. Good/poor SBSOD next-point prediction results (19 good and 11 poor participants)

	Route 1		Route 2	
	Good	Poor	Good	Poor
shortest distance	0.73	0.74	0.74	0.73
minimum deviation	0.69	0.68	0.80	0.76
least complexity	0.67	0.73	0.73	0.71
preference for wider streets	0.65	0.76	0.80	0.75
preference for southern routes	0.55	0.58	0.74	0.78
dynamic real-time predictor	0.74	0.75	0.91	0.89
random baseline	**0.33**	**0.33**	**0.33**	**0.33**

Table 4. Good/poor mental rotation next-point prediction results (25 good and 5 poor participants)

	Route 1		Route 2	
	Good	Poor	Good	Poor
shortest distance	0.73	0.74	0.76	0.66
minimum deviation	0.69	0.65	0.80	0.68
least complexity	0.74	0.66	0.69	0.74
preference for wider streets	0.67	0.76	0.80	0.68
preference for southern routes	0.55	0.59	0.75	0.78
dynamic real-time predictor	0.75	0.73	0.91	0.84
random baseline	**0.33**	**0.33**	**0.33**	**0.33**

5.2   Distribution of Classified Strategies

We used the probabilistic classification described in Section 4.1 to determine wayfinding strategies associated with every point traversed by our 30 participants across the two routes. This gave us a distribution of classified strategies for routes 1 and 2, shown in Figure 3. The first decision point was given, so the classified strategies start at the second decision point, therefore we show results for the second to seventeenth decision points for route 1, and the second to eighth decision points for route 2.

Classified use of the minimize-complexity and minimize-deviation-from-the-goal strategies were significantly different between routes 1 and 2 (Wilcoxon rank-sum test; $p < .001$), as was use of the strategy to minimize distance ($p < .05$). Preference for southern routes was not significantly different between routes 1 and 2, and the preference for wide-streets strategy was not classified as being used at any point during the participants' traversals.

5.3   Graph-Theoretic Metrics

In order to quantify the structure of the graphs of decision points in routes 1 and 2, we calculated the values for each of the graph-theoretic measures shown in Table 5 for each decision point, using the NetworkX library [15] for Python[2]. We also calculated the variances of the values for each metric for routes 1 and 2 and tested the significance of the differences between them. Although seven out of nine metrics investigated showed a higher variance in metric values in decision points for route 2 than route 1, only two of the differences in variance were statistically significant, as examined in the Levene test: eigenvector centrality and current flow closeness centrality.

Table 5. Variance in graph-theoretic metrics (asterisks denote significance)

Metric	Route 2 had higher variance?
PageRank	yes
eigenvector centrality	yes** ($p < .001$)
closeness centrality	yes
betweenness centrality	no
degree centrality	yes
load centrality	no
closeness vitality	yes
current flow closeness centrality	yes* ($p < .05$)
current flow betweenness centrality	yes

To quantify the relationship between classified strategy use and graph-theoretic metrics, we calculated the correlations between averaged graph-theoretic values and the percent use of a strategy for each participant. Results are shown in Figure 4.

---

[2] http://networkx.github.com/

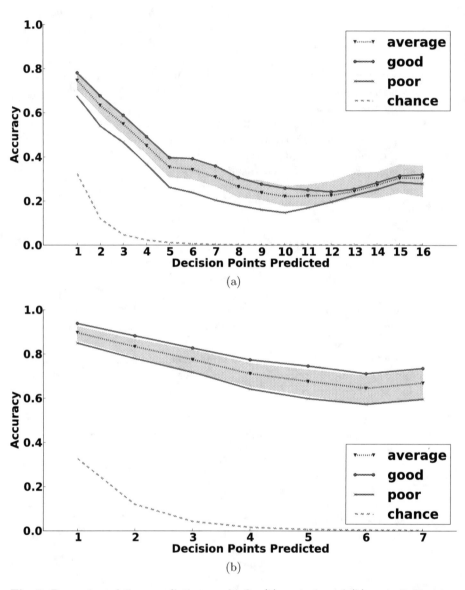

Fig. 2. Dynamic real-time prediction results for (a) route 1, and (b) route 2. Decision points show how many decision points in advance the prediction is for (i.e., 1 is the next decision point, and 7 is the seventh decision point from the current point from which a prediction is being made). Shading denotes the 95% confidence interval for the mean of the predictions for all participants.

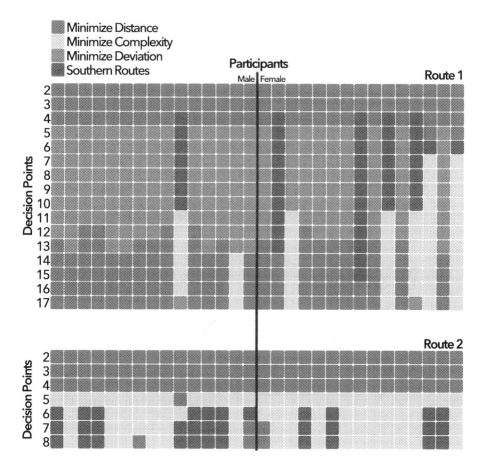

Fig. 3. Classified strategies at decision points for routes 1 and 2. Participants are on the x-axis, whereas decision points traversed are on the y-axis. The colors (or shades of gray in grayscale) denote classified strategies for each decision point traversed. The 15 male and 15 female participants are grouped left and right, respectively.

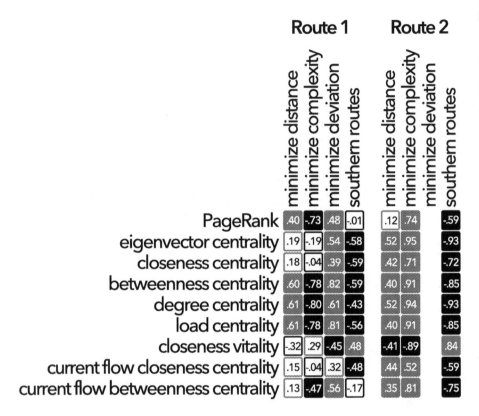

Fig. 4. Correlations between average graph-theoretic metric scores for points traversed by participants and classified strategy use. Solid colors denote significant (Pearson r; $p < .05$) correlations, whereas outlines are non-significant. Red and black are positive and negative correlations, respectively, also shown by numerical values. The wide-streets strategy was not classified for either routes 1 or 2 and the minimize-angular-deviation-to-the-goal strategy was not classified for route 2.

6   Discussion

6.1   Efficacy of Predicting Subsequent Decision Points

Our dynamic real-time route prediction method could predict the next decision point that wayfinders would traverse 82% of the time, and subsequent points after that with accuracy that was significantly higher than chance. Unlike individual strategies, the dynamic predictor was able to predict subsequently traversed decision points with accuracy at least as high or higher than any individual strategy. Although future empirical validation will have to be performed, the results point to the efficacy of using our method across different environments and for wayfinders with varying spatial abilities and wayfinding performances.

Importantly, we were able to predict not only the next point, but also subsequent decision points with accuracy higher than chance ($p < .001$). As shown in Figure 2, although accuracy decreases as the number of points in the prediction chain increased, it is still possible to predict where wayfinders will travel reliably, even seven decision points in advance. Depending on the environment, this could equate to being able to predict where people will be in five to eight minutes.

6.2   Strategy Use and Graph-Theoretic Metrics

For route 1 the dynamic real-time predictor performed only as well as the shortest-distance strategy, whereas for route 2 it greatly outperformed all individual strategies for route prediction. The predictive accuracy, as shown in Figure 2, was also much better for route 2 compared with the second route, and the distribution of classified strategies between routes 1 and 2 were significantly different (Section 5.2). This is an import finding because simply predicting wayfinders using the shortest path to the goal did not perform as well on route 2 as our dynamic predictor that used multiple strategies; for route 2 multiple strategies were a better fit for the behavior of wayfinders than using any single strategy in isolation.

As discussed in Section 5.3, although seven out of nine metrics investigated showed a higher variance for values in decision points for route 2 than route 1, only eigenvector centrality and current flow closeness centrality had significant differences in variance. Eigenvector centrality [3] is a way to measure the influence of a decision point, because points that are themselves important and connected to other important points will have larger values than points that are not as well connected to important points in the graph. We calculated the eigenvector centrality for the decision points in both routes 1 and 2, in order to study differences in the dispersion of influential decision points. Eigenvector centrality variance for route 2 was significantly larger than the variance for route 1 (Levene test; $p < .001$), meaning that differences between the importance and connectedness of important points were more varied in route 2. This may have influenced strategies used by participants, causing a larger separation between different strategies and hurting the efficacy of any single strategy for prediction.

6.3   Classification as a Method for Studying the Use of Route Selection Criteria

Although the actual strategies used by participants to navigate were not recorded, our method of classifying decision points in real-time can enable post-hoc analysis of which wayfinding strategies were most probably associated with a participant's traversed decision points. Since the probability of any decision point being associated with all the strategies can be calculated, the most probable strategies can be inferred for each decision point in a participant's traversal. This can be used to provide insight into how people navigate environments, in the sense that if changes in probable strategies occur, there might be causal factors that can be explored to explain the changes. The stability of wayfinding strategy use is also a topic that should be explored further.

It was surprising that the preference for the wide-streets strategy was not associated with any route traversals in Figure 3. It is possible that our participants preferred to traverse narrower streets, but a more likely explanation may be that our areas of study consisted mainly of very narrow streets in a mostly residential area, with wider streets only on the perimeter. In environments like the ones in our empirical study, a preference for wider streets may not have been a feasible strategy, but it cannot be stated that a preference for wide streets may not be useful in different environments.

As Kato and Takeuchi [24] pointed out, it is not usually possible to know what strategies will work well in an environment a priori, thus "the most important factor affecting wayfinding performance might be the participants' ability to use several strategies in a flexible way in the course of navigation and to select the best ones for the environment in question" (p. 186). Our results of classified strategy use for decision points over time show that strategies associated with wayfinders' traversals did indeed change in the course of navigation, as can be seen in Figure 3. This can suggest that people may be dynamically adapting and responding to the environments that they traverse. Future work should study the stability of route selection criteria that people consciously use as they navigate environments.

Concerning a Southern Route Preference. The poor predictive accuracy of the southern-preference strategy for route 1 (significantly worse than chance for predicting two decision points from the current) and the high accuracy for route 2 can be explained by the placement of the goals for each route, relative to the starting location. The goal was north of the starting location for route 1, whereas the goal for route 2 was south of the starting location. Wayfinders with a preference for southern routes could thus head to the south and then toward the goal for route 2, whereas for route 1 they would have had to go out of their way, resulting in an inefficient traversal, with respect to energy expenditure for locomotion. This suggests that a southern route preference is perhaps implicit when planning routes in some circumstances, but may not be an explicit strategy (cf. [5]).

6.4   Implications for Real-World Applications

The ability to dynamically calculate the probability of wayfinding strategies that wayfinders are using and predict where people navigating an environment will go has implications for many real-world applications. For navigational aids, the display of information can be adapted in real-time during navigation, to provide assistance personalized to the needs of the wayfinder. The need for personalized assistance may not be obvious, because many people may ask, "Why not just display the best route or few best routes from the current location?" However, since mobile navigational devices have been shown to reduce configurational understanding of routes with respect to traditional maps [21], it could be speculated that simply showing a route to a user may adversely affect configurational understanding of routes (see [33]), and thus may make it harder to navigate a return path or a similar route in the future. Also, in some situations, presenting large amounts of visual information is not desirable, as in combat situations where users should not be too distracted by visually presented information [11]. Presenting information aurally has also been shown by Gardony et al. [12] to hurt spatial memory for wayfinders navigating virtual environments, so merely changing the modality of navigational aids is not enough to help wayfinders learn more about their environment; wayfinding aids need to adapt to the individual cognitive needs of users. Finally, the definition of a "best" route is not universal. Even people who plan routes through known environments have been shown ([20]) to take a different route when actually traversing an environment. If navigational aids can determine what strategies wayfinders are using, that information could then be used to display a "best" route tailored for the classified strategy or strategies.

For context-sensitive recommendation systems, the ability to accurately predict subsequent points that will be traversed could allow digital navigational devices or phones to provide contextual information specific to an area that a person is traversing. For example, information about shopping and events could be presented just before people traverse an area, so that they can respond in a desired way, such as visiting a store or attending a concert. Intuitively, wayfinders are more likely to respond to contextual information if they can process it before they visit an area, because wayfinding is a directed activity and thus wayfinders are unlikely to turn around to go out of their way to visit a store that they already passed. Although the current method requires a goal location to be specified, navigational aids such as cellphone map applications already allow goal locations to be inputted, so this is not overly cumbersome.

City design can also take advantage of our method of real-time route prediction. The effect of changes in street networks on different wayfinding strategies could be tested by having humans choose routes through the target environment on a map or in a 3D virtual environment, both before and after a change. Then computer-generated routes could be used to predict the chosen routes and to determine whether the changes increase or decrease predictability of wayfinders between the same start and goal locations. Thus predictability could be a new metric for urban planning, because environmental structure can impact wayfinding performance [37].

For surveillance, our method can be used to find people who deviate from a set of common wayfinding strategies when going toward a known goal. In emergency situations, wayfinders heading toward a known evacuation shelter could be analyzed and where they will traverse can be predicted. Areas to which many wayfinders are predicted to go could be given higher priority for getting cleared, in order to allow more people to pass.

7   Conclusion

Wayfinders have been theorized to use a variety of strategies comprising route selection criteria when traversing environments. Here, five strategies were considered: minimizing distance to the goal, minimizing complexity of traversed routes, minimizing the angular deviation from the goal, having a preference for southern routes, and having a preference for wider streets. We hypothesized that using sets of routes computationally generated according to each strategy, it should be possible to predict subsequent decision points that wayfinders will traverse. We used traversals from 30 participants in an empirical study who navigated two routes to validate our work.

Results show that although individual strategies could be used to predict subsequent decision points that wayfinders traversed, there was no single strategy that was better than all others, across the two routes in our empirical study. Therefore we created a dynamic approach that classified strategies that wayfinders used in real-time and then used the classified strategies to predict decision points wayfinders would traverse. This dynamic real-time prediction approach was able to match or outperform any single strategy for predicting subsequent decision points for wayfinders in our study. Multiple route-selection criteria were better able to model the behavior of wayfinders than any single strategy, including simply minimizing distance. This has implications, in particular, for electronic navigational aids, in that sub-optimal route selections can be predicted and assistance can be offered to wayfinders as they are making important decisions about which way to turn. Prediction information can also be used to prime wayfinders about important features in an environment that they are about to traverse. Future work should empirically test using dynamic real-time prediction to influence wayfinders, and measure the effectiveness of incorporating prediction to improve wayfinding efficiency and survey information of routes travelled.

References

1. Bailenson, J.N., Shum, M.S., Uttal, D.H.: The Initial Segment Strategy: a Heuristic for Route Selection. Memory & Cognition 28(2), 306–318 (2000)
2. Bojduj, B., Weber, B., Richter, K.-F., Bertel, S.: Computer Aided Architectural Design: Wayfinding Complexity Analysis. In: 12th International Conference on Computer Supported Cooperative Work in Design, CSCWD, pp. 919–924. IEEE (2008)
3. Bonacich, P.: Factoring and Weighting Approaches to Status Scores and Clique Identification. Journal of Mathematical Sociology 2(1), 113–120 (1972)
4. Brunyé, T.T., Mahoney, C.R., Gardony, A.L., Taylor, H.A.: North is Up (Hill): Route Planning Heuristics in Real-World Environments. Memory & Cognition 38(6)
5. Brunyé, T.T., Gagnon, S., Waller, D., Hodgson, E., Tower-Richardi, S., Taylor, H.A.: Up North and Down South: Implicit Associations Between Topography and Cardinal Direction. The Quarterly Journal of Experimental Psychology 65(10), 1880–1894 (2012)

6. Dalton, R.C.: The Secret is to Follow Your Nose. Environment and Behavior 35(1), 107 (2003)
7. Dave, K., Lawrence, S., Pennock, D.M.: Mining the Peanut Gallery: Opinion Extraction and Semantic Classification of Product Reviews. In: Proceedings of the 12th International Conference on World Wide Web, pp. 519–528. ACM (2003)
8. Dee, H.M., Hogg, D.C.: Navigational Strategies in Behaviour Modelling. Artificial Intelligence 173(2), 329–342 (2009)
9. Duckham, M., Kulik, L.: "Simplest" Paths: Automated Route Selection for Navigation. In: Kuhn, W., Worboys, M.F., Timpf, S. (eds.) COSIT 2003. LNCS, vol. 2825, pp. 169–185. Springer, Heidelberg (2003)
10. Frankenstein, J., Büchner, S.J., Tenbrink, T., Hölscher, C.: Influence of Geometry and Objects on Local Route Choices during Wayfinding. In: Hölscher, C., Shipley, T.F., Olivetti Belardinelli, M., Bateman, J.A., Newcombe, N.S. (eds.) Spatial Cognition VII. LNCS, vol. 6222, pp. 41–53. Springer, Heidelberg (2010)
11. Garcia, A., Finomore, V., Burnett, G., Baldwin, C., Brill, C.: Individual Differences in Multimodal Waypoint Navigation. In: Proceedings of the Human Factors and Ergonomics Society Annual Meeting, vol. 56, pp. 1539–1543. SAGE Publications (2012)
12. Gardony, A.L., Brunyé, T.T., Mahoney, C.R., Taylor, H.A.: How Navigational Aids Impair Spatial Memory: Evidence for Divided Attention. Spatial Cognition & Computation (to appear, 2013)
13. Geurts, P., Ernst, D., Wehenkel, L.: Extremely Randomized Trees. Machine Learning 63(1), 3–42 (2006)
14. Golledge, R.: Path Selection and Route Preference in Human Navigation: A Progress Report. In: Kuhn, W., Frank, A.U. (eds.) COSIT 1995. LNCS, vol. 988, pp. 207–222. Springer, Heidelberg (1995)
15. Hagberg, A.A., Schult, D.A., Swart, P.J.: Exploring Network Structure, Dynamics, and Function Using NetworkX. In: Proceedings of the 7th Python in Science Conference (SciPy 2008), pp. 11–15 (August 2008)
16. Hegarty, M., Montello, D.R., Richardson, A.E., Ishikawa, T., Lovelace, K.: Spatial Abilities at Different Scales: Individual Differences in Aptitude-Test Performance and Spatial-Layout Learning. Intelligence 34(2), 151–176 (2006)
17. Hegarty, M., Richardson, A., Montello, D., Lovelace, K., Subbiah, I.: Development of a Self-Report Measure of Environmental Spatial Ability. Intelligence 30(5), 425–447 (2002)
18. Hochmair, H., Frank, A.U.: Influence of Estimation Errors on Wayfinding-Decisions in Unknown Street Networks–Analyzing the Least-Angle Strategy. Spatial Cognition and Computation 2(4), 283–313 (2000)
19. Hochmair, H., Karlsson, V.: Investigation of Preference Between the Least-Angle Strategy and the Initial Segment Strategy for Route Selection in Unknown Environments. In: Freksa, C., Knauff, M., Krieg-Brückner, B., Nebel, B., Barkowsky, T. (eds.) Spatial Cognition IV. LNCS (LNAI), vol. 3343, pp. 79–97. Springer, Heidelberg (2005)
20. Hölscher, C., Tenbrink, T., Wiener, J.M.: Would You Follow Your Own Route Description? Cognitive Strategies in Urban Route Planning. Cognition 121(2), 228–247 (2011)
21. Ishikawa, T., Fujiwara, H., Imai, O., Okabe, A.: Wayfinding with a GPS-Based Mobile Navigation System: a Comparison with Maps and Direct Experience. Journal of Environmental Psychology 28(1), 74–82 (2008)

22. Ishikawa, T., Montello, D.R.: Spatial Knowledge Acquisition from Direct Experience in the Environment: Individual Differences in the Development of Metric Knowledge and the Integration of Separately Learned Places. Cognitive Psychology 52(2), 93–129 (2006)

23. Jagadeesh, G.R., Srikanthan, T., Quek, K.H.: Heuristic Techniques for Accelerating Hierarchical Routing on Road Networks. IEEE Transactions on Intelligent Transportation Systems 3(4), 301–309 (2002)

24. Kato, Y., Takeuchi, Y.: Individual Differences in Wayfinding Strategies. Journal of Environmental Psychology 23(2), 171–188 (2003)

25. Krumm, J.: A Markov Model for Driver Turn Prediction. In: Society of Automotive Engineers (SAE) 2008 World Congress. ACM (2008)

26. Krumm, J.: Where Will They Turn: Predicting Turn Proportions at Intersections. Personal and Ubiquitous Computing 14(7), 591–599 (2010)

27. Krumm, J., Horvitz, E.: Predestination: Inferring Destinations from Partial Trajectories. In: Dourish, P., Friday, A. (eds.) UbiComp 2006. LNCS, vol. 4206, pp. 243–260. Springer, Heidelberg (2006)

28. Laasonen, K.: Clustering and Prediction of Mobile User Routes from Cellular Data. In: Jorge, A.M., Torgo, L., Brazdil, P.B., Camacho, R., Gama, J. (eds.) PKDD 2005. LNCS (LNAI), vol. 3721, pp. 569–576. Springer, Heidelberg (2005)

29. Liben, L., Downs, R.: Understanding Person-Space-Map Relations: Cartographic and Developmental Perspectives. Developmental Psychology 29(4), 739–752 (1993)

30. Liu, B.: Web Data Mining. Springer, Berlin (2007)

31. Montello, D.: Navigation. The Cambridge Handbook of Visuospatial Thinking 18, 257–294 (2005)

32. Murakoshi, S., Kawai, M.: Use of Knowledge and Heuristics for Wayfinding in an Artificial Environment. Environment and Behavior 32(6), 756–774 (2000)

33. Parush, A., Ahuvia, S., Erev, I.: Degradation in Spatial Knowledge Acquisition When Using Automatic Navigation Systems. In: Winter, S., Duckham, M., Kulik, L., Kuipers, B. (eds.) COSIT 2007. LNCS, vol. 4736, pp. 238–254. Springer, Heidelberg (2007)

34. Pedregosa, F., Varoquaux, G., Gramfort, A., Michel, V., Thirion, B., Grisel, O., Blondel, M., Prettenhofer, P., Weiss, R., Dubourg, V., Vanderplas, J., Passos, A., Cournapeau, D., Brucher, M., Perrot, M., Duchesnay, E.: Scikit-Learn: Machine Learning in Python. Journal of Machine Learning Research 12, 2825–2830 (2011)

35. Richter, K.-F., Weber, B., Bojduj, B., Bertel, S.: Supporting the Designer's and the User's Perspectives in Computer-Aided Architectural Design. Advanced Engineering Informatics 24(2), 180–187 (2010)

36. Simmons, R., Browning, B., Zhang, Y., Sadekar, V.: Learning to Predict Driver Route and Destination Intent. In: Intelligent Transportation Systems Conference, ITSC 2006, pp. 127–132. IEEE (2006)

37. Takemiya, M., Ishikawa, T.: Determining Decision-Point Salience for Real-Time Wayfinding Support. Journal of Spatial Information Science (4), 57–83 (2012)

38. Takemiya, M., Ishikawa, T.: I Can Tell by the Way You Use Your Walk: Real-Time Classification of Wayfinding Performance. In: Egenhofer, M., Giudice, N., Moratz, R., Worboys, M. (eds.) COSIT 2011. LNCS, vol. 6899, pp. 90–109. Springer, Heidelberg (2011)

39. Takemiya, M., Richter, K.-F., Ishikawa, T.: Linking Cognitive and Computational Saliences in Route Information. In: Stachniss, C., Schill, K., Uttal, D. (eds.) Spatial Cognition 2012. LNCS, vol. 7463, pp. 386–404. Springer, Heidelberg (2012)

An Affordance-Based Simulation Framework for Assessing Spatial Suitability

David Jonietz[*] and Sabine Timpf

Institute for Geography
University of Augsburg
Alter Postweg 118
D-86159 Augsburg, Germany
david.jonietz@geo.uni-augsburg.de

Abstract. In their everyday decision processes, humans depend on their ability to evaluate the suitability of environmental objects for specific actions. Suitability can therefore be understood as an abstract quality that is determined by properties of both the human agent and the environment as well as the action to be performed. The notion of such mutual dependency relationships with regards to action potentials is closely related to the concept of affordances as proposed by Gibson (1977). In this paper, a conceptual framework is proposed for the simulation of human agents assessing spatial suitability. In our model, actions represent a central element and are modeled at different hierarchical levels of abstraction. Finally, a simulation of pedestrian route choice is presented as a case study in order to explain the assessment process.

Keywords: affordances, activity theory, spatial suitability, walkability.

1 Introduction

Evaluating the suitability of environmental objects for specific actions is an everyday task for human beings. We choose the best location to eat, the shortest, quickest or most pleasurable route to walk to the preferred restaurant and, once we have arrived, the optimal table to sit at. Apart from merely identifying an environmental object (restaurant, footpath, table) which affords our planned action, choosing among various alternatives requires a mental assessment of the expected suitability of each object, which then serves as a determining factor for the decision process.

Suitability, however, is not an inherent quality of the environmental object. Thus, it can be expected that other people's perceptions of the optimal restaurant will differ from mine. Similarly, the best route to walk to the destination would probably not be chosen when driving a car. Suitability, therefore, can be defined as an abstract quality which is determined by properties of the environment, the agent as well as the action that will be performed. The notion of such mutual dependency relationships with regards to action potentials is closely related to the core concepts of ecological

---

[*] Corresponding author.

T. Tenbrink et al. (Eds.): COSIT 2013, LNCS 8116, pp. 169–184, 2013.

psychology, a movement in perceptual psychology. With his theory of visual spatial perception, Gibson (1977) set its foundations, describing how the perception of action potentials in the environment, which he termed affordances, depends on the combination of properties of both the environment and the perceiving organism.

Noting this conceptual nearness, a framework for modeling suitability with affordances has been proposed earlier (Jonietz et al. 2013). In this work, we are concerned with using agent-based models (ABM) to simulate the process of human spatial suitability assessment. For this, we build on previous work to define suitability as the degree of correspondence between the properties of both an agent and the environment which are relevant with regards to a particular action. We use ABM because the possibility to model agents with different individual properties is vital to our approach. In our extended framework, actions represent the central elements since they establish the connection between the agent and the environment in complex dependency relationships. In order to model complex actions, we use a basic principle of activity theory, the hierarchical structure of activity (Leontiev 1978, Kemke 2001).

This paper is structured as follows: First we present the relevant background information, including an introduction to the concept of affordances, a description of the affordance-based framework to model suitability and the notion of hierarchically structured actions from activity theory. Then, building on this theoretical basis, we present our affordance-based simulation framework for spatial suitability assessment and further explain it using the practical example of pedestrian route choice. In the last section, we provide our conclusion and discuss future work.

2   Background

This section will provide the theoretical background for our proposed framework, starting with a brief review of the original affordance theory as well as selected work which builds on the initial work by Gibson (1977). In the following, an affordance-based method for modeling spatial suitability, which was proposed earlier, will be described. Finally, the focus will be on activity theory and the concept of hierarchically structured actions.

2.1   Affordances

The affordance-concept stands in the focus of Gibson's ecological theory of visual spatial perception (Gibson, 1977). Investigating human spatial perception, in particular the question how meaning is created from visual observations, he gradually developed his idea of a direct perception of action potentials which are present in the environment. In reference to the verb *to afford*, he termed these action potentials affordances:

> The affordances of the environment are what it offers the animal, what it provides or furnishes, whether for good or ill. (Gibson 1979, p. 127).

Accordingly, as Gestalt psychologist Koffka (1935), whose work is known to have influenced Gibson, expresses it, "each thing says what it is" (Koffka 1935, p. 7). An affordance, however, is in its existence not just determined by attributes of the environmental object, but also by the capabilities of the acting agent (Gibson 1979). Thus, to give an example, a stone can be perceived as liftable by a human agent, but only if its properties such as weight or size match certain of the agent's capabilities such as his or her strength or grasp size. In ecological psychology, this relational dependency is referred to as the principle of agent-environment mutuality, one of the key innovations of Gibson's work (Varela and Thompson, 1991).

In the past, the notion of affordances has served as the theoretical basis for numerous studies from fields such as spatial cognition, agent-based modelling, robotics or geographical information science (GIS). Due to its conceptual vagueness, which is a result of continuous modifications by Gibson himself, there have been several attempts to formalize the affordance-concept as a prerequisite for further applications, sparking discussions about the exact definition of an affordance. While authors such as Turvey (1992) or Heft (2001), for example, allocate affordances to the environment, thereby reducing the agent's role to a merely complementary one, Stoffregen (2003) defines affordances as "properties of the animal-environment system […] that do not inhere in either the environment or the animal" (Stoffregen 2003, p. 123).

Warren (1984) first applied the affordance concept to a practical problem when investigating the perceived climb-ability of stair steps. With his experiment, he could demonstrate that test persons base their assessment not only on the step's height or their own leg length but rather on the ratio between these two properties, with a threshold value of .88 not to be exceeded (Warren 1984). Jordan et al. (1998) create an affordance-based model of place in GIS, claiming that three aspects of affordances must be modelled: the agent, the environment and the task. As an example for a place's semantic meaning to be modelled, the authors refer to a restaurant's suitability for a potential customer, where it is necessary to note not only the agent's capabilities and preferences, but also the actual task, such as eating, socializing or reading (Jordan et al., 1998). Affordances have also been widely used in the context of ABM. For a simulation of wayfinding in airports, for instance, Raubal (2001) bases his epistemological model on an extended concept of affordances, thus enabling agents to interpret the meaning of environmental objects relevant to wayfinding. Another example is Raubal and Moratz (2006), who develop a functional model for affordance-based agents, aiming to enable robots to perceive action-relevant properties of the environment in the context of their own spatio-temporal situation, tasks and capabilities. Several other studies apply affordances to models of local path planning and steering of pedestrian agents, none of which, however, moves beyond basic navigational issues such as obstacle avoidance (Turner and Penn, 2002; Kapadia et al., 2009; Kim et al., 2011).

2.2   Modelling Spatial Suitability with Affordances

We base our framework on an affordance-based method for modeling spatial suitability which was proposed earlier (Jonietz et al. 2013). In accordance with the definition developed by Stoffregen (2003), the authors understand an affordance as a higher-order property of a system $W_{ij} = (agent_i, environmental\ object_j)$ which is in its existence determined by certain agent- and environment-related properties termed capabilities cap_{ij} and dispositions $disp_{ij}$ which are interconnected in complex dependency relationships. Apart from the agent and the environment, the respective action, which the authors refer to as $task_i$, is also mentioned as a critical aspect in accordance with Jordan et al. (1998), but is treated as a subordinate factor in the model.

Similar to the approach by Warren (1984), the authors argue that by creating a ratio in the form of $cap_{ij}/disp_{ij}$, the dependency relationship between the relevant agent- and environment-related properties can be analyzed in order to determine whether an affordance exists in W_{ij}. Extending existing work on affordances, however, it is postulated that, instead of mere binary true/false values, there are various degrees to which an action can be afforded in W_{ij}. Using stair climbing as an example, for instance, although the affordance would be given in both agent-environment-systems, it can be expected that human agents would perceive a person-climbing-stairs system with a ratio close to the critical threshold value of .88 as less suitable compared to another system with a lower ratio value. Consequently, the authors argue that by setting certain agent- and environment-related properties cap_{ij} and $disp_{ij}$ into a relation in the form of $cap_{ij}/disp_{ij}$, and comparing the received ratio values to known critical threshold values, it is possible to derive scaled values for affordances. These values are interpreted as the task-specific $suitability_{ij}$ of $environmental\ object_j$ with reference to $agent_i$.

In order to identify cap_{ij} and $disp_{ij}$ from the total of agent- and environment-related properties $P_{Ai1}...P_{Ain}$ and $P_{Ej1}...P_{Ejn}$, it is necessary to break down the affordance into its constituent sub-affordances, since, as the authors state, actions are complex constructs that cannot be related to just one affordance, but rather have to be modeled as an hierarchical system of several sub-affordances. For instance, an action such as walking can be determined by several affordances such as locomotion, wayfinding, surmounting barriers and so on each of which relate to different properties of both the agent and the environment. Hence, as the authors state, in order to calculate an affordance for W_{ij}, one must identify and further refine the relevant sub-affordances until arriving at the most elementary level of atomic dependency relationships, where each cap_{ij} is confronted by just one $disp_{ij}$, such as leg length and step height in Warren's experiment (Warren 1984). By forming ratios as described above, it is then possible to derive scaled values for the sub-affordances, which can be interpreted as sub-suitability values $suit\_1'_{ij}...suit\_n'_{ij}$. These values can finally be combined to calculate the $suitability_{ij}$ (Jonietz et al. 2013).

2.3   Hierarchical Structure of Actions

As it has been stated in the previous section, in most cases, actions are in fact connected to several affordances. When modeling suitability with affordances, therefore, it is necessary to identify and analyze each of these sub-affordances separately before combining them to the superordinate affordance that is directly connected to the respective action. Instead of assuming the existence of one action with several sub-affordances, however, it is also possible to introduce the notion of an action being constituted by several contributing sub-actions, each of which must be executed to perform the superordinate action and each refer to one particular affordance. For a suitability assessment, it is therefore necessary to identify the sub-actions which contribute to the respective action.

This is closely related to a common problem for researchers dealing with actions that is generally referred to as the action individuation problem, meaning the challenge to unambiguously decide whether an agent's behavior consists of just one or more actions (Trypuz 2008). To state an example given by Kemke (2001), imagine the following three action descriptions: *moving ones finger in a certain way to press the light switch, switching the light on* and *lighting a room*. Depending on the specific viewpoint of the researcher, these actions could either be interpreted as different actions or as different descriptions of the same action. The first position adopts a so-called fine-grained view, in which actions are seen as different if they have different modal, temporal and causal properties (Trypuz 2008). In case of our example, accordingly, one would distinguish between three separate actions which are connected in "by-means-of" relationships, meaning that one action is performed by doing the other (Searle 1983, p. 128). Thus, the *light a room* is done by *switching the light on*, which is in turn accomplished by *moving one's finger in a certain way*. In contrast, a coarse grained view takes all three actions as different descriptions of one and the same action (Trypuz 2008).

Activity theory, a conceptual psychological framework which was developed by psychologists in the former Soviet Union, most notably L. Vygotsky and A. Leontiev, represents a useful approach to cope with the action individuation problem. Thus, one of the basic principles of this conceptual system is the three-level hierarchical structure of activity (Leontiev 1978). According to the author's terminology, the highest hierarchical level is represented by activities which are oriented towards motives that correspond to basic human needs. In order to execute an activity and fulfill a motive, however, it is necessary to perform a number of separate actions which follow subordinate goals and are, in turn, realized by lower-level actions with subordinate goals. When the lowest hierarchical levels of activity are reached, unconscious, automated processes take place. This is the level of operations, which do not follow specific goals, but serve only to implement the corresponding actions (Leontiev 1978).

Until today, activity theory and the notion of hierarchically structured actions has served as a valuable theoretical basis for work for instance in the context of human computer interaction (HCI), ontologies and GIS (Kaptelinin et. al 1995, Kemke 2001, Timpf 2003). In an approach very similar to Leontiev's work, Kemke (2001), for example, differentiates between different levels of abstraction when describing actions:

1. the realization level
2. the semantic level
3. the pragmatic level

On the realization level, an action is described in terms of its physical, motoric realization. For example, *moving ones finger in a certain way* clearly denotes a bodily movement and would therefore be placed on the realization level of abstraction. The semantic level of abstraction refers to action descriptions with a focus on the environmental effect that will be the outcome of the action, such as *switching the light on*. Without giving any information about the actual realization or the motivation that stands behind the action, such a description concentrates on the change that will occur regarding the environment or the agent when the action is carried out. Finally, describing an action on a pragmatic level of abstraction involves a direct reference to the intended goal that is pursued by acting. To *light a room* would be an example for this level of abstraction (Kemke 2001).

In our opinion, when modeling actions, it is necessary to acknowledge their hierarchical structure. A three level structure as proposed by Leontiev (1978) or Kemke (2001) helps to identify the sub-actions which contribute to the execution of a higher-level action. For the evaluation of the suitability of an action, it is therefore necessary to analyze each sub-action separately but also note their hierarchical dependency relationships.

3   Method

In this section, our method is presented. As a first step, it will be explained how actions can be modeled in our framework based on affordances and hierarchical levels of abstraction. This will be followed by a description of our approach for an agent-based simulation of spatial suitability assessment.

3.1   A Model of Hierarchically Structured Actions

For our framework, we build on an affordance-based notion of suitability (Jonietz et al. 2013). As it has been stated in the background section, affordances are action potentials which are determined by certain agent- and environment-related properties. Actions are placed in the center of our framework. In our view, it is the action that connects the agent with the environment, meaning that it determines not only which capabilities and dispositions are involved but also how they are related to each other. In order to identify these agent- and environment-related properties, however, it is necessary to recognize and explicitly describe the relevant sub-actions which contribute to the respective action, for example with the use of ontologies. Admittedly, this poses a challenging task to the modeler, since in realistic situations the number of possible sub-actions can be vast. When dealing with clearly defined scenarios, however, it can be expected that a sufficient set of potential sub-actions can be identified from domain knowledge or the observation of actual human behavior.

Figure 1 illustrates our model of action. We follow the terminology proposed by Kemke (2001) and distinguish between different hierarchical levels of abstraction when describing actions. On the superordinate action level, the relationships between actions on the different levels of abstraction are described. Thus, a pragmatic $action_\alpha$ is carried out by performing one or more semantic $sub\text{-}actions_{\alpha'}$. Actions on the semantic level of abstraction, in turn, must be physically realized by motoric $sub\text{-}actions_{\alpha''}$ on the realization level. Usually, there will be a one-to-many relationship between the actions on the different levels of abstraction. Thus, to refer back to the example used in the previous section, in case of a broken light bulb, for instance, *light* *a room (action$_a$)* could involve the following *sub-actions$_\alpha$: get a new light bulb, replace broken light bulb* and *switch the light on*. Similarly, it is clear that the realization of the *sub-action$_{\alpha'}$ replace broken light bulb* requires a range of *sub-actions$_{\alpha''}$* to be performed. In general, sub-actions can be performed either sequentially, as in the previous example, or simultaneously, such as *hold light bulb* and *screw light bulb in socket*. It can also be the case that sub-actions replace each other to provide alternative procedures for higher-level actions. Thus, *light the room (action$_a$)* could also be carried out by *light a candle (sub-action$_{\alpha'}$)* as an alternative.

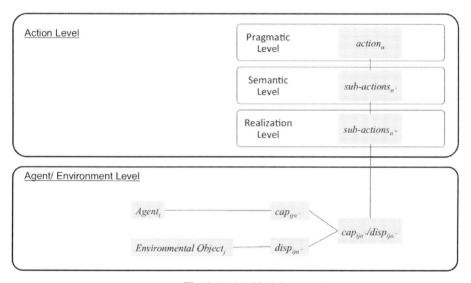

Fig. 1. Action Model

The connection between action, agent and environment is established on the Agent/ Environment Level. Having arrived at the realization level of abstraction, it is possible to describe actions in terms of the specific capabilities *cap$_{ij}$* of *agent$_i$* and environmental dispositions *disp$_{ij}$* of the *environmental object$_j$* that are involved in the action as well as their dependency relationships *cap$_{ij}$/disp$_{ij}$*.

3.2   An Affordance-Based Framework for the Simulation of Spatial Suitability Assessment

Based on the model of actions described previously, it is possible to equip agents with the ability to evaluate the suitability which is afforded by an environmental object with regards to their specific capabilities and the particular action to be performed.

The framework for this assessment process is illustrated in figure 2. We introduce a discretionary system $W_{ij\alpha}$ composed of an *agent$_i$* performing a pragmatic *action$_\alpha$* in an environment or, to be more precise, on an *environmental object$_j$*. We further assume that if the execution of *action$_\alpha$* is possible, implying that there is an *affordance$_{ij\alpha}$* as a higher-order property of the agent-environment-action system $W_{ij\alpha}$, then the *suitability$_{ij\alpha}$* which is afforded in $W_{ij\alpha}$ can be calculated in the form of a standardized numeric value. In order to appropriately assess $suitability_{ij\alpha}$, however, it is necessary to describe the pragmatic $action_\alpha$ in accordance with our model of hierarchically abstracted actions, which has been described in the previous section, meaning that its potentially contributing $sub\text{-}actions_{\alpha'}$ and $sub\text{-}actions_{\alpha''}$ on the lower levels of abstraction must be identified and analyzed separately. The $suitability_{ij\alpha}$ can then be calculated from the received values for $suitabilities_{ij\alpha'}$ and $suitabilities_{ij\alpha''}$, meaning the suitability that is afforded for each $sub\text{-}action_{\alpha'}$ and $sub\text{-}actions_{\alpha''}$ with regards to $agent_i$ and $environmental\ object_j$.

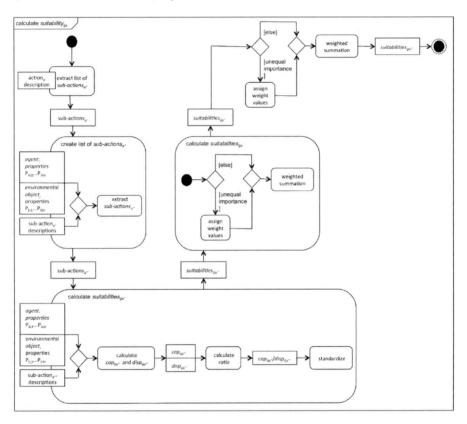

Fig. 2. Simulation Framework for Spatial Suitability Assessment

As a first step, based on the description of the pragmatic $action_\alpha$, all $sub\text{-}actions_{\alpha'}$ on the semantic level of abstraction must be identified. This information can be retrieved from the action description for $action_\alpha$, which, on the pragmatic level, simply lists the semantic $sub\text{-}actions_{\alpha'}$ which are involved in its execution. Additionally, due to the fact that the relation between $action_\alpha$ and its $sub\text{-}actions_{\alpha'}$ can be manifold, it must be determined whether the $sub\text{-}actions_{\alpha'}$ complement each other, either as a sequence or simultaneously, or represent alternatives which could be substituted by others, for example in case of them being not executable since the respective $sub\text{-}affordances_{ij\alpha'}$ are not given. Similarly to the pragmatic level, it is stated that if there are *sub-affordances*$_{ija'}$, then scaled values for the *suitabilities*$_{ija'}$ related to each *sub-action*$_{\alpha'}$ can be calculated, which will eventually be combined to receive *suitability*$_{ija}$.

In order to assess whether the relevant *sub-affordances*$_{ija'}$ exist, and if true, to calculate the *suitability*$_{ija'}$, however, it is necessary to move to the lowest level of abstraction, the realization level. Thus, for each semantic *sub-action*$_{\alpha'}$, the list of *sub-actions*$_{\alpha''}$ on the realization level must be identified, or in other words, it must be defined what must be physically done by *agent*$_i$ in order to achieve a certain environmental effect. Other than on the pragmatic level, however, this no longer depends on the action alone, but involves the specific agent and environmental object. Therefore, in this case, the action description for *sub-action*$_{\alpha'}$, lists potential *sub-actions*$_{\alpha''}$ which may be needed for its execution. Whether these potential *sub-actions*$_{\alpha''}$ are actually relevant with regards to the specific system W_{ija}, however, depends on the properties $P_{Ai1}...P_{Ain}$ of *agent*$_i$ and $P_{Ej1}...P_{Ejn}$ of the *environmental object*$_j$. Thus, for two exemplary agents with different properties or two different environmental objects, it may be necessary to carry out a different set of motoric actions to achieve the same environmental effect. These effects on the calculation process can be implemented in the form of if-else-statements that base the decision of whether a potential *sub-action*$_{\alpha'}$ should be included in the list of *sub-actions*$_{\alpha''}$ on certain property values $P_{Ai1}...P_{Ain}$ of *agent*$_i$ and/or $P_{Ej1}...P_{Ejn}$ of the *environmental object*$_j$.

In contrast to the semantic level, on the realization level, each *sub-action*$_{\alpha''}$ is atomic, meaning that it cannot be broken further down into its contributing sub-actions. In reality, of course, one could refine every motoric action until arriving at a level where every single muscular contraction is taken into account. For most purposes, however, issues of practicality and the limited level of detail of the available data will restrict the level of detail which is used in the analysis. In the model, due to the fact that at this point, the most elementary action level has been reached, it is now possible to relate each *sub-action*$_{\alpha''}$ to one specific pair of a capability *cap*$_{ija''}$ of *agent*$_i$ and an environmental disposition *disp*$_{ija''}$ of the *environmental object*$_j$.

As has been described in section 2.2 of this paper, capabilities and dispositions serve to extract and, if necessary, calculate from the total of properties $P_{Ai1}...P_{Ain}$ and $P_{Ej1}...P_{Ejn}$ of *agent*$_i$ and *environmental object*$_j$ the ones that are relevant to the respective system W_{ija}. These atomic dependency relationships can then be further analyzed by calculating ratio values that, after being standardized, represent the *suitability*$_{ija''}$ for each *sub-action*$_{\alpha''}$ on the realization level of abstraction. In the following steps, first the values for the *suitability*$_{ija'}$ on the semantic level are calculated from these values, before finally, the highest-level *suitability*$_{ija}$ referring to the pragmatic *action*$_{\alpha}$ can be computed by summing the *suitabilities*$_{ija'}$. If required, it is possible to assign individual weights in the course of these calculation processes in order to change the relative importance of certain *suitabilities*$_{ija'}$ or *suitabilities*$_{ija''}$.

In an actual simulation, this evaluation process can be executed just once or repeatedly at every time step. By simply changing the properties of the agent or the environment, it is therefore also possible to address the temporal nature of actions.

Thus, if any of the decisive conditions change, either concerning the environment or the agent, the agent is therefore able to calculate adjusted suitability values and somehow react to these changes.

4   Case Study: Pedestrian Route Choice

In order to further illustrate the ideas which have been discussed so far, we will apply our proposed framework on a simple simulation of pedestrian route choice. We concentrate on the tactical level of travel behavior, which means decisions that are made during the actual trip. For our scenario, we restrict our focus to a particular travel situation, where three pedestrian agents representing a child, an adult and an elderly person move between a fixed origin-destination pair in a hypothetical walking path network. Although the agents know the relative direction to the destination, they possess only limited knowledge of the walking path network. At each decision point, they are able to perceive certain properties of their incident path segments and identify the route alternative which affords the highest suitability with regards to their individual capabilities.

According to our framework, when assessing the suitability of an environmental object (path segment) for an agent (pedestrian) concerning a particular action (walking), it is necessary to describe the action in terms of its potentially contributing sub-actions on the lower levels of abstraction. Put in other words, as a first step, it must be clarified what *walking* actually means for a pedestrian. On this basis, the set of agent- and environment-related factors which are relevant for pedestrian route choice can be identified and dependency relationships determined.

4.1   Modeling the Action *Walking*

Walking involves more than just physical locomotion. In fact, since the mid-1990s, a wealth of empirical research has shown that pedestrians have specific needs and pose particular quality demands on their walking environment. These have often been subsumed under the generic term walkability, or *"the quality of walking environment perceived by the walkers"* (Park 2008, p. 22). Alfonso (2005) classified these walking needs, distinguishing between those related to feasibility, accessibility, safety, comfort and pleasurability.

Transferring these insights into the context of our action-based framework, we can define *walking* as the highest-level action, which involves contributing sub-actions such as *reaching the destination* (accessibility) or *being safe* (safety). According to our definition, these are still actions on the pragmatic level, since they are oriented towards specific goals. In order to identify the contributing sub-actions on the semantic and realization level, and finally recognize the relevant agent- and environment-related properties, it is necessary to refer to the results of empirical studies conducted by researchers from transportation planning, public health or spatial cognition. When focusing on route choice, however, this is not an easy task, since the majority of studies focus on the effects of the walking environment on mode choice, thus investigating what makes people walk. Work on walkability and pedestrian route choice, in contrast, is relatively rare.

We base our model of the action *walking* mainly on the results of two studies, Agrawal et al. (2008) and Özer & Kubat (2007), which both attempted to identify the key variables of the environment that influence pedestrian route choice. The results of the two studies are mainly consistent. Hence, Agrawal et al. (2008) found that pedestrians tend to choose the shortest and safest route, where safety mainly concerns the lowest risk of traffic accidents. The authors operationalized safety as the existence of traffic devices and traffic speed. As secondary factors for the decision process, the authors identify the attractiveness of the route and the quality of the sidewalks (Agrawal et al. 2008). Similarly, Özer & Kubat (2007) found accessibility and safety of the route to be the factors that best explain pedestrian movement patterns. Apart from the distance, they also identified the gradient as a contributing factor for accessibility. Environmental factors related to safety from accidents included pedestrian-vehicle-separation, traffic density, sidewalk width and the design of pedestrian crossings (Özer & Kubat 2007).

In accordance with these findings, and in order to keep the example comprehensible, we restrict the sub-actions to *reaching the destination* (accessibility) and *being safe* (safety) on the pragmatic level of abstraction. Regarding the former, the required environmental effect can be described as *traversing the segment*. Thus, for the goal of reaching the destination to be achieved, the agent must be able to move through one or more path segments. According to the results of both studies on route choice, this is determined by the distance and the gradient, which is why we define *overcoming distance* and *overcoming gradient* as contributing sub-actions on the realization level of abstraction. Concerning the pragmatic action *being safe*, the empirical evidence points to the fact that on the semantic level of abstraction, pedestrians need to *avoid physical contact with motorized means of transportation*. On the realization level, this is done by *keeping a buffer distance to the traffic lanes* and *use appropriate street crossing locations*.

4.2 Modeling the Agents and the Environment

Based on this model of the action *walking* with all its contributing sub-actions, it is now possible to identify the set of agent- and environment-related factors which are related to each sub-action on the realization level.

With regards to the environment, we base our model on the results of the empirical studies mentioned before, therefore including the length of the segment and its gradient for *overcoming distance* and *overcoming gradient* and sidewalk width and the design of pedestrian crossings for *keeping a buffer distance to the traffic lanes* and *use appropriate street crossing locations*. These characteristics are assigned as attributes to each path segment. Pedestrian crossings, however, are modeled as separate nodes in addition to the network links. Here, we distinguish between three different types of crossings regarding their safety, ranging from the least appropriate type without any crossing facilities (type 1) to one with crossing facilities (type 2) and finally one with traffic lights (type 3).

As it has been argued, however, apart from the environmental qualities that have been discussed so far, there are also agent-related aspects that might influence route choice. An overview of functional abilities of pedestrians is given by Vukmirovic (2010). In her work, the author identifies the set of abilities that are related to walking and represent prerequisites for a safe movement, thereby distinguishing between physical, psychomotor, sensory and cognitive abilities. Physical abilities include aspects such as stamina, strength, body coordination and equilibrium or height, whereas psychomotor abilities denote a person's skills to manipulate and control objects, such as reaction time or speed of limb movement. Sensory abilities are related to vision and hearing, whereas cognitive abilities are the skills that are needed for knowledge acquisition and problem solving (Vukmirovic 2010). For our example, we attribute measures of stamina, strength, body coordination, height and vision to the agents. For all properties, we distinguish between 5 ability levels from 1 (low) to 5 (high). As has been mentioned, the three agents are indented to represent a child, an adult and an elderly person. Table 1 shows the exemplary attribute values for these agents.

Table 1. Pedestrian Agents

Agent	Stamina [1-5]	Strength [1-5]	Body Coord. [1-5]	Height [1-5]	Vision [1-5]
1 (child)	3	3	1	1	2
2 (adult)	4	5	4	5	5
3 (elderly)	2	2	2	4	2

According to Vukmirovic (2010), children are especially at risk for traffic accidents since, among other characteristics, they usually have a small height, diminished peripheral vision and often lack body coordination, making them particularly vulnerable when crossing the street but also when walking on insufficient sidewalks too close to the motorized traffic. In our model, this is reflected by low values for height, vision and body coordination. While the adult pedestrian agent is characterized by consistently high capabilities, the elderly often experience some reduction regarding their physical abilities such as stamina and strength, body coordination and vision, which explains the rather low values assigned here (Vukmirovic 2010).

4.3   The Process of Suitability Assessment

Since the relevant sub-actions on the basic realization level of abstraction have been identified, it is now possible to calculate suitability values for each pedestrian-path segment-system. For this, however, the corresponding properties of both agent and environment must be identified for each sub-action. The sub-action *overcoming distance*, thus, can be analyzed by relating the agent's stamina and the path segment's length to each other. Similarly, *overcoming gradient* involves the strength and the

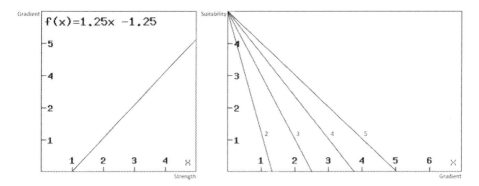

Fig. 3. Suitability assessment using threshold values

gradient. *Keeping a buffer distance to the traffic lanes* is determined by the sidewalk width but involves more than one property of the agent, namely body coordination and vision, just as *use appropriate street crossing locations* refers to the design of pedestrian crossings and vision and height of the agent.

Figure 3 illustrates the process of suitability assessment using the example of *overcoming gradient*. The graph on the left side shows the agent's strength level on the x-axis, ranging from 1 to 5, and the gradient of the path segment on the y-axis in percent. The graph itself denotes the maximum gradients that agents with different levels of strength are able to surmount. In general, a gradient of more than 5% is considered inappropriate for pedestrian traffic, which is why this value is used as a threshold for pedestrian agents with the highest level of strength. Agents with the lowest strength level, in contrast, are assumed to lack the ability to overcome any gradient. Thus, their threshold value is set to 0. For the remaining strength levels, the threshold values can be derived by assuming a simple linear relationship. On the right graph, the suitability values for specific gradients in relation to the strength levels are shown. These are based on the individual threshold values for each strength level, which are assigned the lowest suitability and a linear gradient-suitability relationship for lower gradients. Using known threshold values from the empirical literature on walking, it is possible to assess suitability values for the other sub-actions in a similar way. These are then simply added up to receive a suitability value for *walking* as the highest level action. Table 2 shows the received values for one path segment as an example.

Table 2. Suitability values for an exemplary path segment

Path Segment ID	Gradient	Width	Length	Suitability Agent 1 (child)	Suitability Agent 2 (adult)	Suitability Agent 3 (elderly)
751	1.09	3.91	200	2.50	4.29	-

One can see that, due to generally lower capability levels, much lower suitability values were calculated for the child compared to the adult agent. Regarding the elderly person, this particular path segment would not be accessible at all, which is due to the gradient in relation to the lowest strength level. When located at a decision point, the agents identify all incident path segments which lead in the direction of the destination, calculate their individual suitability values and choose the one with the highest value.

4.4   Results

Figure 4 shows the results of our analysis of pedestrian route choice in an hypothetical walking path network. For this exemplary scenario, apart from the length, the attribute values were randomly assigned to the path segments. A number of crossings with different crossing types were created. The three modeled agents start from the same point of origin in the upper left corner of the network but choose different routes to the destination point in the lower right corner. The deviations in the vicinity of the crossing points can be partly explained by the different sensitivity to more dangerous crossings. Thus, due to low levels of height and vision, it takes longer for agent 1 (child) to find a segment with an appropriate crossing point than for the other agents. While agent 1 (child) and agent 2 (adult) are able to reach the destination point, agent 3 (elderly) is not able to find a suitable path alternative, which is due to high gradients of the path segments, and finally terminates the trip.

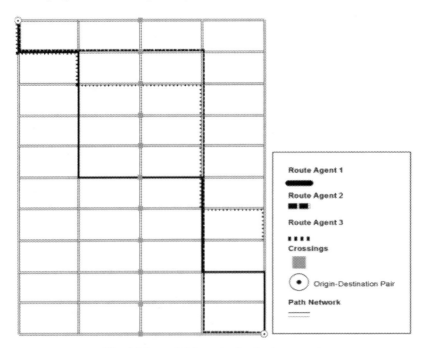

Fig. 4. Most suitable routes for the three agents

5   Discussion

In this paper, an affordance-based simulation framework for spatial suitability assessment was proposed. Building on the concept of affordances by Gibson (1977), spatial suitability was defined as the degree of correspondence between the properties of an agent and an environmental object which are relevant to a particular action. In our framework, complex actions are treated as the central elements which define how the properties of the agent and the environment are interconnected in complex dependency relationships. A model of hierarchically structured actions was developed based on activity theory. Using a case study of pedestrian route choice as an illustrative example, it was demonstrated how the suitability values can be calculated on the basis of known threshold values.

The perceived suitability that arises from a combination of agent, environment and action provides a basis for countless decision processes in our everyday life. With our framework, it is possible to enable agents to evaluate spatial suitability. This can be used to model complex behavior such as searching for an optimal location for one or more intended actions or, on the other hand, adapt behavior to certain characteristics or qualities of the surrounding environment. Modeling actions with hierarchically structured levels of abstraction allows agents to, in case of a low suitability, identify alternative actions on the lower levels which would still lead to the pragmatic action to be executed. In addition, by repeating the evaluation process, the suitability values can be continuously updated. Thus, temporally changing relationships such as, in the context of pedestrian movement, agents getting tired on their way to the destination or blocked sidewalks could be addressed by adjusting the property values such as the agent's stamina or the width of the sidewalk, therefore enabling the agent to react to such changing conditions.

A practical problem is posed by the high level of detail of data required for the environmental as well as the agent and the action model. For complex pragmatic actions, a high number of sub-actions and corresponding properties of the agent and the environment must be identified and analyzed. A validation by the observation of actual human behavior is certainly needed. Another potential source of vagueness is the simplified assumption of a linear relationship between the capabilities of the agent and the properties of the environment. It is imaginable that in many cases, the relation follows a more complex function.

For future work, it is planned to further refine the model and use it as a basis for an agent-based simulation of pedestrian movement in a real urban environment.

References

1. Agrawal, A.W., Schlossberg, M., Irvin, K.: How Far, by Which Route and Why? A Spatial Analysis of Pedestrian Preference. Journal of Urban Design 13(1), 81–98 (2008)
2. Alfonso, M.A.: To Walk or not to Walk? The Hierarchy of Walking Needs. Environment and Behaviour 37(6), 808–836 (2005)
3. Gibson, J.J.: The theory of affordances. In: Shaw, R., Bransford, J. (eds.) Perceiving, Acting, and Knowing: Toward an Ecological Psychology, pp. 67–82. Lawrence Erlbaum, Mahwah (1977)
4. Gibson, J.J.: The Ecological Approach to Visual Perception. Houghton Mifflin Company, Boston (1979)

5. Heft, H.: Ecological psychology in context: James Gibson, Roger Barker, and the legacy of William James's radical empiricism. Lawrence Erlbaum Associates, Mahwah (2001)
6. Jonietz, D., Schuster, W., Timpf, S.: Modelling the Suitability of Urban Networks for Pedestrians: An Affordance-based Framework. In: Geographic Information Science at the Heart of Europe. LNGC, pp. 369–382 (2013)
7. Jordan, T., Raubal, M., Gartrell, B., Egenhofer, M.: An affordance-based model of place in GIS. In: Eighth International Symposium on Spatial Data Handling, July 11-15 (1998)
8. Kapadia, M., Singh, S., Hewlett, B., Faloutsos, P.: Egocentric Affordance Fields in Pedestrian Steering. In: I3D 2009 Proceedings of the 2009 Symposium on Interactive 3D Graphics and Games, February 27- March 1 (2009)
9. Kaptelinin, V., Kuutti, K., Bannon, L.: Activity theory: Basic concepts and applications. In: Blumenthal, B., Gornostaev, J., Unger, C. (eds.) EWHCI 1995. LNCS, vol. 1015, Springer, Heidelberg (1995)
10. Kemke, C.: About the ontology of actions. Technical report mccs- 01-328, Computing Research Laboratory, New Mexico State University (2001)
11. Kim, N., Joo, J., Rothrock, L., Wysk, R., Son, Y.-J.: Human Behavioral Simulation Using Affordance-Based Agent Model. In: Jacko, J.A. (ed.) Human-Computer Interaction, Part I, HCII 2011. LNCS, vol. 6761, pp. 368–377. Springer, Heidelberg (2011)
12. Koffka, K.: Principles of Gestalt psychology. Harcourt Brace, New York (1935)
13. Leontiev, A.N.: Activity, Consciousness, and Personality. Prentice-Hall, New Jersey (1978)
14. Özer, Ö., Kubat, A.S.: Walking Initiatives: A Quantitative Movement Analysis. In: Proceedings, 6th International Space Syntax Symposium, Istanbul (2007)
15. Park, S.: Defining, Measuring, and Evaluating Path Walkability, and Testing Its Impacts on Transit Users' Mode Choice and Walking Distance to the Station. Berkeley: Dissertations, University of California Transportation Center, UC Berkeley (2008)
16. Raubal, M.: Ontology and epistemology for agent based wayfinding simulation. International Journal of Geographical Information Science 15(7), 653–665 (2001)
17. Raubal, M., Moratz, R.: A functional model for affordance-based agents. In: Dagstuhl Seminar Towards Affordance-based Robot Control, Dagstuhl Castle, Germany, June 5-9 (2006)
18. Searle, J.R.: Intentionality: An Essay in the Philosophy of Mind. Cambridge University Press, Cambridge (1983)
19. Stoffregen, T.A.: Affordances as properties of the animal environment system. Ecological Psychology 15(2), 115–134 (2003)
20. Timpf, S.: Geographic Activity Models. In: Duckham, M., Goodchild, M., Worboys, M.F. (eds.) Foundations of Geographic Information Science, pp. 241–254. Taylor & Francis, London and New York (2003)
21. Trypuz, R.: Formal Ontology of Action- A Unifying Approach. Wydawnictwo KUL, Lublin (2008)
22. Turner, A., Penn, A.: Encoding natural movement as an agent-based system: an investigation into human pedestrian behaviour in the built environment. Environment and Planning B: Planning and Design 29, 473–490 (2002)
23. Turvey, M.T.: Affordances and prospective control: An outline of the ontology. Ecological Psychology 4(3), 173–187 (1992)
24. Varela, F.J., Thompson, E.: The embodied mind. MIT Press, Cambridge (1991)
25. Vukmirovic, M.: Functional abilities of humans and identification of specific groups of pedestrians. In: Walk 21 XI International Conference on Walking and Liveable Communities, November 16-19, The Hague, Netherlands (2010)
26. Warren, W.H.: Perceiving affordances: Visual guidance of stair climbing. Journal of Experimental Psychology 105(5), 683–703 (1984)

A Probabilistic Framework for Object Descriptions in Indoor Route Instructions

Vivien Mast and Diedrich Wolter

SFB/TR 8 Spatial Cognition, University of Bremen
{viv,dwolter}@informatik.uni-bremen.de

Abstract. Automatically generated route instructions are common in modern cars, but for indoor environments, such systems hardly exist, possibly due to the fundamental difference of street networks and indoor environments. Indoor environments often lack the clear decision points present in street networks and involve a spatially complex layout. Therefore, good route instructions for indoor environments should provide context by relating instructions to environmental features—they need to describe and locate objects. This paper is concerned with automatic generation of in-advance route instructions for indoor scenarios. Motivated by empirical research, we propose a probabilistic framework for generating referring expressions using vague properties and graded models of spatial relations. We discuss the relevance of this framework for generating indoor route instructions and demonstrate the appropriateness of this approach in case studies.

1 Introduction

In an urban society with increasingly large and complex buildings, indoor wayfinding has become a regular challenge for many people (Carlson et al., 2010; Peponis et al., 1990). For car navigation in street networks, incremental wayfinding assistance based on real-time positioning is well established. For indoor navigation, there are virtually no comparable assistance systems, partly due to limited availability of positioning systems in buildings (see also Gartner et al., 2004).

A viable option for automated indoor wayfinding assistance is offered by in-advance route instructions. For example, the Infokiosk interactive wayfinding assistant (Cuayáhuitl et al., 2010) can give in-advance natural language route instructions in buildings. It uses the route generation system GUARD (Richter, 2008), which is optimized for street network navigation. As indoor environments are structurally and conceptually different from street networks, it is necessary to adapt instruction generation to the requirements of indoor wayfinding. While street networks exhibit a clear structure of decision points with only few options, indoor environments and many environments for pedestrian outdoor navigation often involve a spatially complex layout including open spaces where no clear routes and decision points exist (Rüetschi, 2007). Therefore, good route instructions for indoor environments should provide context by relating instructions to environmental features present—they need to describe and locate objects.

T. Tenbrink et al. (Eds.): COSIT 2013, LNCS 8116, pp. 185–204, 2013.

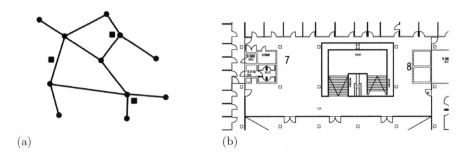

(a)   (b)

Fig. 1. (a) The structure of network space with intersections (circles), paths (lines) and landmarks (squares) (b) A typical scene space area

In this paper, based on an analysis of the structural and conceptual characteristics of indoor wayfinding, we argue that object descriptions and localizations play a crucial role for generating cognitively ergonomic in-advance indoor route instructions. We therefore propose a probabilistic framework for generating referring expressions using vague properties and integrating graded models of spatial relations. Further, we present an implementation of the system and demonstrate its performance in three example scenarios.

2   The Case for Scene-Based, Descriptive In-Advance Indoor Route Instructions

In the following, we will analyze the structural characteristics of indoor spaces and the requirements of in-advance route instructions. We will show that object descriptions play a central role in in-advance indoor route instructions. Further, we will give an overview of empirical research and technical approaches regarding indoor route instructions, referring expression generation (REG) and locative expressions. While important research has been done on adapting the level of granularity in route instructions to the prior knowledge of the user (Klippel et al., 2009; Schuldes et al., 2011), we focus here on detailed instructions for uninformed users.

2.1   Structural Characteristics of Indoor Spaces

Navigation can take place in *network space* (Figure 1 a), which is characterized by clearly identifiable decision points (intersections) connected by paths (streets) (Rüetschi, 2007). The wayfinder is confronted with two types of "which" tasks—firstly, at which decision point an action should be taken, and secondly, which path should be chosen (Moratz and Tenbrink, 2006).

On the other hand, navigation can also involve elements of scene space (Figure 1 b): open spaces that do not contain any clearly identifiable paths or decision points (Rüetschi, 2007; Schuldes et al., 2011). Here, the wayfinder has to navigate wide open spaces, and the wayfinding task involves searching, exploring, and matching. Different strategies such as piloting between landmarks, or oriented search can be used (Allen, 1997). This kind of wayfinding can be considered as a mixed task where the wayfinder needs to deal with both "Which" and "Where" questions, such as "Which door should I take?" and "Where is the door that I need to take?" The relevant cognitive categories in this case are not decision points and turns, but scenes and scene transitions (Rüetschi, 2007). A scene can be defined as "what can be seen by looking around but without significant walking about and forms a coherent and obvious entity in large-scale space" (Rüetschi, 2007, 33). Rüetschi (2007, 45) defines connections between scenes as consciously experienced or unconscious targets of "through" actions, for example doors, pathways or connecting spaces. The distinction between network space and scene space is gradual with different environments and navigation types involving more or less of either. Car navigation is usually strongly dominated by network space while pedestrian navigation in indoor and outdoor built environments contains elements of both scene space and network space.

Additionally, indoor spaces often suffer from a lack of unique landmarks. Landmarks play a central role in wayfinding (Hölscher et al., 2006; Presson and Montello, 1988), and are therefore crucial elements of route instructions (Allen, 1997; Denis, 1997). A central criterion that distinguishes a good landmark is its salience: a good landmark is distinctive, easily recognizable, and therefore easily memorizable (Caduff and Timpf, 2008; Presson and Montello, 1988; Raubal and Winter, 2002). In outdoor environments, there is a high number of diverse landmarks available, making it easier to single out salient landmarks for wayfinding. These can be buildings, atypical intersections, parks and many others. The main landmarks present in indoor environments are corridors, doors, windows, staircases, and elevators, many of which look alike throughout an entire building.

2.2   Requirements for In-Advance Route Instructions

Route instructions can be given *in-advance*, when the wayfinder has not yet started navigation, or *incrementally* during navigation (Richter, 2008). In contrast to incremental instructions, following in-advance route instructions requires the wayfinder to build up and memorize a mental model of the environment that should be traversed and the actions they need to perform (Richter, 2008). Therefore the speaker needs to balance the need for sufficient and exact information against the conciseness required for efficient processing and memorization (Daniel and Denis, 1998; Richter, 2008). As in-advance instructions offer no opportunity for the instruction-giver to correct wayfinding errors en-route, instructions should also be robust against misunderstandings and problems of memorization in order to avoid errors in the first place.

2.3   Descriptive Route Instructions

Due to the requirements discussed above, we argue that descriptive information in the form of object descriptions and locative expressions plays a crucial role in in-advance indoor route instructions. Descriptive elements give information about the spatial relation between mover and landmark, or between landmarks, or about landmark properties. They help the wayfinder build up a representation of an environment (Denis, 1997).

While network space enables action prescriptions tied to very simple landmark terms ("Turn left at the church"), for navigation in scene space the prevalence of "Where" tasks in comparison to "Which" tasks necessitates a larger quantity of descriptive information (cf. Vorwerg and Tenbrink, 2007). Instructions need to present information about the structure of the space, the location of landmarks, and information that directly localizes a required scene transition with respect to a scene description: "... until you reach a big hall area. In front of you there is a staircase. Behind the staircase there are two glass doors. Go through the left door."

This reasoning is supported by empirical research. Denis (1997) and Denis et al. (1999) show that in-advance route instructions for pedestrian navigation contain tightly intertwined descriptive and prescriptive elements that are combined in a highly flexible way, depending on the nature of the route segment described. In an empirical study concerning in-advance indoor route instructions, Mast et al. (2012) show that scene-based, descriptive route instructions are more adequate for scene space areas than network-based, prescriptive instructions. They support mental imagery and memorization of the instructions and reduce wayfinding errors significantly (Mast et al., 2012).

Orienting towards realization of scene-based, descriptive route instructions, Rüetschi (2007) presents a cognitively motivated model of scene space using image schemata. Based on this work, Schuldes et al. (2011) develop a hierarchical architecture for generating combined verbal and visual route instructions for pedestrians at different levels of granularity. The aspect of generating verbal instructions is not detailed enough for evaluation.

Garoufi and Koller (2010) present a system that generates motion instructions and referring expressions as part of an integrated planning module. Their focus is on generating low-level motion instructions that change the environmental conditions systematically in order to enable *simple* and effective referring expressions. We share this approach of creating a unified mechanism that can compare the felicity of messages of different types and complexity, allowing us to generate route instructions with a flexible combination of prescriptive and descriptive information. In contrast to Garoufi and Koller (2010), we are involved with in-advance instructions. Therefore, our focus is on using referring expressions in order to create better motion instructions rather than the other way round. Although the mechanism proposed here is general enough to incorporate motion instructions, in this paper we focus on REG only. In the following, we will give an overview of empirical research and technical approaches to REG and locative expressions before presenting our own approach.

2.4   Referring Expression Generation

As we have seen, the description of objects is crucial for generating cognitively adequate in-advance indoor route instructions. Related research is mainly concerned with the areas of object reference and locative expressions.

Pobel et al. (1988) give an overview of early empirical work on human object reference which we summarize in the following. Reference is context-dependent. Speakers usually aim for *unique identifiability*, i.e. they choose the feature(s) that allow(s) the target object to be distinguished from the range of present distractors (Hermann and Deutsch, 1976; Olson, 1970). If several features enable unique identification, the feature with the largest distance between object and context is chosen (Hermann and Deutsch, 1976). If this distance is identical for several features, humans choose a feature according to learnt preferences (Hermann and Deutsch, 1976). Although some researchers postulate that humans tend to use the shortest distinguishing description (Hermann and Deutsch, 1976; Olson and Ford, 1975), this principle is often violated. A frequent phenomenon is *overspecification* or *redundancy* where speakers use additional non-discriminatory attributes for referring to objects (Mangold and Pobel, 1988; Pechmann, 1984). The reasons for redundant expressions can be summarized thus: speakers overspecify either in order to make production easier for themselves, or to make reference resolution easier for the listener (cf. Pobel et al., 1988). Research on the use of spatial relations in object reference confirms many of the principles found in non-spatial object reference like unique identifiability and conciseness (Tenbrink, 2005), the existence of a preference order (Mainwaring et al., 2003) and flexibility depending on discourse situation (Vorwerg and Tenbrink, 2007).

Early technical approaches on REG focused on crisp properties, i.e. an item was either large or small with no intermediary values. Krahmer and van Deemter (2012) give an overview of the early approaches, the most relevant of which we summarize here. The *full brevity algorithm* focuses radically on the criterion of economy, searching for the shortest uniquely identifying expression. The *greedy heuristic algorithm* is a more efficient approximation of the full brevity algorithm: at each point in time, it chooses the feature that eliminates the most distractors and adds it to the description. The *incremental algorithm* is based on an inherent preference order for the available feature types. It iterates through all possible features in descending preference order. If a feature eliminates at least one distractor, it is added to the description. This algorithm is highly popular due to its computational efficiency. However, if the preference order is badly chosen, the algorithm may generate lengthy, unnatural descriptions. Moreover, this algorithm does not differentiate between a feature that eliminates one distractor and one that eliminates several.

An algorithm for generating referring expressions involving gradable properties is proposed by van Deemter (2006). He treats the meaning and acceptability of graded properties based on perceptual context, i.e. the same door could be conceived as "large" or "small" depending on the other doors in the scene. The system can refer to sets of objects such as "the n large(st) mice" such that every mouse in the target set is larger than every mouse in the distractor set. As van Deemter notes, his approach does not cover combinations of vague adjectives such as "the tall fat giraffe" that are not *pareto-optimal*, i.e. where the giraffe in question is not the largest and the fattest giraffe, but the only one which is large and fat (van Deemter, 2006). He suggests solving this issue with the *Nash arbitration plan*, where a doubly graded description is used when the product of the two acceptability values for the referent exceeds that of all distractors.

2.5   Locative Expressions

Locative expressions are one way to distinguish objects, and they are particularly useful in route instructions as they help the user build up a mental model of the environment. In order to specify the location of an object (target), its spatial relation with respect to a second object (reference object/relatum) needs to be named. In the context of the present work, we focus on the subtype of projective relations ("left", "right", "in front" or "behind").

These relations can be modeled by imposing a co-ordinate system on the reference object from the perspective of a third entity (origin). The prototypical meanings of the projective terms are defined by the axes of this coordinate system.[1] Different reference frames include, amongst others, the *intrinsic* and *relative* reference frame. In a relative reference frame, the reference direction is defined by the line between the origin (usually the viewer) and the relatum, as in "the cup is to the left of the tree (seen from my perspective)". An intrinsic reference frame is given when the origin and the relatum coincide in the same object, and the reference direction is given by the intrinsic orientation of the reference object, as in "the cup is to my right" (Moratz and Tenbrink, 2006).

Gapp (1995b) shows that spatial relations have a broad range of acceptability which is highest for objects on the prototypical axis, and gradually decreases with increasing distance from this axis. While Gapp found a linear decrease, Moratz and Tenbrink (2006) use a smoothing function based on the cosine in their empirically validated model of spatial relations for human-robot interaction.

Apart from the acceptability of the projective term itself, a number of further factors influence the acceptability of using a spatial relation. Barclay and Galton (2008) present a hierarchical influence model for reference object selection with the main factors *locatability of the reference object*, *search-space optimization* and *communication cost*. Gapp (1995a) provides a simpler model using an ordered list of 8 influence factors, including referentiality, mobility and salience of the reference object and the distance between reference object and target. However, the existing sophisticated graded models of spatial relations play a subordinate role in technical approaches to REG, as most systems focus on crisp features.

The planning-based system by Garoufi and Koller (2010) uses spatial position as a crisp unary feature of an object. However, it is capable of using positional

---

[1] See Carlson et al. (2003) for an extensive discussion of axis- and vector-based representation of spatial relations.

adjectives depending on the linguistic and situational context. For example, "the left blue button" does not have to refer to the leftmost button, as long as there are no further blue buttons to its left. An improved version of this system can generate redundant REs for easier understanding, based on a maximum entropy model learnt from a human-human interaction corpus (Garoufi and Koller, 2011).

Dale and Haddock (1991) extend the greedy heuristic algorithm to cover crisp spatial relations between two objects. Kelleher and Kruijff (2005) extend the incremental algorithm to include spatial relations. They diverge from crisp modelling by distinguishing between absolute and relative spatial relations. In absolute spatial relations only one object fits the given relation, for example "the box to the left of the chair" when only one box is to the left of the chair. In relative spatial relations a distractor object is present (there is another box to the left of the chair), but further away from the reference object.

As the empirical research shows, defining spatial relations as crisp categories is problematic. Some way of making fine-grained distinctions in the acceptability of spatial relations is necessary. Humans are willing to accept fairly large deviations from the prototypical axes, depending on the circumstances. If one element is on the prototypical axis of the relation, and a distractor is in a marginally acceptable position, using the relation in question will be sufficient for identifying the target object. The approach by Kelleher and Kruijff (2005) takes this into account, but does not provide any fine-grained way for evaluating the difference.

3   A Probabilistic Framework for REG

In order to generate route instructions that allow a flexible combination of prescriptive and descriptive information, a mechanism is needed that can compare messages of different types and complexity with respect to their adequacy.

In the present paper, we focus on the generation of referring expressions with graded properties including spatial relations. We are aiming for application in real-world scenarios which include target objects that are not easily identifiable by combinations of simple clear-cut features. Our work is motivated by situations where spatial relations with respect to a number of possible reference objects hold to a higher or lower degree, and other features may also be of graded applicability. These scenarios make it necessary to provide a mechanism for combining graded properties in a way that enables fine-grained distinctions of the *discriminatory power* each feature provides. This mechanism should be general enough to be able to include a variety of different types of information. As van Deemter (2006) notes, in such a scenario, unique identifiability is relative rather than absolute. In the following, we will present our general approach and a concrete implementation with a number of object properties and projective spatial relations. We will evaluate this approach with three different exemplary scenarios and discuss how the results differ from previous approaches.

We propose a mechanism of comparison that is at the same time general enough to be able to compare different types of information, and on the other hand modular enough to allow distinct modeling of each specific information type. In order to decide which of a given complex description is the best, we aim to maximize the probability of correctly identifying a referred goal object in a complex referring expression, given limited cognitive resources. For this purpose, the central optimization criterion is the relative difference between two descriptions, and not the absolute probability values.

3.1   Probabilistic Model

Essentially, we aim to identify the description D that maximizes the probability of the listener to identify goal object x correctly. A possible approach could thus be to consider the conditional probability $P(x|D)$ to select the right object, given a description D. This approach would have an undesired side effect of aiming to determine the one description that cannot be interpreted wrongly, i.e., all possible features would be exploited to compile the description, even if they were not very good descriptions for the object per se. Technically speaking, we need to counterweight $P(x|D)$ by a measure of description acceptability. Acceptability of a description can also be approached in a probabilistic manner as the probability $P(D|x)$ that a given description would be accepted [by a human], given the object. This leads to our model:

$$D^* := \arg\max(1 - \alpha)P(x|D) + \alpha P(D|x) \tag{1}$$

Here, D^* stands for the best description and α is a model parameter determining the relative weight of acceptability $P(D|x)$ vs. discriminatory power $P(x|D)$. A higher value for α corresponds to a stronger weight of acceptability. As mentioned above, our goal is to maximize the probability of correctly identifying a referred object in complex object localizations, given limited cognitive resources. Specific probability values may thus be not meaningful. In order to determine the probabilistic evaluation we make use of the well-known Bayes' theorem:

$$P(x|D) = \frac{P(D|x)P(x)}{P(D)} \tag{2}$$

The main factor in our model is $P(D|x)$ which gives the probability that we accept description D for object x. The more similar an object feature is to the prototype of a value referred to in D, the more one is willing to accept the description. We assume that these values are defined for each primitive feature, i.e., for each object and feature we are provided with a number between 0 and 1 that measures the respective feature appropriateness. In this way, we create a modular system that allows different models of features depending on their characteristics. In Section 4, we show how the features used in our current implementation are calculated.

$P(x)$ is the probability of randomly choosing object x in the given scene.[2] In a simple model, this defaults to $P(x) = \frac{1}{N}$, where N equals the number

---

[2] Determining which objects form part of the scene is a complex question in itself which is beyond the scope of this paper.

of objects in the world. In a more sophisticated model, $P(x)$ incorporates the general *Salience* of the object, i.e. the "distinctiveness relative to (perceptual or conceptual) context" (Tenbrink, 2012), as more salient objects will be noticed more easily and therefore have a higher chance of being randomly selected.

$P(D)$ gives the probability that the description D suits an arbitrarily chosen object. The probability of correctly identifying object x is reduced if multiple objects in the scene suit description D. This factor relates to *unique identification* but the probabilistic framework allows for a fine-grained distinction which we term *discriminatory power*.

3.2   Descriptions

In our approach descriptions are sets of tuples that relate a feature dimension (color, size, location, etc.) to respective feature values for an object in the scene. Technically speaking, we say that a description is a set of triples

$$\mathcal{D} := \{(o_1, f_{1,1}, v_{1,1}), (o_1, f_{1,2}, v_{1,2}), \ldots (o_n, f_{n,m}, v_{n,m})\}$$

that relates scene objects $o_1, \ldots o_n$, feature domains $f_{i,j}$ and feature values $v_{i,j}$. A key feature of the probabilistic model is that it can be applied to descriptions of arbitrary complexity, i.e., any size of set of feature dimension/value mappings. To this end, it is helpful to assume that the acceptability of different features is stochastically independent. To give an example, the probability of one accepting that a door is red (color feature) is assumed to be independent of the probability of accepting that the door is large (size feature). Stochastic independency of feature domains allows complex descriptions to be separated by feature domains. For example, $P(D|x)$ can be computed according to $P(D|x) = P(f_0|x) \cdot \ldots \cdot P(f_n|x)$, where each f_i stands for a single feature domain/value pair.

Spatial descriptions may involve multiple objects, for example to identify a door by saying that it is next to a bench. Evaluating a description that involves multiple objects is straightforward when we assume that identification of an object is stochastically independent from identifying another object. This allows the probability of correctly identifying two objects x, y whereby x is described in relation to y to be determined by $P(x|y) \cdot P(y)$.

3.3   Determining the Best Description

In order to determine D^* only the basis acceptance probabilities $P(D|x)$ need to be known for all objects and all combinations of a single feature domain/value. Alternatively, a computational model can be employed to determine required values once they are needed. As domains are finite, the probabilistic evaluation for all descriptions D can be determined in order to select the best description. While this approach is indeed applicable due to the simple mathematical computations required, a more efficient approach is obtained by incorporating search space pruning. We exploit that the evaluation of a description $D = \{(f_1, v_1), \ldots (f_n, v_n)\}$ of n feature domain/value pairs in $P(D|x)$ is monotonically decreasing. If, during computation, the acceptance probability drops below the best description found so far, computation can be aborted and the description candidate discarded.

Table 1. Availability of features in the different scenarios

feature	Scenario 1	Scenario 2	Scenario 3
type	predetermined	predetermined	predetermined
colour	–	predetermined	predetermined
size	local normalization	predetermined	predetermined
corpulence	local normalization	–	–
intrinsic spatial relations	–	left, right, in front of, behind	
relative spatial relations	–	left, right, in front of, behind	
special features	–	–	predetermined

4   Realization and Case Studies

In order to evaluate the capacities of our proposed mechanism, we present three case studies based on two realised variants of the system. Firstly, we present a small example scenario as is common in REG in order to demonstrate the general working of our mechanism and the treatment of combinations of vague features. Secondly, we show a simple spatial scene for demonstrating the integration of spatial relations into the REG mechanism. Finally, we use a spatial setting from a real building as it would be encountered in indoor wayfinding in order to discuss characteristics of the model and potential for improvement.

Table 1 gives an overview of the feature types implemented in the three scenarios, and how the features were modeled. As mentioned above, our approach is compatible with different ways of modeling features, and this is reflected in our implementations. The type feature was fixed in all cases: for each object, the basic-level type was used with an acceptance score of 1. Therefore, although the usage of the type feature was not subject to decision, it did form part of the description and thus influenced the decision process. Simple features were either created by a local normalization procedure based on absolute numerical values (see below), or handcrafted with given values and acceptance scores. For modeling projective terms, a modified version of the empirically founded spatial model by Moratz and Tenbrink (2006) was used.

4.1   Referring Expressions with Vague Terms

The first example scenario is inspired by the "tall fat giraffe" scenario introduced by van Deemter (2006). As discussed in section 2.4, van Deemter's approach does not cover combinations of vague adjectives like "the tall fat giraffe" that are not *pareto-optimal*, i.e. where the giraffe in question is not the largest and the fattest, but the only one which is large and fat (van Deemter, 2006).

We implemented our system for a world of dogs of different height and weight (Figure 2), modeling the features *height* and *corpulence* according to our probabilistic framework. For the realization, it is assumed that an object carries a categorical value for each attribute within the given situation, i.e. a dog is either

1	2	3	4	5
the tall skinny dog	the short skinny dog	the tall fat dog	the tall dog	the short fat dog

Fig. 2. Dogs of different height and weight with descriptions generated by the system

Table 2. Exemplary operationalization of dog sizes—height and corpulence

	physical features			description features			
id	**height** h [cm]	**weight** w [kg]	w/h	**height**	$P(D\|x)$	**corpulence**	$P(D\|x)$
1	47	27	0.57	TALL	0.06	SKINNY	0.98
2	35	20	0.57	SHORT	0.81	SKINNY	1.00
3	55	45	0.82	TALL	0.64	FAT	0.98
4	60	45	0.75	TALL	1.00	FAT	0.44
5	34	26	0.76	SHORT	0.88	FAT	0.55

"fat" or "skinny", but not both. Additionally, an acceptability score between 0 and 1 determines how close the dog in question is to the prototype of this value.

Although different approaches for modeling features are compatible with our general framework, for this case study we used context-based normalization. Table 2 shows the exemplary creation of categorical attribute values with graded acceptability scores for height and corpulence (based on the weight/height ratio). Categories are divided using the mean score of all individuals in the scene as a boundary: all dogs smaller than the mean are considered "short", and all dogs taller than the mean are considered "tall". Acceptability ratings are gained by linear normalization centered around the mean, giving the score that deviates the furthest from the mean an acceptability score of 1, and scores close to the mean ratings close to 0. This process is similar to the z-transformations used by Raubal (2004) in the context of landmark selection. Given this scene without positional information, our system generates the responses shown in Figure 2.

Discussion. As the table shows, our methodology enables combinations of vague attributes that are not *pareto-optimal*. If a dog is both large and fat (3), and there is another dog which may be slightly larger, but less fat (4), the system will call dog 3 "the large fat dog" and dog 4 "the large dog". If dog 4 is made larger (65 cm instead of 60), the system calls dog 3 "the fat dog" only, avoiding confusion with dog 4.

At the same time, our system goes beyond the *Nash arbitration plan* by considering the other dogs in the context. If there are other very fat but smaller dogs in the scene, the attribute "large" is used for dog 3 even if dog 4 is considerably larger. For example if dog 4 is again 65 cm tall, and the weight of dog 5

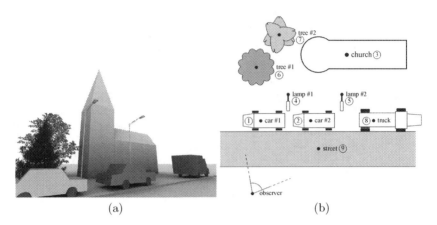

(a)   (b)

Fig. 3. (a) 3D view of street scene example according to Barclay and Galton (2008)
(b) Birds-eye view of the same scene and point abstraction of objects

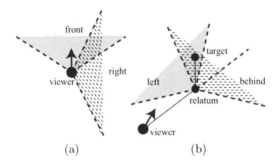

(a)   (b)

Fig. 4. Overlapping acceptance areas for spatial relations (Moratz and Tenbrink, 2006)

is increased to 30 kg, dog 3 is called "the large fat dog", as the attribute "large"
is needed for distinguishing it from the equally fat, but smaller dog 5.

4.2   Simple Spatial Scene

The second scenario was chosen in order to show how spatial relations can be
integrated into our system in a simple spatial setting. We imitate the street scale
scene provided by Galton and Barclay (2008). Figure 3 (a) shows a 3d view of
the scene, and Figure 3 (b) a birds-eye view. As we focus on spatial relations, the
other object features (color and size) and their acceptance values are handcrafted
in this case. Table 3 shows the positions and feature values of the objects used
for the case study. Object types are treated as features and integrated into the
calculation, but each object is obligatorily named with a basic-level type. The
system determines the best description as described in section 3, using $\alpha = 0.1$
(i.e. strong focus on discriminatory power). If several optimal descriptions are
found, it chooses the shortest.

Table 3. Data for the street scene including all features and their acceptance values as well as the referring expression generated by our system

type	position (x, y) [m]	color feature	size feature	description
car	(2,17)	pink (0.8)	small (0.5)	the pink car
car	(15,17)	red (1.0)	small (0.5)	the red car
church	(25,35)	grey (0.5)	large (0.9)	the church
lamp	(9,24)	grey (0.8)	large (0.6)	the lamp to the left of the red car
lamp	(22,24)	grey (0.8)	large (0.6)	the lamp behind the red car
street	(18,11)	grey (0.9)	large (0.7)	–
tree	(2,32)	green (0.8)	large (0.9)	the tree to your left
tree	(5,40)	green (1.0)	large (1.0)	the tree behind the tree to your left
truck	(41,17)	red (0.7)	large (0.7)	the truck

Table 4. Top five descriptions for Objects 3, 7, and 8 of the church scene

rank	Object 3	$P(D\|x)$	Object 7	$P(D\|x)$	Object 8	$P(D\|x)$
1.	the church	1.0000	the tree to your left	0.9529	the tree behind the tree to your left	0.9409
2.	the church in front of you	1.0000	the tree to your front-left	0.9449	the large tree behind the tree to your left	0.9409
3.	the large church	0.9900	the large tree to your left	0.9476	the green tree behind the tree to your left	0.9409
4.	the large church in front of you	0.9900	the green tree to your left	0.9423	the large green tree behind the tree to your left	0.9409
5.	the church behind the red car	0.9524	the large tree to your front-left	0.9404	the tree behind the lamp to the left of the red car	0.9406

With respect to spatial relations, different models can be integrated into our generation system. For the purpose of this paper, we focus on basic projective relations, following an axis-based approach. We assume objects to be point-like, and model prototypes and graded acceptability areas for the relations based on the axes imposed by the reference frame. In doing so, we extend the model provided by Moratz and Tenbrink (2006) for the viewer-intrinsic and the relative reference frame. For the four basic projective relations "left", "right", "front", and "back" we employ the large, overlapping acceptance areas and the prototypes as defined in Moratz and Tenbrink (2006, 89–91). For example, in the intrinsic reference frame, the prototype for "front" is 0°, and the acceptance area ranges from −60° to +60° overlapping that of "right" (Figure 4 a). We also base the graded acceptance values on the formula described there: $cos(\sphericalangle - \sphericalangle^{\text{prototype}})$, which is a smoothing function of the distance to the prototypical angle. We extend the model by including the relative distance between reference object and target. For normalization we compare the distance between target and reference object to the maximal distance between two objects in the scene, yielding the following formula for the acceptability of spatial relations:

$$P(f_{\text{spatial}}|x) = cos(\sphericalangle - \sphericalangle^{\text{prototype}})\left(1 - \frac{dist(o_t, o_{\text{ref}})}{dist_{\text{max}}}\right)$$

Unlike Moratz and Tenbrink (2006), we do not treat combinatory projective terms such as "to the front-right" as simple instances. Rather, we rely on our combinatory mechanism to use either one or two basic spatial relation terms, maximizing discriminatory power. If two basic spatial relations are used (e.g. "front" and "right"), they can be treated as a combinatory projective term for surface realization. Figure 4 (b) shows a target which is contained in the acceptance areas for both "left" and "behind" the relatum, and could therefore be described with either of these relations, or a combination of both.

Further, for the relative reference frame, we use a *principle of same direction* which ensures that the reference object and the target object are in approximately the same direction from the viewer. We again use a cosine function within the accepted maximum of 30° difference in the angle. Table 3 shows the descriptions of the objects in the scene given by the system. Table 4 shows the five best descriptions for objects 3, 7, and 8.

Discussion. As we can see from Table 3 the system generates descriptions without any features as in "the church", with either properties or spatial relations as in "the red car" or "the tree to your left" and combinations of both as in "the lamp behind the red car". In Table 4, we can see that for the church, the option "the church in front of you" is also strongly considered. However, the spatial relation does not add any discriminatory power, and is therefore ignored. The same holds for the first 4 options for object 8: the features "large" and "green" do not further improve identifiability of the tree and are ignored. In the case of object 7 we can see that the same two features minimally reduce discriminatory power—the options "the large tree to your left" and "the green tree to your left" are actually worse than "the tree to your left" as they are not perfect descriptions of the tree.

The set of n-best descriptions gives us a valuable tool both for managing conciseness and for dialogic situations. Firstly, amongst a number of descriptions with (approximately) the same discriminatory power, the system can be set to choose either the shortest description, or one which has x more features. Moreover, in dialogue, variation within the n-best descriptions can be used for responses to clarification requests.

4.3   Realistic Wayfinding Scenario

As the last evaluation scenario, we chose a realistic scenario of an indoor environment as it would occur in a wayfinding task. While our system is currently not supplied with the rich set of spatial features adequate for a task of this complexity, this scenario is interesting to analyze performance of the selection framework itself and to analyze fitness of the spatial models provided. Figure 5 shows the view of the scene as seen by a wayfinder and a floor plan of the area.

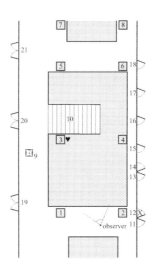

Fig. 5. Indoor test environment inside building GW2 at the University of Bremen

Apart from color and size, two characteristic features were modeled: one pillar had a heart painted onto it, and one door was marked with a large 'X'. Table 5 shows the descriptions generated by the system with $\alpha = 0.1$ and $\alpha = 0.4$, allowing for a maximum chaining of 2 reference objects and 6 features overall (including spatial relations and properties but not object types).

Discussion. For a considerable number of objects, even the limited set of features defined lead to an acceptable description, for example "the door to the left of the door with the X" (object 13) or "the door to the left of the clock" (object 19). But there are also problematic cases which can help to identify problems with our model of spatial relations and to develop refined definitions.

Firstly, issues of surface realization have not been addressed. Usually, one would break down lenghty expressions into several sentences, e.g. for object 1: "There is a door to the left of the clock. The pillar in front of it..."

Secondly, the description for object 7 shows a problem of our model of spatial relations: the subpart "the pillar to the right-behind the pillar with the heart" relates object 5 to object 3, using the combined spatial relation right-behind. However, for a human only the relation "behind" would be acceptable, even though it deviates clearly from the prototypical angle for behind. The reason for this probably lies in principles of *gestalt* perception which lead humans to perceive a row of columns arranged from front to back, leading to a group-based definition of spatial terms.

A similar issue is highlighted by the problematic descriptions of the doors in the right row (objects 14–18). These doors would be hard to describe for humans,

Table 5. Descriptions for the objects in the building scene

object	high discriminatory power ($\alpha = 0.1$)	high acceptability ($\alpha = 0.4$)
1	the pillar in front of the door to the left of the clock	the pillar in front of the door to the left of the clock
2	the pillar in front of the door with the X	the pillar in front of the door with the X
3	the pillar with the heart	the pillar with the heart
4	the pillar to the front-right of the pillar to the behind-right of the pillar with the heart	the pillar to the left of the door to the left of the door with the X
5	the pillar behind the stairs and in front of the door behind the stairs	the pillar behind the stairs and in front of the door behind the stairs
6	the pillar to the right-behind the stairs and to the right of the pillar with the heart	the pillar in front of you
7	the pillar behind the pillar to the right-behind the pillar with the heart	the pillar behind the pillar behind the pillar with the heart
8	the pillar behind the pillar to the right-behind the stairs	the pillar in front of you
9	the clock	the clock
10	the stairs	the stairs
11	the door to the right of the door with the X	the door to the right of the door with the X
12	the door with the X	the door with the X
13	the door to the left of the door with the X	the door to the left of the door with the X
14	the door to your right and to the left of the door to the left of the door with the X	the yellow door
15	the door to the right of the stairs and in front of the door to your front-right	the yellow door
16	the door to your front-right and to the right of the stairs	the yellow door in front of you
17	the door to the right of the stairs and to the right of the pillar with the heart	the yellow door in front of you
18	the door behind the door to the right of the pillar with the heart	the yellow door in front of you
19	the door to the left of the clock	the door to the left of the clock
20	the door to the left of the stairs	the door to the left of the stairs
21	tho door behind the stairs	tho door behind the stairs

but a viable option would be group-based relations using numbers such as "the fourth door on the right". These examples show that group-based relations are important for generating indoor route instructions in realistic scenarios. Principles of *gestalt* perception play a large role and need to be analyzed in detail for a group-based model of relations to work in reality.

Lacking these options, the system produces reasonable alternatives. When α is set to a low value (0.1), the system generates the description with the most discriminatory power, even if this is not a good description of the target object. For example "the door to the right of the stairs and to the right of the pillar with the heart" uses projective terms ranging over long distances in a way that humans cannot distinguish the target from its distractors—an artefact of the graded model of relations. This means the system overestimates the discriminatory power of spatial relations over long distances. With a higher value such as $\alpha = 0.4$ on the other hand, the system produces the best description for the target even if this is not discriminatory ("the yellow door in front of you"), a strategy which is closer to behaviour known from humans but nonetheless not ideal for generating route instructions. The ideal value of α is subject to further research, but we expect it to be approximately at 0.2 to 0.3.

Further, descriptions like "the door to the left of the staircase" show the problem of perspective. Humans tend to use object-intrinsic perspective with objects that have a clearly identifiable front, e.g. the direction one faces when coming into a room from the staircase. The interplay between different frames of reference needs to be considered in a model of spatial relations suitable for realistic scenarios. For example, Carlson-Radvansky and Irwin (1994) show that during interpretation of spatial relations, multiple reference frames are activated and compete. Therefore, relations where reference frames coincide are easier to understand and those where they are in conflict are harder to understand.

5   Summary and Conclusion

In summary, we argue for a scene-based, descriptive approach to generating in-advance indoor route instructions and present a probabilistic framework for comparing and selecting object descriptions. Our approach uses vague models of features and evaluates them based on the principle of discriminatory power.

Our system is able to deal with spatial relations in a more fine-grained way than prior approaches to REG, such as the ones by Garoufi and Koller (2010) or Kelleher and Kruijff (2006). Particularly, by using the concept of *discriminatory power*, our approach takes into account empirical evidence about the relative nature of unique identifiability, and the fact that the relative distance of feature values to those of distractors is an important criterion for feature selection. The potential of this mechanism was demonstrated in three example scenarios. It enables the combination of vague features for cases that are not *pareto-optimal* and enables integration of complex feature models into a general mechanism for REG, as has been shown for spatial relations. In these respects it goes beyond the approach by van Deemter (2006) which only allows simple polar features. Finally, we have demonstrated that our approach is ready for dealing with realistic scenarios. To this end, existing models of spatial relations such as the one by Moratz and Tenbrink (2006) need to be refined and extended. Further, a mature REG system should include group-based and object-intrinsic reference frames, and take into account aspects of gestalt perception.

As our system not only creates object descriptions, but also enables comparisons between them, it also provides a means of selecting those landmarks which can be best described in order to enable simple and effective scene-based route instructions.

While the present work serves to evaluate the overall fitness of this approach, future work will be involved with algorithmically efficient realization. We expect that heuristic search provides the required means. For improving the spatial models, further empirical work will be necessary. Finally, we plan to extend the presented system for dealing with motion instructions to enable the generation of route instructions.

Acknowledgements. This work has been carried out in context of the SFB/TR 8. Financial support by the Deutsche Forschungsgemeinschaft (DFG) is gratefully acknowledged. We thank the I5-[DiaSpace] project group and Daniel Vale for insightful discussions.

References

Allen, G.: From knowledge to words to wayfinding: Issues in the production and comprehension of route directions. In: Hirtle, S.C., Frank, A.U. (eds.) COSIT 1997. LNCS, vol. 1329, pp. 363–372. Springer, Heidelberg (1997)

Barclay, M., Galton, A.: An influence model for reference object selection in spatially locative phrases. In: Freksa, C., Newcombe, N.S., Gärdenfors, P., Wölfl, S. (eds.) Spatial Cognition VI. LNCS (LNAI), vol. 5248, pp. 216–232. Springer, Heidelberg (2008)

Caduff, D., Timpf, S.: On the assessment of landmark salience for human navigation. Cognitive Processing 9, 249–267 (2008)

Carlson, L.A., Regier, T., Covey, E.: Defining spatial relations: Reconciling axis and vector representations. In: van der Zee, E., Slack, J. (eds.) Representing Direction in Language and Space, pp. 111–131. Oxford University Press, Oxford (2003)

Carlson, L.A., Hölscher, C., Shipley, T.F., Dalton, R.C.: Getting lost in buildings. Current Directions in Psychological Science 19(5), 284–289 (2010)

Carlson-Radvansky, L.A., Irwin, D.E.: Reference frame activation during spatial term assignment. Journal of Memory and Language 33(5), 646–671 (1994)

Cuayáhuitl, H., Dethlefs, N., Richter, K.F., Tenbrink, T., Bateman, J.: A dialogue system for indoor wayfinding using text-based natural language. IJCLA 1(1-2), 285–304 (2010)

Dale, R., Haddock, N.: Generating referring expressions involving relations. In: Proceedings of the Fifth Meeting of the European Chapter of the Association for Computational Linguistics, Berlin, Germany (April 1991)

Daniel, M.P., Denis, M.: Spatial descriptions as navigational aids: A cognitive analysis of route directions. Kognitionswissenschaft 7, 45–52 (1998)

van Deemter, K.: Generating referring expressions that involve gradable properties. Computational Linguistics 32(2), 195–222 (2006)

Denis, M.: The description of routes: A cognitive approach to the production of spatial discourse. Cahiers de Psychologie Cognitive 16(4), 409–458 (1997)

Denis, M., Pazzaglia, F., Cornoldi, C., Bertolo, L.: Spatial discourse and navigation: an analysis of route directions in the city of venice. Applied Cognitive Psychology 13(2), 145–174 (1999)

Galton, A.P., Barclay, M.: A scene corpus for training and testing spatial communication systems. In: Proceedings of the AISB 2008 Symposium on Multimodal Output Generation (MOG 2008), pp. 26–29 (2008)

Gapp, K.-P.: Object localization: Selection of optimal reference objects. In: Kuhn, W., Frank, A.U. (eds.) COSIT 1995. LNCS, vol. 988, pp. 519–536. Springer, Heidelberg (1995)

Gapp, K.-P.: An empirically validated model for computing spatial relations. In: Wachsmuth, I., Brauer, W., Rollinger, C.-R. (eds.) KI 1995. LNCS, vol. 981, Springer, Heidelberg (1995b)

Garoufi, K., Koller, A.: Automated planning for situated natural language generation. In: Proceedings of the 48th Annual Meeting of the Association for Computational Linguistics (2010)

Garoufi, K., Koller, A.: Combining symbolic and corpus-based approaches for the generation of successful referring expressions. In: Proceedings of the 13th European Workshop on Natural Language Generation (2011)

Gartner, G., Frank, A., Retscher, G.: Pedestrian navigation system in mixed indoor/outdoor environment - the navio project. In: CORP: 9th International Symposion on Information and Communication Technologies in Urban and Spatial Planning and Impacts of ICT on Physical Space. TU-Vienna (2004)

Hermann, T., Deutsch, W.: Psychologie der Objektbenennung. Huber, Bern (1976)

Hölscher, C., Meilinger, T., Vrachliotis, G., Brösamle, M., Knauff, M.: Up the down staircase: Wayfinding strategies in multi-level buildings. Journal of Environmental Psychology 26, 284–299 (2006)

Kelleher, J.D., Kruijff, G.-J.M.: A context-dependent algorithm for generating locative expressions in physically situated environments. In: Mellish, C., Reiter, E., Jokinen, K., Wilcock, G. (eds.) Proceedings of the 10th European Workshop on Natural Language Generation, SIGGEN. ACL (2005)

Kelleher, J.D., Kruijff, G.-J.M.: Incremental generation of spatial referring expressions in situated dialog. In: Proceedings of the 21st International Conference on Computational Linguistics and the 44th Annual Meeting of the Association for Computational Linguistics, ACL-44, pp. 1041–1048. Association for Computational Linguistics, Stroudsburg (2006)

Klippel, A., Hansen, S., Richter, K.F., Winter, S.: Urban granularities—a data structure for cognitively ergonomic route directions. GeoInformatica, 223–247 (2009)

Krahmer, E., van Deemter, K.: Computational generation of referring expressions: A survey. Computational Linguistics 38(1), 173–218 (2012)

Mainwaring, S.D., Tversky, B., Ohgishi, M., Schiano, D.J.: Descriptions of simple spatial scenes in english and japanese. Spatial Cognition and Computation 3(1), 3–42 (2003)

Mangold, R., Pobel, R.: Informativeness and instrumentality in referential communication. Journal of Language and Social Psychology 7(3-4), 181–191 (1988)

Mast, V., Jian, C., Zhekova, D.: Elaborate descriptive information in indoor route instructions. In: Miyake, N., Peebles, D., Cooper, R. (eds.) Proceedings of the 34th Annual Conference of the Cognitive Science Society. Cognitive Science Society, Austin (2012)

Moratz, R., Tenbrink, T.: Spatial reference in linguistic human-robot interaction: Iterative, empirically supported development of a model of projective relations. Spatial Cognition and Computation 6(1), 63–106 (2006)

Olson, D.R.: Language and thought: Aspects of a cognitive theory of semantics. Psychological Review 77(4), 257–273 (1970)

Olson, D.R., Ford, W.: The elaboration of the noun phrase in children's description of objects. Journal of Experimental Child Psychology 19, 371–382 (1975)

Pechmann, T.: Überspezifizierung und Betonung in referentieller Kommunikation. Ph.D. thesis, Universität Mannheim (1984)

Peponis, J., Zimring, C., Choi, Y.K.: Finding the building in wayfinding. Environment and Behavior 22(5), 555–590 (1990)

Pobel, R., Grosser, C., Mangold, R., Hermann, T.: Zum Einfluß hörerseitiger Wahrnehmungsbedingungen auf die Überspezifikation von Objektbenennungen. Tech. rep., Forschergruppe "Sprechen und Sprachverstehen im sozialen Kontext" Universität Mannheim (1988)

Presson, C.C., Montello, D.R.: Points of reference in spatial cognition: Stalking the elusive landmark. British Journal of Developmental Psychology 6(4), 378–381 (1988)

Raubal, M.: Formalizing conceptual spaces. In: Proceedings of the Third International Conference on Formal Ontology in Information Systems (FOIS 2004), vol. 114, pp. 153–164 (2004)

Raubal, M., Winter, S.: Enriching wayfinding instructions with local landmarks. In: Egenhofer, M., Mark, D.M. (eds.) GIScience 2002. LNCS, vol. 2478, pp. 243–259. Springer, Heidelberg (2002)

Richter, K.-F.: Context-Specific Route Directions—Generation of Cognitively Motivated Wayfinding Instructions, diski edn., vol. 314. IOS Press, Amsterdam (2008)

Rüetschi, U.J.: Wayfinding in Scene Space: Modelling Transfers in Public Transport. phd-thesis, University of Zürich (2007)

Schuldes, S., Boland, K., Roth, M., Strube, M., Krömker, S., Frank, A.: Modeling spatial knowledge for generating verbal and visual route directions. In: König, A., Dengel, A., Hinkelmann, K., Kise, K., Howlett, R.J., Jain, L.C. (eds.) KES 2011, Part IV. LNCS, vol. 6884, pp. 366–377. Springer, Heidelberg (2011)

Tenbrink, T.: Identifying objects on the basis of spatial contrast: An empirical study. In: Freksa, C., Knauff, M., Krieg-Brückner, B., Nebel, B., Barkowsky, T. (eds.) Spatial Cognition IV. LNCS (LNAI), vol. 3343, pp. 124–146. Springer, Heidelberg (2005)

Tenbrink, T.: Relevance in spatial navigation and communication. In: Stachniss, C., Schill, K., Uttal, D. (eds.) Spatial Cognition 2012. LNCS, vol. 7463, pp. 358–377. Springer, Heidelberg (2012)

Vorwerg, C., Tenbrink, T.: Discourse factors influencing spatial descriptions in english and german. In: Barkowsky, T., Knauff, M., Ligozat, G., Montello, D.R. (eds.) Spatial Cognition 2007. LNCS (LNAI), vol. 4387, pp. 470–488. Springer, Heidelberg (2007)

Linked Data and Time – Modeling Researcher Life Lines by Events

Johannes Trame, Carsten Keßler, and Werner Kuhn

Institute for Geoinformatics, University of Münster, Germany
{johannes.trame,carsten.kessler,kuhn}@uni-muenster.de

Abstract. Most datasets on the Linked Data Web impose a static view on the represented entities and relations between them, neglecting temporal aspects of the reality they represent. In this paper, we address the representation of resources in their spatial, temporal and thematic context. We review the controversial proposals for the representation of time-dependent relations on the Linked Data Web. We argue that representing and using such relations is made hard through the direct encoding of inadequate conceptualizations, rather than through inherent limitations of the representation language RDF. Using the example of researcher life lines extracted from curricula vitæ, we show how to model sequences of activities in terms of events. We build upon the event participation pattern from the DOLCE Ultralite+DnS Ontology and show how places and social roles that people play during their careers relate to events. Furthermore, we demonstrate how scientific achievements can be related to events in a career trajectory by means of temporal reasoning.

Keywords: Event patterns, roles, knowledge representation, Linked Data, temporal reasoning.

1 Introduction

Researcher careers are characterized by where and when the researchers have been active. The geographic mobility of researchers plays an important role in expanding scientific knowledge and in forming centers of "scientific gravity" [29]. Not surprisingly, there is also evidence that the mobility of researchers has considerable impact on their access to financial resources and networks [5].

In order to document the careers and achievements of its researchers, the University of Münster has introduced a *centralized research information management system* (CRIS) [19]. CRIS focuses on the representation of information on persons, publications, patents, prices and projects and their interrelation to answer questions such as "Who authored which publications?" The system's query capabilities do not go far beyond such simple questions. Yet, scientific achievements are made while people are engaged in projects and playing a particular role in an organization for a limited and well-defined time. One is therefore often interested in questions of the type "Who did what, when, where (and why)?"

T. Tenbrink et al. (Eds.): COSIT 2013, LNCS 8116, pp. 205–223, 2013.

Table 1. Extract of the curriculum vitæ of Prof. Dr. Andreas Pfingsten, available from
https://www.uni-muenster.de/forschungaz/person/10473

2004–2005	Visiting Professor at University of Illinois at Urbana-Champaign
2001	Visiting Professor at University of Calgary, Canada
1997–1999	Dean and Deputy Dean at Münster School of Business and Economics, University of Münster
since 1994	Professor of Business Administration in Banks and Director of the Institute of Banking at University of Münster
1992	Visiting Professor at University of Graz, Austria
1990–1994	Professor of Economics at University of Siegen

Table 1 shows a career path retrieved from the CRIS database. Assuming that we are able to extract the spatio-temporal information about such a career path by (semi)automatic means,[1] this paper addresses the question how to meaningfully represent the spatial, temporal and thematic references in an academic career on the Linked Data Web. By a meaningful representation we mean in the first place a semantic model that allows for answering the question "when, where and in which role did this person perform what activities during their professional career?", independently of any specific encodings. The model should provide a sound ontological basis for temporal reasoning, so that one can ask which achievements have been made while a person was at a certain place playing a certain role. The model's assumptions regarding space and time need to be anchored in a well-founded ontological theory to improve semantic interoperability with other systems, such as reporting tools or CV generators.

The modeling of researcher career paths serves here as an example to discuss the controversial viewpoints on how time-dependent relations should be modeled on the Linked Data Web and on the role that the Resource Description Framework (RDF) as a knowledge representation language can play. This paper provides an overview of the various existing proposals and of why they fall short of achieving meaningful data integration.

Event-based modeling approaches have attracted attention in several domains and a number of event-centered models have been proposed. While they all highlight the importance of events, they make slightly different assumptions. Additionally, none of them is anchored in an upper level ontological theory. We propose to build on the event participation pattern from the well founded DOLCE+DnS Ultralite Ontology (DUL)[2] instead of inventing yet another event model. We show how to model researcher career trajectories as sequences of events by extending only a few small constructs from the DUL ontology. In this work, we stick to a particular kind of scientific achievement, namely publications. Modeling the act of publishing as an event and separating it from the resulting

---

[1] Automatically extracting entities from textual career descriptions is a research subject in its own right. We have implemented a basic tool to semi-automatically fill our model with real data.

[2] See http://ontologydesignpatterns.org/wiki/Ontology:DOLCE+DnS_Ultralite

publication as an information object provides us with a sound basis for temporal reasoning, which enables us to relate achievements to events in career paths.

The following section reviews different approaches to handle time in knowledge representation languages, followed by an overview and discussion of the different accounts of objects and events in foundational ontologies (Section 3). Section 4 discusses the interplay of events and roles, followed by an overview of our application in Section 5 and conclusions in Section 6.

2 Time in Knowledge Representation

The problem of representing time-varying information on the Semantic Web has led to numerous controversial proposals in hundreds of papers [11]. While there is still no consensus and temporal aspects are largely neglected in practice [35], this section provides an overview of existing proposals.

2.1 Epistemological Knowledge Representation Languages: The Issue of Binary Relations and Time

The Resource Description Framework (RDF) was initially designed to ease the deployment and processing of meta-data about web resources. While RDF has an abstract data model and syntax, providing it with some basic modeling primitives (RDF(S), the RDF Schema extension) and giving it a model theoretic interpretation, it has developed into the de-facto standard language for knowledge representation on the Linked Data Web. The basic abstract concept of the RDF data model is that of a statement in the form of a triple consisting of a subject, predicate and object. The model theoretical interpretation of RDF(S) is strongly monotonic, so that new assertions cannot falsify old ones. Moreover, RDF(S) follows the *Open World Assumption*, which makes no inferences from the absence of a statement.

These properties of RDF(S) make the modeling of time-dependent relationships challenging. If we want to model a person's relation to an organization in the role of an employee, we may introduce two classes, `Person` and `Organization`, and a property `hasEmployee`. Two instances of these classes, *OrganizationXY* and *MisterX*, can be linked by the property `hasEmployee`. Treating the role as a named property or as a subclass of natural type (e.g., modeling `Employee` as a subclass of `Person`) is common practice on the Linked Data Web; however, this solution has some implications, especially concerning the question what happens if the relation changes or if we want to know when and how long the person played that role. In the following, we will review and discuss different approaches treating such time-dependent information in RDF.

2.2 Time as Meta-information

Due to the limitation to binary relations of RDF(S), *RDF reification* and *named graphs* have been proposed as solutions to provide more information about a

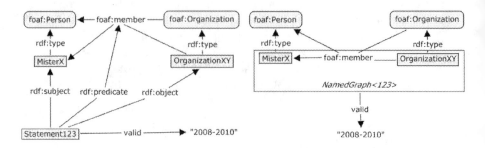

Fig. 1. RDF reification (left) and RDF named graphs (right)

statement (see Figure 1). In the following, we explain these two approaches and argue why they should not be used to account for changes in relationships that reflect changes in the real world.

RDF Reification. While neither the RDF XML syntax nor its model theoretic semantics support direct statements over statements, the RDF reification vocabulary has been introduced to provide meta information about single statements by turning these statements into subjects of other statements.[3] However, RDF reification is one of the most confusing and controversial constructs in the RDF specification, also since its meaning differs from the meaning of reification in conceptual modeling (representing n-ary relationships through binary ones) or linguistics (turning a verb into noun, see Section 3.1). RDF reification blows up the number of statements, makes querying cumbersome, and its model theoretic implications are ambiguous. Consequently, Semantic Web tools generally ignore RDF reification, and its use is discouraged [20].

Listing 1. RDF reification. Example taken from http://spatial.linkedscience.org

```
1  :acmgis/membership840 a rdf:Statement;
2      rdf:object :person666c1cdafc70d38a22b16775df20f004;
3      rdf:predicate foaf:member;
4      rdf:subject :affiliation/affiliation14d23c3ea0e0f08c852c0b329c8e3ee2;
5      dc:date "2008".
```

Listing 1 shows an example of RDF reification in practice [24]. However, in RDF model theory, it does not imply that the person is a member of the organization. Moreover, the property `dc:date` does not have any implications on the model theoretic interpretation of the statement. There is no way to communicate that there is even such information available, nor what it means. The meaning of the reified relation and its attached attribute values have to be defined by the application or by convention. We may query, for example, for all reified statements matching the property `dc:date` and its associated literal "2008":

---

[3] http://www.w3.org/TR/rdf-mt/#Reif

Listing 2. Querying RDF reification in SPARQL

```
1 SELECT * WHERE {
2     ?std a rdf:Statement;
3         rdf:subject ?a;
4         rdf:predicate foaf:member;
5         rdf:object ?b.
6     ?std dc:date "2008".}
```

But then, what does the result mean? Does it mean ?b was member of ?a *during* 2008? or was the statement that ?b is a member of ?a *created* in 2008? or was the statement *valid until* or valid from 2008?

Named Graphs. Because of these semantic problems with RDF reification, metadata on statements is now widely provided through named graphs [6], extending triples to quad-tuples. This development has been driven by the various supersets of RDF syntax and grammar such as N3 or TriG, which extend the triple concept to quad-tuples (or sometimes called context-quads).[4]

Listing 3. Example for a NamedGraph encoded in TriG syntax

```
1 :acmgis/membership840 {
2     :person/person666c1cdafc70d38a22b16775df20f004
3     foaf:member affiliation/affiliation14d23c3ea0e0f08c852c0b329c8e3ee2 .}
4 :acmgis/membership840 dc:date "2008".
```

Named graphs provide a lightweight but useful extension of the initial design encoding and query languages. Named graphs are widely accepted by the community to provide provenance information and they have been incorporated into the W3C recommendation for SPARQL 1.1.[5] Since named graphs are formally subgraphs, they have a clearer syntax and semantics. They are easier to handle and more widely supported by software tools than RDF reification. However, as pointed out by Carroll et al. [6] the semantics are intentionally limited and conform to the basic interpretation of RDF(S) in favor of simplicity. Named graphs are explicitly designed to ease the handling of collections of statements and to attach meta information to those collections, such as provenance or licensing information.

In summary, both RDF reification and named graphs can in principle be used to handle temporally varying information by attaching timestamps to single statements [24] or to (sub)graphs [40]. However, as the temporal dimension stays on a meta-level, it has no implication for the model theoretic interpretation and its existence cannot be stated in the vocabulary. Therefore, named graphs should only be used for temporal information belonging to the meta level, such as provenance data.

2.3 Extensions of RDF(S)

Since both approaches to time are similar to traditional database techniques (time-stamping or temporal versioning), many researchers have taken up that direction,

---

[4] See http://www.w3.org/TeamSubmission/n3/ and
  http://wifo5-03.informatik.uni-mannheim.de/bizer/trig/
[5] See http://www.w3.org/TR/rdf-sparql-query/

trying to provide a model-theoretic account of time. Within the database community, extensive research has been conducted on two notions of time: valid time (when a change occurred in the real world) and transaction time (when a change was entered to the database) [22]. Various proposals adapting the notion of valid time have been made by the Linked Data community, such as temporal RDF graphs (temporal reification vocabulary) [16, 15], multidimensional RDF (extended triple notion) [10], applied temporal RDF (named graphs) [43], stRDF (temporal quad) [25], RDF SpatialTemporalThematic (based on temporal graphs) [34], and temporal quintuples [26].

These approaches have in common that they either extend the RDF syntax or abuse RDF reification or the context quad in order to "label" a triple with a timestamp. The temporal label is then given a model theoretic interpretation modifying the truth value of the statement. This is done by extending the basic entailment rules of RDF(S) or even by moving to a different logic. However, these approaches are in many respects debatable, as they treat the representation of time as a feature of the encoding language. It is unlikely that all existing software implementations will adapt such syntactic and model theoretic extensions. Moreover, time-stamping triples provides no means to share the underlying conceptual model. We can only see changes between different versions, but we fail to explain where these changes in the real world come from. Finally, it is not clear how the notion of valid time interacts with the Open World Assumption, which supports contradicting statements.

2.4   N-Ary Relations

The natural way to deal with the restriction of binary relations in modeling languages is to use the n-ary design pattern. This approach is also known as *conceptual reification* of binary relations. The basic idea is to turn a property into a class and link it to the existing classes via two additional properties. Figure 2 shows an example suggested by the organization ontology.[6] From a practical point of view, n-ary patterns have been criticized because they increase the number of statements and lead to a proliferation of objects. Additionally, the introduction of new (anonymous) individuals through reified relationships causes a maintenance problem and limits the usefulness of RDF and OWL constructs such as domain and range restrictions or inverse property definitions.

The biggest problem, however, is that the potential application scope of the logical n-ary pattern is very broad. The n-ary pattern is often considered as an ad-hoc workaround and thus it is frequently used in an arbitrary fashion, lacking any design rationale. Various modeling solutions are possible, especially since RDF(S), as an epistemological KR language, should be neutral regarding metaphysical and ontological assumptions [13]. One may, for example, treat roles as binary properties or as instances of classes, making ontological commitments that remain implicit [13]. Without any explicit design rationale, reaching even a partial agreement on the conceptual level is unlikely.

---

[6] http://www.w3.org/TR/vocab-org/

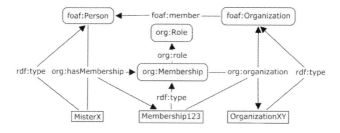

Fig. 2. Example of an n-ary relationship in RDF from the organization ontology

3   Foundational Ontologies: Objects and Events

The use of foundational ontologies has been proposed to make ontological commitments in conceptual models explicit. With its cognitive bias [31], the Descriptive Ontology for Linguistic and Cognitive Engineering[7] (DOLCE), fits the need of the Semantic Web as a frame of reference for building domain ontologies [9]. It organizes particulars and axiomatizes them according to a set of meta-properties [14]. While DOLCE is relatively complex in its full extent, lightweight versions are provided and can be extended in a modular fashion, for example DOLCE-Lite-Plus[7] or DOLCE+DnS Ultralite (DUL).

3.1   Linguistic View on Time

Linguists face the problem of how to capture adverbial modifiers (locative, temporal or instrumental) of action sentences in logic. Sentences such as "Johannes wrote a paper with a pen in the library at midnight" result in predicates of higher arity, such as `write(Johannes,a paper,a pen, the library, midnight)`, where the number of adjuncts for the predicate can become very large. As a solution to this *variable polyadicity* problem of action modifiers, Davidson [7] proposed that action predicates should explicitly range over an additional, normally hidden, event argument: $\exists$e `[write(Johannes,a paper,e)` $\wedge$ `with(e,pen)` $\wedge$ `in(e,library)` $\wedge$ `at(e,midnight)]`. This event argument can then be existentially quantified and bound to the whole sentence, whereas each adjunct will be attached as a separate conjunction clause, thus allowing for an arbitrary number of adjuncts.

Keeping the basic idea of the event argument, the so-called Neo-Davidson approaches have been established as the standard event semantics in linguistics [30]. Parsons [33] advocated to treat the event argument as the only argument of the predicate and link it with a set of thematic roles. Following the proposal by Parson, the structure of our example can be "... replaced by a truly compositional theory of predicates whose fundamental notion is that of event participant and whose fundamental predicates are [thematic roles]" [21, p.444]:

---

[7] See http://www.loa.istc.cnr.it/DOLCE.html

∃e [write(e) ∧ agent(e,Johannes) ∧ location(e,library) ∧ time(e, midnight)]. The decomposition enables one to represent temporal relations directly in first order structures and to reason over them without the need to move, for example, to a temporal or modal logic.

3.2 Ontological Event Theories

A dualistic view, where objects and events complement each other [41], is reflected in most upper-level ontologies such as the Basic Formal Ontology (BFO)[8] or DOLCE [31]. The fundamental distinction is between occurrences and continuants, things that are primarily in space and things that are primarily in time [17, 4]. Objects get then located in time through their participation relation in events and events get located in space through their physical participants. The distinction also helps identifying and separating essential properties from temporally changing ones [38]. Spatial characteristics have an important function for identifying physical objects as well as for defining their unity criteria. Temporal characteristics play the same role for events [4]. We can distinguish two physical objects if they have different spatial locations at the same time. Events are different, if they occur at different times.

Events provide the temporal context in which some relations hold. Roles, in particular, depend on events to come into existence. They carry identity criteria, but do not supply them. Neglecting this may lead to the so-called *isA-overloading*, for example, placing a concept *Employee* under *Person* in the taxonomy) [12]. Extensive work on the representation of roles has been done by Steimann [42], who presented a list of fifteen different features that may apply to roles. In this list, Masolo et al. [32, pp.269–270] identified five characteristics that refer to the dynamic and temporal nature of roles:

(1) an object can play different roles at the same time
(2) an object can change its roles
(3) an object can play the same role for multiple times simultaneously
(4) some roles can be played by different objects, at the same or different times
(5) some roles can only be played by an object depending on the previous roles it played

Loebe and August [28] propose to distinguish between three types of roles, namely relational roles, processual roles, and social roles based on their players and contexts. However, they emphasized that in some cases (for example, for the role *Student*) all three modeling approaches are valid solutions depending on the intention. Masolo et al. [32] argue that the problem with a contextual approach is that the notion of context is still quite fuzzy on its own and one contextual approach may subsume the other. Consequently, they suggest to reify roles and separate them from their specification which accounts for the relational and contextual nature of roles. The object can then be classified for a certain time.

---

[8] http://www.ifomis.org/bfo

This pattern rests on the so called *Descriptions and Situations* (DnS) extension of DOLCE and is also part of the DUL library. While the DnS pattern does not solve all problems of roles, it provides a practical modeling solution that accounts for their dynamic and relational nature.

3.3 Summary

Some researchers see the problem of capturing time in RDF analogously to that of representing time in traditional database systems, since RDF(S) is restricted to binary relations and does not allow to assert when a statement is or was valid in the "real world". Consequently, many proposals have been made that treat time on the meta-level or by extending the syntax and semantics of RDF(S). However, an ontological view reveals that many problems regarding temporal relations result from conceptualizations and from the use of epistemological knowledge representation languages to encode them. Ad-hoc modeling solutions fail to capture the ontological commitments underlying conceptual models.

4 Career Trajectories – Events and Roles

This section focuses on our example problem of how to model researcher career trajectories. The basic idea is to model the career trajectory as a number of events, which relate to certain places and in which researchers play a particular social role. We will shortly discuss existing event models and explain why we do not adapt them, but build upon DOLCE+DnS UltraLite.

4.1 Existing Event Models

Event ontologies have gained attention in recent years. Examples include CIDOC CRM [8], the ABC Ontology [27] or the CultureSampo project [36], where event-based modeling approaches have proven useful to establish a common conceptual reference frame across applications. While these are domain-specific examples, the Event Ontology (EO)[9] aims for more generic event modeling patterns. All these approaches differ slightly in their terminology and regarding their conceptual assumptions. The authors of the LODE ontology, which provides a mapping between different event models, provide a good overview of the differences [39]. While LODE and EO do not account for roles, CIDOC CRM [1] realizes roles as *properties over properties* which cannot be implemented in RDF. Most recently, the Simple Event Model (SEM) [18] has been proposed, as a light-weight model that is neutral with respect to semantic commitments.

However, all concepts and relations in SEM are undefined primitives that are only informally and weakly specified. It is unclear what exactly distinguishes events from objects and whether these concepts are disjoint or not. Knowing this is, however, essential for knowledge sharing, as different conceptualizations

---

[9] http://motools.sourceforge.net/event/event.html

may only be understood by means of a shared understanding of some basic distinctions. Consequently we chose to build on the light-weight version DUL of DOLCE. DUL provides a good combination of flexibility and expressiveness and makes at least the basic distinctions explicit, such as that between events, physical objects and social constructions. In DUL the DOLCE distinction between endurants and perdurants is simplified to a distinction between objects and events, which suffices for most situations and avoids the heavy terminology. Classes have more intuitive names in DUL compared to DOLCE. Moreover, DUL is designed in a modular fashion from which many different ontological content design patterns can be extracted and then combined seamlessly.[10] As we will show, only small extensions are required to DUL to model researcher careers in terms of events. An alignment to DUL provides us with conceptual clarity and serves as a reference frame in order to communicate the basic ontological commitments, which remain ambiguous when using the models mentioned above. Our approach is similar to the F-Event model [37] which also builds on DUL and provides a number of specialized instantiations of descriptions and situations (DnS). By introducing certain types of events and situations, events can be composed to form more complex situations, such as mereological compositions of events or causality relations among events. While this is useful, these patterns add far more complexity than required here. Furthermore, the F-Event model does not suggest how to distinguish among different kinds of participants in an event.

4.2   Basic Design Decisions and Alignment to DOLCE+DnS Ultralite

Events. Events unfold over time, which means that they can be located directly in a temporal region. In DUL, the region values can be directly added to the Event using the hasEventDate[11] property, or the temporal region, such as a TimeInterval, can be separated from the Event and linked via the property isObservableAt, which is a sub-property of hasRegion. While the former approach is simple on the query and application level, the latter provides a "cleaner" and more flexible solution in representing time values from other calendars. Particularly, in the case of temporal intervals we may have a temporal start and end value. These two values belong logically together denoting the boundaries of the interval, rather than the boundaries of the event. Furthermore, by representing the interval explicitly, the model could be extended to account for fuzzy temporal intervals [23].

We follow the second approach here and separate the region values from the event, even if it comes at the cost of simplicity. Since the isObservableAt property in DUL is in the domain of Entity (Objects, Events, Qualities, Abstracts), we introduce a new sub-property eventTime, explicitly linking the Event to its

---

[10] http://ontologydesignpatterns.org/wiki/Category:ContentOP

[11] The property ranges over data values of the XML date schema:
http://www.w3.org/TR/xmlschema-2/#date

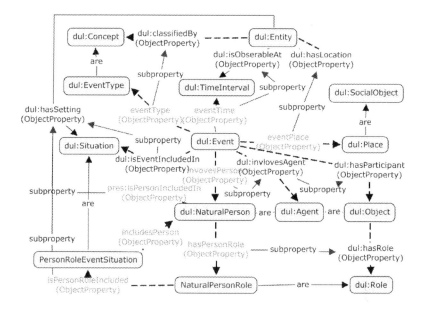

Fig. 3. Overview of the classes and patterns taken from DUL. Extensions to DUL are colored. An RDF(S) encoding is provided online http://vocab.lodum.de/pres/

region `TimeInterval`. In order to link the `TimeInterval` to its region values we introduce the properties `hasIntervalStartDate` and `hasIntervalEndDate` ranging over values from the XML date schema. Both are specializations of the `hasIntervalDate` and `hasRegionDataValue` property in DUL.

Concerning the classification of events, we take a pragmatic approach and link the `Event` to an `EventType` which is a subclass of `Concept`. By this we separate the event from its interpretation, however, we do not provide a situational context. Since the property `isClassifiedBy` is the domain of all entities, we introduce a new sub-property `eventType` with a domain constraint on `Event` and range constraint on `EventTypes`. This allows us to dynamically type event concepts from external vocabularies as `EventTypes`.

Participants in Events. While generally all objects can participate in events via the `isParticipantIn` relation, the primary participants in our case are persons and organizations. Since organizations are hard to extract consistently from CVs, we focus on persons at this point; however, organizations can easily be added later. A `NaturalPerson` in DUL belongs to the category of agentive physical objects because they have intentionality and a proper space region. We specialize the `involvesAgent` property of DUL by introducing a new property `involvesPerson`, with domain constraints on `Event` and range constraints on `NaturalPerson`.

Because we aim to reuse the model in different contexts, we can add further participants as necessary. Participants can then structurally be distinguished based on their functional participation relation (thematic roles). These have already been encoded in the DOLCE Lite extension "Functional Participation Ontology".[12] Thematic roles are realized as sub-properties of the `hasParticipant` property.

Social Roles. We assume that all roles played by researchers during their career belong to the category of social roles. A `Role` belongs to the class `Concept`, as it is externally defined in a description (which we simply assume to exist here). Concepts are social (mental) objects and consequently disjoint from physical agents. The property `classifies` relates the `Role` to any `Entity`. DUL provides a `hasRole` property which constrains the domain to objects. For more domain specific typing, we introduce a new sub-property `hasPersonRole` with domain constraints on `Person` and range constraints on a new sub-class of roles `NaturalPersonRole`. This provides us with a placeholder for the different classes of roles which people may hold, for example, a *Lecturer*.

We introduce a specialized situation called `PersonRoleEventSituation` in order to express that the classification holds in relation to the event, which is quite similar to the *Participant-Role* pattern.[13] Two additional properties are `isPersonRoleIncludedIn` and `isPersonIncludedIn`, so that one can assert OWL cardinality restrictions to the `PersonRoleEventSituation` class.

The way we represent roles is the major difference to existing models like the Simple Event Model (SEM) [18].

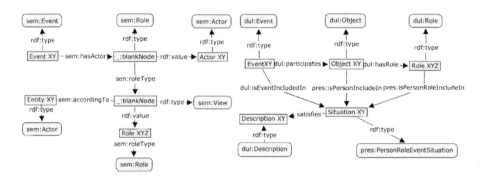

Fig. 4. Comparison of role representations between the SEM model (left) and our role model (right)

SEM shares our idea that social roles of participants in events depend on a particular external view. However, the way SEM models this leads to implementation problems. SEM suggests to replace the actor in an event by a blank node,

---

[12] http://www.loa.istc.cnr.it/ontologies/FunctionalParticipation.owl
[13] http://ontologydesignpatterns.org/wiki/Submissions:ParticipantRole

which can then be typed as a role and links to particular role-type instances, which is again a blank node (see Figure 4). SEM then links from the blank node to the actual participant using the `rdf:value` property, which has no meaning on its own and cannot be constrained with domain and range restrictions. What SEM actually tries is to imitate an n-ary relation, however, by using blank nodes and the `rdf:value` it makes things unnecessarily complicated. Blank nodes are difficult to handle, and using them as an intermediate anonymous node also changes the graph structure. Using the situation setting in DUL provides us with a flexible view on the event, but does not change the logical structure. Furthermore, it clearly separates the event and its participants from the social interpretation.

Place. Since events do not have spatial qualities, they can only be located indirectly through the location of their participants [4]. This can be difficult, however, if participant locations cannot easily be identified. In our example, there is no physical participant that is located inherently. Even if we were able to extract and identify organizations from CV's as participants in an event, the problem would remain that an organization is a social object that by itself does not have a proper space region. Postal addresses might be resolvable to a location in a generic sense (e.g. a political entity). We thus use the social concept `Place` and specialize the generic `hasLocation` by introducing a property `eventPlace`. As the socially constructed concept of place needs to be distinguished from its (approximated) physical region, we introduce a new property `approxGeoLoc`. This is not to say that the place has a region, but that it is approximately located somewhere in a region. While the place might cease to exist, the space region remains.

5   Application

In this section we apply our event model to (semi)automatically extracted data on curricula vitæ of researchers. Exemplary queries show that the model fulfills the basic requirements. Furthermore, we demonstrate how the distinction between objects and events is used to reason about researcher achievements and career events.

5.1   Information Extraction from Researchers' Curricula Vitæ

In the following, we briefly explain how we extracted the relevant pieces of information from the university research database (compare Section 1, Table 1) and how we "filled" our model with this data.

Temporal Information. For each database entry, we create a new event and an interval resource. The person entity to which the database entry belongs is already available through LODUM.[14] We link the event resource to the person resource using the `involvesPerson` property. Since we do not know when specifically a position started or ended, we assume maximal boundaries for the interval (first/last day of the year or month).

Social Roles. Extracting the temporal information is straightforward, but extracting event information from the textual description is more challenging. We implemented some rules according to which we split the description into smaller chunks (punctuations and prepositions). We assume that the role is usually named first and followed by the name and address of the organization or place. For the identification of role concepts, we use DBpedia Spotlight,[15] a tool for automatically annotating textual descriptions with DBpedia.[16] In the case that a role has been identified by DBpedia Spotlight, we take the labels from the DBpedia resource to create new role concepts. Out of the concepts, we create a role resource. At the same time a new instance of the class `PersonEventRoleSituation` is created, which relates the event, the person and the role to each other (see Section 4).

Places. While the Geonames Gazetteer[17] is the most widely used hub for geographic information on the Linked Data Web, its natural language processing capabilities turned out to be too limited for our purpose. Instead, we have used the Google Geocoder API.[18] It is primarily an API for resolving addresses to geographical coordinates, but it also returns a structured list of named entities and their associated types, which have been identified from the input. Out of this list we create a basic place hierarchy, where places are related to each other via the informal `parentPlaceOf` property.

5.2   Querying and Reasoning Capabilities

The core model provides answers to the following questions:

1. In which events does a particular person participate?
2. What are the social roles that a given person plays in an event?
3. Where did the event take place?
4. When did the event start and end?
5. Which concept is assigned to the event?

We formalized and combined several of these questions into one SPARQL query that our prototypical application builds on (see Figure 6).

Since events are time-based and clearly distinguished from objects, we can establish temporal relations between them. Based on Allen's interval calculus [3],

---

[14] Linked Open Data University of Münster; see `http://lodum.de`

[15] `https://github.com/dbpedia-spotlight/dbpedia-spotlight/wiki/Web-service`

[16] `http://dbpedia.org/`

[17] `http://www.geonames.org/`

[18] `https://developers.google.com/maps/documentation/geocoding/`

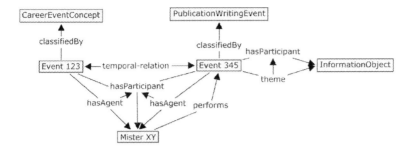

Fig. 5. InformationObjects as functional participants (thematic roles) in events

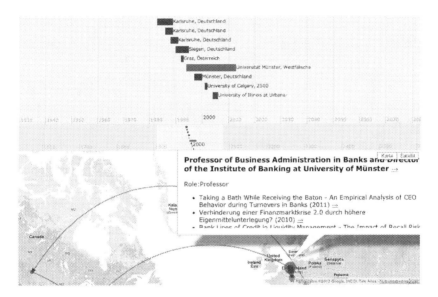

Fig. 6. Prototype application to visualize a researcher career trajectory on a timeline, along with a map of associated places. The circle sizes on the map approximate the number of publications that have been written by the person during the time period the person played a role at a place; see http://data.uni-muenster.de/php/timemap/timemap.html

some basic relations have already been encoded as OWL properties by others.[19] In order to infer these relations automatically, we have formalized them by inference rules using the Jena Rule Syntax.[20,21]

Generally, temporal relations exist between all events, independently of their participants. However, here we are interested in temporal relations between

---

[19] http://www.loa.istc.cnr.it/ontologies/TemporalRelations.owl

[20] http://jena.apache.org/documentation/inference/index.html#RULEsyntax

[21] http://vocab.lodum.de/pres/temporal.rules

events involving the same participant. Therefore, we restrict the inference rules to those situations, reducing the computational overhead. The relations enable simple queries such as "which event precedes this event?" or "which events happen during this event?".

In the context of our use case, this means that we can relate the career events temporally to other kinds of events, for example, to events through which scientific achievements have been made. In order to exemplify this idea, we construct some new events from the meta-data about publications. We call them `PublicationWritingEvents` and create them by making the simplified assumption that an article is written within at most a year before it is published. For representing these events, we apply the basic Event-Participation Pattern. We separate the publication as an `InformationObject` from the event and link it as a participant via the thematic role `theme-of`[12] to the event. The person participates also by a thematic role `performs`[12] in the event.

Temporal relations between the "Career Event" and the "Publication Writing Event" can be derived from the aforementioned set of temporal inference rules. This way we can retrieve, for example, all publications that have been written while a person played a certain role at a certain place. As a demo for the querying capabilities of our model, we have implemented an explorative user-interface which combines a timeline view with a map interface, see Figure 6.

6   Conclusions

We have discussed a variety of proposals regarding the representation of time-dependent relations on the Linked Data Web. Many of these proposed solutions are inspired by the database community, which traditionally treats time at the time stamp level, for example, by assigning validity time stamps to facts. However, doing so fails to provide the means for sharing conceptual assumptions about events with other agents. One can only try to infer real world changes from version changes.

As stated by Allen and Ferguson [2, p.535] "... events are primarily linguistic or cognitive in nature. That is, the world does not really contain events. Rather, events are the way by which agents classify certain useful and relevant patterns of change." This creates all sorts of semantic interoperability problems if event conceptualizations are not made explicit. Treating events as first-class citizens is an essential requirement to make changes traceable [44], as events are the observables in which the semantics of many relationships can be grounded.

As a practical application of our approach, we have shown how researcher career trajectories can be modeled by means of events. We explained how these events relate to places and to social roles that researchers play while participating in events. We built our approach upon the Dolce+DnS Ontology. DOLCE UltraLite does not only provide conceptual clarity, but also servers to communicate basic ontological distinctions (like those between objects and events, or different kinds of event participants) to users of temporal data. These distinctions create a sound basis for temporal reasoning that relates different career events, but also to relate these events to achievements during a career. By extracting relevant data from curricula vitæ and feeding them into our model, we have demonstrated that the event centered modeling approach provides a solution to represent the spatial, temporal, and thematic references in academic life lines meaningfully and usefully. Compared to underspecified event models and workarounds such as RDF reification or named graphs, our solution provides both unambiguous semantics and straightforward querying of time-dependent properties.

The next steps in this research will be to address vague intervals. A clean modeling pattern for such intervals is especially relevant for events where the exact time frame is unknown, but the ordering relations are known – e.g., a paper is always written *before* it is published. On the implementation side, we plan to evolve data annotations in LODUM and spatial@linkedscience [24] to the event model introduced in this paper.

References

[1] Alexiev, V.: Types and Annotations for CIDOC CRM Properties. Digital Presentation and Preservation of Cultural and Scientific Heritage, Report (2011), http://www.ontotext.com/sites/default/files/publications/CRM-Properties.pdf

[2] Allen, J.F., Ferguson, G.: Actions and events in interval temporal logic. Journal of Logic and Computation 4(5), 531–579 (1994)

[3] Allen, J.F.: Maintaining knowledge about temporal intervals. Communications of the ACM, 832–843 (1983)

[4] Borgo, S., Masolo, C.: Foundational choices in DOLCE. In: Staab, S., Studer, R. (eds.) Handbook on Ontologies, 2nd edn., pp. 361–381. Springer, Heidelberg (2009)

[5] Cañibano, C., Otamendi, J., Andújar, I.: Measuring and assessing researcher mobility from CV analysis: the case of the Ramón y Cajal Programme in Spain. Research Evaluation 17(1), 17–31 (2008)

[6] Carroll, J.J., Bizer, C., Hayes, P., Stickler, P.: Named graphs. Web Semantics: Science, Services and Agents on the World Wide Web 3(4), 247–267 (2005)

[7] Davidson, D.: The logical form of action sentences. Essays on Actions and Events (1967), http://ontologics.net/download/dot/Publication/sub_Davi1967a.pdf

[8] Doerr, M.: An Ontological Approach to Semantic Interoperability of Metadata 24(3), 75–92 (2003)

[9] Gangemi, A., Guarino, N., Masolo, C., Oltramari, A., Schneider, L.: Sweetening Ontologies with DOLCE. In: Gómez-Pérez, A., Benjamins, V.R. (eds.) EKAW 2002. LNCS (LNAI), vol. 2473, pp. 166–181. Springer, Heidelberg (2002)

[10] Gergatsoulis, M., Lilis, P.: Multidimensional RDF (Epeaek Ii), pp. 1188–1205 (2005)

[11] Grandi, F.: An Annotated Bibliography on Temporal and Evolution Aspects in the Semantic Web. Tech. rep., Universita di Bologna, Bologna, Italy (2012), http://timecenter.cs.aau.dk/TimeCenterPublications/TR-95.pdf

[12] Guarino, N.: Formal Ontology and Information Systems. In: International Journal of Human-Computer Studies - Special Issue: The Role of Formal Ontology in the Information Technology, pp. 3–15. IOS Press (1998)

[13] Guarino, N.: The ontological level: Revisiting 30 years of knowledge representation. In: Borgida, A.T., Chaudhri, V.K., Giorgini, P., Yu, E.S. (eds.) Conceptual Modeling: Foundations and Applications. LNCS, vol. 5600, pp. 52–67. Springer, Heidelberg (2009)

[14] Guarino, N., Welty, C.: A formal ontology of properties. Knowledge Engineering and Knowledge Management Methods, Models, and Tools, 191–230 (2000)

[15] Gutierrez, C.: Introducing time into RDF. Knowledge and Data Engineering 19(2), 207–218 (2007)

[16] Gutierrez, C., Hurtado, C., Vaisman, A.: Temporal RDF. The Semantic Web: Research and Applications, 167–199 (2005)

[17] Hacker, P.M.S.: Events and Objects in Space and Time. Mind XCI(361), 1–19 (1982)

[18] Hage, W.V., Malaisé, V., Segers, R.: Design and use of the Simple Event Model (SEM). Web Semantics: Science, Services and Agents on the World Wide Web 9(2), 128–136 (2011)

[19] Herwig, S., Höllrigl, T., Ag, A.: All roads lead to Rome: Establishing Best Practices for the Implementation and Introduction of a CRIS: Insights and Experiences from a CRIS Project at the University of Münster. In: Jeffery, K.G., Dvoák, J. (eds.) Proceedings of the 11th Int. Conference on Current Research Information Systems, Prague, Czech Republic, pp. 93–102 (2012)

[20] Hogan, A., Umbrich, J., Harth, A., Cyganiak, R., Polleres, A., Decker, S.: An empirical survey of Linked Data conformance. Web Semantics: Science, Services and Agents on the World Wide Web 14, 14–44 (2012)

[21] Hornstein, N.: Events in the Semantics of English: A Study in Subatomic Semantics. Mind & Language 8(3), 442–449 (1993)

[22] Jensen, C.S., Snodgrass, R.T.: Temporal data management. IEEE Transactions on Knowledge and Data Engineering 11(1), 36–44 (1999)

[23] Kauppinen, T., Mantegari, G., Paakkarinen, P., Kuittinen, H., Hyvönen, E., Bandini, S.: Determining relevance of imprecise temporal intervals for cultural heritage information retrieval. International Journal of Human-Computer Studies 68(9), 549–560 (2010)

[24] Keßler, C., Janowicz, K., Kauppinen, T.: spatial@linkedscience – Exploring the Research Field of GIScience with Linked Data. In: Xiao, N., Kwan, M.-P., Goodchild, M.F., Shekhar, S. (eds.) GIScience 2012. LNCS, vol. 7478, pp. 102–115. Springer, Heidelberg (2012)

[25] Koubarakis, M., Kyzirakos, K.: Modeling and querying metadata in the semantic sensor web: The model strdf and the query language stsparql. The Semantic Web: Research and Applications, 425–439 (2010)

[26] Krieger, H.U.: A Temporal Extension of the Hayes/ter Horst Entailment Rules and a Detailed Comparison with W3Cs N-ary Relations 49(681) (2011)

[27] Lagoze, C., Hunter, J.: The ABC ontology and model. In: DC 2001: International Conference on Dublin Core and Metadata Applications 2001, vol. 2, pp. 1–18. British Computer Society and Oxford University Press (2001)

[28] Loebe, F., August, N.: An Analysis of Roles (2003)

[29] Mahroum, S.: Scientific Mobility: An Agent of Scientific Expansion and Institutional Empowerment. Science Communication 21(4), 367–378 (2000)

[30] Maienborn, C., Wöllstein, A.: Event arguments: foundations and applications, vol. 501. De Gruyter (2005)

[31] Masolo, C., Borgo, S., Gangemi, A., Guarino, N., Oltramari, A., Horrocks, I.: WonderWeb Deliverable D18 Ontology Library, WonderWeb Project (final). Tech. rep. (2003)

[32] Masolo, C., Vieu, L., Bottazzi, E., Catenacci, C.: Social roles and their descriptions. In: Procs. of KR 2004, pp. 267–277 (2004)

[33] Parsons, T.: Events in the Semantics of English. MIT Press, Cambridge (1990)

[34] Perry, M.S.: A framework to support spatial, temporal and thematic analytics over semantic web data. Ph.D. thesis, Wright State University (2008)

[35] Rula, A., Palmonari, M., Harth, A., Stadtmüller, S., Maurino, A.: On the Diversity and Availability of Temporal Information in Linked Open Data. In: Cudré-Mauroux, P., et al. (eds.) ISWC 2012, Part I. LNCS, vol. 7649, pp. 492–507. Springer, Heidelberg (2012)

[36] Ruotsalo, T., Hyvönen, E.: An event-based approach for semantic metadata interoperability. The Semantic Web, 409–422 (2007)

[37] Scherp, A., Franz, T., Saathoff, C., Staab, S.: F-A model of events based on the foundational ontology dolce+ DnS ultralight. In: Proceedings of the 5th Int. Conference on Knowledge Capture – K-CAP, pp. 137–144 (2009)

[38] Semy, S.K., Pulvermacher, M.K., Obrst, L.J.: Toward the Use of an Upper Ontology for U.S. Government and U.S. Military Domains: An Evaluation. Tech. rep. (2004)

[39] Shaw, R., Troncy, R., Hardman, L.: LODE: Linking Open Descriptions of Events. In: Gómez-Pérez, A., Yu, Y., Ding, Y. (eds.) ASWC 2009. LNCS, vol. 5926, pp. 153–167. Springer, Heidelberg (2009)

[40] Sheridan, J., Tennison, J.: Linking UK Government Data.. In: Bizer, C., Heath, T., Berners-Lee, T., Hausenblas, M. (eds.) WWW 2010 Workshop: Linked Data on the Web. CEUR Workshop Proceedings, vol. 628, CEUR-WS.org (2010)

[41] Simon, J.: How to be a Bicategorialist. In: Varzi, A.C., Vieu, L. (eds.) Proceedings of the Third International Conference on Formal Ontology in Information Systems (FOIS 2004), p. 60. IOS Press Inc. (2004)

[42] Steimann, F.: On the representation of roles in object-oriented and conceptual modelling. Data & Knowledge Engineering 35(1), 83–106 (2000)

[43] Tappolet, J., Bernstein, A.: Applied temporal RDF: Efficient temporal querying of RDF data with SPARQL. In: Aroyo, L., et al. (eds.) ESWC 2009. LNCS, vol. 5554, pp. 308–322. Springer, Heidelberg (2009)

[44] Worboys, M.F., Hornsby, K.S.: From objects to events: GEM, the geospatial event model. In: Egenhofer, M., Freksa, C., Miller, H.J. (eds.) GIScience 2004. LNCS, vol. 3234, pp. 327–343. Springer, Heidelberg (2004)

Human Spatial Behavior, Sensor Informatics, and Disaggregate Data

Anastasia Petrenko[1], Scott Bell[1], Kevin Stanley[2], Winchel Qian[2], Anton Sizo[1], and Dylan Knowles[2]

[1] Dept. of Geography and Planning, University of Saskatchewan, 117 Science Place, Saskatoon, SK S7N 5C8, Canada
{anastasia.petrenko,scott.bell,anton.sizo}@usask.ca
[2] Dept. of Computer Science, University of Saskatchewan, 110 Science Place, Saskatoon, SK S7N 5C9 Canada
{kevin.stanley,winchel.qian,dylan.knowles}@usask.ca

Abstract. With the increasing availability of tracking technology, researchers have new tools for examining patterns of human spatial behavior. However, due to limitations of GPS, traditional tracking tools cannot be applied reliably indoors. Monitoring indoor movement can significantly improve building management, emergency operations, and security control; it can also reveal relationships among spatial behavior and decision making, the complexity of such spaces, and the existence of different strategies or approaches to acquiring and using knowledge about the built environment (indoors and out). By employing methods from computer science and GIS we show that pedestrian indoor movement trajectories can be successfully tracked and analyzed with existing sensor and WiFi-based positioning systems over long periods of time and at fine grained temporal scales. We present a month-long experiment with 37 participants tracked through an institutional setting and demonstrate how post-processing of the collected sensor dataset of over 36 million records can be employed to better understand indoor human behavior.

Keywords: Indoor tracking, sensor-based data collection, indoor mobility, indoor movement trajectories.

1   Introduction

A great deal of recent research has leveraged the ability of the Global Positioning System to accurately track human movement through predominantly outdoor environments. These data provide high-level schemas that support the detection and representation of movement patterns [1]. With the growing complexity of urban environments, there is interest in conducting similar analysis indoors that would provide researchers a window on indoor movement behavior. Unfortunately, due to a number of GPS limitations, obtaining accurate positioning or tracking results indoors is rarely possible, making indoor mobility, navigation, and movement analysis challenging.

T. Tenbrink et al. (Eds.): COSIT 2013, LNCS 8116, pp. 224–242, 2013.
© Springer International Publishing Switzerland 2013

Indoor movements are constrained for a variety of reasons: the compact nature of the spaces, the limited number of indoor paths, and the short temporal duration of many indoor movement trajectories. Analysis of pedestrian flow could help architects to design, plan, and build indoor facilities. Building managers and urban planners could monitor indoor human behavior for the purposes of efficient public event management, including determination of visitor densities, waiting times, improving wayfinding experience [2] and localization of possible congestion [3]. Geographers and psychologists would find the large amount of route selection, navigation, and wayfinding information useful for examining how we make spatial decisions during our everyday lives. These types of questions can be answered more efficiently if indoor preferences are better known, which requires the examination of spatio-temporal trajectories over extended periods integrated with the layout of indoor spaces.

A number of research methods have been employed to study human indoor behavior. For example, sketch maps are considered a powerful tool for obtaining information about spatial environments [4-8] and indoor wayfinding strategies [9-13]. Although sketch maps cannot fully depict reality or the actual behavior of individuals, they do give researches an opportunity to extract functional information, omit unnecessary details, and provide a representation of the spatial relations between drawn objects [14]. Alongside traditional visual analysis, a number of approaches exist to interpret the results of sketch drawings. In particular, Skubic [15] proposed a methodology for extracting a qualitative model of a sketched route map; Kopczynski [16] described how sketched location can be identified in a large collection of reference data and developed a framework for solving spatial queries within given spatial reference using sketch maps [17]. However aggregating sketches for many people is difficult when the quality of the route representation varies significantly between individuals. Also, the sketch surface usually limits maps to two dimensions, making research in complex multi-floor environments difficult. Finally, age and background of participants can increase variance and quality of a sketch [18]; moreover mapping skills depends on the participant gender [19] and drawing abilities [20], which can significantly decrease validity and generalizability of collected data.

To overcome these limitations, it is important to incorporate additional tools for monitoring human indoor mobility. Such methods should provide high ecological validity, give an opportunity to perform analysis both on macro and micro spatial scales, as well as assure continuous monitoring for extended period of time. Moreover, collected data should be unbiased, meaning that its representation should be independent of participants' age, origin, or gender. Suitably designed tracking solutions can meet the requirements for representing indoor movement and mobility.

Recently, significant effort has been invested in the development of tracking solutions for indoor human mobility [21-24]. In particular, the deployment of an indoor positioning system in conjunction with a multi-sensor smartphone provides a promising solution for tracking movement indoors and uncovering human spatial behavior. However, the limited accuracy of current indoor positioning solutions requires the incorporation of additional tools to correct erroneous positioning and to provide reasonable interpolation and extrapolation when signals disappear or degrade beyond functional utility. Furthermore, such systems, despite collecting data at the individual level, produced large amounts of data. Over the course of our 30-day tracking experiment we collected between 8 and 11 million pieces of information from some of our highest reporting individuals.

We employed the iEpi system [25] to collect data, SaskEPS [26] to determine position, and the Walkable CentreLINE Network (WCN) [27] to limit the scope of possible positions and infer missing data. We have the following research objectives:

1. Develop a framework for capturing pedestrian movement using a combination of an existing indoor positioning system and a multi-sensor smartphone platform.
2. Demonstrate that by employing an indoor navigational network it is possible to generate accurate trajectories even in a noisy environment that is characterized by low fidelity data.
3. Show that derived trajectories can improve the analysis of space usage and positioning in indoor spaces by displaying/extracting traffic volume that corresponds to the most heavily utilized locations inside a building.
4. Explore the existence of cognitive bias. Cognitive bias is the difference between where a person thinks they spend time and where they actually spend time; such a bias would indicate that some research protocols (sketch maps, interviews, free-recall, etc.) might overlook important aspects of our movement behavior, both indoors and out.
5. Demonstrate that traditional data collection methods can be employed for monitoring of human mobility on the macro level, but that their accuracy and level of detail is insufficient for close spatio-temporal data examination and understanding of the behavioral moving patterns when compared to tracking data.

2   Background

2.1   Positioning System

A number of companies and researchers have introduced additional or alternative positioning solutions that work indoors to replace or enhance GPS systems. Cellular network technology can be leveraged; however cellular signals alone produce positioning with errors up to several hundred meters [28]. Methods using Bluetooth, Cellular, Radio Frequency ID (RFID), and ultra wide band, have been introduced for indoor tracking, but their implementation is often labor intensive, targets a small spatial extent, and requires additional infrastructure [29].

WiFi technology can often leverage existing infrastructure to provide similar spatial resolutions to GPS systems [30], becoming one of the most promising technologies for indoor positioning. WLAN (WiFi) technology is widely installed on devices such as laptops, smartphones, and tablet computers. Given the ubiquity of both fixed beacons (WiFi routers) and mobile receivers (WiFi enabled devices), WiFi-based indoor position solutions can be deployed quickly [31].

Many researchers have experimented with WiFi-based positioning systems [31]. Commercial WiFi-based indoor tracking systems include Google, Apple, Skyhook Wireless, WeFi, and PlaceEngine: all work in a similar manner [28]. An application installed on a WiFi-enabled device records available WiFi signals. These signals are stored on the remote server which sends an estimated location back to the device. Examples include RADAR [32], which was proposed by a Microsoft research group and the COMPASS system [33] that relies on WiFi infrastructure and a digital compass to locate a user. In 2010 researchers from the University of Saskatchewan proposed and produced SaskEPS [26]. Similar to RADAR it uses a WiFi beacons and the trilateration algorithm and supports on-device position calculation. Distances for trilateration are measured according to signal strength from "visible" APs whose precisely calibrated, fixed locations are stored in a database. Bell [26] tested SaskEPS's performance and showed the system provides GPS-like indoor positioning accuracy, although performance was influenced by the arrangement of WiFi routers and building structure [27].

2.2   Limitations of the WiFi-Based Positioning Systems

Signal propagation and attenuation caused by non-line of sight (NLOS) routers, multipath signal attenuation, geometric environmental effects, and noise interference from other devices are common sources of degraded accuracy [31] in WiFi positioning systems. Time of day, the number of people in a local area, and obstacles (such as walls or additional devices) can also influence the performance of the system [34]. There is also evidence that users are sensitive to the context of the positioning results and they are aware that some positioning results are untrustworthy [35].

The accuracy of the system is highly dependent on the algorithms employed for location estimation. For example, the fingerprinting method – which uses pattern matching techniques to locate a user on a grid – is negatively affected by the non-Gaussian distribution of the WiFi signal [36]. Router replacement, antennae orientation changes, changes in router array, or episodic attenuation can require recalibration of the system [37]. A fingerprinting database requires repeated time and effort to maintain an acceptable spatial resolution [36]. Trilateration has problems with determining of the distance to each WiFi point. In particular the trilateration process is sensitive to the arrangement of indoor routers [27].

2.3   Indoor Networks for Improving WiFi Positioning

Many of the methods of spatial analysis employed in outdoor settings can be used inside. For example, the results of long-term indoor pedestrian tracking can be used to improve what we know about path selection through constrained spaces [38]. In particular, the generation of an indoor navigation network to correct tracking data from positioning results can improve the "believability" of positioning results. Such correction can take even relatively accurate positioning results and "snap" them to the logical room or hallway centerline. Almost all GPS-based navigation systems are linked to the underlying road infrastructure in this way, where the road is represented as a linear feature to which the estimated location is matched. The map-matching method [39-41] can be successfully adopted for correction of the WLAN positioning if an indoor navigation networks exists.

Generation of the Indoor Networks. Several corporations provide outdoor navigable networks primarily suited for automobile navigation (such as Navteq and TomTom). Such networks facilitate the presentation and calculation of navigable routes between origins and destinations. However, indoor spaces are quite different from outdoors spaces (even network spaces, like indoor hallways compared to outdoor roadways) [42]. The topological structure of a road network is less complex and diverse than the structure of even a simple building [43]. While roads are easily defined as structures that rarely have more than one instance in a single location, buildings should be regarded as 2.5 [44] or even 3-dimensional, demanding a 3D topological model or graph with an extra degree of freedom in the node labels. The creation of an indoor network model depends on the physical constraints of the building, internal design, presence of other indoor restrictions, and specific indoor properties such as stairwells, doorways, elevators, or escalators [45].

The generation of a comprehensive indoor network requires aspects of existing 2D street network models with additional specifications related to topological relationships between elements in 3D space, summarized in [46] as the following:

1. The indoor network should be a dimensionally weighted topology network that represents indoor route lengths and connections in 3D space.
2. The indoor network should represent the indoor navigable routes that results from decomposition of planar polygons.
3. Space centerlines or medial-axes are an appropriate abstraction for indoor navigation networks.

Indoor network structures should be abstracted as a room-to-room connectivity graph [47]. A traditional 2D node-link structure can be employed, where a room is represented as a node and a link is introduced to connect these nodes. To combine floors, the same node-link method can be applied along level connectors. Werner [48] pointed out that navigation data tailored to a specific query can be generated by hand. However such research is suitable only for small test areas and is problematic and time-consuming for more extensive spaces such as multi-floor buildings. Pu [49] demonstrated that automatically extracting geometry and logic models of a building is a complicated task because nodes and links often have to be created manually.

Example of the Existing Network Model. The Medial Axis Transform [50], Generalized Voronoi Graphs [51], and Straight Medial Axis Transform (S-MAT) [52] are the most commonly employed spatial graph building algorithms. The Geometric Network Model – first introduced by Lee [53] – contains connectivity information and allows for shortest path computation. Later, a Combinatorial Data Model [54] was presented for representing topological relations in 3D. Gillieron and Merminod [42] developed an indoor pedestrian navigation system, where a topological model of navigation spaces for route guidance and map matching was employed.

To correct the erroneous positioning data of SaskEPS using map-matching algorithms, a Walkable CentreLINE Network (WCN) for indoor spaces was developed [27]. The WCN was generated as a graph that includes the center position of hallways, corridors, and rooms of selected buildings in the experimental space.

WCN represents a continuous line that combines a set of points corresponding to the boundary of Voronoi polygons, based on corridor edges similar to the algorithm presented by Wallgrün [51]. All line and point features, integrated in a single network, represent a topologic building structure on the university campus. A graph search algorithm such as Dijkstra's [55] can be used to determine the shortest path between a given node and any other node in a graph. Jung [27] tested the application of the WCN for correcting positioning data collected with SaskEPS and concluded that it reduces positioning errors, particularly when results do not fall within a hallway or room boundary.

3   Positioning and Trajectory Experiment

Empirical datasets that represent individual indoor trajectories provide a basis for human spatial behavior research. However, due to limitations of indoor positioning tools and the complexity of human behavior, it is difficult to collect, analyze, and interpret this data. We carried out an experiment to explore the potential of a network model to improve the utility of indoor positioning data for interpreting indoor mobility through space and time. In particular, we explored whether correcting, or "snapping," WiFi-based positioning results to an indoor network (WCN) can a) be applied to raw positioning data in order to improve accuracy and route generation and presentation, b) be used as a tool to generate movement trajectories that provide a better representation of human spatio-temporal mobility, and c) be used to extract unique spatial behavioral patterns that are more difficult to extract with traditional methods.

3.1   Design and Data Collection

We employed Android Phones programmed the with the iEpi application [25] to collect indoor mobility data (Fig.1).

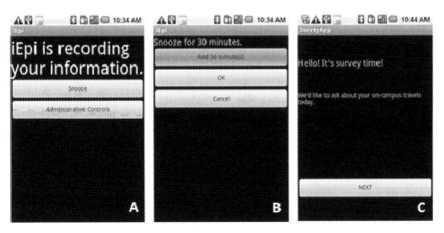

Fig. 1. Interface of the iEpi application. A: Main menu B: Snoozing options C: Survey interface

Thirty seven participants were recruited from a single undergraduate class to ensure that common trajectories between participants would be observed at least twice per week. After receiving study approval from our research ethics board, phones were distributed to participants during orientation sessions; in these sessions participants were shown how to use the phone and completed a pre-survey questionnaire.

As part of the questionnaire, participants were asked to draw a number of sketch maps that indicated their familiarity with the University campus, including the top three on-campus routes they took each day of the week. We used 2D representations of the campus as a map background and did not restrict paths to indoor environments. Example trajectories are presented in figure 2.

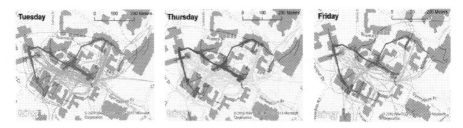

Fig. 2. Aggregate trajectories drawn by participants; darker areas correspond to indoor campus paths (emphasis added)

3.2   Application of Network to Positioning Results

Enhancement of WCN. Initially the WCN [27] was implemented in 2D and contained five campus buildings. To make it suitable for this research, we extended the network model dimensionally (to 2.5D) and expanded it (to the interconnected core of campus, 20 buildings). We used an approach similar to the one presented by Thill [45]. Each of the 20 interconnected buildings represents a multi-level structure subdivided into separate elements representing individual floors. The network consists of the number of nodes and links, where nodes correspond to the central position of the room and decision-points and links represent the medial axis of the hallway polygons that physically connect these rooms and decision points. A similar node-link method is applied along the stairs and elevators to connect building floors. Node and link elements are stored as 2.5D features that allow for the display of differences between various building floors. The 2.5D network model is fully routable. Connections between the floors are allowed only at defined nodes that correspond to the position of stairs, ramps, or elevators (Fig. 3).

Position Estimation. Data collection was consistent with earlier applications of iEpi [25]. To ensure sufficient battery for a full day of monitoring, continuous tracking was not possible. Instead, the phones performed measurements for thirty seconds (referred to as the On Cycle) every two minutes (referred to as the Duty Cycle). Every On Cycle was further divided into 5-second epochs during post-processing. Every Duty Cycle the phones collected 30 seconds of accelerometer records, magnetometer

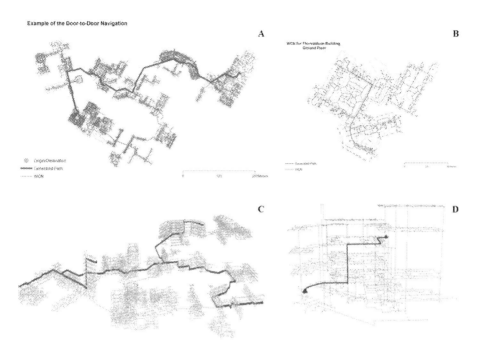

Fig. 3. WCN network with examples of shortest path calculations. A: 2D view entire campus; B: 2D view, ground floor of the Thorvaldson Building; C: 3D view entire campus, path calculation between multiple locations; D: 3D view, Thorvaldson Building

records, Bluetooth contacts, WiFi contacts, and 10 records of battery state. During the 30 day study over 100 million records were made from 6 sensors; magnetometer, accelerometer and Wi-Fi sensors were the most common. Using the University of Saskatchewan secure WiFi network, the phones uploaded data opportunistically. The data was stored initially in flat files and parsed at regular intervals and inserted into a central database. We employed a moving window computation to determine participant location. Within each On Cycle we calculated the position of a participant every second based on the preceding 5 seconds of WiFi router identifiers and received signal strength indicators (RSSI) values (for a total of 25 records per On Cycle). To establish location, this group of unique three-tuples was trilaterated using SaskEPS for all combinations of routers during the 5 second window. The use of duty cycles, while necessary for maintaining measurement longevity, resulted in patchy position updates with up to 25 points available over a 30 second period followed by 90 seconds where no data was available.

Data Post-processing in GIS. To visualize and spatially analyze the output of SaskEPS, the calculated locations were further processed in ArcGIS 10.1. Positioning outliers were removed by identifying the geographic center for a set of recorded points unique for every participant in each one-second interval. Because the data was collected for the first thirty seconds within each two-minute interval, gaps in participant mobility records were evident. To address these gaps, we employed a geometric map-matching algorithm [41] to link filtered SaskEPS points to a corresponding position on the WCN. Prior to snapping, we inserted vertexes along the WCN (with one meter distance between neighboring vertexes) to simplify the snapping process to WCN vertexes rather than segments.

GIS network analysis capabilities allowed for the calculation and generation of a number of shortest/least cost paths that give an appropriate representation of indoor movement trajectories. These algorithms were used to link the strings of known locations calculated by SaskEPS from the 30 second duty cycles. Using ArcGIS 10.1 Network Analyst, we created a tool that generated a shortest path by connecting all daily tracked SaskEPS points recorded for a unique person. This yielded 306 individual paths, which corresponded to an individual's trajectories followed for a single day; these trajectories were later aggregated based on day of the week.

4   Results

The kind of dataset described in this paper can be used to understand human spatial behavior. It represents a compilation of millions of data points for a discrete group of individuals. While the data sizes described here are modest compared to many discussions of Big Data, this is primarily due to the modest N of 37 participants. With the larger study populations possible given the large and accelerating smartphone penetration, data sizes quickly become unwieldy. We believe this type of data represents a unique type of Big Data, one that is disaggregate in form and deep in relevant context. Specifically, we were able to examine route selection, navigation behavior, and collective/shared path choice. After post processing, 180,000 2.5 D location records were available to represent the spatio-temporal trajectories of individuals while on campus.

4.1   Position Tracking Results

During the four weeks of the experiment, participants were tracked in 14 campus buildings (Fig. 4). Figure 4A shows the number of unique times a participant entered a given building, corresponding to visits. Figure 4B shows the number of epochs any participant was found in a particular building and corresponds to an aggregate measure of time spent in the building (in participant-epochs). In general, the location of the participants was consistent with their academic schedule. Most of the identified locations were found in the classrooms and labs of the Arts building and study halls of Murray Library (the main campus library), whereas the Law building and Memorial Hall (unlikely destinations for undergraduates) were the buildings with the fewest, non-zero individual participant visits. Although the Arts and Murray buildings remained in the top two based on the number of counted epochs (meaning that participants were spending considerable time there), results for the Physics building demonstrated that despite the relatively low number of unique participant visits, participants stayed in this building for extended periods of time.

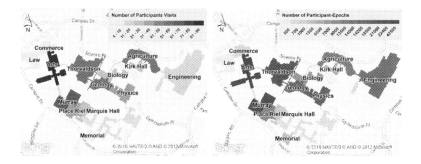

Fig. 4. Results of participants tracking on campus. A: Buildings mostly visited by participants B: Buildings where participants mainly spending time

We performed analysis of the most visited locations separated by floor. In particular, the 1st level of Physics and the 1st floor of the Arts building were the most popular locations among the participants. The graph below displays the ten most visited building floors on campus (Fig. 5).

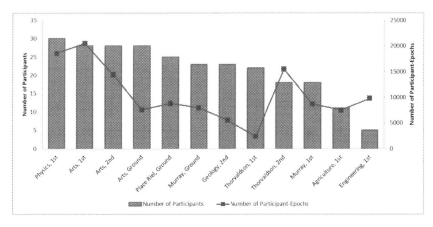

Fig. 5. Most visited building floors on campus

This result allowed us to identify frequently visited locations that were used primarily as transit areas to reach other locations on campus and separate them from destinations at which participants were more likely to linger. The red line on figure 5 indicates such locations. For instance, despite a high number of participants visiting the ground floor of the Arts building, the overall number of epochs is relatively low, implying that participants were crossing this floor to reach their final destination. A similar pattern was evident in the first floor of the Thorvaldson building and the second floor of Geology, which both attach directly to inter-building enclosed skywalks.

4.2   Trajectory Analysis

We selected a single representative day amongst and the top five participants by logged time on campus (Participant IDs = 10, 15, 30, 31, 32). We chose a particular day (not necessarily the same for every participant) where a) the participant had a high number of records and b) she/he visited at least 4 different buildings (Fig. 6).

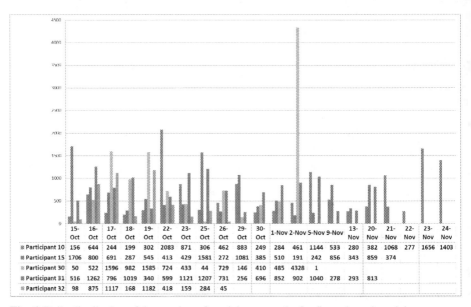

	15-Oct	16-Oct	17-Oct	18-Oct	19-Oct	22-Oct	23-Oct	25-Oct	26-Oct	29-Oct	30-Oct	1-Nov	2-Nov	5-Nov	9-Nov	13-Nov	20-Nov	21-Nov	22-Nov	23-Nov	24-Nov
Participant 10	156	644	244	199	302	2083	871	306	462	883	249	284	461	1144	533	280	382	1068	277	1656	1403
Participant 15	1706	800	691	287	545	413	429	1581	272	1081	385	510	191	242	856	343	859	374			
Participant 30	50	522	1596	982	1585	724	433	44	729	146	410	485	4328	1							
Participant 31	516	1262	796	1019	340	599	1121	1207	731	256	696	852	902	1040	278	293	813				
Participant 32	98	875	1117	168	1182	418	159	284	45												

Fig. 6. Daily distribution of the number of participant-epochs for five selected participants

To better understand an individual participant's mobility we employed a trajectory-based approach and connected corresponding individual locations according to the time stamp of the recording (a time integrated spatial representation) (Fig. 7).

Unfortunately, the results that we initially obtained were not precise enough to perform further analysis. Generated paths were not consistent with building layouts (halls, rooms, entrances, exits, stairs, etc.) in large part because of the intermittent sparseness of the data due to the duty cycling. Higher temporal fidelity of the positioning data would have meant that simple linear interpolation between points was sufficient. However, collecting data at a higher temporal fidelity would have exhausted of the battery in approximately 8 hours. Moreover, most of the generated routes were outside the footprint of campus buildings; such locations are unlikely as WiFi connectivity was recorded, indicating that the participants were indoors or near an indoor location.

Application of map-matching and the WCN significantly improved our results. By linking the tracking position to the corresponded elements on the WCN network, it was possible to reconstruct trajectories (Fig. 8). This also enabled the identification of the most visited locations as well as the temporal duration of these visits.

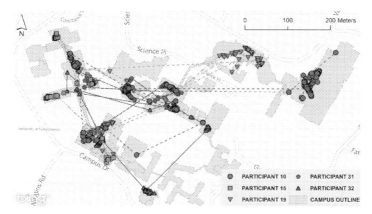

Fig. 7. Raw trajectories before application of map-matching

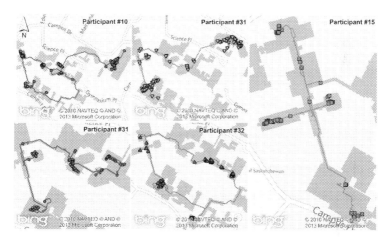

Fig. 8. Adjusted individual trajectories after linking them to WCN

In particular, visualizing the trajectory generated using WCN allowed for the identification of spatio-temporal patterns. For instance, for participant #15 we can easily detect the location of the participant throughout a single day (Fig. 9). In figure 9 bright colors indicate time stamps that correspond to early morning, whereas colors closer to dark red are related to points tracked after 3 pm. By visualizing this data over a base map of campus, it is possible to detect the primarily visited locations, the paths that connect them, and as a result how an individual moves through both time and space. In this case, the trajectory starts at the bus terminal. The participant then moves along the ground floor corridor of the Arts building to reach a classroom located in Commerce. She/he then took the stairs to continue her/his classes on the first floor of Arts. Later, different stairs were used to return to the ground floor of Place Riel and leave campus. It is possible to infer that the first floor of Arts building and ground floor of Commerce were the primary visited locations, whereas the ground floor of Place Riel and Arts building were used as transit corridors to reach the locations of interest.

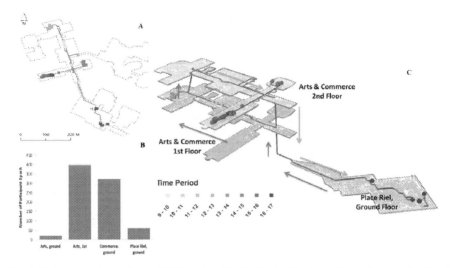

Fig. 9. Trajectory of participant #15 (October 22, 2012). A: 2D view of the trajectory; B: Epoch count for every visited building floor C; 3D view of the trajectories, with the arrows corresponding to the direction of the travel.

4.3   Usage Analysis

Generation of individual trajectories allows for an investigation of traffic flow between various locations on campus. To map the flow, we counted how many times each node of the WCN was visited by participants and represented this count using heat maps. The map in figure 10 shows the aggregated path that participants took on Tuesdays, Thursdays (days when they shared a common class), and Fridays (as a counter example) (Fig. 10). Colors close to light blue represent those campus areas that had fewer visits, whereas colors close to dark blue identify the most popular campus locations. The visual analysis of these three maps demonstrates that the highest number of visits correspond to the location of the hallways that connect various buildings on campus. The skywalks between the Arts and Thorvaldson building and between Biology and Agriculture had the highest number of tracked locations over the three selected weekdays although the number of visits is lower on Fridays.

We compared the trajectories generated by the WCN with the sketch maps that were drawn by participants at the beginning of the experiment. After digitization it was possible to perform an analysis of the sketch maps for various days of the week. From the sketch maps we selected only indoor paths and aggregated them by building. This data can be used to visualize buildings with the highest number of indoor trajectories. The maps in figure 11 show commonly used paths that were identified on sketch maps for Tuesday, Thursday, and Friday with the iEpi/SaskEPS trajectories created from data with the WCN for the same days. Colors in blue correspond to a low number of identified trajectories whereas colors close to red identified the locations visited frequently.

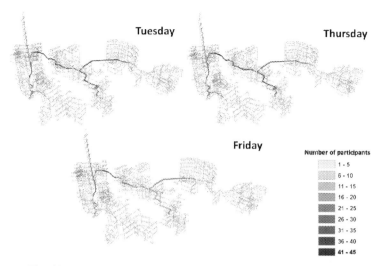

Fig. 10. Usage of campus buildings by participants for different weekdays

By constraining the sketch (as we did here) we could establish common orientation and scale, but we cannot conclude if a drawn path is a simplification of an actual path or the participant's actual or intended path. Many paths appear to be origins and destinations linked by an approximately straight line, others include multiple turns, suggesting greater accuracy or similarity to an actual path. Finally, in the case of our experiment, conducted in a multi-level environment, the sketch maps do not include information regarding floor-to-floor transitions, or floor information of any kind. Likewise, although our sketch maps demonstrate a relatively high number of indoor trajectories in several buildings, it is difficult to identify their trajectories more precisely or detect the vertical transitions of these movements. Moreover, the discrepancy between indoor paths obtained from collected data and trajectories reported by participants' sketch maps demonstrates the existence of cognitive bias. Cognitive bias is the difference between what a person thinks (in this case their knowledge of an environment of their opinion of their movements and behaviors) and what they actually know and do; such a bias would indicate that some research protocols (sketch maps, interviews, free-recall, etc.) might overlook important aspects of our movement behavior, both indoors and out.

In particular, participants did not indicate on their drawn maps, indoor paths that transited intermediary locations such as Thorvaldson and Biology; Place Riel was also rarely identified on the participants' sketch maps; however it corresponded to the location of the Bus Terminal where we tracked a large number of visits. Our results demonstrate a certain spatial variation in the configurations of the generated indoor paths for these locations (Fig. 11 d-f), but when participants drew their indoor routes they failed to include these areas (Fig. 11 a-c). In contrast, the Physics building was marked as one of the most popular locations (Fig. 11 a, b) and corresponded to the

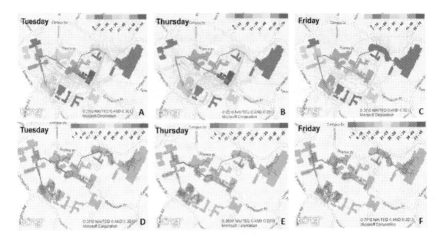

Fig. 11. Usage of the campus for different weekdays. Top three maps: data obtained from sketch maps created by participants; bottom three: paths generated with WCN, 2D view.

location of the class taken by all participants on Tuesday and Thursday. However, this did not completely reflect the actual movement behavior of the participants, despite having to visit this location twice a week, they in fact visited other locations more often; some of these locations were not included on the maps. Although, analysis of the generated trajectories indicated a high number of participants in the Physics building, other areas such as Murray library, Arts building, Agriculture, and the skywalks had similar visit rates, but were not included on sketch maps. Conversely, many participants indicated spending time at the Physical Activities Complex (fitness centre) but were rarely tracked there using the iEpi/SaskEPS tools. It is possible, and likely, that these locations were areas of transition, such as places that were passed through often during the day, or had to be passed through to get to a more important destination (like a class or other meeting location). Capturing transient locations like the skywalks and disambiguating aspirational locations like the PAC is one of the primary advantages of deploying iEpi-like systems for the analysis of institutional spaces. A similar examination of spatial behavior would be more difficult to perform using sketch maps, interview, or environmental observations.

5   Conclusion

Understanding indoor human mobility can significantly improve what we know about spatial behavior and the cognitive processes that underlie it. Such analysis could be performed more efficiently if indoor preferences are better known. This requires the examination of spatio-temporal trajectories integrated with the layout of indoor spaces. However, due to GPS limitations, obtaining movement paths indoors poses a challenge. Our paper outlines the opportunity that multi-sensor and WiFi-based tracking systems offer as a mean for generating accurate indoor trajectories over extended period of time and identifying popular indoor routes.

The results of our study demonstrates a new tool for studying spatial behavior related to navigation and wayfinding as well as results of the correspondence (or lack thereof) between what people think they do and what they are actually doing. It also represents an approach that can generate a trove of spatial information. By connecting the capabilities of iEpi and SaskEPS it is possible to generate accurate indoor movement trajectories. Moreover, we demonstrated that generation of an indoor navigation network (WCN) to correct tracking data from WiFi-based positioning results in an efficient solution for post-processing raw positioning data. We tested our method on the campus of the University of Saskatchewan and showed that the derived movement patterns are consistent with the layout of the building environment, but depicted at a fine-grained spatio-temporal scale. Our approach offers several advantages:

1. Trajectories obtained from automated data collection are not biased by participant recall or embellishment, and can be used to evaluate cognitive bias.
2. The collected data gave us an opportunity to derive locations on campus that are mainly used as transit locations to final destinations (Fig. 5, 9b, 10): areas that are simultaneously characterized by a high number of participant visits and a low number of counted epochs.
3. Derived indoor trajectories can support indoor mobility analysis for various scale-related issues. In particular, our results are accurate enough to depict spatial dynamics either for the entire University campus or within a single building floor.
4. The presented method provides exceptional temporal resolution (Figure 9c).

Overall, the methodology described here will be beneficial for studying different time-scale-dependent indoor patterns, such as congestion, flocking, or hot spots. Results at such fine spatial and temporal scales are unlikely to be identified with sketch maps or observation protocols without substantial effort.

Although our study demonstrated promising results, several shortcomings could be addressed. Our study participants were drawn from a single class section and do not reflect the behavior of the entire University of Saskatchewan population. In the future we will extend our methodology to other, larger populations. Because we had to trade-off battery conservation and data volume, the data was collected for thirty seconds every two minutes, which represented the participant mobility with some systematic level of uncertainty. Future work on improved battery life and adaptive sampling will allow better data acquisition. Additionally, the generation of indoor trajectories in ArcGIS 10.1 can be improved with modification of the algorithm that calculates the shortest path. In our study we did not alter the weighted function and used shortest distance method to generate the trajectories. However with more detailed knowledge of the indoor context and participant preferences it would be possible to introduce additional 'costs' that more accurately interpolate between measured positions. Also, the campus of University of Saskatchewan has dense WiFi coverage; obtaining similar results in the area with lower router density might be not possible. Finally, it will be beneficial to add controlled experiments to our analysis, which will improve the accuracy of the post-processing results and help us better understand the changes in the individual movement behaviour. These limitations should be included as components for improvement in future work.

Acknowledgements. The authors would like to thank Amin Tavassolian, Eishita Farjana, and Wook Jung for assistance in collecting and coding the experimental data, Lindsay Aspen for digitizing the sketch maps, and the University of Saskatchewan

Information Technology Services and Facilities Management Division for sharing access point data. The authors also would like to thank NSERC and the GRAND Network Center of Excellence for providing funding to support this research.

References

1. Laube, P.: Progress in movement pattern analysis. Behaviour Monitoring and Interpretation-BMI-Smart Environments 3, 43–71 (2009)
2. Li, R., Klippel, A.: Wayfinding in Libraries: Can Problems Be Predicted? Journal of Map & Geography Libraries 8(1), 21–38 (2012)
3. Liebig, T., Xu, Z., May, M., Wrobel, S.: Pedestrian Quantity Estimation with Trajectory Patterns. Machine Learning and Knowledge Discovery in Databases, 629–643 (2012)
4. Evans, G.W., Marrero, D.G., Butler, P.A.: Environmental learning and cognitive mapping. Environment and Behavior 13(1), 83–104 (1981)
5. Downs, R.M.: Image & environment: Cognitive mapping and spatial behavior. Aldine (2005)
6. Downs, R.M., Stea, D.: Cognitive maps and spatial behavior: Process and products. Image and Environment 8(26) (1973)
7. Kitchin, R.M.: Cognitive maps: What are they and why study them? Journal of Environmental Psychology 14(1), 1–19 (1994)
8. Tversky, B.: Cognitive maps, cognitive collages, and spatial mental models. In: Spatial Information Theory A Theoretical Basis for GIS, pp. 14–24. Springer, Heidelberg (1993)
9. Harrell, W.A., Bowlby, J.W., Hall-Hoffarth, D.: Directing wayfinders with maps: The effects of gender, age, route complexity, and familiarity with the environment. The Journal of Social Psychology 140(2), 169–178 (2000)
10. Morganti, F., Carassa, A., Geminiani, G.: Planning optimal paths: A simple assessment of survey spatial knowledge in virtual environments. Computers in Human Behavior 23(4), 1982–1996 (2007)
11. Rovine, M.J., Weisman, G.D.: Sketch-map variables as predictors of way-finding performance. Journal of Environmental Psychology 9(3), 217–232 (1989)
12. Moeser, S.D.: Cognitive mapping in a complex building. Environment and Behavior 20(1), 21–49 (1988)
13. Golledge, R.G.: Wayfinding behavior: Cognitive mapping and other spatial processes. JHU Press (1999)
14. Tversky, B.: What do sketches say about thinking. In: 2002 AAAI Spring Symposium, Sketch Understanding Workshop, Stanford University, AAAI Technical Report SS-02-08, pp. 148-151 (2002)
15. Skubic, M., Blisard, S., Bailey, C., Adams, J.A., Matsakis, P.: Qualitative analysis of sketched route maps: translating a sketch into linguistic descriptions. IEEE Transactions on Systems, Man, and Cybernetics, Part B: Cybernetics 34(2), 1275–1282 (2004)
16. Kopczynski, M., Sester, M.: Graph based methods for localisation by a sketch. In: Proceedings of the XXII International Cartographic Conference, ICCC 2005 (2005)
17. Kopczynski, M.: Efficient spatial queries with sketches. In: Proceedings of the ISPRS Technical Commission II Symposium, pp. 19–24 (2006)
18. Passini, R.: Spatial representations, a wayfinding perspective. Journal of Environmental Psychology 4(2), 153–164 (1984)
19. Huynh, N.T., Doherty, S., Sharpe, B.: Gender Differences in the Sketch Map Creation Process. Journal of Maps 6(1), 270–288 (2010)

20. Bell, S., Archibald, J.: Sketch Mapping and Geographic Knowledge: What Role for Drawing Ability? In: Workshop of an Interdisciplinary Approach to Understanding and Processing Sketch Maps in Conjunction with COSIT 2011, Belfast, ME, vol. 42, pp. 1–10 (2011)

21. Stange, H., Liebig, T., Hecker, D., Andrienko, G., Andrienko, N.: Analytical workflow of monitoring human mobility in big event settings using Bluetooth. In: Proceedings of the 3rd ACM SIGSPATIAL International Workshop on Indoor Spatial Awareness, pp. 51–58. ACM (2011)

22. Zhou, Z., Chen, X., Chung, Y.C., He, Z., Han, T.X., Keller, J.M.: Activity analysis, summarization, and visualization for indoor human activity monitoring. Circuits and Systems for Video Technology. IEEE Transactions on Circuits and Systems for Video Technology 18(11), 1489–1498 (2008)

23. Seifeldin, M., Youssef, M.: A deterministic large-scale device-free passive localization system for wireless environments. In: Proceedings of the 3rd International Conference on Pervasive Technologies Related to Assistive Environments, p. 51. ACM (2010)

24. Zhang, D., Liu, Y., Ni, L.M.: Rass: A real-time, accurate and scalable system for tracking transceiver-free objects. In: 2011 IEEE International Conference on Pervasive Computing and Communications (PerCom), pp. 197–204. IEEE (2011)

25. Hashemian, M.S., Stanley, K.G., Knowles, D.L., Calver, J., Osgood, N.D.: Human network data collection in the wild: the epidemiological utility of micro-contact and location data. In: Proceedings of the 2nd ACM SIGHIT International Health Informatics Symposium, pp. 255–264. ACM (2012)

26. Bell, S., Jung, W.R., Krishnakumar, V.: WiFi-based enhanced positioning systems: accuracy through mapping, calibration, and classification. In: Proceedings of the 2nd ACM SIGSPATIAL International Workshop on Indoor Spatial Awareness, pp. 3–9. ACM (2010)

27. Jung, W.R., Bell, S., Petrenko, A., Sizo, A.: Potential risks of WiFi-based indoor positioning and progress on improving localization functionality. In: Proceedings of the Fourth ACM SIGSPATIAL International Workshop on Indoor Spatial Awareness, pp. 13–20. ACM (2012)

28. Zandbergen, P.A.: Comparison of WiFi positioning on two mobile devices. Journal of Location Based Services 6(1), 35–50 (2012)

29. Curran, K., Furey, E., Lunney, T., Santos, J., Woods, D., McCaughey, A.: An evaluation of indoor location determination technologies. Journal of Location Based Services 5(2), 61–78 (2011)

30. Jan, S.S., Hsu, L.T., Tsai, W.M.: Development of an Indoor location based service test bed and geographic information system with a wireless sensor network. Sensors 10(4), 2957–2974 (2010)

31. Gu, Y., Lo, A., Niemegeers, I.: A survey of indoor positioning systems for wireless personal networks. IEEE Communications Surveys & Tutorials 11(1), 13–32 (2009)

32. Bahl, P., Padmanabhan, V.N.: RADAR: An in-building RF-based user location and tracking system. In: Proceedings of the INFOCOM 2000. Nineteenth Annual Joint Conference of the IEEE Computer and Communications Societies, vol. 2, pp. 775–784. IEEE (2000)

33. King, T., Kopf, S., Haenselmann, T., Lubberger, C., Effelsberg, W.: Compass: A probabilistic indoor positioning system based on 802.11 and digital compasses. In: Proceedings of the 1st International Workshop on Wireless Network Testbeds, Experimental Evaluation & Characterization, pp. 34–40. ACM (2006)

34. Chen, Y., Yang, Q., Yin, J., Chai, X.: Power-efficient access-point selection for indoor location estimation. IEEE Transactions on Knowledge and Data Engineering 18(7), 877–888 (2006)

35. Wei, T., Bell, S.: Impact of Indoor Location Information Reliability on Users' Trust of an Indoor Positioning System. Geographic Information Science, 258–269 (2012)
36. Li, B., Salter, J., Dempster, A.G., Rizos, C.: Indoor positioning techniques based on wireless LAN. In: First IEEE International Conference on Wireless Broadband and Ultra Wideband Communications, Sydney, Australia, pp. 13–16 (2006)
37. Ladd, A.M., Bekris, K.E., Rudys, A., Kavraki, L.E., Wallach, D.S.: Robotics-based location sensing using wireless ethernet. Wireless Networks 11(1-2), 189–204 (2005)
38. Harle, R.: A Survey of Indoor Inertial Positioning Systems for Pedestrians, Communications Surveys & Tutorials, pp. 1–13. IEEE (2013)
39. Scott, C.: Improved GPS positioning for motor vehicles through map matching. In: Proceedings of the 7th International Technical Meeting of the Satellite Division of the Institute of Navigation (ION GPS 1994), pp. 1391–1400 (1994)
40. Zhang, X., Wang, Q., Wan, D.: The relationship among vehicle positioning performance, map quality, and sensitivities and feasibilities of map-matching algorithms. In: Proceedings of the IEEE Intelligent Vehicles Symposium, pp. 468–473. IEEE (2003)
41. Quddus, M., Ochieng, W., Zhao, L., Noland, R.: A general map matching algorithm for transport telematics applications. GPS Solutions 7(3), 157–167 (2003)
42. Gilliéron, P.Y., Merminod, B.: Personal navigation system for indoor applications. In: 11th IAIN World Congress, pp. 21–24 (2003)
43. Lorenz, B., Ohlbach, H., Stoffel, E.P.: A hybrid spatial model for representing indoor environments. Web and Wireless Geographical Information Systems, 102–112 (2006)
44. Hölscher, C., Vrachliotis, G., Meilinger, T.: The floor strategy: Wayfinding cognition in a multi-level building. In: Proceedings of the 5th International Space Syntax Symposium, pp. 823–824 (2006)
45. Thill, J.C., Dao, T.H.D., Zhou, Y.: Traveling in the three-dimensional city: applications in route planning, accessibility assessment, location analysis and beyond. Journal of Transport Geography 19(3), 405–421 (2011)
46. Taneja, S., Akinci, B., Garrett, J.H., Soibelman, L., East, B.: Transforming IFC-Based Building Layout Information into a Geometric Topology Network for Indoor Navigation Assistance. In: Computing in Civil Engineering, pp. 315–322. ASCE (2011)
47. Becker, T., Nagel, C., Kolbe, T.H.: A multilayered space-event model for navigation in indoor spaces. 3D Geo-Information Sciences, 61–77 (2009)
48. Werner, M., Kessel, M.: Organisation of Indoor Navigation data from a data query perspective. In: Ubiquitous Positioning Indoor Navigation and Location Based Service (UPINLBS), pp. 1–6. IEEE (2010)
49. Pu, S., Zlatanova, S.: Evacuation route calculation of inner buildings. In: Geo-information for Disaster Management, pp. 1143–1161. Springer, Heidelberg (2005)
50. Lee, D.-T.: Medial axis transformation of a planar shape. IEEE Transactions on Pattern Analysis and Machine Intelligence, 363–369 (1982)
51. Wallgrün, J.O.: Autonomous construction of hierarchical voronoi-based route graph representations. In: Freksa, C., Knauff, M., Krieg-Brückner, B., Nebel, B., Barkowsky, T. (eds.) Spatial Cognition IV. LNCS (LNAI), vol. 3343, pp. 413–433. Springer, Heidelberg (2005)
52. Lee, J.: A spatial access-oriented implementation of a 3-D GIS topological data model for urban entities. GeoInformatica 8(3), 237–264 (2004)
53. Lee, J.: 3D data model for representing topological relations of urban features. In: Proceedings of the 21st Annual ESRI International User Conference, San Diego, CA, USA (2001)
54. Lee, J., Kwan, M.P.: A combinatorial data model for representing topological relations among 3D geographical features in micro - spatial environments. International Journal of Geographical Information Science 19(10), 1039–1056 (2005)
55. Dijkstra, E.W.: A note on two problems in connexion with graphs. Numerische Mathematik 1(1), 269–271 (1959)

Comparing Expert and Non-expert Conceptualisations of the Land: An Analysis of Crowdsourced Land Cover Data

Alexis Comber[1], Chris Brunsdon[2], Linda See[3], Steffen Fritz[3], and Ian McCallum[3]

[1] Department of Geography, University of Leicester, Leicester, LE1 7RH, UK
[2] Department of Geography, University of Liverpool, Liverpool, L69 3BX UK
[3] International Institute of Applied Systems Analysis (IIASA), A-2361, Laxenburg, Austria
ajc36@le.ac.uk, Christopher.Brunsdon@liverpool.ac.uk,
{see,fritz,mccallum}@iiasa.ac.at

Abstract. This research compares expert and non-expert conceptualisations of land cover data collected through a Google Earth web-based interface. In so doing it seeks to determine the impacts of varying landscape conceptualisations held by different groups of VGI contributors on decisions that may be made using crowdsourced data, in this case to select the best global land cover dataset in each location. Whilst much other work has considered the quality of VGI, as yet little research has considered the impact of varying semantics and conceptualisations on the use of VGI in formal scientific analyses. This study found that conceptualisation of cropland varies between experts and non-experts. A number of areas for further research are outlined.

Keywords: Volunteered Geographical Information (VGI), Land Cover, Geo-Wiki, Geographically Weighted Kernel.

1 Introduction

The increase in web-based technologies has resulted in many new forms of data creating and sharing. Individual citizens generate, upload and share a wide range of different types of data via online databases, much of which increasingly has a spatial reference. There is the potential of what is referred to as volunteered geographic information (VGI) [1] to change the nature of scientific investigation as one of its critical advantages of VGI is the potential increase in the volumes of data describing spatially referenced phenomena. Such citizen science activities are supported by the rise of digital, location enabled technologies which offer increased opportunities for the capture of spatial data and for citizens to share information about all kinds of processes and features that they observe in their daily lives.

One of the critical issues to be overcome for crowdsourced data to be included in scientific investigations relates to the quality of the information. In much scientific research data are collected under a formal experimental design which frequently includes consideration of sampling frameworks, quality assurance checks etc. Data

T. Tenbrink et al. (Eds.): COSIT 2013, LNCS 8116, pp. 243–260, 2013.
© Springer International Publishing Switzerland 2013

are collected using well-established methods, by particular instruments or by people with appropriate training and expertise. Thus one of the critical issues in using crowdsourced data relates to the nature of crowd, their familiarity with the domain under investigation and consideration of any impacts of a lack of expertise on the quality of the data that are collected.

This paper generates measures of correspondence between crowdsourced data about land cover / land use and different global land cover datasets. The correspondences between locations identified as being 'cropland' by volunteers was statistically related to measures of the proportions of cropland at those locations from the global datasets. The largest correspondences were used to infer which global dataset best predicted the cropland identified by the crowd. The variation in these correspondences and inferences were examined by considering data contributed by remote sensing experts and by non-experts – ie with contrasting degrees of domain familiarity. In this way the paper explores the impacts of the differences between naïve and expert contributors of VGI and the conceptualisations of landscape features that they hold. This study uses data on land cover, collected as part of the Geo-Wiki project (www.geo-wiki.org), where the registration process captured self-reported measures of contributor expertise. The crowdsourced land cover data captured in this way has a number of potential applications and could, for example, be used to train or to validate statistical classification of remote sensed data. The results of this analysis will inform future uses of VGI by determining whether the differences in expertise about the domain under consideration (in this case remote sensing of cropland land cover) are important and should be considered in such future work.

2   Background

There are many examples of crowdsourced data being exploited that have resulted in novel scientific discoveries such as unravelling protein structures [2], discovering new galaxies [3], reporting of illegal deforestation [4]. The use of VGI is now commonplace in many areas of scientific investigation, from conservation [5] to urban planning [6], and VGI has been found to be particularly useful in endeavours to manage and understand important emerging problems such as ash dieback [7] and post-disaster damage assessments [8, 9]. The latter exemplify the critical advantage of crowdsourced data: the ability to rapidly collect and share large volumes of data describing many kinds of phenomena. These activities are facilitated by ubiquitous ability to capture and share data using many electronic devices (e.g. digital cameras, smartphones, tablets, etc) that are location-enabled, eg through in-built GPS capabilities. The result is an increasing amount of spatially referenced or geo-located data, captured through ubiquitous and low costs citizen owned sensors, that can directly and instantaneously capture and share data of the immediate environment, that are available for formal scientific analysis. This presents a number of challenges associated with use of VGI in formal analyses that relate to questions over data quality and reliability:

- Data are not collected as part of a controlled experiment [10] which may include experimental design, sampling frameworks, validation and error assessments, etc.
- There is no control over what or how information is recorded [11].
- The nature and ability of volunteers can vary greatly in their expertise in subject matter being recorded and may even be malicious [12].

For these reasons consideration of VGI or crowdsourced data quality has attracted increasing attention [11, 13-15] because the usefulness of crowdsourced data for incorporation into scientific analyses depends on its reliability and credibility. In comparison to the designed experiment there is no control over the information that is recorded [14, 16-18] and conflicting information make difficult for analysts to use the information [19] with confidence. One of the frequently employed approaches to overcome a lack of data quality reporting is to use information from a large number of contributors and a number of crowdsourcing projects show a positive relationship between the reliability of contributed data and the number of participants [20], supporting Linus' law [14]. However, other work has shown that providing more data from more contributors may be unhelpful if the volunteers are all similarly confused [21], or if the feature or process being described is associated with alternative conceptualisations, in which case additional contributions may dilute the usefulness of the data. For example, Pal and Foody [22] note that the curse of dimensionality or the Hughes effect [23] is commonly encountered in remote sensing, where the accuracy of a mapping project may decline as the volume of data increases. In order to bridge between these extremes, a number of studies have been undertaken that have sought to develop measures of contributor reliability. These include approaches based on rating data [24], quality assurances based on inputs from trusted individuals acting as gatekeepers [13], random coefficient modeling and bootstrapping approaches to address irregularities in phenology data [10] and the use of control data, where the features under investigation was known, to allow measures of contributor reliability to be generated [11]. More recent research has sought to characterise the reliability of individual volunteers for example through the application of latency measures [25, 26]. These provide an intrinsic measure of contributor quality derived entirely from the data itself and without any additional information such as reference data.

The approaches described above all focus on the *quality* of volunteer and the data they contribute. As yet little research has considered how the background and characteristics of the contributor, citizen or volunteer influences their conceptualisation of landscape features and the impacts of vary conceptualisation on the data this is contributed. It is well know that many landscape features are conceptually uncertain, are conceptually vague or indeterministic in nature, and that different individuals will conceptualise and describe the landscape in different ways depending on their background. Vague interpretations of objects and their boundaries have been considered extensively through the concepts of fiat and bone fide boundaries, corresponding to fiat and bone fide geographic objects [27-29]. More recent research has sought to formalize frameworks for describing and managing alternative conceptualisations of landscape features. These include the use of similarity measures to describe semantics and meanings associated with geographic information retrieval [30] and ontology frameworks [31] for understanding and grounding constructed or differently conceived objects. The latter include mechanisms for integrating different landscape conceptualisations and for describing the associated uncertainty. This research extends consideration of uncertainties associated with varying conceptualisations of the landscape as recorded in crowdsourced geographical information and the impacts on decision making.

3   Case Study: Conceptualisations of Land Cover

Land cover and land cover change have been found to be important variables in understanding land-atmosphere interactions and particularly the impacts of climate changes [32]. A number of different global datasets describe land cover but with considerable disagreement between them in the amount and distribution of different types of land cover features particularly in relation to forest and cropland. This is a long-standing problem, one that has been recognised since the emergence of different global datasets in the early 1990s: they describe significantly different amounts of land cover, for example with differences as great as 20% in the amount of land classified as arable or cropland [33]. The potential errors and uncertainties associated with these products mean that their input into applications such climate models is questionable. They certainly cannot be used to model land cover change. In other research the VGI on land cover as collected through the Geo-Wiki has been used to identify and to validate land cover [34] and to determine which global datasets best correspond to volunteered land cover class labels in different locations – an illustrative example is given in [11] – using spatially weighted kernels [35, 36].

The VGI on land cover can also be used to explore the varying conceptualisations of the land held by different groups or subsets of volunteers who contributed data to the Geo-Wiki project. A recent submission [37] provides a full analysis of the quality of the expert and non-expert data collected through the Human Impact Geo-Wiki campaign. Using control data points, where the actual land cover was known, they analysed the different degrees to which experts and non-expert volunteers correctly identified the land cover and the degree of human impact using a standard statistical data analysis. They found little difference between experts and non-experts in identifying human impact and noted that experts were better than non-experts at identifying the land cover class [37].

The study presented here extends this analysis. It uses VGI on cropland land cover to infer the best global land cover datasets at each location in two study areas. The suggested extension associated with this research, is that it seeks to explore the relative impacts of different levels of expertise rather than absolute ones which would require the use of control data (or ground truth) as in the previous work cited above. The aims of this study were:

- To determine whether the analysis of data contributed by expert and non-expert groups results in different optimal global datasets being selected in different locations.

- To examine the degree to which expert and non-expert groups of contributors hold different conceptualisations of the landscape.
- To explore whether this matters.

These were explored by comparing the data contributed by experts and non-experts and through analysis of cropland land cover and land use relating to agricultural activities as collected by the Geo-Wiki and as recorded in global land cover datasets.

4   Methods

VGI on land cover were collected through the Geo-Wiki project which incorporates a web-based interface that uses Google Earth [38]. Volunteers are invited to record the land cover at different locations. A screen grab is shown in Figure 1 by way of example. The Geo-Wiki project is a directed form of VGI creation in that volunteers are steered towards a structured reporting / scoring environment, they score the land cover in randomly selected global locations and their scores are stored in a databases. This is contrast to other activities such as OpenStreetMap, which provides a direct experiential link between the volunteer and the data being recorded such that a typical Open Street Map volunteer knows the place that is being updated.

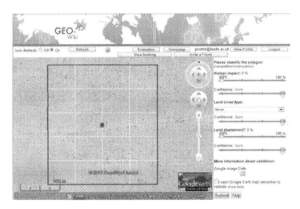

Fig. 1. An illustration of the Geo-Wiki interface

A number of Geo-Wiki campaigns have been initiated to gather different types of information related to specific objectives including agriculture (agriculture.geo-wiki.org), urban areas (urban.geo-wiki.org) and biomass (biomass.geo-wiki.org). The data used in this analysis were collected in the autumn of 2011 in a campaign to collect measures of Human Impact (http://humanimpact.geo-wiki.org). This was developed as a crowdsourcing competition with the aim of collecting sufficient data to validate a spatial dataset of land availability for biofuel production, which was released without any measures of data quality or error and was perceived to misrepresent land availability and thus would implicitly support land grabs in the Global South. As part of their registration, volunteers were asked to provide some background information. Critical for this analysis was, this information included questions about the volunteer's experience, profession and expertise.

Each volunteer was asked to complete an online tutorial in order to demonstrate the process and to provide basic training with explanations of the issues and concepts associated with land cover and land use mapping. The contributors completed as much or as little of the training that they wished to complete. As part of the competition, the top ten participants were offered co-authorship on a paper [39] and Amazon vouchers as incentives [38]. Volunteers were provided with a series of random locations and asked to record what they observed at each location and based on their interpretation of the landscape they assigned each location to one of 10 predefined land cover classes, including a cropland class of *Cultivated and managed*.

These data are considered here for two study areas: South America and Africa. Volunteered information on land cover was captured globally and 17,371 and 6888 data points were recorded in Africa and South America respectively as shown in Figure 2 and Figure 3. Of the data points in Africa, 4562 were recorded as Cultivated and managed and 1390 were similarly recorded in South America. At each location the proportions of cropland land cover as recorded in 5 global land cover datasets were extracted. The global land cover datasets were GlobCover, GLC, MODIS, GeoCover and Hansen and, although other datasets exist (e.g JRC), these datasets were selected to illustrate the analysis this variation in VGI contributed by different groups.

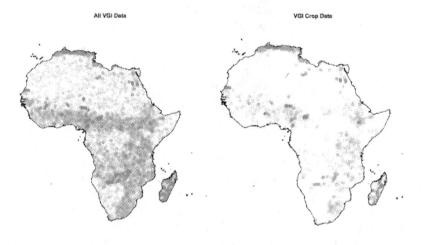

All VGI Data   VGI Crop Data

Fig. 2. VGI data in Africa and the *Cultivated and managed* class used in this study

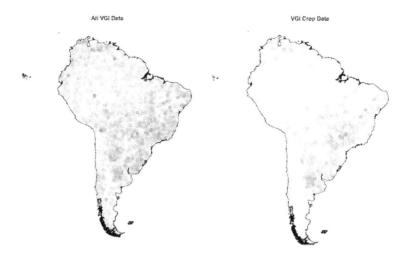

Fig. 3. VGI data in South America and the *Cultivated and managed* class used in this study

Measures of spatially distributed correspondence between VGI cropland and cropland data from the global datasets were generated using the methods described in detail elsewhere [11, 35, 36]. In brief, these approaches use a geographically weighted kernel to generate local correspondence measures calculated at discrete locations on grid covering the study area. This allows local measures of the spatial variations in relationships between the data inputs to be determined. In this case, the analysis considered the presence / absence of VGI cropland against the proportions of cropland cover in the global datasets, when considered simultaneously.

A logistic regression of the binary presence of the VGI cropland class against the proportions of cropland indicated by the global datasets was used to determine which of the different global datasets best predicted the VGI data, using the following calculations.

A *logit* function was defined by:

$$logit(Q) = \frac{exp(Q)}{1+exp(Q)} \tag{1}$$

The logistic geographically weighted regression was then calculated as follows the global land cover datasets:

$$P(y_i = 1) = \text{logit}(b_{0(u_i,v_i)} + b_1 x_{1(u_i,v_i)} \ldots + b_n x_{n(u_i,v_i)}) \tag{2}$$

where $P(y_i = 1)$ is the probability that the VGI cropland cover class, y at location i is correctly predicted, b_0 is the intercept term, $x_{1..n}$ are the proportions of cropland indicated by the 6 global datasets under consideration, $b_{1..n}$ are the slopes and (u_i, v_i) is a vector of two dimensional co-ordinates describing the location of i over which the coefficient estimates are assumed to vary.

The basic idea here is that all of the global dataset proportions are considered as a series of independent variables in the logistic regression. The analyses returns a coefficient for each global dataset, the highest of which indicates the strongest effect in predicting the presence of the VGI data for cropland cover.

5   Results

5.1   VGI to Evaluate Global Datasets

As an initial investigation the relationship between crowdsourced data indicating the presence of the Managed and cultivated class were compared with the proportions of cropland as recorded at those locations in five global land cover datasets. The results of standard logistic regressions for the African and South American data are shown in Table 1 and Table 2.

Table 1. Coefficients of the logistic regression of African crowdsourced cropland data against measures derived from global land cover data

Term	Estimate	Std. Error	z value	Pr(>|z|)
(Intercept)	-1.745	0.025	-69.426	0
GLC	1.807	0.098	18.462	0
GlobCover	0.211	0.074	2.845	0.004
MODIS	1.593	0.103	15.444	0
GeoCover	2.139	0.223	9.578	0
Hansen	1.055	0.07	15.087	0

Table 2. Coefficients of the logistic regression of South American crowdsourced cropland data against measures derived from global land cover data

	Estimate	Std. Error	z value	Pr(>|z|)
(Intercept)	-2.711	0.062	-43.501	0
GLC	1.536	0.161	9.561	0
GlobCover	0.69	0.13	5.294	0
MODIS	1.573	0.176	8.927	0
GeoCover	1.762	0.126	13.954	0
Hansen	0.635	0.152	4.172	0

These show that for both Africa and South America the proportions of cropland recorded by each of the different global land cover datasets are highly significant (<0.01) predictors of volunteered data on cropland (Cultivated and managed). Of interest are the varying degrees to which the different datasets correspond to the VGI: in both cases GeoCover has the strongest relationship. In Africa GlobCover has the weakest relationship and in South America is with the Hansen data and if we were to chose one dataset that best reflects the actual land cover on the ground, then GeoCover would be selected from this set of five global datasets.

However it is also possible to examine how these relationships vary spatially. A geographically weighted logistic regression was used to model the variation in the degree to which the individual global datasets best predicted the VGI data (ie had the highest coefficient) at different locations. In the geographically weighted approach [40], local regression models are estimated at discrete locations throughout the study area, in this case defined on 100km grid. At each location the local model is calculated from the data points under a kernel and their distance from the location under consideration weights their individual contribution to the model. The size of kernel can be predefined or can be automatically determined. In this case kernel sizes were specified to incorporate 0.01 (all cropland data) and 0.02 (expert and non-expert groups) of the data points. The application of these techniques in the context of the VGI on land cover is described in detail elsewhere [11].

The results of using a geographically weighted approach, i.e. extending the ordinary and stationary logistic regression, are shown in Figure 4. These show which of the global datasets best reflect the conceptualisation of the volunteers in each location (100km grid cell). By examining the maps in Figure 4, it is evident that the GLC data is best predicts cropland in areas with little cropland – the absence of cropland – and that the results in areas of cropland indicate that other datasets more closely correspond to the conceptualisations of the land held by volunteers.

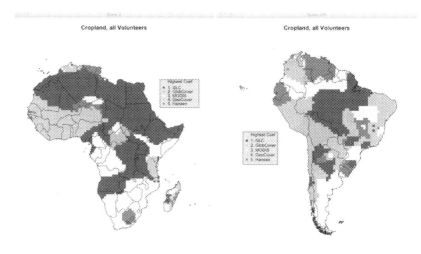

Fig. 4. The global datasets that most strongly correspond to the landscape conceptualisations held by all volunteers in different locations

5.2   Comparing Expert and Non-expert Evaluations of Global Datasets

The analyses described above were rerun, but this time after subsetting the data in order to compare Experts with Non-experts. The results are shown in Figures 5 and 6.

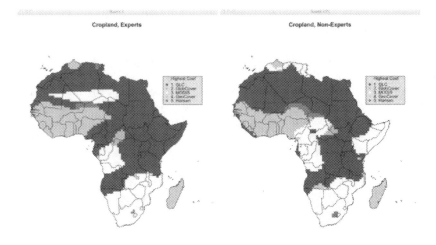

Fig. 5. The global datasets that most strongly correspond to the landscape conceptualisations held by experts and non-experts in Africa

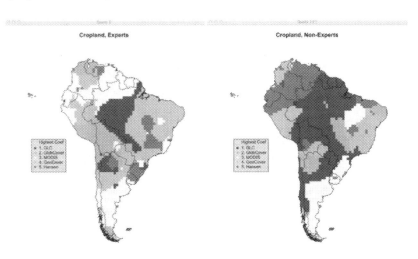

Fig. 6. The global datasets that most strongly correspond to the landscape conceptualisations held by experts and non-experts in South America

These show distinguishable differences in the spatial distribution of the global land cover datasets that most strongly corresponded with the crowdsourced data cropland / Cultivated and managed. The locations of difference are mapped in Figure 7 and it is noticeable there are much greater differences between expert and non-experts and the global land cover datasets they infer in South America than in Africa. This may be due to a number of factors:

- The nature of the land cover. The cropland and cultivated land cover may be more confusing or more difficult to discern in South America.
- The quality of the global land cover data in those areas. There are well known and large variations in the amounts of land cover features reported in different global datasets.

Or, the observed differences may be due to alternative landscape conceptualisations held by the 2 groups. The differences between them can be more formally investigated by considering the sample locations for which land cover was recorded by an expert and a non-expert: 1165 data locations in Africa and 494 in South America.

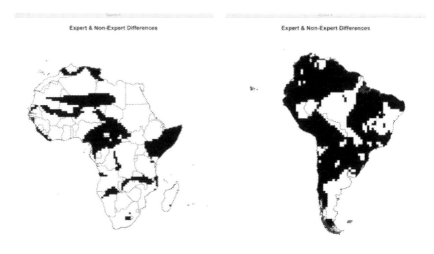

Fig. 7. Differences between the global datasets that best match crowdsourced data on cropland contributed by experts and non-experts

Tables 3 and 4 show correspondence matrices and the per class measures that can be derived from them for the case study areas. If the assumption is that the experts are correct then the equivalent of Type I and Type II error rates can be calculated on a per class basis.

In this context, Type I errors relate to errors of omission (exclusion) for cropland by the non-experts relative to the experts. These indicate the relative probabilities of areas being identified as cropland by experts and not being identified as such by non-experts. These are low (~25%) for Africa and high (~45%) for South America. Considering the Type II errors, the errors of commission (inclusion) by the experts, it is evident that there are similar levels for cropland in both the study areas (~25-30%) which indicates that the experts are including many false positives in their scoring. Of course, it is important to remember that these are *comparative* statements in the way that 'error' is being interpreted, rather than *absolute* statements about error.

Table 3. Correspondence between experts and non-experts in South Africa with the cropland class highlighted

| | | Expert | | | | | | | | | | |
		1	2	3	4	5	6	7	8	9	10	Type II
Non-Expert	1	111	50	26	2	29	2	0	0	0	0	0.505
	2	25	75	22	6	14	1	0	0	10	0	0.490
	3	6	9	28	3	9	3	0	0	9	0	0.418
	4	0	4	4	215	71	0	0	0	3	0	0.724
	5	4	8	5	69	168	0	0	1	2	1	0.651
	6	0	0	0	0	0	0	0	0	1	2	0
	7	0	0	0	0	0	0	1	0	0	0	1.000
	8	1	0	0	0	0	0	0	0	3	0	0
	9	1	6	11	4	1	0	0	1	138	0	0
	10	0	0	0	0	0	0	0	0	0	0	0
	Type I	0.750	0.493	0.292	0.719	0.575	0	1.000	0	0	0	

Table 4. Correspondence between experts and non-experts in South Africa with the cropland class highlighted

| | | Expert | | | | | | | | | | |
		1	2	3	4	5	6	7	8	9	10	Type II
Non-Expert	1	128	19	12	4	19	3	0	1	0	0	0.688
	2	3	12	12	2	3	0	0	0	2	0	0.353
	3	1	8	18	5	12	2	0	0	8	1	0.327
	4	2	0	3	59	19	0	0	1	0	0	0.702
	5	7	1	2	31	59	1	0	0	2	0	0.573
	6	2	0	1	0	0	1	0	0	0	0	0.250
	7	0	0	0	0	1	0	2	0	0	0	0.667
	8	1	1	1	0	0	0	0	2	0	0	0.400
	9	0	1	3	0	1	0	0	1	8	0	0
	10	1	0	0	0	0	0	0	0	0	5	0
	Type I	0.883	0.286	0.346	0.584	0.518	0.143	1	0.400	0	0	

5.3   Volunteer Confidence

The Geo-Wiki on Human Impact also asked volunteers to record their confidence in the land cover class that they allocated to each sample location. Four levels of confidence were available, *Sure*, *Quite sure*, *Less sure* and *Unsure*, and the analysis described below examines whether the differences between experts and non-experts observed persisted when their confidence was Sure. Although other work [37] has shown that experts to be better than non-experts in identifying land cover particularly herbaceous, cropland and mosaic classes, the confidence associated with those class allocations has not been considered. The expert and non-expert data described above were further subsetted to extract the data points that were allocated a Sure confidence attribute. These were then analysed in the same way as above to determine which global dataset best matched the certain VGI cropland data in the two study areas, and maps of difference were created.

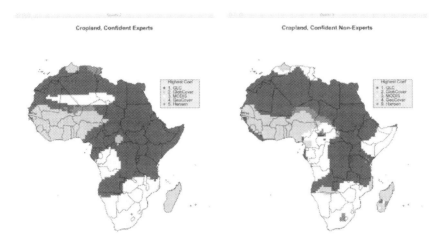

Fig. 8. The global datasets that most strongly correspond to the landscape conceptualisations held by *Confident* experts and non-experts in Africa

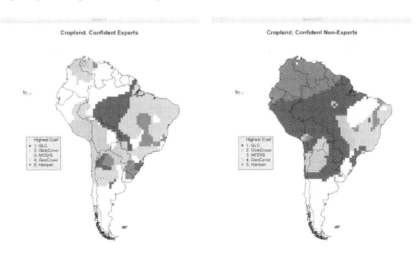

Fig. 9. The global datasets that most strongly correspond to the landscape conceptualisations held by *Confident* experts and non-experts in South America

Figures 8 and 9 show the mappings of the global datasets that most strongly corresponded to the different subsets of VGI data and Figure 10 shows the differences between the expert and non-expert groups. Different levels of confidence were not considered. Interestingly, when these results and figures are compared with Figures 5-7 it is evident that, for Africa, extracting the crowdsourced data where the volunteers were confident (sure) resulted different global datasets being selected by experts and non-experts. In South America it did not.

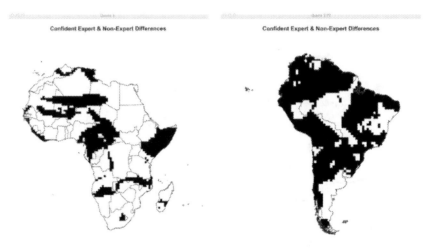

Fig. 10. Differences between the global datasets that best match crowdsourced data on cropland contributed by *Confident* experts and non-experts

5.4   Summary of Results

The aim of this analysis was to explore the varying conceptualisations of cropland land cover held volunteers. The example application was the use of VGI to select from a range of different global land cover datasets, whose quality is known to vary.

Evaluating Global Datasets
For both study areas, the proportions of cropland recorded by each of the global land cover datasets were found to be significant predictors of the presence of crowdsourced data of Cultivated and managed.

GeoCover was found to have the strongest relationship with volunteer conceptualisation of cropland with the GlobCover and Hansen datasets having the weakest relationships in Africa and South America respectively.

The GLC data was found to best predict the absence of cropland being recorded by volunteers.

Comparing Expert and Non-Expert Conceptualisation of Global Datasets
The differences between expert and non-experts conceptualisations of land cover and global datasets measures of cropland were much greater in South America than in

Africa, potentially indicating the impacts of different conceptualisations (eg for data selection) of this feature in these areas.

If the expert data are considered as the referent then low (~25%) errors omission (exclusion) occurred in Africa with high levels (~45%) for South America. Errors of commission (inclusion) were similar in both the study areas (~25-30%).

Volunteer Confidence
The impacts of considering only Sure / certain crowdsourced data on cropland were much greater for the selection of the best global dataset in Africa than in South America.

6   Discussion

There is much interest in methods and applications to take advantage of the large volumes of crowdsourced data that are potentially available to scientific research. The ubiquity of location-enabled devices that attach a geographic reference or geo-tag to such information that is easily shared via various portals and networks provides spatial information scientists with a number of challenges. Not least of these is the need to understand how the concepts embedded in the data that are contributed relate to the intended use of that data. This research has shown that conceptualisations between expert and non-expert groups vary and that decisions made using crowdsourced data (in this global dataset selection) will also vary.

There has been much recent research to understand the quality and reliability of crowdsourced geographical data particularly in relation to the data collected by the Geo-Wick project and methods have been developed for generating error and uncertainty measures [11, 14, 15, 25, 26]. However, as yet little consideration has been given to different conceptualisations of landscape features held by different contributors and the impacts of those. This research has highlighted those differences and their impacts on decisions and the need for further investigation into formal structures to allow such differences to be modelled and reasoned with.

Participation in VGI-related activities may empower citizens to create spatial information and to represent *their* world [41, 42], but may also enhance social disparities across the digital divide as "the favored few "(with digital access) are allowed to exploit "the mediocre many" (without) (quotes from [43]). As spatial data creating and mapping is an inherently politically and socially mediated activity [44], the digital divide results in disparities in the opportunity to create information. Globally, a narrow subset of the world's population has the opportunity to describe what *is there*. For instance 6.2% of total internet use in 2011 was in Africa despite it accounting for 14.7% of the global population[1], and within developed countries the digital divide reflect an urban bias in access to broadband technologies [45].

---

[1] http://www.newmediatrendwatch.com/world-overview/
   34-world-usage-patterns-and-demographics

This suggests a number of areas appropriate for further research that are being developed in on-going research by the authors:

- Analysis of the impacts of volunteer characteristics such as their background, training and socio-economic contexts on their perceptions and what they record.
- Consideration of the digital context within which information is volunteered, the impacts of the digital divide on participation, both in terms of contribution and the extent to which participation has the unintended effect of increasing the digital divide.
- Analysis of the degree to which spatial knowledge is shaped by identity, power, and socioeconomic status and the extent to which is spatial data handling through VGI is socially and politically mediated.

References

1. Goodchild, M.F.: Citizens as sensors: the world of volunteered geography. GeoJournal 69, 211–221 (2007)
2. Khatib, F., DiMaio, F., Group, F.C., Group, F.V.C., Cooper, S., et al.: Crystal structure of a monomeric retroviral protease solved by protein folding game players. Nature Structural & Molecular Biology 18, 1175–1177 (2011), doi:10.1038/nsmb.2119
3. Timmer, J.: Galaxy Zoo shows how well crowdsourced citizen science works (2010), http://arstechnica.com/science/news/2010/10/galaxy-zoo-shows-how-well-crowdsourced-citizen-science-works.ars
4. Nayar, A. (2009), Model predicts future deforestation. Nature News, http://www.nature.com/news/2009/091120/full/news.2009.1100.html
5. Newell, D.A., Pembroke, M.M., Boyd, W.E.: Crowd Sourcing for Conservation: Web 2.0 a powerful tool for biologists. Future Internet 4, 551–562 (2012)
6. Brabham, D.C.: Crowdsourcing the public participation process for planning projects. Planning Theory 8, 242–262 (2009)
7. MacLean, D., Yoshida, K., Edwards, A., Crossman, L., Clavijo, B., Clark, M., Swarbreck, D., Bashton, M., Chapman, P., Gijzen, M., Caccamo, M., Downie, A., Kamoun, S., Saunders, D.G.O.: Crowdsourcing genomic analyses of ash and ash dieback – power to the people. GigaScience 2(2) (2013), doi:10.1186/2047-217X-2-2
8. Goodchild, M.F., Glennon, J.A.: Crowdsourcing geographic information for disaster response: a research frontier. International Journal of Digital Earth 3, 231–241 (2010)
9. Zook, M., et al.: Volunteered Geographic Information and Crowdsourcing Disaster Relief: A Case Study of the Haitian Earthquake. World Medical & Health Policy 2(2), 7–33 (2010)
10. Brunsdon, C., Comber, A.J.: Experiences with Citizen Science: Assessing Changes in the North American Spring. Geoinformatica (2012), doi:10.1007/s10707-012-0159-6
11. Comber, A., See, L., Fritz, S., Van der Velde, M., Perger, C., et al.: Using control data to determine the reliability of volunteered geographic information about land cover. International Journal of Applied Earth Observation and Geoinformation 23, 37–48 (2013), doi:10.1016/j.jag.2012.11.002

12. Vuurens, J., de Vries, A.P., Eickhoff, C.: How Much Spam Can You Take? An Analysis of Crowdsourcing Results to Increase Accuracy. In: Proceedings of the SIGIR 2011 Workshop on Crowdsourcing for Information Retrieval (CIR), Beijing, China (2011)
13. Goodchild, M.F., Li, L.: Assuring the quality of volunteered geographic information. Spatial Statistics 1, 110–120 (2012), doi:10.1016/j.spasta.2012.03.002
14. Haklay, M., Basiouka, S., Antoniou, V., Ather, A.: How Many Volunteers Does it Take to Map an Area Well? The Validity of Linus' Law to Volunteered Geographic Information. The Cartographic Journal 47, 315–322 (2010)
15. Foody, G.M., Boyd, D.S.: Using volunteered data in land cover map validation: Mapping tropical forests across West Africa. In: 2012 IEEE International on Geoscience and Remote Sensing Symposium (IGARSS), pp. 6207–6208 (2012)
16. Hudson-Smith, A., Batty, M., Crooks, A., Milton, R.: Mapping for the masses. Accessing Web 2.0 through crowdsoucring. Social Science Computer Review 27, 524–538 (2009)
17. Li, L., Goodchild, M.F.: The Role of Social Networks in Emergency Management: A Research Agenda. In: International Journal of Information Systems for Crisis Response and Management, vol. 2(4), 11 pages (2010), doi:10.4018/jiscrm.2010100104
18. Wiersma, Y.F.: Birding 2.0: Citizen science and effective monitoring in the web 2.0 world. Avian Conservation and Ecology 5, 13 (2010),
    http://www.ace-eco.org/vol5/iss2/art13/,
    http://dx.doi.org/10.5751/ACE-00427-050213
19. Voigt, S., et al.: Rapid damage assessment and situation mapping: learning from the 2010 Haiti earthquake. Photogram. Eng. and Rem. Sen. 77, 923–931 (2011)
20. Snow, R., O'Connor, B., Jurafsky, D., Ng, A.Y.: Cheap and fast—but is it good?: evaluating non-expert annotations for natural language tasks. In: 2008 Proceedings of the Conference on Empirical Methods in Natural Language Processing, pp. 254–263 (2008)
21. Flanagin, A., Metzger, M.: The credibility of Volunteered Geographic Information. GeoJournal 72, 137–148 (2008)
22. Pal, M., Foody, G.M.: Feature Selection for Classification of Hyperspectral Data by SVM. IEEE Transactions on Geoscience and Remote Sensing 48(5), 2297–2307 (2010)
23. Hughes, G.F.: On the mean accuracy of statistical pattern recognizers. IEEE Transactions on Information Theory 14(1), 55–63 (1968), doi:10.1109/TIT.1968.1054102
24. Raykar, V.C., Yu, S.: Ranking annotators for crowdsourced labeling tasks. Advances in Neural Information Processing 24 (2011)
25. Foody, G.M., Boyd, D.S.: Exploring the potential role of volunteer citizen sensors in land cover map accuracy assessment. In: Proceedings of the 10th International Symposium on Spatial Accuracy Assessment in Natural Resources and Environmental Science (Accuracy 2012), Florianopolis, Brazil, pp. 203–208 (2012)
26. Foody, G.M., See, L., Fritz, S., Van der Velde, M., Perger, C., Schill, C., Boyd, D.S., Comber, A.: Accurate attribute mapping from volunteered geographic information: issues of volunteer quantity and quality. Paper to be submitted to Cartographic Journal (in preparation)
27. Smith, B.: On drawing lines on a map. Spatial Information Theory A Theoretical Basis for GIS, 475–484 (1995)
28. Smith, B.: Fiat Objects. Topoi 20, 131–148 (2001)
29. Smith, B., Mark, D.: Geographical categories: an ontological investigation. International Journal of Geographical Information Science 15, 591–612 (2001)
30. Janowicz, K., Kessler, C.: The Role of Ontology in Improving Gazetteer Interaction. International Journal of Geographical Information Science (IJGIS) 22(10), 1129–1157 (2008)

31. Kuhn, W.: Semantic engineering. Research Trends in Geographic Information Science, 63–76 (2009)
32. Feddema, J.J., Oleson, K.W., Bonan, G.B., Mearns, L.O., Buja, L.E., Meehl, G.A., Washington, W.M.: The importance of land-cover change in simulating future climates. Science 310, 1674–1678 (2005)
33. Fritz, S., See, L., McCallum, I., Schill, C., Obersteiner, M., et al.: Highlighting continued uncertainty in global land cover maps for the user community. Environmental Research Letters 6, 044005 (2011), doi:10.1088/1748-9326/6/4/044005
34. Fritz, S., McCallum, I., Schill, C., Perger, C., See, L., et al.: Geo-Wiki: An online platform for improving global land cover. Environmental Modelling & Software 31, 110–123 (2012), doi:10.1016/j.envsoft.2011.11.015
35. Comber, A.J.: Geographically weighted methods for estimating local surfaces of overall, user and producer accuracies. Remote Sensing Letters 4(4), 373–380 (2013), doi:10.1080/2150704X.2012.736694
36. Comber, A., Fisher, P.F., Brunsdon, C., Khmag, A.: Spatial analysis of remote sensing image classification accuracy. Remote Sensing of Environment 127, 237–246 (2012)
37. See, L., Comber, A.J., Salk, C., Fritz, S., Van der Velde, M., Perger, C., Schill, C., McCallum, I., Kraxner, F., Obersteiner, M.: Comparing the quality of crowdsourced data contributed by expert and non-experts. Paper submitted to PLOSOne (submitted, February 2013)
38. Perger, C., Fritz, S., See, L., Schill, C., Van der Velde, M., et al.: A campaign to collect volunteered geographic Information on land cover and human impact. In: Jekel, T., Car, A., Strobl, J., Griesebner, G. (eds.) GI_Forum 2012: Geovisualisation, Society and Learning, pp. 83–91. Herbert Wichmann Verlag, Berlin (2012)
39. Fritz, S., See, L., Van der Velde, M., Nalepa, R.A., Perger, C., et al.: Downgrading recent estimates of land available for biofuel production. Environ. Sci. Technol. 47, 1688–1694 (2013), doi:10.1021/es303141h
40. Brunsdon, C.F., Fotheringham, A.S., Charlton, M.: Geographically Weighted Regression - A Method for Exploring Spatial Non-Stationarity. Geographical Analysis 28, 281–298 (1996)
41. Tulloch, D.A.: Is VGI participation? From vernal pools to video games. Geojournal 72(3-4), 161–171 (2008)
42. Coleman, D.J., Georgiadou, Y., Labonté, J.: Volunteered geographic information: The nature and motivation of producers. International Journal of Spatial Data Infrastructures Research 4, 332–358 (2009)
43. Sui, D.Z.: The wikification of GIS and its consequences: Or Angelina Jolie's new tattoo and the future of GIS. Environment and Urban Systems 32, 1–5 (2008)
44. Harvey, F., Chrisman, N.: Boundary objects and the social construction of GIS technology. Environment and Planning A 30, 1683–1694 (1998)
45. ONS, Internet Access - Households and Individuals (2012), http://www.ons.gov.uk/ons/dcp171778_275775.pdf

The Meanings of the Generic Parts of Toponyms: Use and Limitations of Gazetteers in Studies of Landscape Terms

Curdin Derungs[1], Flurina Wartmann[1], Ross S. Purves[1], and David M. Mark[2,3,*]

[1] Geocomputation Unit, Geographic Institute, University of Zurich, Switzerland
{curdin.derungs,flurina.wartmann,ross.purves}@geo.uzh.ch
[2] Department of Geography, University at Buffalo, NY, USA
[3] National Center for Geographic Information and Analysis (NCGIA),
University at Buffalo, NY, USA
dmark@buffalo.edu

Abstract. Are the contents of toponyms meaningless, as it is often claimed in linguistic literature, or can the generic parts in toponyms, such as hill in *Black Hill*, be used to infer landscape descriptions? We investigate this question by, firstly, linking gazetteer data with topographic characteristics, and, secondly, by conducting analysis of how the use of landscape terms might have changed over time in a historic corpus. We thus aim at answering a linguistic, and ethnophysiographic, research question through digital input data and processing. Our study area is Switzerland and our main focus is on geographic eminences, and in particular on the use of the terms *Spitze*, *Horn* and *Berg*. We show that most prominent generic parts in toponyms show expected topographic characteristics. However, not all generic parts strictly follow this rule, as in the case of *Berg*. Some generic parts have lost their meaning in standard language over time (e.g. *Horn*). We therefore put a cautionary note on the use of generic parts in toponyms in landscape studies, but point out that the subtle details of these differences provide rich topics for future research.

Keywords: toponyms, proper names, generic parts, landscape terms, GIR, ethnophysiography, gazetteers.

1 Introduction

Recently, research has accelerated on the topic of how landscape and its elements are conceptualized, on how these conceptualizations are expressed in human natural language, and on how and why the conceptualizations vary across languages and cultures. Mark and Turk coined the term ethnophysiography for this topic (Mark and Turk, 2003a, 2003b). A major development was the publication of a special issue of Language Sciences (Burenhult, 2008), followed by a book entitled "Landscape in Language" (Mark et al., 2011). Burenhult and Levinson listed some key research questions in this domain:

---

* Corresponding author.

T. Tenbrink et al. (Eds.): COSIT 2013, LNCS 8116, pp. 261–278, 2013.

How are landscape features selected as nameable objects ('river', 'mountain', 'cliff')? What is the relation between landscape terms (common nouns) and place names (proper nouns)? How translatable are landscape terms across languages, and what ontological categories do they commit to? (Burenhult and Levinson, 2008, p. 136)

The detailed research agenda, and the potential of finding fundamental explanations for spatial particularities in conceptualizations, are a key strength of ethnophysiography. On the other hand, ethnophysiography up to now has concentrated on small samples of speakers, usually through ethnographic interviews or focus group discussions, and thus data collection is thus very time intensive.

In this context, Geographic Information Retrieval (GIR) can be considered another line of research. GIR is usually defined as a combination of methodologies from Geographic Information Science and Information Retrieval, where unstructured data, often in the form of natural language, is parsed for geographic information. Jones and Purves list some of the following tasks as being most relevant in GIR:

Detecting geographical references in the form of place names and associated spatial natural language qualifiers within text documents and in users' queries; disambiguating place names to determine which particular instance of a name is intended; geometric interpretation of the meaning of vague place names, such as the 'Midlands' and of vague spatial language such as 'near'; (Jones and Purves 2008, p. 220)

The two fields of research, namely GIR and ethnophysiography, can be usefully combined by using digitized input data to explore ethnophysiographic research questions. In this paper we analyze the use of landscape terms as parts of Swiss toponyms, as in for example the generic landscape term *Berg* (mountain) in the toponym *Uetliberg*, and their association with the use of landscape terms in everyday language. By incorporating data from gazetteers and digitized language corpora we aim to contribute to the ethnophysiographic research question:

What is the relation between landscape terms (common nouns) and place names (proper nouns)? (as outlined in Burenhult and Levinson 2008, p. 136)

We first introduce relevant theoretical background on proper names from different disciplinary angles including psychology, linguistics, semantics, and geography. We then provide an overview of the methods used. We investigated properties of generic parts in toponyms through a topographic analysis, i.e. geomorphometry (e.g. Pike et al. 2008), and explored change over time in the use of landscape terms in a digital text corpus and user generated content (UGC) (Goodchild 2007) through textual analysis and association of generic parts with regions.

In our study we therefore develop a framework that allows us to explore ways in which toponyms are used. In so doing, we demonstrate that caution is needed when using gazetteers as a corpus and, in particular, when using proper names as sources of generic information to describe landscapes. Moreover, we establish different approaches that exploit digital input data, automatic data processing, and thus, quantitative evaluations, in order to explore ethnophysiographic research questions, which so far have been primarily addressed by empirical ethnographic studies in the field. We argue that this combination of approaches has considerable potential, especially if we wish to broaden the spatial scope of ethnophysiographic research.

2   Theoretical Background

Domains of Proper Names. Proper names are used to refer to individual entities mainly in two domains of human experience: people, and places. It is important for our purposes to note that toponyms, also known as place names, are one of the most important subclasses of proper nouns or proper names. Recently, Levinson (2011) has pointed out that people and places, the main referents for proper nouns, are also two domains with specialized processing areas in the brain (faces and places).

The Psychology of Proper Names. Valentine et al. (1996) reviewed cognitive and psychological differences in how people process proper names, compared with how people process common nouns and noun-phrases. However, few rigorous psychology experiments with geographic proper names have actually been conducted. One exception is the work by Hollis and Valentine (2001) on effects of priming and associations between pictures and names, and between auditory priming and visual recognition of names. They found that the priming improved recognition for people names and landmark names (e.g. Levinson 2011), but not for country names which were processed in a way similar to common nouns.

In general proper names are often said to be pure referencing expressions, and therefore not to have semantic meaning or sense. Yet, as Hollis and Valentine point out, toponyms, at least in English, often have semantically-meaningful components, that is, parts of the names that at least *appear to make sense*:

> *Landmark names often contain a greater degree of meaning compared with people's names and country names that can be considered arbitrary. For example, the Eiffel Tower is a tower and Tower Bridge is a bridge, next to the Tower of London.* (Hollis and Valentine 2001 p, 113)

However, as an example, of the 157 toponyms called "Mount Pleasant" in the US Geological Survey's GNIS database (USGS, 2013), 76 percent are populated places or locales, rather than being topographic eminences[1,2]. The sense in toponyms thus appears to be neither absolutely true or absolutely false, but rather partial or probabilistic, to vary with circumstances, and to be different for different languages and cultures. Coates' (2006) article on properhood provides a basis for understanding this.

Properhood. Coates (2006) provides a very detailed account of how "properhood" works in language. Coates reviews previous efforts to define and distinguish proper names, mainly by philosophers and linguists, and to show how they differ from ordinary language. He then provides a compelling alternative theoretical treatment. Coates concludes that proper nouns are not a lexical category (type of noun), but rather are "a type of referring that discounts the sense of any lexical items (real or

---

[1] In this paper, we use the term *eminence* in its older landscape sense, as a superordinate term for a part of the landscape that stands above its surroundings.

[2] We leave it to the reader to judge which of these places are "pleasant"!

apparent) in the expression that is being used to do the referring" (Coates 2006 p. 378). Coates notes that the view that proper names have "no sense in the act of reference" does not exclude the possibility that they have senses that are "accessible during other (meta)linguistic activities" (Coates, 2006, p. 356). Conventions for understanding proper names license "probabilistic implications, nothing logically stronger, even though the probability in a given case could be extremely high—the implied categorization should always be taken as falsifiable in principle even if not yet falsified" (Coates 2006, p. 365).

Toponyms. Toponyms have referential meaning, that is, they denote individual geographic features. Toponyms are often compiled to form gazetteers, typically by government agencies, but sometimes as projects within volunteered geographic information (VGI), a current 'hot' topic in Geographic Information Science (Goodchild 2007).

Since, following from Coates' theory of properhood, the non-referential 'content' of every toponym is to some extent 'encapsulated', that is, it may or may not "make sense", or follow Grice's maxims (Grice 1975), it should not be a surprise that the generic parts of toponyms, if present, may be from a different vocabulary than the general landscape terms of a language. This is an important point, since many researchers (including one of the present authors, e.g. Feng and Mark 2012) have uncritically treated lists of toponyms as corpora for studying general landscape terms. For the English language, the possibility that toponyms are not "normal language" is masked by the fact that, except for names of populated places, the generic parts of many toponyms in English involve the same terms as those that are used as landscape terms to denote classes of landscape features (e.g. hill, valley, lake, etc.).

One illuminating exception for natural landscape features in English is the word 'Mount', which is common in toponyms denoting mountains or other eminences, but which in contemporary English would not be used as a common noun outside of toponyms (i.e., "*I live near a mount"; "*what is the name of that mount?")[3].

2.1   Case Study of Toponyms for Mountains in Swiss German

Motivation. This research project began with the observation that equations of the form:

"the Eiffel Tower" is a "tower"

are not valid for many names of mountains in German, such as *Zugspitze* or *Matterhorn*. The following sentences, relating specific peaks to the generic parts of their names (*Spitze* and *Horn*[4]):

"*Die Zugspitze ist eine Spitze" or "*Das Matterhorn ist ein Horn"

---

[3] Note: Following the practice in linguistics, an asterisk [*] is placed in front of an expression to indicate that it would not be considered to be a valid or well-formed utterance by speakers of the language

[4] The primary meaning of Horn in German is the same as in English, namely a projection on the head of an animal, while the primary meaning of Spitze is similar to the meaning of the English word *point*, such as in "pencil point".

are not well-formed sentences in standard German, because *Spitze* and *Horn* on their own refer to parts, but not whole, topographic features. However, *Spitze* and *Horn* are commonly used nouns in standard German, in contrast to *Mount*, *-ville* or *–ton*, popular generic terms in English and American toponyms, which do not form useful nouns in their own right. Furthermore, the meanings of the words in standard German suggest proximate association with topographic features, for instance in referring to topographic shape.

Coaz (1865) pragmatically explains the issue of *Horn* and other generic parts in toponyms in an article on place names in the Swiss Alps ("Ueber Ortsbenennung in den Schweizeralpen") published by the Swiss Alpine Club.

> *From the canton of Bern, to the borders of the Canton of Vaud and the Oberwallis, most mountain tops have the generic name "Horn", and, in fact, this part of the Alps is characterized by numerous pyramidal, rocky mountain tops.* (Coaz 1865, p. 467, translated from German by the authors)

Coaz emphasizes that toponyms containing the generic part *Horn* belong to the feature class Berg (mountain), and that *Horn* is related to the summit of the mountain. This could also be true for *Spitze*, since the sentence "Die Spitze des Berges" makes perfect sense for German speakers. However, this would indicate that *Horn* and *Spitze* only refer to parts of a mountain, whereas the toponym, as for instance "Matterhorn", certainly refers to the mountain as a whole. In the following, we will look at this divergence more closely.

In standard German, the association of *Spitze* and *Horn*, with *Berg*, is reflected by the high acceptance of the sentence:

"Die Zugspitze und das Matterhorn sind Berge"

(e.g. de.wikipedia.org/wiki/Zugspitze). However, and this makes the issue much more complex, *Berg* is also prominently used as a generic part in Swiss toponyms. Nonetheless, the sentence:

"Zürichberg, Hönggerberg, Chäferberg und Uetliberg sind Berge"

would probably not be accepted by locally knowledgeable, native speakers. All four toponyms, although referring to geographic eminences, would not be considered as mountains, resembling the Matterhorn or Zugspitze, but would rather be classified as *Hügel* (hill), as is indeed the case in the feature classification of the gazetteer used in this study.

Problem Statement and Research Questions. From a small set of specific examples, we find the relation between landscape terms and toponyms, and generic parts in toponyms in particular, to be of a complex nature. *Spitze* and *Horn* are frequently used as generic parts in toponyms, but they do not seem to represent independent geographic feature types. Furthermore, these two generic parts are not productively used in language to refer to features that have these generics in their names. Instead, such features would rather be called *Berg*. However, a certain association of *Spitze* and *Horn* with geographic eminences is suggested by the use of the words in standard German. Derungs and Purves (2007), for instance, found evidence that *Spitze* was often associated with *Berg*. *Berg*, on the other hand, refers to a basic level geographic eminence type in standard German, but its occurrence as a generic part in toponyms appears to have little topographic overlap with prototypical mountains, such as the *Zugspitze* or *Matterhorn*. Toponyms containing *Berg* as a generic part seem to refer to geographic features that would not be called *Berg* in standard language.

From these observations three research questions emerge:

- Can generic parts in toponyms be associated with particular topographies?
- Are generic parts in toponyms fossilized, meaning that the productive use in standard German has changed over time?
- Is the use of generic parts in Swiss toponyms arbitrary, such that there is no meaning within toponyms, or, can we, for instance, deduce spatial relations or settings?

Topographic and Corpus Based Study. To shed light on the relation between generic parts in Swiss mountain toponyms and their associated forms, we will link the locations of toponyms which contain generic parts to topographic characteristics, thus allowing comparison of topographic similarity between different classes of generic parts as used in toponyms.

In a corpus based study, we investigate the apparent mismatch of concepts between the generic part *Berg* and its underlying referent, as well as the use of *Horn* and *Spitze*, by analyzing whether the use and meanings of these terms has changed over 1975time. Our hypothesis is that change is of a different nature for *Horn/Spitze* than it is for *Berg*. In the case of *Horn* and *Spitze*, we suppose that these terms were productively used as category norms for geographic eminences in the past, and that they have become fossilized over time and are nowadays no longer productively used in standard language. Thus, the use of the generic parts has changed over time. In the case of *Berg* on the other hand we assume that the meaning itself has changed and that over time, different types of topographies have been referenced with the same generic part in their toponyms.

If, in both cases, change over time is not a useful explanatory variable, this indicates that, at least in the Swiss context, and examples from other European languages point in the same direction, generic parts in toponyms do not necessarily have full normal meaning.

In the following sections of this paper we present the set of input data and methods that we used for comparing topographic characteristics of generic parts in toponyms. Secondly, we report on the analysis on change of meaning over time, by using a large historic corpus on Swiss alpine landscape descriptions. We then present the results for these two empirical case studies and discuss them in view of the three research questions outlined above.

3   Methods

3.1   Topographic Characteristics

Input Data. The input data consists of a digital elevation model (DEM) and a gazetteer. The digital elevation model we used in our analysis has a spatial resolution of 25m and its spatial coverage includes Switzerland in its entirety. Vertical accuracy in the Swiss Alps varies between 3 and 8 meters. This DEM is provided by the Swiss Federal Office of Topography Swisstopo (http://www.swisstopo.admin.ch/).

As a gazetteer we use "Swissnames", the most extensive compilation of Swiss toponym data that provided by Swisstopo. It consists of more than 190,000 individual toponyms and the geographic coordinates of the locations that are associated with them. The georeferencing of toponyms reflects the location where the respective label is drawn on topographic maps from the National Mapping series (scales 1:25,000 and 1:50,000). Swissnames contains a feature classification that differentiates toponyms into types of geographic features. Roughly half of all toponyms refer to geographic features of natural feature type, with numbers of toponyms for different types of geographic eminences listed in Table 1. We also added separate columns for the generic parts *Berg*, *Spitze* and *Horn* and indicated their relative frequencies.

Table 1. Count of different types of geographic eminences in Swissnames and relative frequencies of *Berg, Spitze* and *Horn* as generic parts distributed over feature types

Feature type	Description/ Example	Count	*Berg*	*Spitze*	*Horn*
Huegel	hill, e.g. *Uetliberg, Hönggerberg*	2543	81	16	4
KGipfel	small mountain peak, e.g. *Lochberg, Märenspitz*	4414	17	68	72
GGipfel	big mountain peak, e.g. *Allalinhorn, Breithorn*	866	1	14	19
HGipfel	main mountain peak, e.g. *Matterhorn, Weisshorn*	165	0	1	5
Massiv	mountain range, e.g. *Engelhörner, Churfirsten*	143	0	0	0

Analysis. By considering values of elevation and slope from a set of three buffer zones (200m, 400m and 2000m) around each toponym location, we make a simple association between toponym location and topographic properties. From the distribution of elevations within each buffer zone we store relief (the maximum difference between the elevation of two raster cells within the buffer zone) and standard deviation in elevation (which is related to surface roughness). From the distribution of slopes we retain mean slope and standard deviation. Since both types of measurements are computed for all three buffer zones, we generate 12 attributes that represent the topographic characteristics for each toponym location. The three buffer sizes can be seen as a very simplistic form of multi-scale analysis, which could be extended by, for example, identifying and generating term vectors of morphometric classes (e.g. pit, peak, pass, etc.) for each buffer at multiple scales (e.g. Fisher et al. 2004).

The topographic characteristics of neighborhoods of toponyms can thus be represented as a vector of features. Similarity between toponyms can then be compared either quantitatively, for example using a measure such as cosine similarity (e.g. Bayardo et al. 2007), or qualitatively, by reducing the multi-dimensional vector space to a two dimensional map (a "Self Organizing Map", or SOM) (Mark et al. 2001; Skupin and Esperbé 2011). SOMs are a form of artificial neural networks, where training data is used to reduce the multidimensionality of feature vectors through a mapping to two dimensional space, whilst preserving the similarity between feature vectors as neighborhood relationships (Kohonen 1995).

To compute SOMs we used the *kohonen* package, as implemented in R (Wehrens and Buydens 2007). We selected, additionally to *Berg*, *Spitze* and *Horn*, sets of semantically comparable generic landscape terms from toponyms found in Swissnames that either have comparable meaning in different languages (*Pass*, *fuorcla* and *col*, all meaning pass in German, Rumantsch and French respectively or *champs* and *Acker* meaning agricultural field in French and German respectively) and generics with differing spelling across dialects (*Acker*, *Acher*), and extracted feature vectors for all toponyms containing these generic parts.

3.2   Change of the Meaning of Generic Parts in Toponyms over Time

Input Data. Historic uses of terms related to landscape features are investigated using Text+Berg (Volk et al. 2009), a large corpus (collection) of Swiss alpine landscape descriptions, published as yearbooks by the Swiss Alpine Club and covering the last 150 years. These books have been digitised and linguistically parsed (Sennrich et al. 2009), such that they can be automatically queried for the occurrence of particular terms. The descriptions cover all official Swiss languages, with a majority of the articles written in standard German. Articles cover a wide variety of topics ranging from scientific reports on the state of Swiss glaciers to reviews of newly introduced outdoor products. However, most common are reports on mountaineering trips in the Swiss mountains, where landscapes are often described in great detail. The authors of the articles are not necessarily residents from the regions that are being described, but can be mountaineers from other parts of Switzerland and the world.

In order to examine uses of landscape terms and their spatial distributions in a contemporary context, we use georeferenced Flickr photographs (www.flickr.com) that have been uploaded and tagged with different terms by users. For the bounding box of Switzerland we gathered a total number of about 4 million items using the Flickr API. In this case study we focus on analyzing the user-generated tags that were applied to describe the content or context of pictures.

Analysis. We split the analysis in two parts. First we relate the generic parts *Berg*, *Horn* and *Spitze* to regions. Using these initial results we analyze the relatedness of these terms to each other. In a second step, we investigate the occurrences and uses of these terms in a historic corpus.

Spatial Footprints. We approximated footprints for generic parts in toponyms by mapping density peaks, calculated from all associated locations (e.g. Hollenstein and Purves 2010). We then compared these footprints with footprints gathered from more recent uses of landscape terms, outside of toponyms, by calculating spatial footprints from georeferenced Flickr photographs tagged with landscape terms, such as *Berg*. *Spitze* and *Horn* occur rarely as generic terms in their own right, and thus we did not generate spatial footprints. By comparing these footprints we approximate the relatedness of these concepts in terms of their spatial overlap.

Occurrences of Terms in Corpus. In order to explain patterns of relatedness, we compare occurrences of *Spitze* and *Horn* as nouns in early (prior to the year 1870) and recently (after 2000) published yearbooks from the Text+Berg corpus, with a particular focus on change of use over time. Since word sense disambiguation, for instance in detecting if the term *Spitze* refers to mountain, part of mountain, or pencil point, is challenging, we relied on manual annotation. To detect change in meaning over time in the use of *Berg*, we again analyzed early published yearbooks of the Text+Berg corpus, this time with a particular focus on the types of toponyms that *Berg* is referring to. Through this approach we can investigate the use of generic parts in standard language outside toponyms.

4   Results

4.1   Topographic Characteristics

First, we report on the topographic characteristics of the three generic parts *Horn*, *Spitze* and *Berg* and show the vector similarities of within-, and across-group comparisons. We then present the SOM of a set of nine generic parts that frequently occur in Swiss toponyms. These terms do not exclusively represent topographic eminences, but cover a broad selection of landscape terms.

SOM. We analyzed SOMs derived from feature vectors representing the topography of toponyms containing generic parts that frequently ($n>200$) occur in Swissnames (Fig. 1). These nine generic parts represent very different topographies. The top row of Fig. 1 contains the three generic parts discussed in the introduction. We arranged the comparable generics in horizontal rows so that similarities in their SOM representations are clearly visible. The SOMs of *Spitze* and *Horn* are auto-correlated, and almost congruent, whereas the SOM of *Berg* shows a quite diffuse pattern, which is very different from *Spitze* and *Horn*. Equally, the terms describing passes through mountains, *Pass, fuorcla* and *col* are all rather similar and have some overlap with the feature space of *Horn* and *Spitze*. Finally, the generic parts describing cultivated fields (*champs, Acker, Acher*) occupy a different region of the SOM, interestingly overlapping slightly with the region occupied by *Berg*.

Cosine Similarities. Cosine similarities between all nine generics, computed from topographic vectors, are represented in Fig. 2. The grey boxes delineate groups of generic parts considered semantically similar, such as passes or cultivated fields that were arranged in comparable rows in Fig. 1. In general, generic parts that describe similar features have high cosine values in Fig. 2.

The (self-)similarity of *Horn* and *Spitze* is also reflected by cosine similarities, within and across groups (Fig. 2, eminences box). Toponyms containing *Horn* or *Spitze* have median cosine values of around 0.8, in both, within- and across-group comparisons, indicating topographic relatedness. *Berg* toponyms, on the other hand, are unrelated to other *Berg, Horn* or *Spitze* toponyms, manifested by cosine values of around 0.

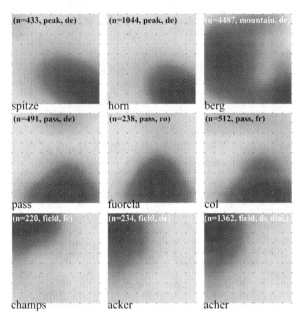

Fig. 1. SOM representation of the topographic vectors for nine frequent generic parts in Switzerland

4.2   Change of the Meaning of Generic Parts in Toponyms over Time

Spatial Footprints. We calculated the spatial footprints for the generic parts *Berg*, *Horn* and *Spitze* within toponyms (Fig. 3). For comparison, we also included the spatial footprints of the use of *Berg* and *mountain* as user-generated tags in georeferenced Flickr photographs. The footprint of the term *Berg*, used as a generic part (Fig. 3 upper right), is markedly different to the use of *Berg* (solid line), and *mountain* (dashed line), as a tag in Flickr (Fig. 3 upper left). The Flickr footprints reflect a spatial distribution of the term *Berg* and *mountain* that overlaps with what we would expect from its use in everyday language and "naïve" geography (Egenhofer and Mark 1995), covering Swiss alpine regions (Valais, the Bernese Oberland, parts of central Switzerland, Toggenburg, the Albula region and Oberengadin). On the other hand, toponyms containing *Berg* (Fig. 3 upper right) are mainly located in the *Swiss Mittelland*, an extensive plain. This does not imply that *Berg* toponyms refer to plains or valley floors: *Berg* toponyms do refer to geographic eminences. However, these features containing the term *Berg* in the Swiss Mittelland are topographically different from prototypical mountains, such as *Matterhorn* or *Zugspitze*.

Occurrences of Terms in the Corpus. *Spitze* and *Horn* are both prominently used in Text+Berg, with about 5,500 references to *Spitze* and 750 to *Horn*. These terms, however, are semantically ambiguous, with a significantly smaller number of references referring to actual geographic features. *Spitze* is used more often compared to *Horn*, presumably because *Spitze* is also used to refer to different types of objects having a peaky shape (e.g. pencil points). We found that *Horn*, as a noun occurs more often in older yearbooks. A comparison between the 10 oldest and the 10 most recent yearbooks resulted in 34/mw (i.e. number of occurrence per million words) occurrences in old, compared to 10/mw occurrences in new yearbooks. This relation is even more marked for *Spitze*, with 500/mw in old compared to only 27/mw occurrences in the most recent yearbooks.

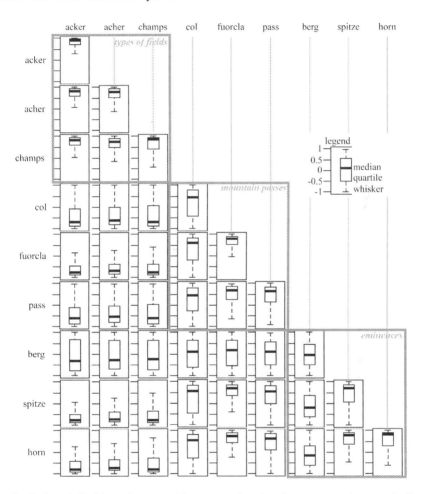

Fig. 2. Cosine similarities of topographic vectors for nine frequent generic parts in Swiss toponyms. Related generic parts are delineated (grey boxes) and labeled.

In a reading of early yearbooks, *Horn* appears to be most often used in reference to mountains, and mostly in reference to whole mountains. However, the mountain toponym does not necessarily need to contain the generic part *Horn*. For instance, in an article from 1865 (Weilenmann 1865), *Piz Rosegg* is referred to as 'a *Horn* that shrinks, the more one approaches it'. In five yearbooks we found 15 cases where *Horn* refers to a whole mountain and only 5 cases where *Horn* refers to the top part of a mountain.

By contrast, in recent yearbooks *Horn* is very rarely used to refer to mountains. In 10 yearbooks we only found three cases where *Horn* is used in this context, where in two cases it was used to reference a whole mountain.

For *Spitze*, we found few occurrences in early yearbooks that refer to whole mountains. The *Dufourspitze*, for instance, is called a *Spitze* in 1864 (Studer 1864), where the author regrets that this *Spitze* was not named after its first ascendant. Indeed, *Spitze* was, and is, prominently used to refer to the top part of a mountain. In a mountaineering context, for instance, it is well accepted to state: "Ich näherte mich der *Spitze*" (I approached the summit). Most interesting are the early occurrences, mainly of *Horn*, but also of *Spitze*, that refer to whole mountains. This suggests that is was acceptable to refer to a mountain by using the word *Horn*, and thus, the sentence:

"Das Matterhorn ist ein Horn"

was at some point in time valid. Although use of *Spitze* has decreased, its meaning in both old and new yearbooks appears to have primarily referred to the summit regions of a mountain.

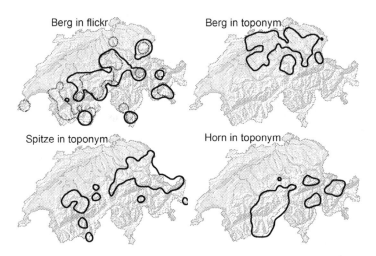

Fig. 3. The density contours enclosing 20% of the volume of Flickr photographs containing the tag *Berg* (upper left, solid line) and mountain (upper left, dashed line), and toponyms that contain the generics *Berg*, *Spitze* and *Horn*.

In the case of *Berg*, we did not detect a change in meaning over time using Text+Berg, such that *Berg* used to refer to topographic eminences in the *Swiss Mittelland* in early yearbooks, but now rather describes an eminence in the *Alps*, such as might be suggested by Fig. 3. In the very first yearbooks we find references to *Berg* where it is used to refer to prototypical mountains, such as the *Finsteraarhorn*, referred to as the "mächtige Berg" (majestic mountain, Lindt 1864), or, the *Aletschhorn*, which is called a "herrlicher Berg" (magnificent mountain, Fellenberg 1864). This indicates that the concept of *Berg* has not significantly changed over the last 150 years, at least in terms of its use in Text+Berg. One explanation for the mismatch between our observations in the corpus and the comparison of gazetteer data with UGC, may be that variations in the use of *Spitze* and *Horn* occurred within the 150 year lifespan of Text+Berg, whereas changes in the use of *Berg* date further back.

We found support for this hypothesis by searching for first occurrences of toponyms in historic documents. For three out of four *Berg* toponyms listed in section §1.2 we found historic references, dating back to the early 15[h] century or earlier, in a local, very fine grained gazetteer (www.ortsnamen.ch). *Hönggerberg* first occurs in 1432, *Zürichberg* is first mentioned in 1188, and the first reference to *Chäferberg* dates back to 1420. We could not find any historic reference to *Uetliberg*.

5   Discussion

We aimed to investigate an ethnophysiographic research question, namely the relation between toponyms and landscape terms, by exploiting a range of digital input data including gazetteers, text corpora, UGC and elevation models. To explore such data, we used a variety of methods allowing us to characterize regions in terms of geomorphometry, to calculate similarities between generic parts of toponyms and to identify parts of speech for a qualitative reading of a text corpus.

We argue that the combination of approaches used has considerable potential, especially if we wish to broaden the spatial and temporal scope of ethnophysiographic research. In the following, we discuss our findings with respect to the research questions set out in §1.2.

RQ1: Can Generic Parts in Toponyms Be Associated with Particular Topographies?

For nine different generic parts that are frequently found in Swiss toponyms we used geomorphometric measurements of toponym locations as represented by relative drop and slope in different buffer zones. We have shown, by reducing feature vectors to two dimensions with SOM, that groups of toponyms with identical generic parts such as *Spitze* or *Horn* exhibit specific topographic patterns (Fig. 1). For instance, the patterns for *Spitze* and *Horn* were self-similar and similar to one another, while both differed from patterns exhibited by toponyms with generic terms related to agricultural fields and mountain passes, which we expect to have correspondingly different toponyms. Interestingly, we observed this pattern even for generic parts in toponyms in different languages (*Pass, col* and *fuorcla* for mountain pass in German, French and Rumantsch). Thus, our method can be used to compare landscape terms and their relation to morphometry across languages – thus potentially contributing to generating taxonomies and ontologies (Scheider et al. 2009). Using a quantitative measurement in the form of cosine similarity produced similar results (Fig. 2). These findings are in accordance with our expectation that semantically similar terms are also topographically similar as measured by their SOMs and cosine similarities.

However, we did find an important exception to the more general pattern described above. The SOM of toponyms containing the generic part *Berg* showed a much more diffuse pattern than other generic parts, which was not similar to any other SOM of the terms investigated (Fig. 1). Furthermore, cosine similarity for *Berg* showed low self-similarity. This indicates that toponyms containing *Berg* include varying topographies that differ from the topographies associated with *Spitze* or *Horn* toponyms. This is, at first glance, puzzling because *Spitze*, *Horn* and *Berg* are all used to refer to topographic eminences (e.g. Zugspitze, Matterhorn, Chäferberg). In the following paragraph, we discuss the difference of the use of generic terms in toponyms as compared to the use of the terms in standard language.

RQ2: Are Generic Parts in Toponyms Fossilized, Such That the Productive Use in Standard German Changed over Time?
In order to investigate how generic parts in toponyms such as *Berg, Horn* and *Spitze* are spatially distributed in Switzerland we computed separate footprints for the three generic terms from the Swissnames gazetteer. The footprints for *Horn* and *Spitze* show that these terms are both most often used in the Swiss Alps, with slightly different distributions. *Spitze* is used more in Eastern Switzerland, while *Horn* seems to be primarily used in the Valais and the Central Alps. If we follow the argument that generic parts in toponyms have meaning (Hollis and Valentine 2001), this would be an indication that the topography in Eastern Switzerland is different from the topography in the Valais such as that *Spitze* and *Horn* are used to denote features with differing topographies. However, our results from SOM and cosine similarities clearly shown that *Spitze* and *Horn* indeed share similar topographies and both their spatial footprints overlap with mountainous terrain. Thus, we suggest that the spatial differences in the occurrence of toponyms with *Spitze* and *Horn* are probably more directly related to language use in dialect regions of Switzerland. For the generic term *Berg* the spatial footprint covers the *Swiss Mittelland*, but not the Alpine regions in Switzerland. This finding is counter-intuitive, as we would expect that the generic term *Berg* ("mountain") is also used where the geographic features of mountains exist, i.e. in the Alps, which was not the case. However, this finding is consistent with our previous observations that the cosine similarities for *Berg* are distinct from *Horn* and *Spitze*, and the generic terms containing *Berg* have a more diffuse topography as represented by the SOM. Thus, through the spatial footprints we found a pattern that supports our first results, where *Spitze* and *Horn* are similar topographically, even though they do not have exact overlapping footprints, and *Berg* shows a distinct pattern.

In order to look more closely at this peculiarity of the use of *Berg*, we compared its use as a generic part in toponyms with its use in current standard language. The current use of the term *Berg* is, for example, encapsulated in user-generated and georeferenced Flickr tags. These two footprints of *Berg* show almost no spatial overlap (Fig. 3 upper row), which strongly suggests that the generic part *Berg* as used in toponyms is different from the use of *Berg* in contemporary standard German. We thus suggest caution when using gazetteers to mine for contemporary uses of landscape terms.

Since we found that current use in standard language differed from use of generic terms, we looked at the historic uses of these terms in a corpus of Swiss Alpine yearbooks. Our results indicate that in a historic context, both *Horn* and *Spitze* were also used to refer to entire mountains. This was unexpected, since such an association of *Horn* and *Spitze* with an entire mountain is not reflected in the current use in standard German where *Spitze* and *Horn* are used as partonyms (names of parts) in reference to mountains. However, in some Swiss dialects, this association appears to still be found, but with our methodological approach, we could not investigate the details of this relationship, which should be examined using questionnaires or ethnographic methods.

For the use of the term *Berg* we could not find any evidence in our historic corpus that indicates a change of meaning of the term over time. On the contrary, many early occurrences of *Berg*, which date back to the late 19th century, refer to features that would still be classified as prototypical mountains. In this case, either the use of *Berg* in toponyms has always been independent from the use of *Berg* outside toponyms, or this change in meaning over time happened earlier than the time that our Text+Berg corpus covers. If we look at first historic references of a set of *Berg* toponyms dating back to the early 15th century, they support the assumption of earlier coining of toponyms with the term *Berg*. However, we currently lack more historic data to compare the use of the generic term and the use in standard language of the term *Berg* at the time these toponyms were coined.

In sum, we found that *Spitze* and *Horn* were indeed once used to refer to whole mountains in standard language as well, which may explain their high frequency as generic parts in toponyms associated with mountains, but that the use of these terms has changed over time. The case for *Berg* is different, since we found no change of meaning in the historic corpus and suggest that if a change has happened, it would date back further, to the time   when the toponyms containing *Berg* were coined.

RQ3: Is the Use of Generic Parts in Swiss Toponyms Arbitrary, Such That There Is No Meaning in Toponyms, or, Can We, for Instance, Deduce Spatial Relations or Settings from Generic Parts?

Do generic parts in toponyms have meanings? Based on our results that we discussed in the context of the first research question we argue that there is a relation of semantically similar groups of generic parts with topographic patterns. This result supports the finding of Hollis and Valentine (2001), namely that names for landmarks seem to have a greater degree of meaning (sense), compared to other types proper names. For toponyms denoting natural feature types, there seems indeed to be meaning in the generic parts; however, this meaning does not seem to be the meaning of that same term in a non-toponyms context, but rather an older meaning. Also, even within our small sample of terms, we found exceptions. For example by comparing the use of terms inside and outside of toponyms we found that *Spitze*, used in general language, refers to the summit of a mountain, whereas *Spitze* as a generic part of a toponym, as in Zugspitze, refers to the whole mountain. Another exception is *Horn*, which changed over time, such that its original meaning inside toponyms can be considered fossilized. The examples of *Spitze* and *Horn* show that the relation between landscape terms, inside and outside toponyms, sometimes is complex, which is in accordance with Coates (2006) claim on "properhood", that there is no *lexical* trust in the meaning of proper names beyond their function as merely referential.

The example of *Berg* has shown that with our methodological approach in some cases no clear-cut explanation can be found for how the generic part in toponyms relates to the use of *Berg* in general language, and how the unexpected and diffuse topographic patterns can be explained (Fig. 1 and Fig. 2). In such cases we need to rely on additional information and historic background of specific toponyms. For instance in the case of *Uetliberg* we found historic accounts that the etymology of *Berg* goes back to *Burg* (castle). However, we refrain from considering this a general pattern, as for instance *Zürichberg* is translated from the Latin name *monte Turegicum*, and thus already contains the reference to mountain (www.ortsnamen.ch).

6   Conclusions and Future Work

Through our approach we aimed to contribute to the ethnophysiographic research question on the relation of generic parts in toponyms and landscape terms in general language. Current approaches to such research range from theorizing axiomatic frameworks, to empirical investigations and ethnographic enquiries. In our study, we applied a new palette of data and methods, and were able to fruitfully extend the state of the art in ethnophysiography. The strength of our approach is the large spatial and temporal coverage and the comparison of large groups of generic terms, whereas we are limited in finding explanations for specific instances, as we showed for the case of *Uetliberg*. Our results may thus contribute to the existing discussions on proper names in domains such as linguistics, geography and psychology, and provide a means of hypothesis generation for more detailed empirical studies.

Several future research questions arise from our work on the relation of generic parts in toponyms with general landscape terms, which used a combination of approaches from ethnography and GIScience. In the following, we only discuss a set of issues that are relevant in the context of the ethnophysiographic research question of the relation between toponyms and landscape terms.

The results we reported in this paper have made it clear that more detailed ethnographic studies are needed in the future. One such case study would be to investigate the process of fossilization, and in particular if fossilization of generic parts in toponyms means that such terms are not productively used for coining 'new' toponyms anymore. We did a small thought-experiment by thinking of Antarctica as a tabula rasa for contemporary toponym coining. We found that around 1940, during the time of German expeditions in Antarctica and the discovery of *Neuschwabenland*, many new toponyms were coined, containing the generic parts *Spitze* (e.g. *Buckinghamspitze* or *Cladonspitze*) and *Horn* (e.g. *Schnidrighorn*) (Brunk 1986). In future research we aim to investigate this topic in more detail through field studies in Switzerland.

Our results show that both generic parts of *Spitze* and *Horn* have similar topographic characteristics (Fig. 1 and Fig. 2), however, the individual spatial footprints are, to large extents, different. Are these differences caused by the influence of language patterns, such as different dialects, or is there a physical parameter, other than topography, which is distinct for both these generic parts of toponyms? The underlying reasons for this fine granular difference, if there is indeed any, is a question that remains to be answered.

The question of whether the meaning of *Berg* has changed over time, which could not be fully answered in this paper, could be investigated by using a georeferenced corpus of historic landscape descriptions. In a recent paper two of the authors presented an approach for automatically gathering landscape specifications from natural language descriptions, which might be adapted for answering the present issue with *Berg* (Derungs and Purves, 2013).

Finally, it would be interesting to extend this analysis to different languages or settings. We would assume that in Europe, where place names have a long history, it is more likely that the generic parts in toponyms have become fossilized, or that there has been a change in the meaning of these generic parts, compared to the use in standard language. In other contexts, however, where colonization has replaced place names more recently, (e.g. United States, Latin America), fossilization might only have limited impact. However, the interpretation of New World toponyms is complex, since in many cases they were borrowed from the Old World, rather than being coined in more recent times.

Acknowledgments . The research reported in this paper was funded by the project FolkOnt supported by the Swiss National Science Foundation under contract 200021-126659. We would like to thank the Text+Berg team for giving us access to a pre-processed digital version of the corpus. We would also like to thank Werner Kuhn for valuable discussions of this topic.

References

Bayardo, R.J., Yiming, M.A., Srikant, R.R.: Scaling up all pairs similarity search. In: Proceedings of the 16th International Conference on World Wide Web, pp. S. 131–S. 140 (2007)

Brunk, K.: Kartographische Arbeiten und deutsche Namengebung in Neuschwanden, Antarktis. Geschichte und Entwicklung der Geodäsie 24(1) (1986)

Burenhult, N. (ed.): Language and Landscape: Geographical Ontology in Cross-linguistic Perspective. Language Sciences, vol. 30(2/3). Elsevier, Amsterdam (2008)

Burenhult, N., Levinson, S.C.: Language and landscape: a cross-linguistic perspective. Language Sciences 30(2/3), 135–150 (2008)

Coates, R.A.: Properhood. Language 82(2), 356–382 (2006)

Coaz, J.: Ueber Ortsbenennung in den Schweizeralpen. In: Jahrbuch des S.A.C von 1865, pp. 461–478. S.A.C Verlag, Bern (1865)

Derungs, C., Purves, R.S.: Empirical experiments on the nature of Swiss mountains. In: Proceedings of GISRUK 2007, Maynooth, Ireland, pp. 348–352 (2007)

Derungs, C., Purves, R.S.: From text to landscape: Locating, identifying and mapping the use of landscape features in a Swiss Alpine corpus. International Journal of Geographic Information Science (2013)

Egenhofer, M.J., Mark, D.M.: Naive geography. In: Frank, A.U., Kuhn, W. (eds.) Spatial Information Theory: A Theoretical Basis for GIS, pp. 1–15. Springer, Berlin (1995)

Fellenberg, E.: Das Aletschhorn. In: Jahrbuch des S.A.C von 1864, pp. 183–210. S.A.C Verlag, Bern (1864)

Feng, C., Mark, D.M.: Exploring NGA GEOnet Names Server Data for Cross-Linguistic Research on Landscape Categories: A Case Study of Some Toponyms in Malaysia and Indonesia. In: GIScience 2012, Columbus, Ohio, September 19-21 (2012) (extended abstracts)

Fisher, P., Wood, J., Cheng, T.: Where is Helvellyn? Fuzziness of multi-scale landscape morphology. Transactions of the Institute of British Geographers 29(1), 106–128 (2004)

Goodchild, M.F.: Citizens as sensors: the world of volunteered geography. GeoJournal 69(4), 211–221 (2007)

Grice, P.: Logic and conversation. In: Cole, P., Morgan, J. (eds.) Syntax and Semantics. 3: Speech Acts, pp. 41–58. Academic Press, New York (1975)

Hollenstein, L., Purves, R.S.: Exploring place through user-generated content: Using Flickr tags to describe city cores. Journal of Spatial Information Science 1, 21–48 (2010)

Hollis, J., Valentine, T.: Proper-Name Processing: Are Proper Names Pure Referencing Expressions? Journal of Experimental Psychology: Learning, Memory, and Cognition 27(1), 99–116 (2001)

Jones, C.B., Purves, R.S.: Geographical information retrieval. International Journal of Geographical Information Science 22(3), 219–228 (2008)

Kohonen, T.: Self-Organizing Maps. Springer, Berlin (1995)

Levinson, S.C.: Foreword. In: Mark, D.M., Turk, A.G., Burenhult, N., Stea, D. (eds.) Landscape in Language: Transdisciplinary Perspectives, pp. ix–x. John Benjamins, Amsterdam (2011)

Lindt, R.: Das Finsteraarhorn. In: Jahrbuch des S.A.C von 1864, pp. 273–312. S.A.C Verlag, Bern (1864)

Mark, D.M., Skupin, A., Smith, B.: Features, Objects, and Other Things: Ontological Distinctions in the Geographic Domain. In: Montello, D.R. (ed.) COSIT 2001. LNCS, vol. 2205, pp. 489–502. Springer, Heidelberg (2001)

Mark, D.M., Turk, A.G.: Landscape categories in yindjibarndi: Ontology, environment, and language. In: Kuhn, W., Worboys, M.F., Timpf, S. (eds.) COSIT 2003. LNCS, vol. 2825, pp. 28–45. Springer, Heidelberg (2003a)

Mark, D.M., Turk, A.G.: Ethnophysiography. Paper presented at: Workshop on Spatial and Geographic Ontologies (Prior to COSIT 2003), September 23 (2003b) (published on web site)

Mark, D.M., Turk, A.G., Burenhult, N., Stea, D. (eds.): Landscape in Language: Transdisciplinary Perspectives. John Benjamins, Amsterdam (2011)

Pike, R.J., Evans, I., Hengl, T.: Geomorphometry: A Brief Guide. In: Hengl, T., Hannes, I. (eds.) Geomorphometry - Concepts, Software, Applications. Series Developments in Soil Science, vol. 33, pp. 3–33 (2008)

Scheider, S., Janowicz, K., Kuhn, W.: Grounding Geographic Categories in the Meaningful Environment. In: Hornsby, K.S., Claramunt, C., Denis, M., Ligozat, G. (eds.) COSIT 2009. LNCS, vol. 5756, pp. 69–87. Springer, Heidelberg (2009)

Sennrich, R., Schneider, G., Volk, M., Warin, M.: A New Hybrid Dependency Parser for German. In: Chiarcos, C., de Castilho, R.E., Stede, M. (eds.) Proceedings of Biennial GSCL Conference 2009, pp. 115–124 (2009)

Skupin, A., Esperbé, A.: An Alternative Map of the United States Based on an n-Dimensional Model of Geographic Space. Journal of Visual Computing 22(4), 290–304 (2011)

Studer, G.: Das Mattwaldhorn. In: Jahrbuch des S.A.C von 1864, pp. 226–232. S.A.C Verlag, Bern (1864)

USGS, Geographic Names Information System (GNIS) (2013),
http://geonames.usgs.gov (accessed January 31, 2013)

Valentine, T., Brennen, T., Brédart, S.: The Cognitive Psychology of Proper Names: On the Importance of Being Ernest. Routledge, London (1996)

Volk, M., Bubenhofer, N., Althaus, A., Bangerter, M.: Classifying Named Entities in an Alpine Heritage Corpus. Künstliche Intelligenz 4, 40–43 (2009)

Wehrens, R., Buydens, L.M.C.: Self- and Super-organising Maps in R: the kohonen package. Statistical Software 30(6), 1–19 (2007)

Weilenmann, J.J.: Der Piz Roseg. In: Jahrbuch des S.A.C von 1865, pp. 86–128. S.A.C Verlag, Bern (1865)

Creating a Corpus of Geospatial Natural Language

Kristin Stock[*], Robert C. Pasley, Zoe Gardner, Paul Brindley, Jeremy Morley, and Claudia Cialone

Nottingham Geospatial Institute, University of Nottingham, United Kingdom
kristin.stock@nottingham.ac.uk

Abstract. The description of location using natural language is of interest for a number of research activities including the automated interpretation and generation of natural language to ease interaction with geographic information systems. For such activities, examples of geospatial natural language are usually collected from the personal knowledge of researchers, or in small scale collection activities specific to the project concerned. This paper describes the process used to develop a more generic corpus of geospatial natural language.

The paper discusses the development and evaluation of four methods for semi-automated harvesting of geospatial natural language clauses from text to create a corpus of geospatial natural language. The most successful method uses a set of geospatial syntactic templates that describe common patterns of grammatical geospatial word categories and provide a precision of 0.66. Particular challenges were posed by the range of English dialects included, as well as metaphoric and sporting references.

Keywords: corpus linguistics, geospatial, natural language.

1   Introduction

The content and nature of language describing the location of geographic features, people, events and socio-economic phenomena relative to actual places and to other phenomena (for example, spatial relations) or in absolute terms (for example, cardinal directions) are of interest for a number of research activities in disciplines including linguistics, geography, computer science and cognitive science. For example, such language (referred to herein as *geospatial natural language*) may give insight into the nature of underlying cognitive structures [12], may reflect more general syntactic, grammatical and semantic patterns and thus provide information about linguistics, and is also relevant for the development of computerized natural language systems.

This paper addresses the research question of how to create a corpus of geospatial natural language that contains a large number of geospatial natural language clauses across a range of application domains and dialects, but that excludes clauses without any geospatial content. The motivation in creating this corpus was to support the authors' ongoing research in natural language geospatial querying as an alternative to traditional spatial querying which is often poorly understood by non-expert users [30],

---

[*] Corresponding author.

T. Tenbrink et al. (Eds.): COSIT 2013, LNCS 8116, pp. 279–298, 2013.
© Springer International Publishing Switzerland 2013

but there are many other potential applications for the data [24], including numerical analysis of the use of particular language constructions (e.g. frequency and structure) for linguistic studies; the development and evaluation of natural language based ontologies and languages [2, 29]; the identification of contextual use of location expressions (for example, with different geographic feature types) and data mining to establish common geospatial expressions to support natural language interpretation, and as a basis for the design of empirical studies.

The scope of the corpus is limited to language describing the location, movement or shape of geographic features, people, events and socio-economic phenomena, relative to actual places, relative to other geographic features or absolutely (north etc). The content was harvested from a range of different web sites, with the aim of covering different types of language and different application domains. The first release contains 10,146 clauses totaling 233,115 words, harvested from 46 different web sites. A monitor corpus approach [24] is planned and new content is being continuously harvested.

Examples of text from the geospatial corpus include:

- *The wall crossed Baldon Gate and turned in a south west direction to Carter Gate.*
- *The main visitor entrance to Pompeii ruins is right outside Pompeii Scavi station, across the road and to the right.*
- *Still continuing in a straight line cross the next field towards the end of a line of trees and then follow the field edge a short way to cross a wooden style...*

The paper is structured as follows: Section 2 provides a review of relevant literature; Section 3 describes the scope of the corpus with respect to how geospatial language is defined for the purposes of the corpus; Section 4 explains the method adopted for identifying geospatial natural language, including a set of geospatial syntactic templates that form the basis of the method and the evaluation of the method against a manually selected evaluation set; Section 5 describes the procedure for harvesting content using the approach described in Section 4 by crawling selected web sites and Section 6 discusses issues arising from the approach. The corpus content can be searched at http://geospatiallanguage.nottingham.ac.uk/.

2   Background

A linguistic corpus is a large body of examples of language use that may be studied and analysed, usually with computational methods. A sampling frame is required to allow representative and balanced examples to be drawn from the total of all language. Corpora can be annotated or un-annotated, and may include written and spoken language [24].

2.1   Existing Geospatial Corpora

A large number of general machine-readable corpora have been developed, including the Brown Corpus, the British National Corpus (BNC) and the International Corpus of English (ICE). Others are based on some form of restriction such as the Reuters corpus of news. More specialised still are domain specific corpora (for example, the Wolverhampton Business English Corpus and GENIA for biomedical text [9]).

Corpora dedicated to geospatial natural language are very limited. Some small collections of spoken language relating to spatial relations [41] and paths [5] have been collected, but these contain only a small number of examples of geospatial natural language collected from individuals. The HCRC Map Task Corpus contains descriptions of map routes and studies geospatial language and also considers aspects such as eye contact and familiarity of communicators [1]. Written language collections include [38], extracted from four textual descriptions of European regions and cities from web sources; [44], extracted from a large set of web pages containing route descriptions (identified by the presence of postcodes) [15] particularly focused on complex spatial language extracted from the web and [19], drawn from respondent descriptions of movement patterns.

These examples were developed for specific project purposes. The work described in this paper intends to go beyond this and provide a source of examples of a range of geospatial natural language that can be used for multiple purposes. To this end, the work aimed to extract a wide range of geospatial natural language from around the English speaking world, from a range of different domains. One of the biggest challenges in this endeavour was recognising and extracting geospatial natural language (as opposed to natural language without a geospatial component).

2.2   Methods for Extraction of Geospatial Natural Language

The majority of the work addressing the extraction of geospatial natural language from text focuses on the extraction of actual location information from web pages [27, 39], the alteration of those locations by natural language modifiers (for example, 'northeast of Washington DC'), and how natural language spatial relation expressions can be converted into locations on the ground [3], or be used to help disambiguate locations [28] and define their boundaries [32]. Gregory and Hardie [14] combine linguistic and geographic approaches by extracting place names from generic text corpora using a combination of part of speech tagging with place name listings in a gazetteer. Another recent approach [4] describes a system for geospatial mining using spatial language descriptions, focusing on geospatial path understanding (the latitude/longitude coordinates of a path given a natural language description of that path). None of these approaches attempts to detect geospatial natural language in a range of different forms.

Zhang at al. [45] adopt a broader approach and attempt to extract spatial information from Chinese text in a manually constructed unstructured text corpus created from the geography subset of the encyclopaedia of China. They initially use a rule based approach, in which they manually develop syntactic rules for different parts of geospatial speech to automatically extract spatial clauses. In later work, Zhang et al [46] adopt a Support Vector Machine approach for parsing and classifying complex spatial expressions. This work bears some similarity to the work presented in this paper, particularly with its manual syntactic rules. However, the focus of Zhang et al's work is narrower in that they are only interested in clauses that relate place names. [39, 29] also discuss simple syntactic structures for spatial expressions that define places.

2.3   Generic Extraction Methods

There is an established vein of research within Information Extraction (IE) [2] which aims to extract related entities. This is often done using Regular Expressions or patterns which contain wildcards that are filled by the entities [6, 7]. Many of the methods commonly used in Information Extraction were considered unsuitable for the purposes of this research because they relied on the availability of a large training set (for example, by automatically mining patterns) which was not available, or because they assumed the full range of geospatial natural language was known at the outset, limiting the ability to go beyond this and collect a wide range of diverse language.

3   Scope

The aim of the research was to create a corpus of geospatial natural language, geospatial natural language being defined (for the current purposes, as distinct from that specified by [21]) as language that describes the location in space or movement through space of geographic features, people, events or socio-economic phenomena relative to places, to other geographic features or in absolutely terms (for example north etc). However, even within this definition, the identification of geospatial natural language also proved vague, and the choice about whether a particular clause qualified as geospatial or not was often difficult.

Specifically, the examples of geospatial natural language encountered can be considered to fall into two categories:

1. those that describe the locations of static geographic objects (features, places, events or phenomena), either absolutely (*the mountain in the north*) or more commonly, relative to other geographic objects (*the mountain beside the bay; the mountain near Lake Taupo*); and
2. those that describe the movement of objects (usually people, animals or vehicles) relative to static geographic features (*she crossed the bridge*) including people's walking paths, bus and van routes, etc.

Note that objects referred to using language in the first category are sometimes described using language that is more akin to that in the second category as per Talmy (2000), in which static objects are described using fictive paths or patterns (for example, *the road followed the cliff from Innisfail to the headland*).

Our broad definition is further refined with the following:

1. Language describing location and movement at the scale of a person's body or within a room or small building was excluded. Geospatial natural language is defined as language relating to space larger than a room or small building (although locations in some large public buildings are included), within the scope of the planet earth, hence the reference to geospatial, rather than spatial language. The scale scope corresponds roughly with environmental and geographical scales using Montello's [26] typology, although this typology is more focused around perception than absolute scale in the way that we have defined it.
2. Sporting language and language describing the weather were excluded on the basis that these two areas in particular (which were commonly encountered) often use

unusual and specialised jargon that does not apply generically to the ways in which people describe geographic object locations and movements in space. However, language describing disaster events (floods, tsunamis) that result from the weather was included, as these are often reported in mainstream news and blogs.

3. Metaphoric uses of location or movement of geographic objects were excluded (for example, *his business is on the road to nowhere*) where they appear without accompanying non-metaphoric geospatial natural language, but some clauses combine the two, in which case they are included. As a rule of thumb in detecting metaphors, whether or not it would be possible to draw a plan/map/diagram of the clause (which would not be possible with the *road to nowhere* example) was used.

4. Addresses were excluded. While these are certainly descriptions of location, it may be debated whether they qualify as geospatial natural language. Nevertheless, addresses have their own particular format (depending to some extent on the administrative jurisdiction) that has been studied in detail [22], and it was not our goal to study and analyse address formats in this work.

The corpus contains only clauses (sentences or parts of sentences), not entire documents as is usually the case with text corpora (in order to allow text to be studied in context). The reason for this is that an entire document may contain one or a small number of geospatial references within a document that is not primarily geospatial in nature, and the actual geospatial natural language may be difficult to extract from within the larger document. However, with each clause, the document source is stored (all of which are publicly available web pages), so it is possible to view the entire document as long as the source web site remains with the content in place.

The linguistic scope of the corpus is English broadly, and is not confined to British English or any other English dialect to allow cross-dialect examination of geospatial natural language. American English in particular has some spatial relation expressions that are not commonly found in other dialects of English (for example, *in back of*), and the corpus aims to include these. The corpus also contains language in a range of different registers and styles.

4   Identifying Geospatial Natural Language

The biggest challenge in creating a corpus of geospatial natural language involves automation of the identification of geospatial natural language within documents that contain both geospatial and non-geospatial natural language.

4.1   Method

The development of a method for automatic identification of geospatial natural language was driven by two goals:

1. to identify the broadest possible range of geospatial natural language, avoiding reliance on a pre-existing list of all words that might be considered spatial, as this would not permit the discovery of new language that was not on the list and

2. to develop a method that could ultimately be extended to include other languages.

However, these goals were only partially achieved – see below.

The identification methods that were used to populate the corpus involved identifying different words and phrases within a clause that are likely to indicate that it is geospatial. The clauses were scored according to the presence of geospatial words/phrases[1], and the most highly scoring clauses were considered more likely to be geospatial. In order to do this, a series of lists of potentially geospatial words/phrases were created. In the most basic (and least successful) of the four scoring methods, a simple, unstructured list was used, while in the other three methods, a series of lists of different categories of words/phrases were combined with syntactic templates describing how the different categories of geospatial words/phrases might be combined in a geospatial clause. As discussed below, the approach was adapted to ensure that unexpected geospatial natural language would still be identified using these approaches, as described in goal 1 above.

Three categories of geospatial words and phrases were used.

Geographic Features. Geographic features are an important component of many clauses that describe locations in geospace. Examples include *the **camping ground** beside the **river**, the **road** skirts the **bay*** or *the **library** is beside the **supermarket***.

A list of geographic features was created by combining concepts from selected components of GEMET[2], SWEET[3], the hydrology, buildings and places ontologies from Ordnance Survey[4] and the geonames ontology[5]. A selective approach proved more successful than a more generic approach in which the widest possible range of geographic feature types from a large number of ontologies were included.

A measure for weighting the geographic feature words/phrases was developed that was an adaptation of the well-known term frequency-inverse document frequency (tf-idf) weighting statistic [42]. We refer to this measure as sf-icf, and is calculated as the number of times a given word/phrase appears in the evaluation set of 2455 geospatial clauses manually identified from the British National Corpus (see Section 4.2): the spatial frequency (sf), divided by the number of times the same word/phrase appears in the entire British National Corpus: the inverse corpus frequency (icf). This measure is intended to give higher weight to those words/phrases that are found to appear in clauses that are known to be geospatial, offset by the overall frequency of the word/phrase, to remove the effect of common words like 'the'.

$$sf - icf = \frac{word\ count\ in\ spatial\ phrases}{word\ count\ in\ BNC} \tag{1}$$

Place Names and Vernacular Names. Geonames[6] was used as a source of place names, since it is one of the most comprehensive worldwide gazetteers available.

---

[1] While most of the entries in the lists were individual words, some place names and some members of other word categories were actually phrases (for example, *as far afield as*).

[2] http://www.eionet.europa.eu/gemet

[3] http://sweet.jpl.nasa.gov/ontology/

[4] http://www.ordnancesurvey.co.uk/oswebsite/ontology/

[5] http://www.geonames.org/ontology/documentation.html

[6] http://www.geonames.org/

However, the full list of over 10 million names proved difficult to manage from a processing point of view, so the list was confined to only those place names in English speaking countries, or countries that were common places of interest or discussion for English speakers (usually due to tourism or prominence in events that were of particular relevance to English speakers, like areas of military involvement by English speaking countries).

The list of place names from Geonames was augmented with a list of vernacular sub-settlement (suburb and neighbourhood) names for the UK only, as the comprehensiveness of sub-settlement place names within Geonames was questionable. The sf-icf weighting measure was also used for place names, although it is has the disadvantage of being biased towards British place names due to the use of British National Corpus to calculate sf-icf (see Section 6).

Words/Phrases Describing Location and Movement in Space. While place names and geographic features give some indication of geospatial natural language, on their own, they are insufficient to identify whether text is geospatial or not. Firstly, lists of place names are commonly found on web sites, and these were not of interest for the geospatial corpus. Secondly, and most importantly, large amounts of text may mention place names and geographic features and discuss many aspects of those places or features, without describing locations or movement in space. For this reason, words/phrases that actually describe that location and movement are required to be included in the method.

However, most words/phrases that describe location and movement on their own do not provide sufficient evidence of spatial language. For example, most spatial relation words/phrases (especially spatial prepositions) are used for numerous non-geospatial purposes in English (for example, *she was beside herself; the mystery surrounding Stonehenge; John is very close to his mother*). Space and spatial words and phrases are commonly used to describe many aspects of human experience [20]. For this reason, the simple inclusion of a list of spatial words/phrases is not very successful in identifying geospatial clauses, even when combined with place names and geographic features. The ways in which spatial words/phrases are combined is an essential component of identifying geospatial natural language.

To resolve this problem, several categories of geospatial words/phrases were identified (Table 1), along with a set of geospatial syntactic templates (Table 2) to describe commonly identified patterns of geospatial natural language based on a simple grammatical categorisation of geospatial natural language. The templates were developed manually, through examination of a set of example geospatial clauses from the British National Corpus. The process involved identification of the 'constructions' or patterns by applying a technique developed in linguistics research [13,17] to examine language clauses. [35] provides more details of some lower level constructions, from which the syntactic templates were distilled (as well as some additional constructions not described in [35]).

Table 1. Geospatial Word Categories

Category	Examples	Description
Generic Specifier (GS)	one, 29, this, your	An indication of the thing or things described, either a numerical quantifier or a determiner (this, that, your).
Spatial Noun (SN)	Nottingham, hill	A place name (including vernacular place names) or a geographic feature type.
Location Specifiers (LS)	the top of, the coast of, the north of	Location specifiers define a particular part of the location. In our categorization, location specifiers must relate to a geographic feature or place name in order to be considered spatial (for example, *the top of the hill*).
Spatial Verb Satellites (SVS)	across, near, at	In English, and other satellite-framed languages, spatial relations may be expressed using verb satellites (prepositions and postpositions) as opposed to in verb-framed languages in which they are expressed by specialised verbs (*the road goes across the park/the road crosses the park*)[7].
Spatial Relation Qualifiers (SRQ)	entirely, just	Adverbs that qualify a spatial relation (*the road goes just outside the park*).
Spatial relation verbs (SRV)	encircles, avoids	Verbs that describe the spatial relationship between two objects, whether static or moving and that can be used alone without other verbs to express a spatial relation (*the road crosses the park*).
Location and movement verbs (LMV)	goes, runs,	Verbs that describe actual location, movement or action, referring to either a static object fictively (as per Talmy (2000)), or an actual movement.
Directions (D)	left, right, north, south	Adjectives indicating direction, combined with a reference object either explicit or implied.
Measurement Reference (MR)	walk, drive	A measure that is used as a reference for movement (*three minutes walk; 5 hours drive*)
Quantity Unit (QU)	miles, minutes, feet	The unit used to specify movement (*three minutes walk; 5 hours drive*)

Table 2. Geospatial Syntactic Templates

Definition	Examples
Spatial Noun Phrase = geographic feature | place name | (SN + SN) | (GS + Spatial Noun Phrase) | (LS + 'of' + Spatial Noun Phrase) | (D + LS + 'of' + Spatial Noun Phrase)	*the lake; Nottingham; Lambley Church; this river; end of the lake; uphill side of your road*
Spatial Verb Satellite Phrase = SVS | (SRQ + SVS) | (Spatial Verb Satellite Phase + SVS)	*right beside; just near to*
Direction Phrase = (Spatial Verb Satellite Phrase + GS + D) | (SpatiaVerb Satellite Phrase + GS + D + Spatial Verb Satellite Phrase + Spatial Noun Phrase) | (D + Spatial Noun Phrase) | (D + Spatial Verb Satellite Phrase + Spatial Noun Phrase)	*on your left; in the north; to the right of the police station; uphill from the well; north-east London*
Measurement Phrase = (GS + MR) | (GS + QU + MR) | (SVS + Measurement Phrase) | (Measurement Phrase + Direction Phrase)	*after 3 minutes walk; in a short distance; three miles from; twenty-nine minutes walk north*
Spatial Verb Phrase = SRV | (SRV + (Spatial Verb Satellite Phrase | Direction Phrase)) | (LMV + (Spatial Verb Satellite Phrase | Direction Phrase)) | (LMV + Spatial Verb Phrase) | (SRV + Spatial Verb Phrase)	*runs along; goes to the left; is passed by; rises beside; crosses*
Spatial Clause = Spatial Noun Phase + Spatial Verb Phrase | Spatial Noun Phase + Spatial Verb Satellite Phrase | Spatial Clause + Measurement Phrase	*the road goes to the left; the house is nearby*
Spatial Sentence = Spatial Noun Phrase + Spatial Clause	*the road crosses the river*

In these templates, the + symbol indicates combinations of categories that would normally be adjacent (logical AND). **The + symbol is not order specific, it merely represents adjacency, not order of appearance,** although the templates presented here describe the most common order of appearance in English. The | symbol indicates alternative definitions (logical OR).

---

[7] In the case of motion events, this distinction is referred to as manner languages vs. path languages [37].

For each of the geospatial word/phrase categories, an extensible list of words/phrases was created, and these lists were used together with the templates to score clauses for evidence that they contained geospatial natural language. The list of spatial relation verbs was initially populated from multiple sources including Wordnet [25] (GIS spatial operator words were used as seeds and semantic links were followed to create a larger list) and selected spatial words/phrases from the English prepositions in Wikipedia, [10,23,33,37], and manual selection from a set of geospatial clauses from the British National Corpus. The other lists were populated from manually identified geospatial clauses from the British National Corpus.

New items were added to these lists as they were found, in order to ensure that the widest possible range of geospatial natural language was identified. The lists themselves contain the lemma (canonical form) for each word/phrase, and different variations (plurals, or conjugations in the case of verbs) were generated programmatically.

The syntactic templates in Table 2 express the most common ways in which the different categories of geospatial words/phrases may be combined to produce geospatial clauses. This is a high level set of templates, based on broad categories, and is intended to be generous. That is, it allows more than it prohibits. There may be cases within particular categories that may not validly combine with cases from other categories, but this is used merely as a mechanism to determine how spatial a clause is, not as any attempt towards a Chomskian generative grammar [8]. Also, these templates do not address the details of language use within categories, as included for example in image schemata [18,20], or individual members of closed-class grammatical categories and their linguistics treatment [10,36]. Such detail was not considered to be warranted for our purposes. Other attempts to create a categorization of natural language expressions of movement and path (for example, [16]) do exist. However these do not summarise the geospatial syntactic structures as required here.

Scoring Methods. Four different scoring methods were tested. The first method (naïve scoring), does not use the geospatial syntactic templates, but the remaining three methods use the templates and apply scoring in different ways. For all of the methods that used the syntactic templates, a small number of intervening words that are not explicitly listed in the geospatial word/phrase categories (for example, *the big road crossed the narrow estuary*, rather than *the road crossed the estuary*) were permitted. The approach proved most effective if a maximum of 1 intervening word was permitted in most cases, although for some word combinations up to three were permitted, based on manual examination of geospatial natural language combined with the results returned. For all methods, lists that contain only place names (with no other geospatial word) were excluded.

Naïve Scoring. The most basic method for scoring a clause (included for comparison purposes) uses the following formula:

$$score = \frac{(number\ of\ words\ that\ appear\ in\ the\ spatial\ word\ megalist)}{(total\ number\ of\ words\ in\ the\ sentence)} \qquad (2)$$

The spatial word/phrase megalist is a list of words/phrases from all of the different sources, including the geographic feature ontologies, the place names lists and the words/phrases in all of the geospatial word/phrase categories. All words/phrases are weighted equally. This scoring method does not give any consideration to the structure of the clause, but considers simply the **density** of spatial words/phrases.

Simple Aggregation Scoring. The Simple Aggregation method scores a clause according to the total number of spatial components, at both a phrasal and intra-phrasal level, but does not include spatial clauses and spatial sentences. For each of the different types of phrases listed in Table 2, a basic value indicating the frequency with which that particular phrase type appeared in a geospatial clause was calculated using the BNC evaluation set (see Section 4.2) with the following formula:

$$value = \frac{(count\ of\ phrase\ type\ in\ BNC\ evaluation\ set\ of\ geospatial\ phrases)}{(count\ of\ phrase\ type\ in\ full\ BNC\ evaluation\ set)} \tag{3}$$

The calculated values are shown in Table 3. The value was then multiplied by the number of elements within the clause that were present out of the 'normalised'[8] total. The final score for the clause was calculated using the formula:

$$S = \sum_{i=1}^{n} x_i^{value} * x_i^{elementweighting} \tag{4}$$

Table 3. Simple Aggregation Scoring Method

Element/Phrase Type	Value	Weighting
Simple Aggregation Scoring Method		
Spatial Noun Phrase	Value calculated from weighting of geographic feature or place name using the sf-icf method (Section 4.1.1.1)	Number of elements/3 ('of' is ignored)
Spatial Verb Satellite Phrase	0.242	Number of elements/2
Direction Phrase	0.522	Number of elements/4
Measurement Phrase	0.176	Number of elements/3.
Spatial Verb Phrase	0.224	Number of elements/2. (Each phrase is counted as one element).
Enhanced Aggregation Scoring Method		
Spatial Clause	(Sum of value * weighting of component element phrases)	Number of elements/2
Spatial Sentence	(Sum of value * weighting of component element phrases)	Number of elements

Elements were only counted once: if they appeared in a phrase, then they were counted as part of that phrase. If they appear individually, they are counted individually.

Enhanced Aggregation Scoring. The Enhanced Aggregation method uses the same approach to scoring individual base-level phrases as for the Simple Aggregation

---

[8] Many rules theoretically allow an infinite number of components because they are self-referencing. The normalised total is the maximum number of element without recursion. Where the number in the actual sentence exceeds this, the element weighting > than 1.

method, but also considers spatial clauses and spatial sentences. The final score for the clause was calculated using the formula:

$$S = \sum_{i=1}^{n} x_i{}^{value} * x_i{}^{elementweighting} \tag{5}$$

Exponential Scoring. The final method applies exponents to the basic weights for different phrase types in order to determine whether an exponential rather than an aggregation approach gives better results. In addition, this method builds the weight-ings of the component elements into the weight of the higher level element, rather than simply aggregating the components within the clause that is being evaluated. Table 4 shows the method that was used to score each element within the overall clause that was being evaluated, with the score of the highest level element that was present in the overall clause being given to the clause as its score. The numbers shown as the base for the weighting of each element type were calculated using for-mula 3.

Table 4. Exponential Scoring Method

Grammatical Construct	Weighting
Spatial Noun	As for Simple Aggregation Method
Spatial Noun Phrase	$w_{SNP} = w_{SN}{}^{1/(1 + quantity\ of\ elements)}$
Direction Phrase;	$w_{DP} = 0.522$
Measurement Phrase;	$w_{MP} = 0.176$
Spatial Relation Verb;	$w_{SRV} = 0.248$
Location and Movement Verb;	$w_{LMV} = 0.206$
Spatial Verb Satellite Phrase	$w_{SVSP} = 0.242^{1/(quantity\ of\ SVSs\ +\ quantity\ of\ SRQs)}$
Spatial Verb Phrase	$w_{SVP} = w_{highest\ weighted\ component}{}^{(1 - summed\ weight\ of\ all\ components\ in\ the\ phrase)}$
Spatial Clause	$w_{SP} = w_{SNP}{}^{(1 - combined\ weight\ of\ all\ components\ in\ the\ clause)}$
Spatial Sentence	$w_{SS} = w_{SP}{}^{(1 - combined\ weight\ of\ all\ components\ in\ the\ sentence)}$

Spatial Word/Phrase Propagation. One of the main objectives of the work was to develop a method to identify spatial clauses without restriction to a particular set of words/phrases or terms. However, it proves difficult to get satisfactory results without recourse to a list of relevant words/phrases, so the lists were instead extended as new words/phrases were found using two different approaches. The first approach was completely manual and involved adding new words that were identified from har-vested content into the word/phrase lists. The second approach involved the use of a search engine to actively search for new words/phrases using the geospatial syntactic templates combined with content already harvested. The spatial elements within the evaluation set clauses were replaced with wildcards and set up as http requests, ready to send as quoted search terms to a specific search engine. Thus, if a clause such as *the road goes by the park* was passed as a search request (*the road goes * the park*), the search engine might also return references such as *the road goes next to the park* or *the road goes around the park*. The first iteration of this approach during testing and development of the method resulted in an increase in the spatial verb satellites category of 13%, but after subsequent additions, only occasional new words/phrases were found. Furthermore, all of the methods are designed in such a way as to minimize any absence of single geospatial words/phrases from the word/phrase lists, in that they are weighted according to the presence of multiple geospatial words/phrases.

Multilingual Aspects. Another main objective of this research was to create an approach that could be applied to different languages. For this reason, methods that involved machine learning based on particular syntactic combinations (for example, identifying syntactic rules attached to a particular spatial relation term) were avoided. The initial approach was to develop methods involving content from multilingual ontologies/thesauri (for example, GEMET) and place name lists that could be directly applied to document sources in different languages. However, these approaches proved inadequate (see Section 4.2), hence the consideration of syntactic patterns (as expressed in the geospatial syntactic templates). In order to apply this method in other languages, a similar set of templates for the language in question would need to be derived. For some languages, these templates are likely to vary widely.

4.2   Evaluation

Defining the Evaluation Set. In order to evaluate the methods developed, an evaluation set of 13,953 clauses was created, 2455 of which were considered geospatial, the remainder not. The clauses were extracted from the British National Corpus (BNC) using basic methods designed to be weighted towards selection of geospatial clauses for practical purposes, and manually flagged as geospatial or not, according to the criteria presented in Section 3.

The size of the evaluation set was tested by examining the well-known precision, recall and f-measure [31] for different sized evaluation sets using the Simple Aggregation scoring approach. These methods were applied in the context of this work using the following formulae:

$$precision = \frac{num\ of\ top\ n\ ranked\ clauses\ that\ are\ geospatial}{n} \tag{6}$$

$$recall = \frac{number\ of\ geospatial\ clauses\ in\ the\ top\ n\ ranked\ clauses}{number\ of\ geospatial\ clauses\ in\ evaluation\ set} \tag{7}$$

$$f - measure = 2.\frac{precision.recall}{precision + recall} \tag{8}$$

Figure 1 shows the figures for evaluation sets of sizes ranging from 1000 up to 13,911, with n values ranging from 100 to 900.

The results show that the graphs level out for precision at an evaluation set size of 7000, and for recall at an evaluation size around 12000. On this basis, the evaluation set size was considered adequate to select the best method for harvesting of content, and to provide an indication of the quality of the automated methods. The complete set was used for the evaluations.

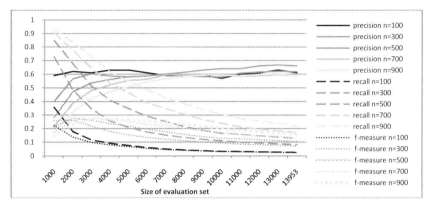

Fig. 1. Precision, Recall and f-measure for different scoring methods

Evaluation sets with the same number of geospatial clauses (2455) and varying numbers of non-geospatial clauses (as distinct from randomly selecting the set of the required size from the 13,953) were also tested, and showed a similar pattern.

Evaluation Results. The evaluation set was used to evaluate the four methods presented in Section 4.1. However, the evaluations were run on the British National Corpus, and the web sites that were used for harvesting were different in nature in some cases (depending on the web site) from the BNC and there was wide variation between web sites. For example, the BNC contains mainly (but not exclusively) British content, which is particularly relevant because of the reliance of our methods on place names. The evaluations described here were run using a set of British place names, while in the final harvesting, we included place names from a wide range of countries to ensure that non-British language was also detected. Figure 2 shows the results from the evaluation.

Testing of statistical significance showed that the differences between all three measures for all of the pairs of methods were statistically significant at $p >= 0.005$, with one exception: the difference in precision values between the Simple Aggregation and Exponential methods was not confirmed to be significant ($z = 1.94$), while recall and f-measure difference for the same pairing were significant. Statistical significance for all measures and pairings were calculated using the Wilcoxson signed-rank test for non-parametric distributions (the non-normal nature of the distribution was confirmed using the Shapiro-Wilk test).

All of the methods that used the syntactic templates provided significantly better results than the Naïve method, which was very poor. The internal structure of the clauses is therefore important. The Simple Aggregation Scoring method gave the best results overall, although the Exponential and Enhanced Aggregation methods were superior for very small values of n. The difference between the Simple Aggregation and Exponential methods was very small, with only slightly better results for the former. As shown in Section 4.1.2.3, the Enhanced Aggregation method only augments the Simple Aggregation method with consideration of spatial clauses and spatial sentences, and this deteriorates the performance of the method. This is thought to be because incorrectly identified lower level phrases (for example, spatial noun phrases, spatial verb phrases) are further emphasised by their consideration as spatial clauses and spatial sentences, meaning that while the augmented approach increases the score of geospatial clauses, it also increases the score of false positives. The Exponential method also considers spatial clauses and spatial sentences, but performs better than the Enhanced Aggregation method. It uses an exponential model to increase the cumulative effect of multiple spatial phrases of different types (relative to the simple and enhanced aggregation methods, which use linear aggregation) within a given spatial clause or sentence, but still does not perform as well as the Simple Aggregation method. A test of the Exponential method with spatial clauses and spatial sentences excluded was also performed, with minimal effect on the results for the Exponential method, so this was not pursued further.

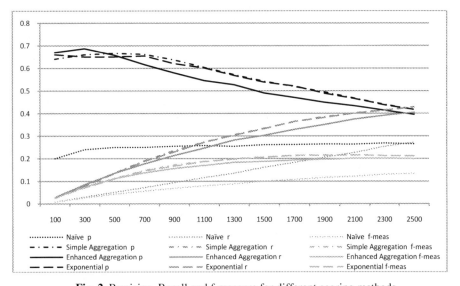

Fig. 2. Precision, Recall and f-measure for different scoring methods

The shape of the graphs, combined with the values for the three evaluation measures, were used to select an initial threshold score for content harvesting. The value of n=900 was selected as a point at which precision had just started to decrease (0.66), while recall was still increasing (0.23), and the f-measure (combining the two) had started to flatten off. In the evaluation set, this value of n coincided with a score of 0.7 using the Simple Aggregation method, which was used as a threshold for automatic harvesting. For comparison, for complex multi-word queries (which most closely match the complexity of the corpus retrieval task), Google's precision and relative recall[9] have been assessed as 0.73 and 0.80, while Yahoo's were found to be 0.89 and 0.20 respectively [39]. While we achieved a precision level approaching that of the Google, our recall was significantly lower (although higher than that of Bing).

---

[9] A measure of the degree of overlap in recall between the two search engines.

However, we were less concerned with recall than precision, as our goal was to collect a range of language examples, many of which are similar or repeated in different sources, rather than find every relevant document.

5   Populating the Corpus

5.1   Procedure

Corpus harvesting was implemented using a series of steps: (1) Web sites were crawled using a crawler based on the websphinx[10] java library, confined to only the relevant domain name; (2) The html tags from the web site content were stripped using lynx[11], then cleaned up to produce text without any tags; (3) The GATE[12] suite of software [11] was used to parse the text from the web pages to establish sentence boundaries which were then loaded into a mysql database; (4) Java was used with GATE's JAPE rules to annotate the text clauses with tags corresponding to the geospatial word/phrase categories and the geospatial syntactic templates in Table 2; and to calculate scores. Clauses were ranked in order of their score.

Following automated content harvesting, clauses were manually filtered to remove clauses that were not geospatial. The resulting corpus data set contains the clause, together with a unique identifier, the date of harvesting and web URL.

5.2   Sources

Corpus content was harvested from selected web sites. Sites were selected because they were expected to be rich in geospatial natural language, or to provide an important cross-section of language. The aim in harvesting content was intentionally skewed towards an attempt to collect as wide as possible a range of examples of geospatial natural language in selection of the web sources, in preference to statistical representativeness and balance [24]. The selected sites do not aim or claim to be a statistically representative sample of all sites on the web, and nor do the clauses in the corpus aim or claim to be a statistically representative sample of all geospatial natural language. Instead, they aim to provide a diverse range of geospatial natural language expressions, and as the corpus grows, the diversity and the representativeness of the sample will grow [43].

The aim of diversity was considered both in selection of the sites and in the filtering strategy. In regard to selection of sites, the selection aimed to include a range of:

- written language styles, ranging from official reports (for example, world news sites, automobile association reviews) to blogs (for example, travel blogs);
- country of origin from around the English speaking world, aiming to capture different English dialects and
- subject matter, including news, travel, local history and creative writing.

---

[10] http://www.cs.cmu.edu/~rcm/websphinx/
[11] http://lynx.isc.org/
[12] http://gate.ac.uk/ie/

However, the goal of achieving a wide range of geospatial natural language was balanced against the practicalities of content harvesting and the need to identify sources with high geospatial content for pragmatic reasons. Some web sites proved to return so few geospatial clauses that the time taken to harvest and filter the content could not be justified. In particular, it was difficult to establish a wide cross-section of subject matter as most geospatial natural language was found on certain types of sites, particularly travel and tourism and local history. On news web sites, most geospatial natural language related to reports on military operations and natural disasters.

The aim of collecting a wide cross-section of geospatial natural language was also addressed with the manual filtering strategy. In particular, many examples of very common expressions of geospatial location are included in the corpus, but in some extreme cases, inclusion was limited to avoid dominance of these very common constructions at the expense of a wider range of language (for example, the use of 'in' to describe geospatial location). Secondly, very short clauses, often photo captions, (for example, sunset at Bowen) were excluded because they provided insufficient context to determine their geospatial semantics. Finally, the quantity of clauses returned from the harvesting that were manually examined for potential inclusion in the corpus was dictated by a combination of:

- the proportion of the clauses examined that were geospatial, for practical reasons;
- the diversity of the source (in subject matter, language style etc.) against content already in the corpus;
- the score threshold: in most cases, clauses were examined down to a score of about 1.2, since across a number of web sites, this turned out to be roughly the score at which about a third of the clauses examined were confirmed as geospatial.

Ultimately, this selection process was subjective, as it involved a judgment based on knowledge of the existing content, as well as the content itself.

The use of the web itself as a representative source of language has been called into question [24] due in part to the quality of the language (sometimes mistakes, typographic or grammatical errors are included). However, the inclusion of these errors in the corpus of geospatial natural language were not considered problematic, as they might also represent the kinds of errors found in other natural language sources, because common errors tend to become part of normal accepted language over time and because examination of frequencies in a corpus would tend to exclude single errors.

6   Discussion

The results returned by the automated harvesting approach identified some particular shortcomings with the approach. Most of these shortcomings related to incorrect identification of words/phrases as members of one of the geospatial word/phrase categories, which then had a cascading effect.

Firstly, in some cases, words/phrases were incorrectly identified as place names, because they were found in the place names list that came from GeoNames. In order to correctly find clauses that relate to locations, the largest possible set of place names is ideal. However, the larger the set of place names used, the more likely it is that false positives will be identified (an example of the well-known trade-off between precision and recall). These often include people's names (for example, *Sydney*), or generic words that are also place names (for example, *Bath*). The sf-icf method was designed to reduce this possibility by giving words that were not commonly included in confirmed geospatial clauses in the evaluation set a lower score in the scoring methods, but the benefit of this is reduced when place names from around the world are included, as the evaluation set used to calculate sf-icf scores was based on British National Corpus content which is heavily skewed towards the UK. However, an evaluation set to cover world wide names would need to be very large (and must be manually created), which is impractical. However, as the corpus grows, we propose to use the corpus content as a source for the sf-icf method, thus improving results.

The method is also deficient at detecting metaphors. Spatial metaphors are very common in generic language, and in many cases were scored highly. While the inclusion of geographic features and place names reduces this risk, if the clause contains a number of metaphoric spatial references, this is sometimes not sufficient to avoid these clauses. Methods for the detection of metaphors have been developed [34], and could be applied to improve the method.

Examples of sporting language occasionally scored highly (although much less commonly than for metaphors). In order to detect these, a method for the automated identification of sporting language would need to be developed, requiring a study of language relating to a particular sport. The likelihood of success in doing this is high, given that a set of specialized sporting language is normally attached to a particular sport or group of sports, but more work is required in this area.

While each of these issues is addressable, it is thought likely that manual intervention in the process will always be necessary, due to the complexity of language itself.

7   Conclusions

This paper has described the methods and process used to address the research question of how to create a corpus of geospatial natural language, designed to contain a wide range of examples of geospatial natural language in English in order to support geospatial natural language research.

The process adopted was semi-automated, requiring manual filtering of results returned by an automated scoring method. While good returns can be achieved for text from some sites that have high proportions of geospatial natural language, it proved much less efficient to collect geospatial natural language from more generic web sites, or sites containing a range of different subject matter.

Acknowledgements. This work was funded under the European Union FP7 Euro-GEOSS project; the Horizon Doctoral Training Centre at the University of Nottingham (RCUK Grant No. EP/G037574/1) and the RCUK's Horizon Digital Economy Research Institute (RCUK Grant No. EP/G065802/1). The comments of the anonymous reviewers resulted in significant improvements to the paper and are gratefully acknowledged.

References

1. Anderson, A., Bader, M., Bard, E., Boyle, E., Doherty, G.M., Garrod, S., Isard, S., Kowtko, J., McAllister, J., Miller, J., Sotillo, C., Thompson, H.S., Weinert, R.: The HCRC Map Task Corpus. Language and Speech 34, 351–366 (1991)
2. Bateman, J.A., Hois, J., Ross, R.J., Tenbrink, T.: A Linguistic Ontology of Space for Natural Language Processing. Artificial Intelligence 174, 1027–1071 (2010)
3. Bitters, B.: Geospatial Reasoning in a Natural Language Processing (NLP) Environment. In: Proceedings of the 25th International Cartographic Conference (2011)
4. Blaylock, N., Swain, B., Allen, J.F.: Tesla: A tool for annotating geospatial language corpora. In: HLT-NAACL (Short Papers), pp. 45–48 (2009)
5. Blaylock, N., Swain, B., Allen, J.: Mining Geospatial Path Data from Natural Language Descriptions. In: ACM QUeST 2009, Seattle, November 3 (2009)
6. Brill, E.: A simple rule-based part of speech tagger. In: Proceedings of the Third Conference on Applied Natural Language Processing, ANLC 1992, pp. 152–155. Association for Computational Linguistics, Stroudsburg (1992)
7. Califi, M.E., Mooney, R.J.: Relational learning of pattern-match rules for information extraction. In: Proceedings of AAAI Spring Symposium on Applying Machine Learning to Discourse Processing, Standford, CA, pp. 6–11 (1998)
8. Chomsky, N.: Three models for the description of language. IRE Transactions on Information Theory 2, 113–124 (1956)
9. Cohen, K.B., Fox, L., Ogren, P.V., Hunter, L.: Corpus design for biomedical natural language processing. In: Proceedings of the ACL-ISMB Workshop on Linking Biological Literature, Ontologies and Databases: Mining Biological Semantics, pp. 38–45 (June 2005)
10. Coventry, K.R., Garrod, S.C.: Saying, Seeing and Acting: The Psychological Semantics of Spatial Prepositions. Psychology Press, East Sussex (2004)
11. Cunningham, H.: GATE, a General Architecture for Text Engineering. Computers and the Humanities 36, 223–254 (2002)
12. Evans, V., Green, M.: Cognitive Linguistics: An Introduction. Edinburgh University Press, Edinburgh (2006)
13. Goldberg, A.: Constructions at Work: The Nature of Generalization in Language. Oxford University Press, Oxford (2006)
14. Gregory, I., Hardie, A.: Visual GISting: bringing together corpus linguistics and Geographical Information Systems. Literary and Linguistic Computing 26, 297–314 (2011)
15. Hirtle, S., Richter, K.-F., Srinivas, S., Firth, R.: This is the tricky part: When directions become difficult. Journal of Spatial Information Science 1, 53–73 (2010)
16. Hornsby, K.S., Li, N.: Conceptual Framework for Modeling Dynamic Paths from Natural Language Expressions. Transactions in GIS 13, 27–45 (2009)
17. http://www.nottingham.ac.uk/~lgzwww/contacts/staffPages/kristinstock/documents/PatternsinGeospatialNaturalLanguagev1.0.pdf
18. Hunston, S., Francis, G.: Pattern Grammar: A Corpus-Driven Approach to the Lexical Grammar of English. John Benjamins Publishing Co., Amsterdam (2000)
19. Johnson, M.: The body in the mind: the bodily basis of meaning, imagination, and reason. University of Chicago Press, Chicago (1987)
20. Klippel, A., Xu, S., Li, R., Yang, J.: Spatial event language across domains. In: Workshop on Computational Models for Spatial Language Interpretation and Generation, CoSLI-2 (2011)

21. Lakoff, G.: Women, fire, and dangerous things: what categories reveal about the mind. University of Chicago Press, Chicago (1990)
22. Landau, B., Jackendoff, R.: "What" and "Where" in spatial language and spatial cognition. Behavioral and Brain Sciences 16, 217–265 (1993)
23. Law, M.: Guide to Worldwide Postal Code and Address Formats. WorldVu LLC (2010), http://www.worldvu.com (accessed May 22, 2013)
24. Mark, D.M., Egenhofer, M.J.: Topology of Prototypical Spatial Relations Between Lines and Regions in English and Spanish. In: Proceedings of the Auto Carto 12, Charlotte, North Carolina, pp. 245–254 (1995)
25. McEnery, T., Hardie, A.: Corpus Linguistics: Method, Theory and Practice. Cambridge University Press, Cambridge (2012)
26. Miller, G.A.: Wordnet: A lexical database for English. Communications of the ACM 38, 39–41 (1995)
27. Montello, D.R.: Scale and multiple psychologies of space. In: Campari, I., Frank, A.U. (eds.) COSIT 1993. LNCS, vol. 716, pp. 312–321. Springer, Heidelberg (1993)
28. Morimoto, Y., Aono, M., Houle, M.E., McCurley, K.S.: Extracting spatial knowledge from the web. In: SAINT 2003: Proceedings of the 2003 Symposium on Applications and the Internet, pp. 326–333. IEEE Computer Society, Washington, DC (2003)
29. Morton-Owens, E.: A tool for extracting and indexing spatio-temporal information from biographical articles in Wikipedia. Masters Thesis. New York University (2012)
30. Pustejofsky, J., Moszkowics, J., Verhagen, M.: ISO-Space: The Annotation of Spatial Information in Language. In: Proceedings of the Sixth Joint ISO - ACL SIGSEM Workshop on Interoperable Semantic Annotation, Oxford, UK (2011)
31. Riedemann, C.: Naming Topological Operators at GIS User Interfaces. In: 8th AGILE Conference on Geographic Information Science, Estoril, Portugal, pp. 307–315 (2005)
32. Saracevic, T.: Evaluation of evaluation in information retrieval. In: Proceedings of the 18th Annual International ACM SIGIR Conference on Research and Development in Information Retrieval. Special Issue of SIGIR Forum., pp. 138–146 (1995)
33. Schockaert, S., De Cock, M., Kerre, E., Smart, P., Abdelmoty, A., Jones, C.: Mining topological relations from the web. In: Bhowmick, S.S., Kung, J., Wagner, R. (eds.) DEXA 2008. LNCS, pp. 652–656. Springer (2008)
34. Schwering, A.: Evaluation of a semantic similarity measure for natural language spatial relations. In: Winter, S., Duckham, M., Kulik, L., Kuipers, B. (eds.) COSIT 2007. LNCS, vol. 4736, pp. 116–132. Springer, Heidelberg (2007)
35. Semino, E., Hardie, A., Koller, V., Rayson, P.: A computer-assisted approach to the analysis of metaphor variation across genres. In: Barnden, J., Lee, M., Littlemore, J., Moon, R., Philip, G., Wallington, A. (eds.) Corpus-based Approaches to Figurative Language, pp. 145–153. University of Birmingham School of Computer Science, Birmingham (2005)
36. Stock, K.: NaturalGeo Project: Identifying Patterns in Geospatial Natural Language (2012) (accessed on May 22, 2013)
37. Talmy, L.: Toward a Cognitive Semantics. MIT Press, Cambridge (2000)
38. Tellex, S.: Natural Language and Spatial Reasoning. PhD Thesis, Massachusetts Institute of Technology (2009)
39. Tomai, E., Kavouras, M.: Where the city sits? Revealing Geospatial Semantics in Text Descriptions. In: 7th AGILE Conference on Geographic Information Science, pp. 189–194. Association of Geographic Information Laboratories for Europe, Heraklion (2004)
40. Usmani, T.A., Pant, D., Bhatt, A.K.: A Comparative Study of Google and Bing Search Engines in Context of Precision and Relative Recall Parameter. International Journal on Computer Science & Engineering 4, 21–34 (2012)

41. Vasardani, M., Winter, S., Richter, K.-F.: Locating place names from place descriptions. International Journal of Geographical Information Science (2013)
42. Wang, X., Matsakis, P., Trick, L., Nonnecke, B., Veltman, M.A.: A study on how humans describe relative positions of image objects. In: Ruas, A., Gold, C. (eds.) Headway in Spatial Data Handling, Proceedings of SDH 2008, 13th Int. Symposium on Spatial Data Handling, pp. 1–18. Springer Publications (2008)
43. Wu, H.C., Luk, R.W.P., Wong, K.F., Kwok, K.L.: Interpreting tf–idf term weights as making relevance decisions. ACM Transactions on Information Systems 26, 1–37 (2008)
44. Xiao, R.: Corpus Creation. In: Indurkhya, N., Damerau, F.J. (eds.) The Handbook of Natural Language Processing, 2nd edn., pp. 147–165 (2010)
45. Xu, S., Klippel, A., MacEachren, A., Mitra, P., Turton, I., Zhang, X., Jaiswal, A.: Exploring regional variation in spatial language - a case study on spatial orientation with spatially stratified web-sampled documents. In: Spatial Cognition Conference – Poster Session, Mt. Hood, Portland Oregon (2010)
46. Zhang, C., Zhang, X., Jiang, W., Shen, Q., Zhang, S.: Rule-Based Extraction of Spatial Relations in Natural Language Text. In: International Conference on Computational Intelligence and Software Engineering, CiSE 2009, pp. 1–4 (2009)
47. Zhang, X., Zhang, C., Du, C., Zhu, S.: SVM based Extraction of Spatial Relations in Text. In: Proceedings of the IEEE International Conference on Spatial Data Mining and Geographical Knowledge Services 2011, Fuzhou, China (2011)

From Descriptions to Depictions:
A Conceptual Framework

Maria Vasardani[1], Sabine Timpf[1,3], Stephan Winter[1], and Martin Tomko[2]

[1] Department of Infrastructure Engineering
[2] Faculty of Architecture, Building and Planning University of Melbourne,
Parkville, VIC, 3010, AU
{mvasardani,winter,tomkom}@unimelb.edu.au
[3] Department of Geography, University of Augsburg, 86135, Augsburg, DE
sabine.timpf@geo.uni-augsburg.de

Abstract. People use verbal descriptions and graphical depictions to communicate spatial information, thus externalizing their spatial mental representations. In many situations, such as in emergency response, the ability to translate the content of verbal descriptions into a sketch map could greatly assist with the interpretation of the message. In this paper, we present an outline of a semi-automatic framework enabling seamless transition between verbal descriptions and graphical sketches of precinct-scale urban environments. The proposed framework relies on a three-step approach: NL parsing, with spatial named entity and spatial relation recognition in natural language text; the construction of a spatial Property Graph capturing the spatial relationships between pairs of entities; and the sketch drawing step where the identified entities are dynamically placed on a canvas in a manner that minimizes conflicts between the verbalized spatial relationships, thus providing a plausible representation of the described environment. The approach is manually demonstrated on a natural language description of a university campus, and the opportunities and challenges of the suggested framework are discussed. The paper concludes by highlighting the contributions of the framework and by providing insights for its actual implementation.

Keywords: NL place descriptions, spatial information extraction, dynamic sketching.

1   Introduction

People often use verbal and graphical (sketch) language to communicate information about spatial scenes in a way that externalizes the mental images they have created through direct or indirect experience with the environment. An intuitive spatial human-computer interaction should accommodate both modes of communication. For example, in emergency response, fast automatic sketching of the affected environment from a caller's verbal input can be critical, in cases where direct georeferencing is not possible. By asking for additional natural language (NL) input from the caller, thus initiating a dialog, the service could refine or disambiguate the

T. Tenbrink et al. (Eds.): COSIT 2013, LNCS 8116, pp. 299–319, 2013.
© Springer International Publishing Switzerland 2013

sketch. In an effort to enable seamless switching between the two modes of communication, and since different groups already work on generating spatial NL expressions from map sketches, this research focuses on generating spatial sketches from unrestricted NL.

An important aspect of both verbal and graphical spatial language is the extended use of spatial relations to communicate the various absolute and relative locations of features in the descriptions. In verbal language the expression of spatial relations is done using abstract linguistic terms. Such terms enable the understanding of the infinite variability of the world, while revealing aspects of the environment with some functional relevance [1]. And sketch maps are themselves visually simple, two-dimensional representations of features in an environment and the spatial relations among them. Sketch maps can serve as an intuitive user interface for geospatial computer applications (e.g., [2–4]). In this paper, we demonstrate the conceptual outline of an algorithm for the (semi) automatic construction of plausible, two-dimensional sketch maps, based on the extraction of spatial relations and features from text-based natural language (NL) descriptions.

Both verbal descriptions and sketches differ substantially from how computers and geospatial services generally process and present geographic information. While work from various communities is being done separately on NL processing for extracting geospatial information (e.g., [5]), on comparing NL descriptions and sketches that express the same cognitive image (e.g., [6]), and on assessing the information presented on a man-made sketch map (e.g., [7]), there is no effort so far—to the best of our knowledge—on assessing whether the information extracted from NL processing can result in a plausible, two-dimensional sketch-like topological representation of the described environment. This transformation forms a substantial challenge, acknowledging that NL descriptions can be notoriously underspecified and ambiguous.

This paper is a first take on this challenge of going from descriptions to depictions. Our hypothesis is that spatial information extracted from NL place descriptions can be (semi) automatically represented in a topologically conflict-free sketch map. Such a sketch can reveal how the person who expressed the NL description conceptualizes the environment. Positive evidence would support Tversky's suggestion that the meanings of linguistic elements can be mapped one-to-one onto the meanings of depicted elements and vice versa [8, 9].

The conceptual algorithm we are proposing in this work actually comprises three steps: 1) a NL parsing step, 2) a graph-producing step, and 3) and a sketch-drawing step. We are expecting unrestricted NL descriptions of a specific, urban environment. Thus, implementation of the parsing step depends on ongoing work on part-of-speech tagging, spatial named entity recognition, and toponym resolution for a parser that would be able to identify the geographic features in a description and the spatial relations among them (cf. [10–16]). For a plausible automated placement of geographic features on the plane, spatial relations are categorized according to formal models of topological, cardinal, projection and orientation relations ([17–20]). During the second phase, the identified spatial features and relations are placed on a spatial Property Graph (sPG). The sPG strips away the cognitive details of the NL place descriptions and provides the basic spatial information in the form of reference objects (RO), locata (L)—objects to be located in relation to the RO—and the identified spatial relations (*r*) between them. The ROs and Ls are the nodes in such graphs, while the *r*s are labeled edges. During the final phase, all ROs and Ls are arranged on the plane, in a sketch that represents a plausible representation of the described environment, according to the extracted spatial information. By plausible here, we mean a topologically conflict-free representation that is most probably one of many such plausible representations that one can derive from the descriptions. The conceptual design of the proposed algorithm is based on a case study of NL descriptions from memory of the layout of a university campus, from volunteer graduate students.

The remainder of the paper is structured as follows: Section 2 discusses background work on concepts and tools necessary for the design of our framework, and our approach is presented in Section 3—the three steps of the conceptual algorithm. Section 4 discusses the case study and an example of (manually) applying the algorithm to produce the sPG and corresponding sketch(es). In Section 5 we study the results and provide an overall discussion of the important aspects of our conceptual design. We conclude this paper and discuss future work in Section 6.

2 Related Work

In this section we discuss previous work that is related to the three steps of our approach, namely the NL parsing step, the property graph production, and the sketch drawing, and point out the differences with comparable approaches.

2.1 Extracting Spatial Information from NL Descriptions

We are interested in an algorithm that can produce a plausible two-dimensional representation of the place described in an unrestricted NL narrative. Some earlier work had focused either on the extraction of specific and application-dependent spatial information from restricted language [10, 21, 22], or on conceptualizations from specific communication contexts, for example route descriptions [23]. Most computational aspects of spatial information do not consider the intricate linguistic concepts that lead to under- or over-specification when mapping NL to formal representations of spatial information, and which can be regarded responsible for the lack in domain-independent applications. Bateman *et al.* [24] argue extensively about this shortage, especially when it comes to formal models of spatial relations which are based on the logic of human spatial cognition. Such models do not take into consideration the linguistic constructs and the way that people actually verbalize spatial relations. In [25] the need for semantics and a linguistically-oriented spatial ontology that can facilitate mapping between NL and spatial calculi, such as the Generalized Upper Model (GUM) ontology [26], is supported. Recently, however, work on parsers that can process unrestricted NL for spatial information and thus come closer to the requirements of our algorithm has surfaced, and is described here.

Kordjamshidi *et al.* [27] introduce *spatial role labeling* (SpRL), a method for assigned the roles of ROs, locata and *spatial relations* roles to terms. NL sentences are tagged according to the *holistic spatial semantic* theory in [28], and later mapped onto formal spatial relations frameworks, to facilitate subsequent spatial reasoning. The authors propose machine learning methods to cope with the various sources of ambiguity of spatial information in NL. In [11] they employ the proposition project (TPP) used in SemEval-2007 [29] to disambiguate the prepositions' spatial meanings and enhance their SpRL technique. Evaluation of the spatial role labeling task is done using the General Upper Model (GUM) spatial ontology [26].

Another group, interested in the generation of rendered 3D images from NL scene descriptions, created a dependency parser, which labels all terms in a sentence according to their direct or indirect relation with the (finite) verb [13] . The first version of the system (Words-Eye) handled 200 verbs in an *ad hoc* manner with no systematic semantic modeling, while their current system utilizes the Scenario-Based Lexical Resource (SBLR) [14]. SBLR consists of an ontology and lexical semantic information extracted from WordNet [30] and FrameNet [31]. With such parsers, the goal of labeling terms of unrestricted NL descriptions with specific spatial roles becomes much more feasible, and they are considered a necessary prerequisite for the conceptual design of the proposed framework.

2.2   Property Graphs

In our conceptual framework, there is an intermediate step between the input NL processing and the two-dimensional representation of the layout—the creation of a spatial Property Graph (sPG). Property graphs [32–35] are a special type of graph data models introduced whenever the data complexity exceeds the capabilities of the relational model as in e.g., transport networks [36] or spatially embedded networks [37]. Property graphs are directed, labeled, edge- and node-attributed multi-graphs, with nodes representing entities, and edges the relations between them [35]. We call the property graph constructed in the suggested framework a *spatial* property graph, emphasizing the fact that the labeled directed edges between nodes represent spatial relations with (possible) properties, frequently belonging to formal models.

2.3   Mental Maps

Producing a sketch map from text descriptions cannot be a direct action, but requires first understanding of the process by which people built mental models [38]. This process, called cognitive mapping, comprises a series of psychological transformations by which an individual acquires, codes, stores, recalls, and decodes information about the relative locations and attributes of phenomena in their everyday spatial environment" (p.9, [39]). Sketching the spatial relations from a mental model is called mental mapping and the end results, i.e. mental maps, are a prominent source for researching human's understanding and feelings about geographic places [40, 41].

The properties of mental maps exhibit all of the processes humans go through by first cognitive mapping and then mental mapping: they are highly simplified, schematized, selective and distorted [42]. They are qualitative, thus reflecting humans' thinking about geographic space [43]. Topological relations are usually correctly preserved, as are order and sequence relations [44, 45]. Previous research indicates that information represented in a mental map is preferentially North-Up oriented [46]. From this discussion, it should become clear that mental maps have specific properties that this research intends to exploit in the sketching process.

2.4   Comparable Approaches

With a focus on spatial relations inference to extract new information, Wiebrock *et al.* [15] produce a visualization from descriptions of spatial layouts. They also produce a graph of objects and spatial relations between them, which resembles the sPG suggested here, in that their graph is also a labeled, directed graph where the edges are spatial relations. The authors then use multiplication of transformation matrices, constraint propagation, and verification to realize inference. Apart from the focus on inference of new spatial information alongside the visualization of the spatial layout, their approach differs from ours in that input to their system is not unrestricted NL descriptions, but a sequence of propositions from instructed agents. They also consider only table-top objects and room environment descriptions, with a main interest in human-robot communication for indoor robot movements, instead of relations between features in urban environments, which are what we are mainly focusing on. Finally, while for robot navigation a representation of indoor environments must be consistent with reality at every location, we require our representations to be plausible (i.e., free of topological conflict), but they need not be consistent with reality.

In another approach, a route description from search engines is used to produce a map of the route called LineDrive [47]. In this work, spatial cognition insights are used to schematize the route map, i.e. linearizing the streets, straightening angles between route segments, emphasizing the route and de-emphasizing (up to disappearance) the context information, resulting in a very clear and simple spatial representation similar to sketch maps. Among other fundamental differences, input information for the sketch comes from an already interpreted set of route instructions that were produced by a database search, whereas we focus on the generation of plausible sketches from unrestricted NL descriptions of a spatial layout.

Finally, in an another effort to generate sketch maps from route descriptions, Fraczak [48] attempts to directly map NL expressions to a so called graphic code, which in turn is supposed to generate a map. From the literature it is not clear if this mapping was ever completely implemented. Again, the author focuses on NL descriptions of routes rather than spatial layouts. In addition, Fraczak's goal is to derive an interpretation of the route description that is consistent with reality, whereas we stress the dynamic nature of the sketching process in this framework that allows for several plausible solutions.

3   Approach

The general approach we are proposing consists of three steps:

1. An unrestricted NL parser processes the descriptions of a specific environment's layout and labels the terms in each sentence according to whether they carry a spatial role or not.
2. Algorithm 1 takes the parser's output and produces a spatial Property Graph (sPG).
3. Algorithm 2 processes the sPG graph and produces a plausible 2D representation of the described environment's layout, in the form of a sketch.

3.1   Parser Output

The first step in the conceptual framework is feeding the parser with an unrestricted NL description of the spatial layout of an environment. The parser labels the terms in each sentence according to a spatial role set, consisting of some of the basic spatial semantic concepts that are defined as 'universals' in the spatial semantics literature [28], also used in the annotation scheme developed in [11], and contains the core roles: *reference object* (RO), *locatum* (L), *spatial relation* (*r*), and *none*—if the word provides no additional spatial information. The *spatial relation* can be a *static spatial indicator* or a *motion indicator*. We decided to employ this already defined spatial role set because it is commonly used in the current spatial semantics parsers' literature (cf. [12, 13, 15]), with some of the roles having alternative names such as *trajectory* for the L, or *landmark* and *relatum* for the RO (cf. [28]). The roles of RO and L are identified for features on level 3-building and level 4-street of the granularity schema developed in [49], as we are focusing on spatial relations and features of an urban environment.

The RO is the reference object in relation to which the location or the trajectory of motion of the locatum L is indicated. The L is the entity whose location or trajectory is of relevance, and it can be static or dynamic. The *r* explains the type of spatial relation between RO and L and is usually a preposition (e.g., "house *on* the lake"), but can also be a verb (e.g., "buildings *surrounding* the lawn"), or a noun (e.g., "the road to *the east*), or even implicit (e.g., "Paris, France", implying *in*). In GUM ontology [26] it is defined as *spatial modality*, and is the main axis of the spatial relation. It can also be an indicator of motion, e.g., in the form of *motion verbs* [11].

The desired and required final output of the parser is a list of ordered triplets of the form in Equation 1,

$$<L \; r \; RO>i, \text{ with } i=1..n \tag{1}$$

where *n* is the number of triplets, and *i* is the index of each triplet according to the order with which it was identified in the NL description. At this point, words labeled as *none* are not considered any further. It should be made clear here, that the same spatial elements may participate in multiple triplets. For example, from the sentence

"You can also find NAB and Commonwealth bank branches near the Union House", the parser should produce the following triplets: <NAB *near* Union House>, <Commonwealth bank *near* Union House>. In other words, 'Union House' is the RO for both Ls, 'NAB' and 'Commonwealth bank'. For this reason, and to assist with the algorithm's design, the parser assigns to each RO a unique numerical identifier RO_ID. This identifier is later used to: a) help in recognizing how many Ls are related to the same RO, and b) to help resolve synonymy issues, i.e., if two ROs with different RO_IDs at first are recognized as the same entity, but with different names (e.g., one official and one shorter for brevity), then they get assigned the same RO_ID. In a similar manner, synonymy is resolved between locatums (Ls).

Apart from relations in the form of prepositions, adjectives, nouns, or implied ones, the parser is required to recognize the more complex, path indicating relations such as *across*, *along*, *down the hill from*, etc. which are identified as *PathRepresentingInternals* in the GUM ontology [26]. Motion verbs, when conveying spatial information should also be detected as such. For example, in "You may continue north, past the bakery to reach the market", by interpreting 'may continue north', 'past' and 'to reach' the parser should produce the triplet <market *north* bakery> recognizing that there is a path that leads from one feature to next, in a sequence. Finally, the parser is able to resolve anaphora and other co-references (cf. [27]). For example, in "...of the old buildings. These buildings are..." the parser is able to identify that 'these buildings' refer to the 'old buildings' previously mentioned, and therefore are the same spatial entity.

Interpretation Rules. Certain rules are imposed on the parser for the interpretation of the spatial semantics, for our conceptual framework to work properly.

Rule 1: Each description corpus falls entirely into a single top-level container. It is initially assumed that with input from the analyzer, the described spatial environment would be identified as the main container and reference system of each description. For example in our case study, the term 'campus' was recognized as such, and identified ROs and Ls were positioned in it. Use of superlative in descriptions then indicates the outer bounds of such reference system, and is picked out by the parser. For example, in "The Arts School is furthest north on campus", the output triplet should be <Arts School *furthest north on* campus>, instead of just <Arts School *north on* campus>.

Rule 2: A sequence of Ls with no additional spatial information produces inherited relations to the same RO. In the example "From the intersection by continuing north you can get to the museum, the library and then reach the market.", the expected output triples are <museum *north* intersection>, <library *north* museum> and <market *north* library>. Hence the relation *north* is inherited for all triplets, the L of each triplet becomes the RO in the next, and orientation does not change (see Rule 3).

Rule 3: Unless explicitly stated otherwise, the initial orientation identified does not change – this is a general phenomenon when dealing with the orientation of a description. However, when a change in direction is stated, and when there is no obvious RO after the change, then the notion of a corner-point (CP_i) is introduced.

The CP$_i$ acts as the RO for the next bit of the description when the direction has changed, but also as the L for the previous part of the description. So, the output triplets in the example "The second path from the south entrance takes you a little bit to the west and then switches to the north. From this path [...] you can reach the University House by continuing north" should be <CP$_1$ *west* south entrance> and <University House *north* CP$_1$>.

Rule 4: Verbs that imply a container, such as 'contains' or 'houses' are interpreted as the containment relation *in*, and the container feature is then considered the RO of the relation. In the example "The Union House is containing many cafes and restaurants..." the output triplets are <cafes *in* Union House> and <restaurants *in* Union House>. In these cases, if the RO is involved in another relation, then the relation also holds for the contained features as well, however, the reverse is not true. In the example "[...] in Baretto's in the Alan Gilbert building, across Grattan St. is the medical building", then apart from 'Alan Gilbert building', also 'Baretto's' is considered as a possible locatum across the street from the medical buildings.

Rule 5: Named streets are assigned a role of either RO or L, whereas paths, which are usually unnamed or inferred, are only expressed through the spatial relations of the spatial elements they connect.

Rule 6: Complex spatial relationships including two or more ROs or Ls are broken into binary relationships. Complicated descriptions require extra inference capabilities from the parser. Prepositions such as *around*, require usually an orientation for locating features. In the example "Around South Lawn you would see (from left to right) Graduate House, the MBS [...]" the parser should be able to assign the relation *around* between South Lawn and all the mentioned locatums, and also assign the relation *to the right of* between the locatums in the mentioned sequence, i.e. <Graduate House *around* South Lawn>, <MBS *around* South Lawn>, <MBS *to the right of* Graduate House> and so on. In addition, the desired parser output triplets include spatial relations that are binary in nature, Therefore, with ternary relations such as 'A *between* B and C', and 'A *across the street from* B', the parser is required to break them down into simple binary relations that fit with the rest. *Between* is decomposed into <A *between* B> and <A *between* C>, while *across* into <street *between* A> and <street *between* B>. This way, we maintain uniformity in the required output of binary relations' triplets.

3.2 Spatial Property Graph Algorithm

In this section, the second step of the suggested framework is described; a graph-producing algorithm that creates the spatial-Property Graph (sPG). Input is the ordered list of triplets produced by the NL parser at the first step. A property graph has a set of nodes and a set of edges. Each node has a unique identifier, a set of incoming and a set of outgoing edges, and a collection of node properties defined as key/value pairs. An edge has a unique identifier, an outgoing and an incoming node, a label denoting the type of spatial relationship between the two nodes and a collection of properties defined as key/value pairs. These key/value pairs make it possible to describe in detail the relationship between two entities. In our framework, the nodes represent ROs and Ls and the edges the spatial relations between them. In the case of a spatial PG, the types of relations may belong to different formal frameworks of relations such as topological, directional, or cardinal relations. The direction of the edge points from the L to the RO, with the exception of *between* relations that are represented with bi-directional edges, and additionally decomposed into pairs of directed edges, showing the existence of one L and two ROs (see Rule 6). There can be multiple directed edges between the same pair of RO and L (i.e., combined with a logical AND in a formal interpretation).

Below, the sPG constructing algorithm, *Algorithm 1*, is outlined and an example graph from the case study is presented in Section 4 (Fig. 2).

***Algorithm 1*: Constructing the spatial Property Graph**

Input = n triplets in the ordered list, each assigned with a sequence index i:
   For i=1...n,
    a. If not already created, create nodes for the RO and the L in $<L\ r\ RO>_i$
    b. Create a directed edge E_j pointing to RO and originating from L, labeled with r

   Next i

3.3   Sketch Drawing Algorithm

The last step in the text-to-sketch conversion framework is the generation of a sketch guided by the graph structure. In the current state, this process is performed by a person—it is not (yet) expect it to work unsupervised. The expected outcome is a plausible representation of the relationships among spatial objects encoded in the graph, in a topologically conflict-free sketch, called a *topological sketch*.

Related literature suggests that NL place descriptions are sequential and clustered [50]. In all the case study examples analyzed for this paper, people followed the communication maxims, i.e., provided information that was of the best assumed quantity, quality, relevance and manner, avoiding unnecessary wordiness or ambiguity [51]. The descriptions are not just a collection of facts, but they reveal each person's preference for what is important, relevant, and in which sequence it should be presented. In their text-based descriptions, volunteers started from a certain part of the environment, described it in sequence of spatially related elements, and when done, moved on to the next, possibly neighboring part. The steps in our algorithm are designed with this characteristic in mind: The parser's output triplets are ordered according to the order with which ROs, Ls, and rs are introduced in each description. The algorithm places all ROs and Ls in the sketch, one at a time, starting with the first RO in the sPG.

McNamara *et al.* [45] identified region membership as an important property of spatial environments. Therefore, (small) regions are the main spatial feature to be placed on the sketch canvas. However, in this framework, this idea is taken further than in McNamara *et al.* by giving any entity in the graph the shape of a rectangle that can be dynamically enlarged to include other entities, shrunk, or elongated as needed. Furthermore, a general reference frame is introduced in the form of a double cross of absolute cardinal directions [19], with the true North.

Basic Process. The first entity to be placed on the blank canvas is the container feature for all named entities in the description, in our case study the 'campus'. This entity is treated as the reference frame for all subsequent relations. In this bounding feature, a double cross is introduced as reference frame for absolute cardinal directions (Fig. 1a). Next, find in the sPG the reference object RO_*i* with the smallest sequence number and place its rectangle in the double cross—if no cardinal information with respect to the reference frame is given, the rectangle shall be placed in the center region of the double cross. If a cardinal relation is given, then the rectangle is sketched in the appropriate place on the canvas (Fig. 1b).

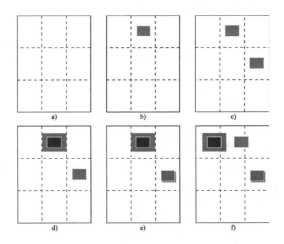

Fig. 1. Dynamic generation of topological sketch example

Once the RO_*i* is placed, a spatial relation of the type cardinal direction is selected from the property graph, that relates a L to that RO_*i*. The rectangle for the corresponding L is placed on the canvas in the given cardinal direction relation at a fixed distance from the RO_*i* (Fig. 1c). This is repeated for all relations containing cardinal directions relating to RO_*i*. When no other cardinal direction relations with RO_*i* remain, one of the other relations with RO_*i* shall be selected for inclusion in the sketch, and each corresponding L is placed on the canvas in the given spatial relation. If an L already exists from a previous relation, its position is checked for consistency and readjusted if necessary, maintaining the sketch conflict free. This is repeated until no spatial relations with RO_*i* remain in the graph. Next, RO_*i* +*1* is selected and its rectangle sketched on the canvas; the procedure is repeated until all ROs and their Ls are placed on the canvas. The outline of the sketch algorithm, *Algorithm 2*, can be found below.

Defining the Heading. One characteristic of natural language descriptions of places is the switch of perspective or reference frame. Following Meilinger [52], who proposes to allow different reference frames for each location, we assign each rectangle a heading, which forms one axis of the reference frame. The heading of a rectangle is defined as the direction in which the entrance of a building lays, or in the case of other entities such as the campus or the park, the direction to the North. This rule is justified by recent findings by Meneghetti *et al.* [46], who showed that the mental representation derived from spatial descriptions is North-Up oriented. In the absence of any other orientation information we will thus default the heading value to "North". More complex rules could be made here, e.g. making the heading dependent on the features already placed on the canvas or on the type of the entity. We will however leave this refinement for future work.

***Algorithm 2*: Generating a topological sketch**

1. If known, draw the RO that is considered the main container and introduce a double cross reference system in it. Else, introduce the double cross and consider it the main reference frame.

2. For $i=1\ldots n$, where n is the number of unique RO_IDs,
   a. Place RO_i as a rectangular feature on the double cross. If there is information available for its absolute cardinal position place it in the appropriate cell; otherwise place it in the central cell.
   b. For all the <L r RO_i> triplets,
      i. Identify all cardinal directions relations r and, if not already there, place the corresponding L on the frame.
      ii. For all other relations, if not already there, place the corresponding L relative to RO_i according to r on the frame.
   Next i

Dynamically Adjusting the Feature Placement. All rectangles can be enlarged, shrunk, or moved to accommodate a new rectangle on the canvas. For example, if there exists a spatial relation "IN" between two rectangles then the reference rectangle is enlarged to make space for the new feature (Fig. 1d). On the canvas, the "IN" relation is represented as the topological relation "RO contains L". In the special case of 3D spatial relations, e.g., spatial relation "L underneath RO", two rectangles are allowed to overlap partially, i.e. showing the topological relation "RO overlaps L" (Fig. 1e). Rectangles that represent streets or other linear features (e.g., rivers) are allowed to stretch alongside several other objects when information is available or may be inferred (see for an example Section 4 – case study).

During the placement process it can happen that a rectangle cannot be put on the canvas, because doing so would violate one or several of the relations that already exist. In this case the existing rectangles can also be moved as long as their original relations are not violated. For example in Fig. 1f, the containing rectangle is moved a little to the west, to leave room for the new rectangle that is placed to its east, to satisfy the sequence of information, as this is revealed by the ordered triplets. When a rectangle cannot be placed without producing a conflict with existing features, past decisions on how and where to place features are revisited, aiming to find another solution that still satisfies all existing spatial relations. For this process to work we introduce a data structure called a *placement decision tree*.

Placement Decision Tree. The placement decision tree records incrementally the history of decisions of where features were placed and their potential alternative placements. For example, when placing a rectangle on the canvas using the relation "entity1 is in the northern part of the campus", three placements of that rectangle could be viable, since a system with eight cardinal directions is used. In the placement decision tree, three new links (north, north-east, north-west) with new leaf nodes are added, one of which (the one attached to the link "north") contains entity1. If at any point in the dynamic sketching process a conflict arises out of a potential placement, then the last decision is revisited and changed, i.e., up in the decision tree, and then the new feature is placed again.

The decision on a particular placement thus corresponds to a default hypothesis if alternative solutions exist. By recording not only the decision itself, but also the alternative solutions, we construct incrementally a search space in which we keep track of our hypotheses. In contrast to [53], our data structure is lazy, i.e., we only compute a new solution if a conflict is identified. The data structure reflects the under-specificity of the place description and can be used as a quality measure.

4   Case Study

The case study for testing our conceptual framework consists of a set of four NL descriptions of a university campus. The descriptions were submitted by a group of graduate students with varying degrees of familiarity with the campus. Students were asked to produce the descriptions from memory, as if explaining to a new student the layout of the university environment. No further directions or expectations of the descriptions were suggested. Below is a campus description (*Narrative A*) to which the suggested framework is applied and Table 1 contains the ordered set of triplets expected from the NL parser. As the parser is still not implemented, the triplets were manually produced and ordered according to the sequence of ROs and Ls in the description. Triplets 3 and 4 present an example of interpreting relation *across* as two *between* relations (Rule 6). After extracting the ordered list of triplets, Algorithm 1 (Section 3.2) was manually implemented on the list, and produced the sPG in Figure 2. Double-directed edges correspond to the *between* relationships discussed above, to show the difference between binary and ternary relations, represented also as binary.

Narrative A

"We're sitting in Baretto's in the Alan Gilbert Building, across Grattan street is one of the medical buildings. Down the hill along Grattan street the new building being constructed is the Peter Doherty Institute and diagonally across the road (Royal Parade) is Melbourne Hospital. In the other direction the open area is University square (there is a carpark underneath) at the city end of University square is the law building. Across Grattan street from University square there is an entrance to the campus, straight ahead is an overpass building and to the right are the various Engineering buildings. The road goes in a big loop around the campus you can either go left towards the Medical buildings or right past the Engineering buildings then head North (away from University square). One other area you may want to explore is South Lawn which you can get to by going underneath the overpass building directly in front of you when you enter the campus."

Table 1. Expected parser output – ordered list of triplets <L *r* RO>i

<Barreto's *in* Alan Gilbert Building>1
<Medical Building *across (Grattan St)* Alan Gilbert Building>2
<Grattan Street *between* Alan Gilbert Building>3
<Grattan Street *between* Medical Building>4
<Peter Doherty Institute *down the hill* Alan Gilbert Building>5
<Peter Doherty Institute *along* Grattan Street>6
<Melb. Hospital *diagonally across (Royal Parade)* Peter Doherty Institute>7
<Royal Parade *between* Peter Doherty Institute>8
<Royal Parade *between* Melbourne Hospital>9
<University Square *in the other direction from PDI* Alan Gilbert Building>10
<carpark *underneath* University Square>11
<law building *at the city end of* University Square>12
<entrance to the campus *across (Grattan Street)* University Square>13
<Grattan Street *between* University Square>14
<Grattan Street *between* entrance to the campus>15
<overpass building *straight ahead* entrance to the campus>16
<engineering buildings *to the right* entrance to the campus>17
<medial buildings *left* entrance to the campus>18
<overpass building *directly in front* entrance to the campus>19
<overpass building *between* entrance to the campus>20
<overpass building *between* South Lawn>21

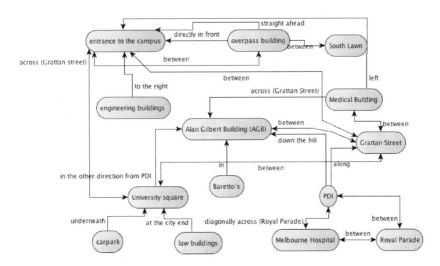

Fig. 2. The sPG produced by applying Algorithm 1 on the example campus description

Finally, Algorithm 2 (Section 3.3) is applied on the sPG and (at least) two plausible sketch maps are derived (Fig. 3-4). As is evident, there may be many topological sketches that can come out of one description. In the example above, it is not specified which side of 'Alan Gilbert Building' to place 'Peter Doherty Institute'; therefore, a decision needs to be made and recorded in the placement decision tree. Figure 2 shows the outcome topological sketch when first choosing one side of 'Alan Gilbert Building', while Figure 3 shows the outcome when choosing the other. Due to the description being underspecified, and since the rest of the spatial relations can be realized without conflict, both topological sketches are valid.

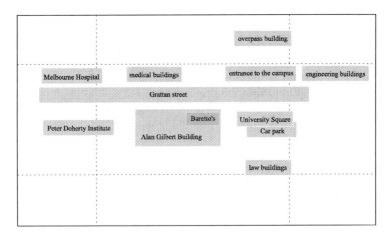

Fig. 3. A plausible sketch map produced from the description by applying Algorithm 2

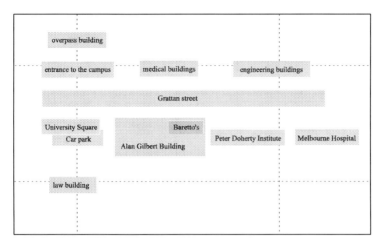

Fig. 4. Another plausible sketch map produced by applying Algorithm 2

The suggested framework was also applied in the same manner to the rest of the NL descriptions in the set and produced sPGs and plausible topological sketches for all of them. The fact that plausible sketches could, in fact, be generated supports our hypothesis, considering that all descriptions were very different from each other and varied in style, even within the same description.

5   Discussion

The complexity of putting unrestricted NL onto a sketch map stems from the under-specificity of NL. Even the fact that we can have a multitude of plausible topological sketches is a direct effect of this observation. However, we believe that the suggested framework is a first step toward accommodating this language characteristic.

There are a few observations that can be immediately made from the NL place descriptions we collected for our case study. First of all, people have the tendency to mix different styles of spatial descriptions, alternating between survey and route views, or changing the reference object and switching from an allocentric to an egocentric description and vice versa. Even though this poses additional challenges to the extraction of the appropriate spatial roles (L, r, RO), there has been found no contradiction, and the suggested structure of the triplets is flexible enough to accommodate these different styles. For example, in the sentence "[…] the overpass building directly in front of you, when you enter the campus", the implied co-location of the person and the 'entrance of the campus' is enough to suggest that the 'entrance' can be used as the RO in the triplet. Thus, the structure allows for any type of description, and for the co-existence of different types of spatial relations. The interpretation of the relations follows then certain rules (Section 3.1) that make a topological sketch possible.

A second observation verifies related literature that suggests that NL place descriptions are sequential and clustered [50]. In all our examples, we found that people like to start from a certain part of the environment, describe it in sequence of spatially related elements, and when done, move on to the next, possibly neighboring part. There is evidence that people move through their mental representation when giving directions or describing places and regions. This is especially true for descriptions using a route perspective, but seems also to be true for a survey perspective [54]. The simplicity of the roles assigned in our framework makes it easy to identify the common spatial elements, which then act as the bridges between the various parts of described environment.

A third observation is the fact that people like to refer to the same things with different names or expressions (e.g., abbreviations), which adds another layer of difficulty when assigning spatial roles to the NL expressions. A different, but related problem is the identification of synonyms that are in different parts of the description. In the current approach, this identification is carried out by a person (who can consult gazetteers and other available knowledge sources), and is expected to be addressed in natural language processing.

The choice of the double cross of absolute cardinal directions to represent the main reference system is motivated by the need to delineate and structure the canvas in a uniform way. Cardinal relations provide the most information when trying to place things on a blank canvas, and are thus used first in our sketch producing algorithm whenever available. In this process, the double cross allows for suggested areas where elements can be placed relative to each other. When absolute relative cardinal relations are missing, we start with the assumption that the first mentioned RO is placed in the center, and things are located around it according to the rest of the available spatial relations.

We need to emphasize here that the production of the topological sketch is an incremental and dynamic sketching process. Therefore, spatial elements are allowed to shrink, expand, and move around in order to avoid topological conflict, as long as they do not violate previously established relations. For example, in the description, "the Arts building is to the north of the Union House", if we later on find out that there is another spatial object between the Arts building and the Union House toward the North, it can still be placed between them, by moving for example the Arts building further North, to make room for the new object. Also, in the case study example (Fig. 3-4), since there is available information about the span of "Grattan St." along certain objects from the extracted triplets (Table 1), the rectangle representing the street is stretched alongside the corresponding objects. Additionally, spatial objects must occupy an area (which can be resized accordingly), and they are not allowed to overlap, unless specifically stated so (e.g., "carpark *underneath* South Lawn").

Finally, by using the term 'plausible' we refer to the fact that the outcome sketches must be free of topological conflict. This means that from a single NL description there may be multiple sketch map interpretations—homomorphic representations are all 'plausible' in this sense. A mental model represents one possibility, capturing what is common to all the different ways in which the possibility may occur [55]. The spatial objects can be placed on the plane in a way that agrees with the incremental revelation of information from the ordered triplets. If, by revisiting the placement decision tree, conflict with a previously established relation cannot be resolved after examining all possible decision points in decision placement tree, we then come to the conclusion that there is an error in the description. And even after all conflicts are resolved, there is no guarantee that the resulting sketch-map accurately represents what the narrator meant.

6   Conclusions and Future Work

The main goal of this paper was to show that a plausible topological sketch can be produced from text-based natural languages descriptions of places, such as a campus, without the use of additional knowledge sources. There are three processing steps in our approach. In the first step, textual descriptions are parsed for spatial relationships and the result is presented in an ordered list of binary spatial relations between objects, i.e., a list of triplets $<L\ r\ RO>i$. In the second step, objects in this list and the binary relations are stored in a spatial property graph. In the third and final step, the information contained in the property graph is used to dynamically construct a topological sketch, keeping track of the placement decisions made when the spatial relation is underspecified in the placement decision tree. We allow for the fact that some of the placement decisions need to be default solutions. Such default solutions may be inconsistent with reality. This is not critical to our goal, which is to determine if a plausible, rather than the actual topological layout, can be found from the NL descriptions.

We have shown in this paper that our approach to (semi-) automatically produce a topologically plausible sketch from natural language description is feasible, by applying the framework to an example (*Narrative A*) from the case study, and producing plausible sketches (Fig. 3-4). We also manually applied the algorithm to the full set of NL descriptions and could derive plausible topological sketches for all of them. We have given pseudo-algorithms to describe the necessary processing steps, where an automatic solution and thus implementation is intended. We clarified the rules we impose to arrive at a feasible solution. Such rules mostly stem from research in spatial cognition and psychology. We also identified where more research is needed to fully automate the process, i.e. mostly in the parser and in identifying additional rules needed for the sketching process.

One of the strengths of the approach lies in the fact that we accept the under-specification of given spatial relations and provide a cognitively motivated default interpretation in order to sketch them. We do not force quantification on the qualitative descriptions, rather allowing for a re-interpretation of previously sketched relations in the dynamic sketching process by shrinking, enlarging or moving features as necessary. To achieve that, we populate the topological sketch by following the order with which people reveal the spatial information in their descriptions, exhausting all information for a part of the description, before moving to the next. This dynamic sketching process reveals how each person wanders in her own mental representation of the environment to produce a place description. We thus exploit and apply results from research in spatial cognition and natural language processing. By producing a list of requirements for what needs to be provided by a spatial parser, we can guide research in NL processing of spatial descriptions.

Another strength of our approach is that we can extract a very simple structure of spatial information (ordered triplets) from unrestricted NL place descriptions. The simple structure <L r RO>i lends itself to a straightforward production of a property graph and a spatial topological sketch, by following the provided rules. An important characteristic of this structure is that it allows for different types of spatial relations and different styles of narrations to be accommodated together and to contribute synergistically to the production of a conflict free sketch.

In our current solution, the input of an intelligent agent (i.e., a person) is still needed at the parsing step. We do realize that we are building our process on a parser that is fairly sophisticated. However, our in-depth study on previous work in NL interpretation shows that an automatic parsing of spatial expressions as we require it, may not actually be that far in the future. Another point of consideration are the rules for sketching, i.e. the default interpretations of spatial relations. The rules we apply were motivated by research in spatial cognition. We will continue to check and refine these rules as more research becomes available.

In the future, apart from actually fully automating the three steps that make up our approach and testing them on a different set of NL place descriptions, we want to enable processing of multiple descriptions of the same place. In our current approach the input is a single description. We have already discussed the resolution of place name synonyms (Section 3.1). The same approach, expanded to accommodate identifying vernacular synonyms from context, can be used to point to identical entities mentioned in different descriptions. This will allow for combining parser outputs into a single list of triplets that contain the extracted information from all descriptions. We are expecting that abundance and redundancy of the combined information will lead to more detailed positioning of elements on the topological sketch, coming closer to enabling an 'actual' representation of the described environment. By actual representation we mean a topological sketch that excludes some homomorphisms as not resembling the layout of the described environment.

So far, sketch evaluation is only about plausibility. Other evaluations can offer different insights to our approach. For example, a comparison of an output sketch with a map would allow to ground-truth our approach. Since ground truth was not an issue for the current research, we did not touch on the potential of using gazetteers and spatial databases for geocoding the known elements in a description and thus enabling alignment of sketches, but this is certainly something to consider for future versions of the algorithm.

Moreover, some additional information can be extracted from our sPGs. For example, the number of incoming edges of a single node in the graph can be considered as a measure of 'popularity' of the element, since it reveals the number of times the element was used as a reference object. Especially in the case where the descriptions of multiple people are combined, such popularity measure would reveal *landmark* candidates. Similarly, counting the number of leaves in the decision tree can be considered as a measure of the quality of the description, where higher number of decisions implies less specific descriptions that can lead to a higher number of plausible topological sketches. Finally, in the current framework, processing is restricted to buildings and street level [49]. Future extensions will deal with other granularity levels such as cities or rooms.

References

1. Sjoo, K., Pronobis, A., Jensfelt, P.: Functional Topological Relations for Qualitative Spatial Representation. In: Proceedings of 15th International Conference on Advanced Robotics (ICAR 2011), pp. 130–136. IEEE, Tallinn (2011)
2. Egenhofer, M.J.: Query Processing in Spatial-Query-by-Sketch. Journal of Visual Languages and Computing 8, 403–424 (1997)
3. Blaser, A.D.: User interaction in a sketch-based GIS user interface. In: Hirtle, S.C., Frank, A.U. (eds.) COSIT 1997. LNCS, vol. 1329, pp. 505–505. Springer, Heidelberg (1997)
4. Blaser, A.D., Egenhofer, M.J.: A visual tool for querying geographic databases. In: Proceedings of the Working Conference on Advanced Visual Interfaces, pp. 211–216. ACM, New York (2000)
5. Tappan, D.: Knowledge-Based Spatial Reasoning for Scene Generation from Text Description. In: Proceedings of Association for the Advancement of Artificial Intelligence, Chicago, IL (2008)
6. Denis, M.: The description of routes: A cognitive approach to the production of spatial discourse. Cahiers de Psychologie Cognitive 16, 409–458 (1997)
7. Wang, J., Schwering, A.: The Accuracy of Sketched Spatial Relations: How Cognitive Errors Influence Sketch Representation. In: Tenbrink, T., Winter, S. (eds.) International Workshop on Spatial Information: Granularity, Relevance, and Integration, In Conjunction with COSIT 2009, pp. 40–47 (2009)
8. Tversky, B.: What do Sketches say about Thinking? AAAI Spring Symposium, Sketch Understanding Workshop, AAAI Technical Report, Stanford University (2002)
9. Tversky, B., Lee, P.U.: Pictorial and Verbal Tools for Conveying Routes. In: Freksa, C., Mark, D.M. (eds.) COSIT 1999. LNCS, vol. 1661, pp. 51–64. Springer, Heidelberg (1999)
10. Tappan, D.A.: Knowledge-Based Spatial Reasoning for Automated Scene Generation from Text Descriptions. PhD Thesis, New Mexico State University, Las Cruces, New Mexico (2004)
11. Kordjamshidi, P., Van Otterlo, M., Moens, M.-F.: Spatial role labeling: Towards extraction of spatial relations from natural language. ACM Trans. Speech Lang. Process. 8, 4:1–4:36 (2011)
12. Kordjamshidi, P., Frasconi, P., Van Otterlo, M., Moens, M.-F., De Raedt, L.: Relational Learning for Spatial Relation Extraction from Natural Language. In: Muggleton, S.H., Tamaddoni-Nezhad, A., Lisi, F.A. (eds.) ILP 2011. LNCS, vol. 7207, pp. 204–220. Springer, Heidelberg (2012)
13. Coyne, B., Sproat, R., Hirschberg, J.: Spatial Relations in Text-to-Scene Conversion. In: Ross, R.J., Hois, J., Kelleher, J.D. (eds.) Proceedings of 1st Workshop on Computational Models of Spatial Language Interpretation (COSLI 2010), pp. 9–16. Mt Hood, Portland (2010)
14. Coyne, B., Rambow, O., Hirschberg, J., Sproat, R.: Frame semantics in text-to-scene generation. In: Setchi, R., Jordanov, I., Howlett, R.J., Jain, L.C. (eds.) KES 2010, Part IV. LNCS, vol. 6279, pp. 375–384. Springer, Heidelberg (2010)
15. Wiebrock, S., Wittenburg, L., Schmid, U., Wysotzki, F.: Inference and Visualization of Spatial Relations. In: Freksa, C., Habel, C., Brauer, W., Wender, K. (eds.) Spatial Cognition II. LNCS (LNAI), vol. 1849, pp. 212–224. Springer, Heidelberg (2000)
16. Leidner, J.: Toponym Resolution in Text: Annotation, Evaluation and Applications of Spatial Grounding of Place Names. PhD Thesis, Institute for Communicating and Collaborative Systems, School of Informatics, University of Edinburgh (2007)

17. Egenhofer, M., Herring, J.: Categorizing Binary Topological Relations between Regions, Lines and Point in Geographic Databases. Department of Surveying Engineering. University of Maine, Orono (1990)

18. Randell, D.A., Cui, Z., Cohn, A.G.: A Spatial Logic based on Regions and Connection. In: Proceedings of 3rd International Conference on Principles of Knowledge, Representation and Reasoning (KR 1992), pp. 165–176. Morgan Kaufmann, San Mateo (1992)

19. Frank, A.U.: Qualitative spatial reasoning: cardinal directions as an example. International Journal of Geographical Information Systems 10, 269–290 (1996)

20. Hernández, D.: Relative Representation of Spatial Knowledge. In: Mark, D., Frank, A. (eds.) Cognitive and Linguistic Aspects of Geographic Space, pp. 373–385. Kluwer Academic, Dordrecht (1991)

21. Kelleher, J.D.: A perceptually based computational framework for the interpretation of spatial language (2003)

22. Li, H., Zhao, T., Li, S., Zhao, J.: The Extraction of Trajectories from Real Texts Based on Linear Classification. In: Proceedings of the 16th Nordic Conference of Computational Linguistics (NODALIDA 2007), pp. 121–127. University of Tartu, Tartu (2007)

23. Richter, K.-F., Klippel, A.: Before or After: Prepositions in Spatially Constrained Systems. In: Barkowsky, T., Knauff, M., Ligozat, G., Montello, D.R. (eds.) Spatial Cognition 2007. LNCS (LNAI), vol. 4387, pp. 453–469. Springer, Heidelberg (2007)

24. Bateman, J.A., Hois, J., Ross, R., Tenbrink, T.: A linguistic ontology of space for natural language processing. Artificial Intelligence 174, 1027–1071 (2010)

25. Bateman, J.A.: Language and Space: a two-level semantic approach based on principles of ontological engineering. Int. J. Speech Technol. 13, 29–48 (2010)

26. Bateman, J., Tenbrink, T., Farrar, S.: The Role of Conceptual and Linguistic Ontologies in Interpreting Spatial Discourse. Discourse Processes: A Multidisciplinary Journal 44, 175–212 (2007)

27. Kordjamshidi, P., van Otterlo, M., Moens, M.-F.: From language towards formal spatial calculi. In: Ross, R.J. (ed.) Proceedings of 1st Workshop on Computational Models of Spatial Language Interpretation (COSLI 2010), pp. 17–24. Mt.Hood, Portland (2010)

28. Zlatev, J.: Spatial Semantics. In: Cuyckens, H., Geeraerts, D. (eds.) The Oxford Handbook of Cognitive Linguistics, pp. 318–350. Oxford University Press (2007)

29. Litkowski, K., Hargraves, O.: SemEval-2007 Task 06: Word-Sense Disambiguation of Prepositions. In: Proceedings of the 4th International Workshop on Semantic Evaluations (SemEval), pp. 24–29. Association for Computational Linguistics (2007)

30. Fellbaum, C. (ed.): WordNet: An Electronic Lexical Database. MIT Press, Cambridge (1998)

31. Baker, C., Fillmore, C., Lowe, J.: The Berkeley FrameNet Project. COLING-ACL (1998)

32. Hidders, J.: Typing Graph-Manipulation Operations. In: Calvanese, D., Lenzerini, M., Motwani, R. (eds.) ICDT 2003. LNCS, vol. 2572, pp. 394–409. Springer, Heidelberg (2002)

33. Güting, R.H.: GraphDB: Modeling and Querying Graphs in Databases. In: Proceedings of the 20th International Conference on Very Large Databases (VLDB), Santiago, pp. 297–308 (1994)

34. Kuper, G.M., Vardi, M.Y.: A new approach to database logic. In: Proceedings of the 3rd ACM SIGACT-SIGMOD Symposium on Principles of Database Systems, pp. 86–96. ACM, New York (1984)

35. Rodriguez, M.A., Neubauer, P.: Constructions from dots and lines. Bulletin of the American Society for Information Science and Technology 36, 35–41 (2010)

36. Mainguenaud, M.: Modelling the network component of geographical information systems. International Journal of Geographical Information Systems 9, 575–593 (1995)
37. Freeman, L.C.: The Development of Social Network Analysis: A Study in the Sociology of Science. Empirical Press (2004)
38. Tversky, B.: Cognitive maps, cognitive collages, and spatial mental models. Presented at the Spatial Information Theory: A Theoretical Basis for GIS (1993)
39. Downs, R.M., Stea, D.: Cognitive Maps and Spatial Behaviour: Process and Products. In: Dodge, R., Kitchin, R., Perkins, C. (eds.) The Map Reader, pp. 312–317. John Wiley & Sons, Ltd. (2011)
40. Gould, P., White, R.: Mental Maps. Routledge, New York
41. Lynch, K.: The Image of the City. MIT Press (1960)
42. Tversky, B.: Visualizing Thought. Topics in Cognitive Science 3, 499–535 (2011)
43. Freksa, C.: Using Orientation Information for Qualitative Spatial Reasoning. In: Frank, A.U., Formentini, U., Campari, I. (eds.) GIS 1992. LNCS, vol. 639, pp. 162–178. Springer, Heidelberg (1992)
44. Hirtle, S., Jonides, J.: Evidence of hierarchies in cognitive maps. Memory & Cognition 13, 208–217 (1985)
45. McNamara, T.P., Hardy, J.K., Hirtle, S.C.: Subjective hierarchies in spatial memory. Journal of Experimental Psychology Learning Memory and Cognition 15, 211–227 (1989)
46. Meneghetti, C., Pazzaglia, F., De Beni, R.: The Mental Representation Derived from Spatial Descriptions is North-Up Oriented: The Role of Visuo-spatial Abilities. In: Stachniss, C., Schill, K., Uttal, D. (eds.) Spatial Cognition 2012. LNCS, vol. 7463, pp. 262–278. Springer, Heidelberg (2012)
47. Agrawala, M., Stolte, C.: Rendering effective route maps: improving usability through generalization. In: SIGGRAPH 2001: Proceedings of the 28th Annual Conference on Computer Graphics and Interactive Techniques. ACM Request Permissions (2001)
48. Fraczak, L.: Generating "mental maps" from route descriptions. In: Olivier, P., Gapp, K.-P. (eds.) Representation and Processing of Spatial Expressions, pp. 185–200. L. Erlbaum Associates Inc., Mahwah (1998)
49. Richter, D., Vasardani, M., Stirling, L., Richter, K.-F., Winter, S.: Zooming In - Zooming Out: Hierarchies in Place Descriptions. In: Proceedings of the 9th Sympo-sium on Location Based Services (LBS 2012), Munich, Germany (2012)
50. Brunyé, T.T., Mahoney, C.R., Taylor, H.A.: Moving through imagined space: Mentally simulating locomotion during spatial description reading. Acta Psychologica 134, 110–124 (2010)
51. Grice, P.: Logic and Conversation. Syntax and Semantics 3, 41–58 (1975)
52. Meilinger, T.: The Network of Reference Frames Theory: A Synthesis of Graphs and Cognitive Maps. In: Freksa, C., Newcombe, N.S., Gärdenfors, P., Wölfl, S. (eds.) Proceedings of the International Conference on Spatial Cognition VI: Learning, Reasoning, and Talking about Space, pp. 344–360. Springer, Heidelberg (2008)
53. Wallgrün, J.O.: Qualitative spatial reasoning for topological map learning. Spatial Cognition & Computation (2010)
54. Barsalou, L.W.: Grounded Cognition. Annual Review of Psychology 59, 617–645 (2008)
55. Johnson-Laird, P.N., Byrne, R.M.J.: Conditionals: A theory of meaning, pragmatics, and inference. Psychological Review 109, 646–678 (2002)

Reading Geography between the Lines: Extracting Local Place Knowledge from Text*

Clare Davies

University of Winchester, Dept. of Psychology (HJB), King Alfred Campus,
Sparkford Road, Winchester SO22 4NR, United Kingdom
ccdavies@ccdavies.co.uk

Abstract. The computational linguistics tool Latent Semantic Analysis (LSA) can approximately map out geographic placenames within spatially unreferenced web-based text. This paper discusses LSA's practical potential to help discover and locate local vernacular (unofficial) geographic names. It may also highlight people's cognitive distortions and biases, creating hypotheses for human experiments, and hinting at identical mechanisms for semantic and spatial memory at these scales. Some successes, problems and methodological issues are illustrated and discussed, with reference to a sample dataset of one area of England. Previous findings were replicated and surpassed, but the space may show more topological than metric tendencies.

1 Introduction

Modern-day humans talk and write every day about places such as villages, towns and city districts, because we spend most of our lives within them or travelling between them. When we talk or write we usually assume that we – both the communicator and the listeners or readers – broadly know and agree about the locations and extents of those places. The variability and the non-Euclidean geometry of our spatial knowledge is well documented in the spatial cognition and human geography literature, yet we somehow manage to communicate useful information about it most of the time.

This state of affairs has long been tacitly acknowledged in cartography by the inclusion of (often) unofficial placename text labels, with no accompanying fixed boundaries. However, in recent years it has become a serious challenge for geographic information (GI) science, as GI is used increasingly on programmed computers and mobile devices whose data models do not allow for vagueness [1].

Socio-economic factors have also increased the frequency with which people without local geographic knowledge have to access and interpret GI for a given area. Across various countries, the discrepancies between local people's geographies and those of official or commercial data have created controversial, expensive and sometimes life-threatening situations, frequently leading to well-meaning but possibly futile attempts to produce rigid official definitions (e.g., [2]

---

* This research was funded by Ordnance Survey of Great Britain.

T. Tenbrink et al. (Eds.): COSIT 2013, LNCS 8116, pp. 320–337, 2013.

[3] [4]). Yet humans seem to create and tolerate both vagueness and inconsistency in their knowledge of place locations and extents, and in multiple names for the same feature or place, as well as differing from one another. These phenomena have been collectively labelled 'vernacular geography' [1].

As previously suggested and tested ([5] [6]), from a cognitive perspective vernacular geography phenomena bear a striking resemblance to the well-established empirical evidence about how people store concepts and categories in their semantic memory. Traditionally semantic and spatial memory ('what' and 'where') were viewed as linked but largely distinct areas of human cognition. However, the above evidence (and the contrasting heavily integrated theorising in human geography, e.g. [7]) both suggest that places may be better considered as semantically rich concepts.

Meanwhile, in recent years within cognitive science, 'embodied cognition' theories have proposed that we appear to employ some of the same neural resources in our brains that we use for physical spatial tasks (in close personal space), even when we perform apparently non-spatial tasks such as thinking about moods or morality, or understanding language (e.g., see [8]; [9]). Such theories suggested that (close-up) spatial processing would fundamentally underpin many more abstract cognitive tasks. They have more recently been challenged and broadened, however, by authors presenting evidence and arguments for far more widespread brain reuse [10]. Evidence suggests that many parts of the brain (especially the cerebral cortex) are used in a very much wider variety of cognitive tasks than was previously thought, and that its interactions with the rest of the body are also more complex than previously assumed. At the same time, the emphasis in neuroscience itself has shifted further away from studies of individual brain areas and their potential function, to the notion of semi-specialised 'circuits' of information flow around the brain [11].

Thus it appears now that any given cognitive task, such as recalling the location of a local settlement from what we used to call our 'spatial memory', may involve a range of resources that we would use similarly to think about many other entities. If we happen to reuse the same apparatus to think about places as we do about, say, categories of animals or birds (the traditional examples in the literature on semantic memory), then this fits better with current notions of an apparently highly flexible, circuit-based, endlessly resource-recycling brain, than with traditional assumptions in cognitive (neuro)science. However, as social psychologists and human geographers are well aware, obtaining and modelling such conceptual cognitive models of place is painstakingly difficult when we have to depend upon asking human participants about them.

2   Latent Semantic Analysis and Placenames

2.1   Description of LSA

Latent Semantic Analysis (LSA) is a computational technique to extract meaning information from large text corpora [12]. Controversially for some linguists, it does this through a crude 'bag of words' approach. Rather than painstakingly tagging and interpreting words (terms) within individual sentences, every term in every document is simply counted, creating a term-document matrix (table) of counts. The values for each term are then adjusted with weightings according to factors such as how unusual it is locally within the document, and globally within the corpus as a whole.

The LSA process then uses Singular Value Decomposition (the mathematics of which need not concern us here) to transform the matrix. The vector within it for each term is not based on simple co-occurrences of that term with others, but more on how much it seems to fit into a similar context (i.e., where texts that have similar content tend to mention both term A and term B – but not necessarily together). Thus it is even possible for two terms which never co-occur within any document in the corpus, to still be associated closely because of a common textual context of use. This allows LSA to effectively disambiguate two meanings of the same word (e.g. 'bank'), based on the different contexts in which they appear – although the term itself will still only appear once in the resulting semantic 'map', lying between its two clusters of related terms. The technique can also, more importantly, yield useful associative information about terms which occur very sparsely in the texts and yet are of great interest to the researcher.

It has long been claimed, by the inventors and users of LSA, that it reflects and models people's cognitive associations between concepts (noting, of course, that this implies shared associations among different individual writers, wherever the corpus reflects multiple authors). Thus LSA has been put forward as a tool for cognitive science to disambiguate and reveal hidden semantic representations, beyond its established use for information retrieval and modelling [13] [14] [15].

The LSA output which is of most interest in this respect is the final term-term matrix. This is a diagonally symmetrical matrix usually consisting of cosine values, each calculated to measure the similarity between each pair of terms. In order to visualise and interpret the matrix outputs of an LSA, it is common to subject the vectors of this eventual term-term cosine matrix to multidimensional scaling (MDS), which tries to optimally reduce the data down to a 2D map of the terms. Coordinates of points within this 'semantic map' reflect the relative position of each term as compared to all the others. Depending on the application, the clusters of related terms can be visually identified, subjected to a clustering algorithm, or compared statistically to an alternative representation.

2.2   LSA and Geographical Placenames

If the concepts represented by terms were actually geographic places, it is reasonable to test the assumption that the 2D post-LSA+MDS 'semantic map' for such places would show greater proximity among places which are geographically closer. Put more simply, it would be very useful if the places that we *say* together also *lay* together. Is it reasonable to expect this? In any text corpus there will be many instances where a placename is mentioned alongside others that are *not* physically close – e.g., when comparing major cities or travel destinations. However, over a large corpus containing many placename occurrences, we can hypothesise that neghboring places of a similar magnitude will be mentioned together more often than those separated by many other places (of the same magnitude).

To test this, over the past few years LSA has been applied to geographic (placename) information within various text corpora, mostly by Max Louwerse's team at the University of Memphis [15] [16] [17] [18]. For both city names within large countries cited in sources such as national newspapers [16] [17], and fictional places within a fantasy world described in a single large novel (i.e. Tolkien's Middle Earth in *The Lord of the Rings*, [18]), fairly high correlations have been found between geographic location in real space and the coordinates of points in a 'semantic' post-LSA +MDS space. In other words, at least for these quite constrained data sources and sets of places, the answer to the above question is a tentative 'yes'.

2.3 Local-Scale Vernacular Geography

These recent findings hold exciting potential in two ways. First, at a practical level, if these findings can be shown to hold at more local geographic scales then we may be able to identify, and perhaps even (at least roughly) to locate, previously unknown local vernacular placenames, without complex querying or dependence on existing gazetteers. Previous efforts to identify vernacular names via web trawling have usually involved direct placename querying of the text (or of a structured online database)–sometimes identifying extra mentioned names that indicate related places–and then 'grounding' the names to known spatial coordinates either from the text or a gazetteer (e.g. [19] [20] [21] [22]).

However, most other projects to capture (or to use or develop better query methods for) geographic web-based information have relied upon gazetteers, to find information about places whose names were already documented. Such research has nevertheless often extended and enriched the gazetteer information through various techniques that increase its relevance to vernacular understanding of the placenames. The SPIRIT project, for instance, used co-occurrences of landmark names or smaller places, that apparently lay within a larger 'vague' place (e.g. a region or locality), to roughly demarcate its location and extent [23].

The traditional surveyor's alternative to such 'grounded' desk-based techniques – directly asking local people for such information – is considered too expensive nowadays on any scale above a single city, and is also prone to error and biased sampling. Victorian surveyors tended to rely on middle-class professionals rather than the peasantry, and nowadays even volunteered online resources often show a bias towards the views of a few highly engaged and (usually) technically literate individuals, whose views may not be representative of the whole population, and who may overlook or underspecify certain geographic areas [24] [25] [26]. Thus automatic web-based corpus building and analysis sounds extremely tempting as an alternative, yet the participation biases should warn us that the choice of sources, and the degree to which they truly represent local vernacular usage, are still likely to affect results.

This leads us to the second potential use for the analysis, the research question: to what extent does the 'geography' revealed by LSA reflect the qualities of people's cognitive representations about the places in the space? We can, for now, leave aside the strongest Whorfian claims of language effectively being, or determining, the cognition itself. The less dramatic, but still strong, claim made for LSA (and similar tools, summarised by McNamara [14]) is that their semantic spaces reflect the way that people's underlying cognition is structured and biased. After all, where else do texts come from?

These two research angles have opposing aims in one respect. If the LSA accurately represents the underlying geography, such that every already-known place falls into its correct relative position in the semantic map, then the texts appear to reflect reality without any obvious cognitive bias - but they are of far more immediate use for capturing new vernacular geography. On the other hand, if the semantic maps of local-scale spaces prove to be far messier than the neater ones derived so far for wider geographic spaces (real and fictional), we may have more work to do to locate new placenames, but we gain many interesting questions about the biases and distortions appearing in the data. As Louwerse has pointed out [15], "LSA does not measure distances, or rather, measures more than distances. Chicago and New Orleans may not be close in space, but are close in jazz music." In any case, techniques to compare Louwerse's findings to the 'real' space have so far led to an r (correlation) value of only around 0.5, explaining only 25-30 per cent of the variance in the data - which is a little too low to be of practical use without further improvement or triangulation with other techniques.

So the initial research question is whether Louwerse's interesting findings can be replicated at all for local-scale data, especially when gathered from multiple sources across the internet rather than from more internally coherent sources. Newspapers and books are unlikely to reflect everyday local people's geographies: journalists tend to use more official placenames and spatial referents, and obviously a single-author source (even when of a real environment rather than Middle Earth) captures only one person's idiosyncratic knowledge. A second question is whether, with better attention to placename-specific issues and optimising techniques, the geographic accuracy and relevance can be improved.

The following section outlines the corpus building and analysis process, as applied to a small region of the county of Hampshire in southern England. Along the way, it will highlight certain methodological choices whose implications will be discussed at the end of the paper.

3   Method

3.1   Geographical Area

For the present sample, a 432km^2 rectangular study area was selected covering the city of Southampton and the immediate area around it, on the south coast of England.[1] This topographically complex, but largely quite flat, area includes a mixture of urban city localities and small outlying suburbs and settlements (but avoiding nearby large towns, so that Southampton was the only major settlement). It also includes part of a National Park with much more sparse population density, some rivers and a major dividing water feature (Southampton Water) with no bridge across it. It was thought that these features might be seen to influence the spatial-semantic mappings, especially between places on either side of the Water, which are separated by only two miles but more than half an hour's driving distance.

Using an Ordnance Survey 1:100,000 map (via the OS 'GetAMap' service, getamap.ordnancesurvey.co.uk), all of the 59 visible placenames within the defined oblong area were used as 'seeds' (search query terms) for retrieving the web corpus described below. It was hypothesised that the corpus itself would then include, and the LSA could roughly locate, many other local names not shown within this official map. The logic here was that wherever one local place is mentioned, others are quite likely to be included too, even though each non-seeded name might occur relatively rarely. (By including the map-labelled names as seeds, it was guaranteed that at least some of those would occur multiple times within the corpus, although of course searches can also produce null results.)

3.2   Procedure: Assembling Corpus

For assembling the example web-based corpus analysed below, the freeware program 'BootCaT' was used [27]. At the time of writing, this currently uses Microsoft's 'Bing' API to retrieve, clean up and collate text results from a series of pre-specified queries. Based on its authors' experience of successful corpus building, the user is encouraged to submit a set of structured queries consisting of fixed-length combinations ('tuples') of a small set of key search terms.

Traditionally, and in some (but not all) of Louwerse's studies, the corpus to which LSA is applied has consisted of at least several million tokens[2]. For the present analysis, a web corpus was collated using a list of query tuples in which each 'seed' placename was included alongside its wider local government authority area (Hampshire or the city of Southampton, as appropriate). Two- or three-word names were enclosed in quotation marks. The BootCaT developers encourage greater recall and precision by creating longer tuples from combinations of two seeds, so their script for this was used to create a much longer list of 2051 query tuples such as:

"Emery Down" Lyndhurst Hampshire
Woolston Swaythling Southampton

---

[1] OS grid square SU, eastings 280-540 (approx. WGS84 longitude: -1.60406 degrees to -1.23250 degrees), northings 000-180 (approx. WGS84 latitude: 50.79889 degrees to 50.95890 degrees).

[2] 'Token' is the linguistics term for any occurrence of any word, regardless of which word; for example, the title of this paper contains 11 tokens, and would do even if two of the words were the same.

BootCaT retrieves a maximum of 50 results per query tuple, and then eliminates all duplicate URLs before retrieving the actual text and attempting to clean all HTML tags and other junk from it. Some filtering of 'adult' content (by including a short 'blacklist' of unwanted terms), and the use of a 'whitelist' of high-frequency words and common spatial terms (especially prepositions), increased the relevance of the search results. The resulting raw corpus, before further cleaning (see below), contained around 13.7 million tokens in roughly 15,000 documents. A quick relevance check by searching for the name 'New Hampshire', which had been by far the most common inappropriate inclusion in previous smaller-scale trials of this method, found 264 occurrences - implying over 98 per cent precision.

However, using this method with a focus on placenames raises some specific issues, previously recognised but not addressed in Louwerse's work. The first of these is placename duplication: the same name being used for more than one place, often in another area or even country. In the present analysis, the use of the local authority name did seem very effective in focusing results on the UK and the specific local area. Exact duplication within the local area, though rarer and largely only applying to small suburbs or hamlets, was harder to avoid. A possible future solution for this issue will be outlined in the Discussion section below.

The second phenomenon is the tendency of many placenames to be homonymous with ordinary words (e.g. Bath, Rugby, Fleet, Woodlands). In the present analysis this was minimised as far as possible by identifying, and adding a 'zz' to, all capitalised instances of such placenames in the corpus. In the interests of potential future automation, the names to which this applied can be identified automatically by running the Aspell spell-checking algorithm on the placenames list, using a British English dictionary that omitted proper names. Obviously, tagging capitalised instances is not perfect since people do not always correctly capitalise placenames (nor avoid capitalising ordinary nouns), especially in casual online writing such as discussion forums. Furthermore, sometimes the ordinary meaning of the word would be capitalised legitimately in a title or at the start of a sentence (and, of course, this technique could not be applied with a language such as German). Nevertheless, in English, such relatively unusual occurrences may be 'ironed out' when considering a very large corpus such as the present one.

The third issue with placenames is the frequency of two- or three-word names, sometimes hyphenated but often not (e.g. New York, Stratford-upon-Avon). As the term-document matrix takes a term to be a single word token, this can lead to confusions and irrelevant data. In the present analysis, this was resolved as far as possible by removing the white space, apostrophes and hyphens from all instances of such names in the corpus to create single tokens, as part of the general 'cleaning' process.

The cleaning process also included identifying and deleting around two hundred documents (i.e., downloaded web page texts) that were either extremely short (just a URL or a few additional words), or had been replaced entirely by an error or crawler-blocking message. The eventual corpus contained 14,231 documents.

The R implementation of LSA tends to be extremely memory consuming, and can hang with a large corpus. It was necessary to perform some final corpus cleaning that included removing extremely common 'stop' words (it, a, the, etc.), to reduce the number of tokens to around 7.6 million. The most extremely sparse terms (more than 99 per cent sparsity) were then also removed from the text matrix of counts, since sparse matrices are particularly inefficient to process. This removed a lot of non-word terms and other junk, but also all other terms occurring in less than 1 per cent of documents (i.e., appearing in less than 142 out of the 14,231). Although this did involve losing a few very rarely mentioned names of places and related landmarks (including three of the original seed names), making it a non-ideal step if a larger-scale cloud facility was available, it yielded a far more manageable matrix of just 4062 terms (linguistically, 'types'). These thus formed a matrix of around 58 million term-document frequency counts, of which just over two million were non-zero.

3.3 Procedure: Deriving Semantic Space

The LSA procedure was run using R, following the same process as used in Louwerse's papers to date [15] [16] [17] [18]. This sets the number of calculated dimensions at 300, and uses default settings for local and global weightings (see [16] for more details).

The cosine matrix for the original placenames was calculated from the resulting LSA space, and then Euclidean distances were derived from these[3]. The resulting term-to-term distance matrix was then submitted to a multidimensional scaling procedure.

The coordinates of the places in the resulting 2D 'map' were then compared with grid coordinate data extracted from a gazetteer (U.K. OS grid references, expressed as approximate two-digit 'easting' (longitude) and 'northing' (latitude) values representing 1km intervals within each 100km2 grid square), to examine how well the MDS space matched the local geographic layout. This was done initially with simple correlations between the dimensions of the two spaces, and then with Tobler's bidimensional regression procedure (see [28] [29] [30]).

4 Results

4.1 Seed Placenames Alone

Louwerse's work previously used non-metric MDS to map the cosine-based distance matrix among the placenames of interest, presumably on the basis that

---

[3] While this may seem an odd approach, the cosines represent the angles between term vectors in the initial 300-dimension semantic space, ignoring the vectors' lengths. If raw distances are calculated from that space instead, they are mainly influenced by the vector lengths which, in themselves, depend upon the raw frequency of each term's occurrence in the original corpus. Taking cosines first, and then deriving distances from that cosine matrix, focuses on relatedness rather than frequency.

the data could not be assumed to be evenly distributed ratio data. Applying the same process to the present data, a two-dimensional non-metric MDS on the cosines-based Euclidean distance matrix for the seed placenames gave r-squared = 0.968, after converging in only two iterations with a stress of 0.180.

When the two dimensions from this MDS were correlated against the two dimensions of the geographic coordinates for the seed places, the results were unclear. Ideally, this analysis would show a clear correlation between one MDS dimension and the latitude, and between the other MDS dimension and the longitude. On this occasion, however, this was not the case - one of the MDS dimensions did not correlate strongly with anything, while the other correlated above 0.4 with both the easting (longitude) and the northing (latitude).

A bidimensional regression on the dimensions suggested that either a Euclidean or an affine transformation would work moderately well in transforming one set of points into the other, although the affine transformation was significantly better $(F(2,106) = 3.34, p <0.05)$. The r value for this transformation was 0.562 - similar to Louwerse's best findings, as might have been expected since the same procedure was followed (see Discussion for suggested future enhancements which may improve on it).

4.2   'Discovered' Placenames

Complete Dataset. As mentioned earlier, one interesting aspect of this technique lies in its potential to 'discover' and roughly locate local vernacular placenames that are not currently present in existing geographic data. A second analysis was undertaken to examine this possibility, while avoiding an extensive process of named entity extraction to truly 'discover' them in the data, which is beyond the scope of this paper. Thus this second analysis extracted cosines from the LSA results not only for the originally seeded places, but also for extra local placenames which were identified via an unofficial gazetteer. It was hoped that alongside the seeded placenames, other local ones would be picked up as LSA terms and could thus be included in another MDS analysis, to see if they too would at least be located within appropriate clusters of places.

The gazetteer used to identify these extra places (see www.gazetteer.co.uk) was developed for Great Britain by volunteer members of a group called The Association of British Counties (ABC), whose aim is to capture and integrate historic and modern place information, particularly with respect to the disparities between historic (but often still mentioned in everyday use) and modern administrative geographic units, and different spellings of placenames. For the purpose of the current analysis, it thus offered a sort of halfway house between truly vernacular data (i.e., obtained from local residents' everyday speech or writing), which even when available tends to be incomplete and poorly referenced, and the official data from which the original seeds were extracted.

By contrast to the 59 'official' names originally identified from the small-scale cartographic map, the ABC gazetteer listed 205 placenames within the specified test area. This list was 'cleaned' in the same way as previously discussed, for three purposes: to tag names that corresponded to dictionary words, to reduce multi-word names to single ones, and to remove inadvertent duplicates (since the gazetteer often had the same place listed both as e.g. "Swanwick, Lower" and "Lower Swanwick"), which reduced the list to 181, of which 58 had been seeds (one seed was not in the gazetteer). Thus there were 181 - 58 = 123 extra placenames sought in the data, to see whether placename discovery was feasible in this way.

Despite the earlier removal of extremely sparse entries, and the presence in the gazetteer of some rather obscure names which are quite likely to be obsolete, 48 of the non-seeded 123 names (39 per cent) were identified within the output from the LSA. Along with the seeds already identified, this total of 104 placenames was resubmitted to the nonmetric MDS and bidimensional regression procedures outlined above.

This time the nonmetric MDS converged in just two iterations, with stress = 0.188 and r-squared = 0.965. The output dimensions correlated more cleanly with the easting or longitude (MDS dimension 2, r = -0.51) and the northing or latitude (MDS dimension 1, r = 0.53). Since the angle between the west-east and south-north axes has to be fixed at -90 degrees (i.e., to make geographic sense, the dimensions need to positively correlate), this implied that the data would fit better if this dimension was reflected. Since the MDS dimensions based on cosines tend to include values on either side of zero, this was done simply by multiplying them all by -1. To make the dimensions entirely positive, since earlier work had made the author question how well the scaling factor can work in bidimreg when one dataset includes negative numbers and the other does not, 2 was added to the values on both MDS dimensions (which, being distances between cosine values, inevitably ranged from -2 to +2).

A bidimensional regression showed that again an affine transformation would work best, with an r value of 0.535: comparing it to a Euclidean transformation, $F(2,202) = 39.6$, p <0.0001.

Seeds Subset. Reducing the data down to just the seed placenames again, to compare this data more directly with the previous analysis on just that data, the two coordinate dimensions produced by the MDS were again correlated with the lat-long ('easting and northing') coordinates for the actual places, as obtained from the gazetteer. Interestingly, this time the results were much clearer and stronger - implying that including the data from the extra places had improved the results for the original seeds.

The first MDS output dimension had a Pearson correlation r = 0.673 with the 'easting' (longitude), but the second correlated r = -0.556 with the 'northing' (latitude). Once again, the latter dimension was reflected as above due to the negative sign.

Submitting (again just the seed) data from this to a bidimensional regression, to compare with the earlier one, it was clear that once again an affine transformation mapped the two coordinate sets far better than a Euclidean one. This time the data gave noticeably stronger r values: r (affine)=0.637, $F(4,106)=18.08$, p <0.0001. Comparing the two directly, $F(2,106) = 34.1$, p <0.00001. In terms of explaining variance, the r-squared value for the affine transformation was 0.406, i.e., the bidimreg could model 40 per cent of the difference between the real and the corpus-derived coordinate data.

Figure 1 is a map comparing the 'real' seed place coordinates to their post-MDS coordinates (after transformation according to the parameters from this last, stronger, bidimensional regression). Lines from each real-world place location (filled circle) link it to the position predicted by the corpus data. The approximate coastline and local rivers have been outlined in grey, to delineate natural boundaries and barriers in the local topography.

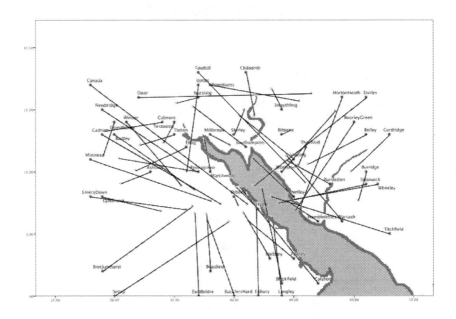

Fig. 1. Plot of the 'seed' placename coordinates in geographic and predicted 'semantic' space

It will be observed that although there are some very large displacements, none actually crosses Southampton Water, suggesting as predicted that this is a major cognitive boundary in local mental geography as reflected in written language. In general, the map gives the impression that the bidimensional regression may not have been effective enough in rescaling the data: most places seem displaced towards the centre of the map. There is also a sense of clustering in the data which may be worth investigating further in future work: lines from neighboring places often appear to be moving in the same direction to stay broadly near one another. In general, though, the map suggests that at local level the topology may be preserved more effectively than the metric spatial layout. This is what we might expect, of course, not only from the fact that the former is easier to describe in language, but also from many years of studies of human spatial cognition itself, dating from Lynch's seminal work [36] up to more recent experiments [37] and neuroscience investigations [35].

4.3   Examining City Influence

The sense of central 'pull' in the above map led to one final analysis, in which the name of Southampton itself was deliberately omitted from the MDS and bidimreg to see if it reduced the amount of central tendency in the data. All the other placenames (not just the seeds) were again included in the MDS, since this appeared to improve the modelling of the key 'seed' locations in the subsequent bidimreg last time. With 103 placenames in the analysis, the non-metric MDS again converged in just two iterations, with stress = 0.186 and r-squared = 0.965. The bidimensional regression was then run once more on all of the places. Similarly to the previous analysis, an affine transformation was preferred (F(2,200) = 40.1, p <0.0001), with a similar r value of 0.539. When repeated with just the seed placenames, however, the r for the affine transformation was now 0.652; this transformation was still considerably more effective than a Euclidean one (F(2,104) = 36.1, p <0.0001).

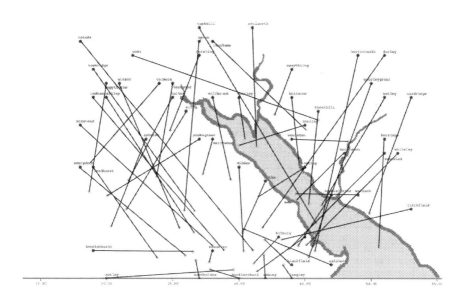

Fig. 2. Plot of the 'seed' placename coordinates as above, after removing Southampton from the analysis

Plotting this last output on a map as in Figure 2, we see that in reality there is still some central tendency but it no longer seems to point to Southampton nor on Southampton Water; instead, the centre of gravity of the representation has moved south towards the main sea coast and the New Forest. Despite the similar r value to last time, the map appears even less ordered and has some even longer displacements. However, clusters of nearby settlements still seem to be displaced as groups in the same direction. Also, although this time it looks as though some eastern settlements have crossed the water, in reality they largely still do not cross the lines of those on the west side: it is more as though the displaced line of the water would be too thin and slightly misplaced. (This may of course be a partial consequence of the waterway also having Southampton in its name.)

Overall, this analysis shows that exclusion of major nodes from the scaling and mapping analysis can affect results, even where their names had been included in the LSA and the subsequent calculation of cosines and Euclidean distances. On a positive note, it again seems to hint (as the cognitive literature leads us to expect) at a topological, more than topographical, representation of local-scale space within the text data, and at a cognitive clustering of places in a similar pseudo-hierarchical way to that of concepts in semantic memory [39]. (However, as in the case of semantic concepts, this notion of hierarchy is likely to be far looser and more flexible than was once thought [38].)

5   General Discussion

The above analyses have demonstrated that it is possible to use these techniques to explore the geographic component of large-scale semantic networks derived from text, and their potential to reflect cognitive geographies (particularly, in this case, the apparent importance of local topology in creating separate categories of places, with a clear boundary between those on either side of Southampton Water). They have also shown that additional placenames, discovered in the text either via knowledge gained from other sources or perhaps via 'named entity' extraction, can be identified and, when added to the analysis, appear to enhance its spatial accuracy in mapping the more frequently occurring 'seed' placenames (and hence, presumably, others too).

A wide range of options and improvements could take this work further - and indeed are necessary, if the data is to be useful for establishing vernacular geography. For a start, the initial corpus cleaning, and the attempts to solve ambiguous placename issues, could be enhanced still further.

However, as mentioned above, it is almost impossible to disambiguate placenames that are duplicated *within* a local area without using a more sophisticated and iterative technique, such as statistically identifying places which appear to have two (or more) related clusters that are unusually distanced from each other. LSA, as mentioned above, can reveal such clusters but cannot actually separate the term into its two separate occurrences. It may be possible to use such cluster results in a co-occurrence search of a corpus (to identify the instances where the name occurred with members of one cluster, as opposed to members of the other), tag the instances differently, and re-run the LSA to separate the two tagged versions of the name. This may well be achievable through further research.

Secondly, although the present analyses kept Louwerse's parameters for the LSA and MDS processes for the sake of replication, it seems worth questioning them. In particular, the oft-quoted 300 dimensions may well not be the optimum number to choose for this kind of LSA application. The number appears to have been based on studies in which an LSA was applied to more semantic domains, such as automatically scoring student essays and comparing the results to human markers. It seems unlikely that the same number of domains will be optimal for such a different application as this one. LSA experts frequently stress, apparently against the general trend in its usage, that one should avoid using an arbitrary number of dimensions, instead basing the choice on theoretical or mathematical reasoning [40].

Thirdly, the apparent importance of local topology in the map shown above - where despite distortions of distance, the barrier of Southampton Water was preserved in the data, and the loss of a major local settlement altered the outcome for all the smaller ones surrounding it - suggests the same kind of issues as in the existing literature on large-scale spatial cognition. Once the results have been optimised for a given set of known places, the remaining distortions of LSA outputs compared to the real (cartographic) geography can be more confidently explored as possible cognitive 'distortions' in writers' mental geography of the local area. There is clearly rich potential here, since the geography itself only explained 40 per cent of the variance in the post-LSA mappings at best, suggesting that much else is going on to underpin people's understanding of places.

We may find, for instance, that the road distance or more especially travel time between places is a better bidimreg match to the LSA output (via MDS) than direct distances, since the essentially episodic experience of travel between places seems to be an essential part of the way that our brain tracks space (see, e.g., [35]). We may also need to try to factor in the extent to which places cluster partly on semantic similarity rather than solely geographic proximity. If we can do this in a way that makes a text-based model match well with the results of human studies, then this kind of technique may have very useful predictive potential for people's local spatial behaviour and knowledge.

Such 'distortions' also raise interesting hypotheses for human-participant studies, to see how far the results from LSA studies reflect people's performance on other place-related tasks (such as categorisation, map drawing, word associations etc.). If we are right to see places as essentially semantic categories with a spatial dimension, then studies of people's cognitive task performance should match the findings for more conventionally studied categories such as animals and trees. Indeed, there is already some experimental evidence to support this [6].

This also reminds us that the collation and nature of the initial corpus might be expected to make the greatest impact on the eventual place mappings. As mentioned earlier, the present corpus was constructed to be as large as possible, while maintaining a high degree of precision (relevance). However, as also mentioned earlier, representativeness of the general population may be more critical for useful vernacular geography (and, indeed, for valid psychological inference and testing) than simply capturing every possible name and every possible usage of it.

The present analysis stuck to a predefined procedure, in order to try to replicate previous results without fiddling. As a result, we have seen that even the default and most commonly used parameters for these analyses do reveal some meaningful underlying geographic knowledge, even with more complex and messy local geographies and sources than those previously tried. Nevertheless, the series of methodological decisions taken during the LSA (choices of weighting calculations as well as the number of dimensions) and also during the MDS (choice of method, parameters, distance measures, treatment of the data as ordinal or interval, and many more besides: see [31]) would surely affect the outcome of the modelling. Indeed, the singular value decomposition (SVD) performed within an LSA also underpins, and has been described as a specific form of, MDS [41] – *and* apparently underlies the calculations in bidimensional regression [42]. So there might be a more integrated and effective way to approach this kind of statistical analysis. This would be preferable, computationally more efficient and methodologically more transparent, than performing rather similar statistical tricks three times over.

Since this is really a modelling rather than a hypothesis-testing approach, it would obviously be best if methodological choices in future studies were driven by explicit theoretical considerations. The tweaking of some of the parameters might also resolve the (relatively few) encounters with very severely misplaced individual placenames. However, a more refined approach to the names themselves (duplicates, lexical and multi-word names, etc.) may be more fruitful as discussed above.

Meanwhile, it remains to be seen whether an even more local scale of geography – where vernacular names arise for individual local features such as parks, pubs, alleyways and street corners – might also be usefully explored using these techniques. Far less data would be available online for such local landmarks, of course, creating an even sparser matrix in the first place (which would require efficient large-scale computing techniques and facilities), but that may be the only limiting factor in terms of scale.

True vernacular placename discovery, of course, requires us to find and locate names that are not yet in any gazetteer. This requires some pre-analysis of a text corpus, to identify such names and ensure they are extracted for modelling after the LSA process. Several algorithms exist (e.g. [32] [33]) for such Named Entity extraction, as it is known in the information science literature. However, when trialled on some text samples from the current corpus these were found to be seriously lacking, failing to pick up many placenames which would be obvious from the linguistic context to any human viewer. As mentioned earlier, this aspect of vernacular name discovery is beyond the scope of the current paper, but it requires further investigation if the technique is to be used for discovering new and entirely undocumented local names.

Even without this, however, the present paper has shown that to some extent, clues as to the location of any newly identified placenames may be gained from the refinement and use of these entirely language-based techniques, even in the complete absence of explicit spatial information. LSA's most common application, known as latent semantic *indexing* (LSI), allows one to compare a new query to the existing dataset (say, to retrieve relevant documents whose vectors in the space are similar to the query vector) - which is why it is used by major search engines. Thus 'querying' an established, well-constructed dataset about a newly identified placename may also be possible in future, without re-running the entire LSA-MDS process.

Nevertheless, the problems encountered when handling placenames representing apparently multiple scales and categories suggest that we still have a lot of Louwerse's 'jazz' in the data. This leaves us plenty of scope for experimenting with differently categorised places (such as those falling into different local authority areas or socioeconomic communities), to see if treating them separately in the MDS approach can produce more geographic fidelity. In other words, as we factor out potential biases, the geometric accuracy of the map should improve. Or, conversely, if we keep reaching a limit to that accuracy no matter how well the issues outlined above are handled, it may be due to the unavoidable biases of human cognition.

Furthermore, to reconcile the two apparently opposing research agendas discussed earlier, it is likely that we can use cognitive theories and possibly even computational models of semantic memory (e.g., [34]) to hypothesise how such category or hierarchy biases (and perhaps some others) might influence the results. In this way we may be able to produce optimised results for vernacular geography extraction, while simultaneously testing the limits to the application of semantic memory theories to place.

References

1. Davies, C., Holt, I., Green, J., Harding, J., Diamond, L.: User Needs and Implications for Modelling Vague Named Places. Spatial Cognition and Computation 9, 174–194 (2009)
2. Levitz, Dena: The Awkward Art of Neighborhood Naming. The Atlantic Cities (January 3, 2012), http://www.theatlanticcities.com/neighborhoods/2012/01/awkward-art-neighborhood-naming/843/
3. Beaumont, Lucy: Secret Suburbs. The Age (October 3, 2007), http://www.theage.com.au/articles/2007/10/03/1191091194789.html
4. British Broadcasting Corporation: PCSOs 'did not watch boy drown'. BBC News (2007), http://news.bbc.co.uk/1/hi/england/manchester/7007081.stm
5. Davies, C.: Are Places Concepts? Familarity and Expertise Effects in Neighborhood Cognition. In: Hornsby, K.S., Claramunt, C., Denis, M., Ligozat, G. (eds.) COSIT 2009. LNCS, vol. 5756, pp. 36–50. Springer, Heidelberg (2009)
6. Davies, C., Battye, E., Engelbrecht, P.: Places as Fuzzy, Partly Visuospatial Categories. Journal of Experimental Psychology: Applied (under review)

7. Cresswell, T.: Place: a Short Introduction. Blackwell, Oxford (2004)
8. Foglia, L., Wilson, R.A.: Embodied Cognition. Wiley Interdisciplinary Reviews: Cognitive Science (2013), doi: 10.1002/wcs.1226
9. Anderson, M.L., Richardson, M.J., Chemero, A.: Eroding the Boundaries of Cognition: Implications of Embodiment (1). Topics in Cognitive Science 4, 717–730 (2012)
10. Anderson, M.: Neural Reuse: a Fundamental Organizational Principle of the Brain. The Behavioral and Brain Sciences 33, 245–266 (2010)
11. Sporns, O., Tononi, G., Kötter, R.: The Human Connectome: A Structural Description of the Human Brain. PLOS Computational Biology 1(4), e42 (2005), doi:10.1371/journal.pcbi.0010042
12. Landauer, T.K., McNamara, D.S., Dennis, S., Kintsch, W.: Handbook of Latent Semantic Analysis. Lawrence Erlbaum, Mahwah (2007)
13. Landauer, T.K.: LSA as a Theory of Meaning. In: Landauer, T.K., McNamara, D.S., Dennis, S., Kintsch, W. (eds.) Handbook of Latent Semantic Analysis, pp. 3–34. Lawrence Erlbaum, Mahwah (2007)
14. McNamara, D.S.: Computational Methods to Extract Meaning from Text and Advance Theories of Human Cognition. Topics in Cognitive Science 3, 3–17 (2011)
15. Louwerse, M., Cai, Z., Hu, X., Ventura, M., Jeuniaux, P.: Cognitively Inspired NLP-Based Knowledge Representations: Further Explorations of Latent Semantic Analysis. International Journal on Artificial Intelligence Tools 15, 1021–1039 (2006)
16. Louwerse, M.M., Zwaan, R.A.: Language Encodes Geographic Information. Cognitive Science 33, 51–73 (2009)
17. Louwerse, M.M., Hutchinson, S., Cai, Z.: The Chinese Route Argument: Predicting the Longitude and Latitude of Cities in China and the Middle East Using Statistical Linguistic Frequencies. In: Proceedings of the 34th Annual Conference of the Cognitive Science Society
18. Louwerse, M.M., Benesh, N.: Representing Spatial Structure Through Maps and Language: Lord of the Rings Encodes the Spatial Structure of Middle Earth. Cognitive Science 36, 1556–1569 (2012)
19. Hollenstein, L., Purves, R.: Exploring Place Through User Generated Content: Using Flickr tags to Describe City Cores. Journal of Spatial Information Science 1, 21–48
20. Leidner, J.L., Sinclair, G., Webber, B.: Grounding Spatial Named Entities for Information Extraction and Question Answering. In: Proceedings of the HLT-NAACL 2003 Workshop on Analysis of Geographic References, vol. 1, pp. 31–38. Association for Computational Linguistics (2003)
21. Pasley, R.C., Clough, P.D., Sanderson, M.: Geo-Tagging for Imprecise Regions of Different Sizes. Proceedings of the 4th ACM Workshop on Geographical Information Retrieval, pp. 77–82 (2007)
22. Twaroch, F.A., Jones, C.B., Abdelmoty, A.I.: Acquisition of a Vernacular Gazetteer from Web Sources. In: Proceedings of the First International Workshop on Location and the Web, pp. 61–64. ACM (2008)
23. Jones, C.B., Purves, R.S., Clough, P.D., Joho, H.: Modelling Vague Places with Knowledge from the Web. International Journal of Geographical Information Science 22, 1045–1065 (2008)
24. Whittaker, S., Terveen, L., Hill, W., Cherny, L.: The dynamics of mass interaction. In: Proceedings of the 1998 ACM Conference on Computer Supported Cooperative Work, pp. 257–264. ACM (1998)
25. Purves, R., Edwardes, A., Wood, J.: Describing Place Through User Generated Content. First Monday 16(9) (September 5, 2011), firstmonday.org

26. Haklay, M.: Nobody Wants to Do Council Estates: Digital Divide, Spatial Justice and Outliers. In: Presented at the 108th Annual Meeting of the Association of American Geographers, New York, USA (February 25 2012),
http://poversham.wordpress.com/2012/03/05/

27. Baroni, M., Bernardini, S.: BootCaT: Bootstrapping Corpora and Terms from the Web. In: Proceedings of LREC, pp. 1313–1316 (2004), bootcat.sslmit.unibo.it

28. Tobler, W.: Bidimensional Regression. Geographical Analysis 26, 186–212 (1994)

29. Friedman, A., Kohler, B.: Bidimensional Regression: A Method for Assessing the Configural SImilarity of Cognitive Maps and Other Two-Dimensional Data. Psychological Methods 8, 468–491 (2003)

30. Carbon, C.C.: BiDimRegression: Bidimensional Regression Modeling Using R. Journal of Statistical Software (accepted pending minor revision, 2013)

31. Borg, I., Groenen, P.J.F.: Modern Multidimensional Scaling: Theory and Applications, 2nd edn. Springer, New York (2005)

32. Cunningham, H., Maynard, D., Bontcheva, K., Tablan, V.: GATE: A Framework and Graphical Development Environment for Robust NLP Tools and Applications. In: Proceedings of the 40th Anniversary Meeting of the Association for Computational Linguistics (ACL 2002), Philadelphia (July 2002), gate.ac.uk

33. Finkel, J.R., Grenager, T., Manning, C.: Incorporating Non-local Information into Information Extraction Systems by Gibbs Sampling. In: Proceedings of the 43nd Annual Meeting of the Association for Computational Linguistics (ACL 2005), pp. 363–370 (2005), nlp.stanford.edu

34. Rogers, T.T., McClelland, J.L.: Semantic Cognition: A Parallel Distributed Processing Approach. MIT Press, Cambridge (2004)

35. Burgess, N., Maguire, E.A., O'Keefe, J.: The Human Hippocampus and Spatial and Episodic Memory. Neuron 35, 625–641 (2002)

36. Lynch, K.: The Image of the City. MIT Press, Cambridge (1960)

37. Stankiewicz, B.J., Kalia, A.A.: Acquisition of Structural Versus Object Landmark Knowledge. Journal of Experimental Psychology: Human Perception and Performance 33, 378–390 (2007)

38. Murphy, G.L., Hampton, J.A., Milovanovic, G.S.: Semantic memory redux: an experimental test of hierarchical category representation. Journal of Memory and Language 67, 521–539 (2012)

39. McNamara, T.P., Hardy, J.K., Hirtle, S.C.: Subjective Hierarchies in Spatial Memory. Journal of Experimental Psychology: Learning, Memory, and Cognition 15, 211–227 (1989)

40. Quesada, J.: Creating Your Own LSA Spaces. In: Landauer, et al., op. cit

41. Bartell, B.T., Cottrell, G.W., Belew, R.K.: Latent Semantic Indexing is an Optimal Special Case of Multidimensional Scaling. In: Proceedings of the 15th Annual International ACM SIGIR Conference on Research and Development in Information Retrieval (ACM), pp. 161–167. ACM (1992)

42. Platt, J.: On Statistical Approximations of Geographical Maps. In: Proceedings of the 8th International Conference on GeoComputation (2005),
http://www.geocomputation.org/2005/Platt.pdf

Modeling Spatial Knowledge
from Verbal Descriptions

Lamia Belouaer, David Brosset, and Christophe Claramunt

Naval Academy Research Institute,
BP 600, 29240, Brest Naval, France
{lamia.belouaer,david.brosset,christophe.claramunt}@ecole-navale.fr

Abstract. Over the past few years, several alternative approaches have
been suggested to represent the spatial knowledge that emerges from natural environments. This paper introduces a rule-based approach whose
objective is to generate a spatial semantic network derived from several
humans reporting a navigation process in a natural environment. Verbal
descriptions are decomposed and characterized by a graph-based model
where actions and landmarks are the main abstractions. A set of rules
implemented as first order predicate calculus are identified and applied,
and allow to merge the common knowledge inferred from route descriptions. A spatial semantic network is derived and provides a global and
semantic view of the environment. The whole approach is illustrated by
a case study and some preliminary experimental results.

Keywords: navigation knowledge, verbal description, spatial semantic
network.

1 Introduction

Spatial cognition is one of the most fundamental experiences of humans interacting in their environment [6]. All of us are in our every day life navigating in
known or unknown environments, and trying to infer new spatial knowledge that
will help us to act and behave appropriately. Amongst many spatial concepts,
maps have been long used as valuable references for structuring a representation
of our environment. Maps can be modeled in different forms: either qualitative
or quantitative, metric or topological. The representations that emerge from
maps are often sufficient for many applications, but in some cases they might
not be appropriate. This is especially the case for human navigating with a low
knowledge of their environment, particularly in natural and poorly structured
environments. The spatial knowledge that can be inferred from human perception of such environments cannot be directly described using conventional map
representations. In fact, navigation in these environments should be structured
enough to infer new spatial knowledge that will allow humans to navigate and
reach their destinations [9]. Moreover, when perceived and described appropriately, navigation can generate a useful representation of the environment. The
spatial knowledge that emerges might be even analyzed in order to infer a generalized representation of the environment. Of course, humans perceive different

T. Tenbrink et al. (Eds.): COSIT 2013, LNCS 8116, pp. 338–357, 2013.

landmarks and relations that nature offers to derive a cognitive and structured representation [15].

Human navigation emerges from either path planning or displacement along a specified route in a given environment. Navigation can be modeled as a sequence of locations and actions from an initial location to a target location. It has been shown that verbal descriptions are well suited to the description of human navigations [7,8,14]. Several recent studies have attempted to model navigation knowledge from human verbal descriptions [12,16]. The research challenge relies not only on the modeling of the cognitive abstractions that help to model such displacements, but also to map these concepts to appropriate linguistic constructs that can be in turn mapped towards a structured representation [13]. In particular, a close relationship between what is perceived and what can be logically described should be investigated and formalized. Amongst many approaches developed so far, ShopTalk is an example of a proof-of-concept wearable system designed to assist visually impaired shoppers with finding shelved products in grocery stores [11]. Using synthetic verbal route directions and descriptions of a store layout, ShopTalk leverages the everyday orientation and mobility skills of independent visually impaired travelers to direct them to aisles with target products. A natural language route description can be inferred from a set of landmarks and a set of natural language route descriptions, whose statements have been tagged with landmarks from a landmarks set. However, and despite the fact that humans are effective to process and transmit such knowledge, there is still a conceptual gap between these descriptions and map representations. Other approaches in indoor or outdoor environments have been suggested to model verbal routes descriptions [8,14].

The research presented in this paper introduces an approach whose objective is to generalize the spatial knowledge inferred from several descriptions of navigation routes. The approach is applied and experimented in the context of poorly structured environments and a panel of humans navigating in such environments, and where every panelist is asked to move between an initial and target locations. Each human is given appropriate navigation instructions; and the route performed is recorded and verbally described after each navigation. More formally, each route is modeled as a sequence of locations and actions as introduced in [2]. Locations and actions are interactively identified at the expert level. All route descriptions are manually mapped to the most appropriate landmarks and spatial relations. Navigation is modeled as a graph where nodes represent locations, and edges actions between these locations. This experimental setup provides several navigation descriptions where each generated graph corresponds to the verbal description of a human route. The nodes are labeled by their names and types, actions according to a categorization that qualifies the spatial terms and relations identified. The main objective of the modeling approach is to reconcile as much as possible the different navigation routes generated. We define a rule-based approach where several operators are applied. Those operators allow to evaluate the semantic matching between nodes in different route descriptions. When their semantics match these nodes are merged.

The iterative application of this process generates a spatial semantics network that gives a synthetic view of the environment.

The rest of the paper is organized as follows. Section 2 introduces the modeling background of our approach, while Section 3 develops our model and defines the set of rules for the spatial semantic network generation. Section 4 presents the experimental evaluation conducted. Finally, Section 5 discusses the potential of our approach and outlines further work.

2   Modeling Approach

The objective of our modeling approach is to generate a global and semantic view of the environment as depicted by several humans navigating in that environment. The approach first introduces a graph-based model whose principles are described in Section 2.1, while Section 2.2 develops the ontology support of the modeling approach.

2.1   From Verbal to Formal Descriptions

A navigation route is modeled using four complementary constructs as suggested in [10]:

- an action models the displacement behavior of a human acting in the environment. Actions are materialized by verbs. They represent the dynamic component of a human navigation. An action can be associated to a starting position and a target position;
- a landmark denotes a decision point in the environment, or assimilated to a decision point;
- a spatial entity models a thing of interest in the environment;
- a spatial relation models a relation between a spatial entity and the environment that can be associated to the description of an action.

Let us introduce the modeling concepts that support the formal representation of a route. A navigation route is represented by a directed graph where actions, landmarks and spatial entities are the main primitives:

- a *node* models either the start or the end of an action in geographical space. It can be characterized by the presence or not of a landmark or a spatial entity in a route description. A node (l, t) is labeled by its name l and its type t whose values are unknown when no landmarks or spatial entities are given at this node. A node label is a noun given by the wayfinder in the verbal description. It often plays the role of a unique identifier for a landmark or a spatial entity. Node types are defined according to a hierarchy, noted $\mathcal{T}$, where each node is a type (e.g., 111 hill, 301 lake) derived from IOF (International Orienteering Federation) map legends.[1]

---

[1] http://orienteering.org/wp-content/uploads/2010/12/
Control-Descriptions-2004-symbols-only1.pdf

– an *edge* models an action, that is, an elementary displacement described by a direct path between two nodes. Actions can be described by additional relations such as cardinal orientation (O_{r_C} = {North, South, East, West, North East, North West, South East, South West}), relative orientation (O_{r_R} = {Right, Left, Forward, Backward}) and/or elevations (E = {Up, Down}) and/or metric or qualitative distances (D_Q = {close, ..., far}). The semantics of an action can be also related to a spatial entity. An action starts and terminates at a location that can model a landmark. Formally, O_r is the union of these relations, the null element is added to the set to denote the absence indication of the orientation of the action in a verbal description: $O_r = O_{r_R} \cup O_{r_C} \cup E \cup D_Q \cup \{null\} \cup \mathbb{R}^+$.

Let us consider the following verbal description v_{d_1}: "*From the eighth control point climb the knoll going to the north to arrive at the electric pole. Then go to the footpath. Turn to the right and you must see the ninth control point*". In order to model the verbal description v_{d_1}, nodes and edges are extracted:

1. nodes:
   – the eighth and the ninth control points are modeled as the start and end nodes, respectively, of that verbal description;
   – two landmarks are identified: electric pole (517) and footpath (507). They are modeled as nodes labeled with appropriate types.
2. edges:
   – *From the eighth control point climb the knoll going to the north to arrive at the electric pole* gives an edge that begins with the node representing the eighth control point and ends with the node representing the landmark electric pole. The action *climb the knoll* is generalized as an elevation relation Up, while the action *going to the north* corresponds to the cardinal relation North. Therefore, this edge is first interpreted and labeled with the spatial relation: North ∧ Up;
   – from the electric pole to the footpath there is an action, but it is not qualified so the edge corresponding to that action has no explicit label;
   – *Turn to the right and you must see the ninth control point* is composed of an action *Turn to the right* that begins with the landmark footpath and ends with the ninth control point. Therefore, the edge is labeled with the relative orientation: Right.

Fig. 1. Graphic representation of $f_{d_1} = \{Er_0, Er_1, Er_2\}$

A route is defined as a sequence of elementary routes $\{Er_0, Er_1, ..., Er_n\}$. An elementary route Er_i is defined as an action between an origin and a destination. It is modeled as a 3-tuple $Er_i(p_i, a_i, p_{i+1})$ where p_i is the starting node of the action, a_i the action, p_{i+1} the termination node of the action. Let us consider the verbal description v_{d_1}. The formal route description (f_{d_1}) contains four nodes and three edges (Fig. 1). Formally, f_{d_1} is composed by three elementary routes:

1. Er_0 =(Start, North $\wedge$ Up, 517), Er_0 models the expression *From the eighth control point climb the knoll going to the north to arrive at the electric pol*;
2. Er_1 =(517, $\emptyset$, 507), Er_1 models the expression *Then go to the footpath*;
3. Er_2 =(507, Right, End), Er_2 models the expression *Turn to the right and you must see the ninth control point.*

2.2   Spatial Semantic Network Generation

The ontology developed provides a set of formal abstractions that allow to refine the semantics that categorize and relate locations and actions. Moreover, the ontology verifies the consistency of the knowledge identified, and derives and enriches the whole description (Fig. 2).

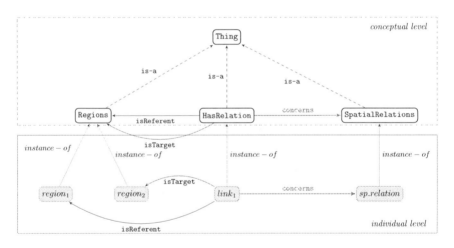

Fig. 2. Ontology instantiation

This ontology consists of two parts. The first is devoted to the representation of geographic features (concept `Regions`). The second part presents the spatial relations (`SpatialRelations`). The concept `HasRelation`, relates through a spatial relation, a reference region to one or more target regions. The concepts (`Regions`, `SpatialRelations`, `HasRelation`) and relations (`is-a`, `isReferent`, `isTarget`) are extracted from *SpaceOntology* [1]. Figure 2 illustrates the instantiation of this terminology. The result of the instantiation is a directed graph with two levels called *spatial semantic network*:

1. the first level, called *conceptual level*, is a directed graph where nodes denote concepts and arcs link between these concepts;

2. the second level, called *individual level*, is a directed graph where nodes
   denote individuals of concepts and arcs denote links between these nodes.

Formally, the ontology instantiation is as follows. From a graphical representation of a formal description, each node is an instance of the concept Region and each arc is an instance of the concept SpatialRelations. The direction of the arc defines the reference and target regions. Let us consider the description f_{d_1} presented in Figure 1.

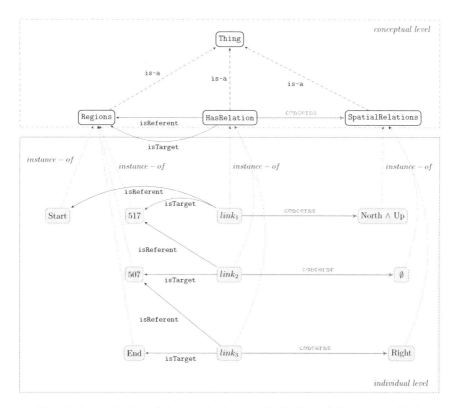

Fig. 3. Instantiation of the terminology with the formal description (f_{d_1})

The nodes *Start, 517, 507* and *End* that define the formal description f_{d_1} are instances of the concept Region (Fig. 3). The edges that define the formal description f_{d_1} are instances of the concept SpatialRelations (Fig. 3). All these elements are modeled as node at the instance level. The link between these elements is defined by the concept HasRelation. In fact, the formal definition of the concept HasRelation denotes a spatial relation between regions as a 3-uplet $\langle Rg_{tgt}, Rel, Rg_{ref} \rangle$, where Rg_{tgt} is a set of target spatial entities, Rg_{ref} is a reference spatial entity, while Rel denotes the spatial relations that relates these entities. Finally, an elementary route implements the ontology by instantiating the concept HasRelation, where: (1) p_i and p_{+1} are two instances of the concept Region, (2) p_i is a reference and p_{+1} is the target and (3) a_i is an individual of the concept SpatialRelations. Let us consider the elementary route Er_0 =(Start, North $\wedge$ Up, 517) (Fig. 1). Er_0 models the expression *From the eighth control point climb the knoll going to the north to arrive at the electric pole*. Er_0 implements the concept HasRelation, where: (1) Start and 517 are two instances of the concept Region, (2) Start is the reference and 517 is the target, (3) North $\wedge$ Up is an instance of the concept SpatialRelations (Fig. 3).

3   From Formal Descriptions to a Spatial Semantic Network

The route model represents the navigation descriptions generated by several humans navigating in a given environment. Our modeling approach relies in the (1) generation of a spatial network derived from all navigations formally represented as routes and as introduced in the previous section (2) identification and modeling of semantic rules whose objective is to merge part-of routes that share a common knowledge and application of those rules (3) and generation of a semantic spatial network that provides a global semantic and spatial view of the environment.

3.1   Spatial Network Generation

The spatial network is derived from the fusion of the graphs that describe the route descriptions (actions, landmarks, spatial entities, spatial relations). Algorithm 1 generates a spatial network after application of the fusion mechanisms applied to the verbal descriptions as modeled by our approach.

Algorithm 1. Constructspatialnetwork(F_d)

Require: $F_d = \{f_{d_0}, f_{d_1}, \ldots, f_{d_n}\}$
1: $Sp\mathcal{N} \longleftarrow \oslash$, $f_d \longleftarrow f_{d_0}$ {*a randomly chosen formal description*}
2: **for** $i \in [1..n]$ **do**
3:     $f_d \longleftarrow$ merge(f_d, f_{d_i})
4: **end for**
5: **return** $Sp\mathcal{N} \longleftarrow f_d$

Algorithm 1 takes as an input a set of formal descriptions and returns a spatial network (denoted $Sp\mathcal{N}$). The function merge takes two formal descriptions and returns a merged formal description. At the i-th step, this algorithm merges the formal description produced by the (i-1)-th step (f_d) with a formal description randomly selected in the i-th step(f_{d_i}) from the set $\{f_{d_i}, \ldots, f_{d_n}\}$. Algorithm 2 implements the function merge, it takes as input two formal route descriptions (as illustrated in Fig. 4) and returns a union of routes (Fig. 5) where common

Algorithm 2. Merge(f_{d_0}, f_{d_1})

Require: f_{d_0}, f_{d_1}
1: **for** i from 0 to $|f_{d_0}|$ **do**
2:    **for** j from 0 to $|f_{d_1}|$ **do**
3:       $Similar_p(p_i^{f_{d_0}}, p_j^{f_{d_1}}) \longleftarrow false$ {$p_i^{f_{d_0}}$ position in f_{d_0}, $p_j^{f_{d_1}}$ position in f_{d_1}}
4:       apply 1
5:       **if** $Similar_p(p_i^{f_{d_0}}, p_j^{f_{d_1}})$ **then**
6:         $p \longleftarrow$ fusionp($p_i^{f_{d_0}}, p_j^{f_{d_1}}$)
7:         $edges_{p_i^{f_{d_0}}} \longleftarrow$ getEdges($p_i^{f_{d_0}}$)
8:         $edges_{p_j^{f_{d_1}}} \longleftarrow$ getEdges($p_j^{f_{d_1}}$)
9:         updateEdges($p, edges_{p_i^{f_{d_0}}}, edges_{p_j^{f_{d_1}}}$)
10:     **end if**
11:   **end for**
12: **end for**
13: **return** f_{d_1}

nodes are merged whenever possible. The implementation of Algorithm 2 assumes that (1) the input set of formal descriptions is valid; namely, initial and target destinations are similar (Fig. 2) and (2) no empty description.

Let us consider two formal descriptions f_{d_0} and f_{d_1} (Fig. 4). $p_i^{f_{d_k}}$ denotes the i-th node in f_{d_k} and $action_i^{f_{d_k}}$ denotes the i-th edge in f_{d_k}.

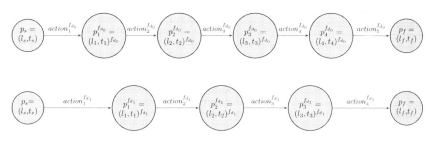

Fig. 4. Route examples f_{d_0} and f_{d_1}

Figure 5 presents the fusion of similar nodes from f_{d_0} and f_{d_1} towards one node common to those elementary routes, by taking all possible connections between them (execution of Algorithm 2). At the i-th step, the similarity between two positions $p_i^{f_{d_0}}$ and $p_j^{f_{d_1}}$ ($Similar_p(p_i^{f_{d_0}}, p_j^{f_{d_1}})$) is checked by applying Rule 1. If these positions are similar they can be merged (fusionp($p_i^{f_{d_0}}, p_j^{f_{d_1}}$)).

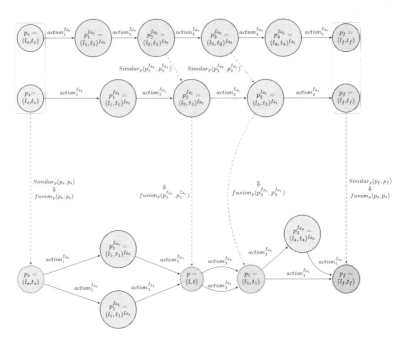

Fig. 5. Merge of two route descriptions: f_{d_0} and f_{d_1}

Rule 1 (Nodes Similarity)

$$\bigwedge_{(f_{d_1},f_{d_2})\in FD/f_{d_1}\neq f_{d_2}} \bigwedge_{p_i^{f_{d_1}}\in f_{d_1},p_j^{f_{d_2}}\in f_{d_2}} \bigwedge_{\exists p_k^{f_{d_1}}\in f_{d_1},p_l^{f_{d_2}}\in f_{d_2}/k\in[0,i[,l\in[0,j[}$$

$$\left(\begin{bmatrix} computeTypeDist(p_i^{f_{d_1}}.type,p_j^{f_{d_2}}.type)\leq\alpha \\ \wedge \\ computeSim((p_k^{f_{d_1}},p_i^{f_{d_1}}),(p_l^{f_{d_2}},p_j^{f_{d_2}}))\leq\beta \end{bmatrix} \Rightarrow Similar_p(p_i^{f_{d_1}},p_j^{f_{d_2}})\right)$$

Rule 1 is defined as a formula of first-order predicate calculus. Two nodes $p_i^{f_{d_1}}$ and $p_j^{f_{d_2}}$ from two different formal descriptions (f_{d_1}, f_{d_2}) are considered as similar when:

1. $p_i^{f_{d_1}}$ and $p_j^{f_{d_2}}$ are of similar type ($computeTypeDist(p_i^{f_{d_1}}.type, p_j^{f_{d_2}}.type) \leq \alpha$). The function $computeTypeDist$ computes the distance between two types, when this distance is below a given threshold α, then the two nodes are considered as similar;

2. and there is a sequence in f_{d_1} which ends by $p_i^{f_{d_1}}$ similar to a sequence in f_{d_2} which ends by $p_j^{f_{d_2}}$ ($computeSim((p_k^{f_{d_1}}, p_i^{f_{d_1}}), (p_l^{f_{d_2}}, p_j^{f_{d_2}})) \leq \beta$). The function $computeSim$ computes the similarity between two sequences. When this distance is below a given threshold β, then the two nodes are considered as similar.

These two parameters (α and β) give to the expert the possibility to control the errors contained in the descriptions or differences in perception. Their respective roles are described in Section 3.2.

3.2   Rule-Based Approach

The generation of a spatial network is performed by the application of Rule1. Rule1 encapsulates two functions: *computeTypeDist* that returns the distance between two types, and *computeSim* that returns the similarity measure between two no-empty positions.

Distance between Types: The function *computeTypeDist* computes the semantic distance between two types. This distance is given by the step-distance of the two types in the hierarchy denoted $\mathcal{T}$ as defined in [3]. The semantic distance allows us to consider cases where landmarks are described by different terms, but whose semantics is relatively close. Each term in that hierarchy is labeled by an IOF code. For example, depression (115), small depression (116) and pit (117) are considered as equivalent in the following case study. The function *computeTypeDist* computes the distance between two nodes in the graph $\mathcal{T}$, where:

- $computeTypeDist(t_i, t_j) \in \mathbb{N}$;
- $computeTypeDist(t_i, t_j)$ with $t_i \neq \emptyset$ and $t_j \neq \emptyset$;
- $computeTypeDist(t_i, t_j) = 0$, means that t_i and t_j are the same node in $\mathcal{T}$.

α is an integer that denotes a threshold. According to Rule 2, two positions $p_i^{f_{d_1}}$ and $p_j^{f_{d_2}}$ are considered as not similar if the semantic distance between two types $p_i^{f_{d_1}}.type$ and $p_j^{f_{d_2}}.type$ is greater than α.

Rule 2 (Type-Distance Similarity)

$$
\bigwedge_{(f_{d_1}, f_{d_2}) \in FD/f_{d_1} \neq f_{d_2}} \bigwedge_{p_i^{f_{d_1}} \in f_{d_1}, p_j^{f_{d_2}} \in f_{d_2}} \left(\begin{array}{c} [computeTypeDist(p_i^{f_{d_1}}.type, p_j^{f_{d_2}}.type) > \alpha] \\ \Rightarrow \neg Similar_p(p_i, p_j) \end{array} \right)
$$

Distance between Routes: The function *computeSim* computes the similarity distance between two set of elementary routes: $(p_k^{f_{d_1}}, p_i^{f_{d_1}})$ and $(p_l^{f_{d_2}}, p_j^{f_{d_2}})$. A sequence of elementary route is represented by a graph as a formal description. The similarity distance between two sequences corresponds to the length of the longest common sequence between their respective graphs. Therefore, *computeSim* is valued as suggested in [5]:

$$
computeSim((p_k^{f_{d_1}}, p_i^{f_{d_1}}), (p_l^{f_{d_2}}, p_j^{f_{d_2}})) = 1 - \frac{|MaxSeq((p_k^{f_{d_1}}, p_i^{f_{d_1}}), (p_l^{f_{d_2}}, p_j^{f_{d_2}}))|}{max(|(p_k^{f_{d_1}}, p_i^{f_{d_1}})|, |(p_l^{f_{d_2}}, p_j^{f_{d_2}})|)}
$$

$computeSim(p_k^{f_{d_1}}, p_i^{f_{d_1}}), (p_l^{f_{d_2}}, p_j^{f_{d_2}}))$ computes the longest subsequence common to two given sequences $(p_k^{f_{d_1}}, p_i^{f_{d_1}})$ and $(p_l^{f_{d_2}}, p_k^{f_{d_2}})$. $MaxSeq(p_k^{f_{d_1}}, p_i^{f_{d_1}})$ denotes the longest common sub-sequence of $(p_k^{f_{d_1}}, p_i^{f_{d_1}})$ and $(p_l^{f_{d_2}}, p_k^{f_{d_2}})$, where $|(p_k^{f_{d_1}}, p_i^{f_{d_1}}), (p_l^{f_{d_2}}, p_j^{f_{d_2}}))|$ respectively denote the maximum number of nodes of the two sequences, and $|(p_k^{f_{d_1}}, p_i^{f_{d_1}})|$ and $|(p_l^{f_{d_2}}, p_j^{f_{d_2}})|$ respectively denotes the number of nodes of the two sequences $(p_k^{f_{d_1}}, p_i^{f_{d_1}})$ and $(p_l^{f_{d_2}}, p_j^{f_{d_2}})$. This measure fulfills the following properties [5]:

- $0 \leq computeSim((p_k^{f_{d_1}}, p_i^{f_{d_1}}), (p_l^{f_{d_2}}, p_j^{f_{d_2}})) \leq 1$;
- $computeSim((p_k^{f_{d_1}}, p_i^{f_{d_1}}), (p_l^{f_{d_2}}, p_j^{f_{d_2}})) = computeSim((p_l^{f_{d_2}}, p_j^{f_{d_2}}), (p_k^{f_{d_1}}, p_i^{f_{d_1}}))$
- $computeSim((p_k^{f_{d_1}}, p_i^{f_{d_1}}), (p_l^{f_{d_2}}, p_j^{f_{d_2}})) \leq computeSim((p_k^{f_{d_1}}, p_i^{f_{d_1}}), (p_m^{f_{d_2}}, p_n^{f_{d_2}})) +$
  $computeSim((p_m^{f_{d_1}}, p_n^{f_{d_1}}), (p_l^{f_{d_2}}, p_j^{f_{d_2}}))$

Rule 3 evaluates the equivalence of two sequences, that is, two sequences are considered as equivalent if their origin and destination nodes are similar. According to Rule 4, two sequences are not equivalent if there are no equivalence of origin and destination nodes.

Rule 3 (Strong Equivalent Sequences)

$$\bigwedge_{(p_k^{f_{d_1}}, p_i^{f_{d_1}}) \in f_{d_1}, (p_l^{f_{d_2}}, p_j^{f_{d_2}}) \in f_{d_2}} \left(\begin{array}{l} [Similar_p(p_k^{f_{d_1}}, p_l^{f_{d_2}}) \wedge Similar_p(p_i^{f_{d_1}}, p_j^{f_{d_2}})] \\ \Rightarrow computeSim((p_k^{f_{d_1}}, p_i^{f_{d_1}}), (p_l^{f_{d_2}}, p_j^{f_{d_2}})) = 0 \end{array} \right)$$

Rule 4 (Non Equivalent Sequences)

$$\bigwedge_{(p_k^{f_{d_1}}, p_i^{f_{d_1}}) \in f_{d_1}, (p_l^{f_{d_2}}, p_j^{f_{d_2}}) \in f_{d_2}} \left(\begin{array}{l} [\neg Similar_p(p_k^{f_{d_1}}, p_l^{f_{d_2}}) \vee \neg Similar_p(p_i^{f_{d_1}}, p_j^{f_{d_2}})] \\ \Rightarrow computeSim((p_k^{f_{d_1}}, p_i^{f_{d_1}}), (p_l^{f_{d_2}}, p_j^{f_{d_2}})) = 1 \end{array} \right)$$

By applying Rule 5, Rule 6 and Rule 7, the similarity between two given nodes $p_i^{f_{d_1}}$ and $p_j^{f_{d_2}}$ requires the existence of two equivalent paths: in f_{d_1} ending with $p_i^{f_{d_1}}$, and f_{d_2} ending with $p_j^{f_{d_2}}$. The reasoning is a recursive chain backward.

Rule 5 (Recursive Nodes Similarity)

$$\bigwedge_{p_i^{f_{d_1}} \in f_{d_1}, p_j^{f_{d_2}} \in f_{d_2}} \bigwedge_{\exists p_k^{f_{d_1}} \in f_{d_1}, p_l^{f_{d_2}} \in f_{d_2} / k \in [0,i[, l \in [0,j[} \left(\begin{array}{c} \left[\begin{array}{l} Similar_p(p_k^{f_{d_1}}, p_l^{f_{d_2}}) \wedge computeTypeDist(p_i^{f_{d_1}}.type, p_j^{f_{d_2}}.type) \leq \alpha \\ \wedge computeSim((p_k^{f_{d_1}}, p_{i-1}^{f_{d_1}}), (p_l^{f_{d_2}}, p_{j-1}^{f_{d_2}})) \leq \beta \\ \wedge computeSim((p_{i-1}^{f_{d_1}}, p_i^{f_{d_1}}), (p_{j-1}^{f_{d_2}}, p_j^{f_{d_2}})) \leq \beta \end{array} \right] \\ \\ \Rightarrow \\ Similar_p(p_i^{f_{d_1}}, p_j^{f_{d_2}}) \end{array} \right)$$

Rule 6 (Recursive Equivalent Sequences)

$$
\bigwedge_{(p_k^{f_{d_1}}, p_i^{f_{d_1}}) \in f_{d_1}, (p_l^{f_{d_2}}, p_j^{f_{d_2}}) \in f_{d_2}} \quad \bigwedge_{\exists p_k^{f_{d_1}} \in f_{d_1}, p_l^{f_{d_2}} \in f_{d_2} / k \in [0,i[, l \in [0,j[}
$$

$$
\left(\begin{array}{c}
\left[\begin{array}{c}
computeSim((p_k^{f_{d_1}}, p_{i-1}^{f_{d_1}}), (p_l^{f_{d_2}}, p_{j-1}^{f_{d_2}})) \leq \beta \\
\wedge \\
computeSim((p_{i-1}^{f_{d_1}}, p_i^{f_{d_1}}), (p_{j-1}^{f_{d_2}}, p_j^{f_{d_2}})) \leq \beta
\end{array} \right] \\
\Rightarrow \\
computeSim((p_k^{f_{d_1}}, p_i^{f_{d_1}}), (p_l^{f_{d_2}}, p_j^{f_{d_2}})) \leq \beta
\end{array} \right)
$$

Rule 7 (Action-Distance and Equivalent Sequences)

$$
\bigwedge_{p_i^{f_{d_1}} \in f_{d_1}, p_j^{f_{d_2}} \in f_{d_2}}
$$

$$
\left(\begin{array}{c}
\left[\begin{array}{c}
Similar_p(p_{i-1}^{f_{d_1}}, p_{j-1}^{f_{d_2}}) \\
\wedge computeTypeDist(p_i^{f_{d_1}}.type, p_j^{f_{d_2}}.type) \leq \alpha \\
\wedge \\
computeActionDist((p_{i-1}^{f_{d_1}}, p_i^{f_{d_1}}), (p_{j-1}^{f_{d_2}}, p_j^{f_{d_2}})) \leq \gamma
\end{array} \right] \\
\Rightarrow \\
computeSim((p_{i-1}^{f_{d_1}}, p_i^{f_{d_1}}), (p_{j-1}^{f_{d_2}}, p_j^{f_{d_2}})) \leq \beta
\end{array} \right)
$$

The function $computeActionDist$ computes the distance between two actions, and respects the following properties:

- $computeActionDist(a_i^{f_{d_1}}, a_j^{f_{d_2}}) \in [0,1]$;
- $computeActionDist(a_i^{f_{d_1}}, a_j^{f_{d_2}}) = computeActionDist(a_j^{f_{d_2}}, a_i^{f_{d_1}})$;
- $computeActionDist(a_i^{f_{d_1}}, a_j^{f_{d_2}}) = 0$, expresses that $a_i^{f_{d_1}}$ and $a_j^{f_{d_2}}$ denote the same action;
- $computeActionDist(a, null) \leq \gamma$, where $a \in O_r$.

Distances between relative or cardinal orientations are valued according to their conceptual proximity, and within the unit interval $[0,1]$. For example the distance between North and South is valued to 1, while the distance between North and North East is valued to 0.25. Therefore, we define the value of the function $computeActionDist(a_i^{f_{d_1}}, a_j^{f_{d_2}})$ for each pair of actions $(a_i^{f_{d_1}}, a_j^{f_{d_2}})$ according to their properties. Let γ an integer that denotes a threshold. Two actions $a_i^{f_{d_1}}$ and $a_j^{f_{d_2}}$ are considered as equivalent if the action distance between them is smaller than γ.

3.3   From Spatial Network to Spatial Semantic Network

The spatial semantic network is derived from the spatial network. Let us consider the spatial network illustrated by Figure 6. Figure 7 develops the matching process between the abstractions of the spatial network and the underlying spatial ontology.

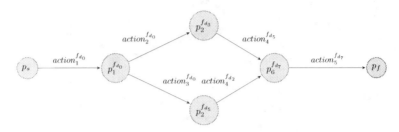

Fig. 6. Spatial network example

A spatial network is the union of several elementary routes. The matching process decomposes the input spatial network into a set of elementary routes as follows: $\{(p_s, action_1^{f_{d_0}}, p_1^{f_{d_0}}), \ldots, (p_6^{f_{d_7}}, action_5^{f_{d_7}}, p_f)\}$ (Fig. 7). Then, for each element e_i in the tuple $(p_i, action_i, p_{i+1})$ a correspondence to concepts of the considered ontology is established.

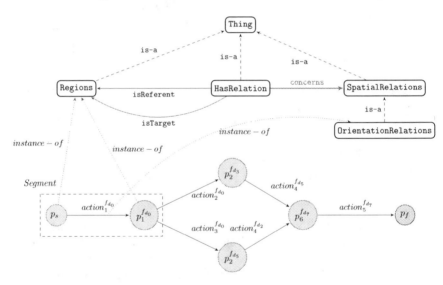

Fig. 7. Spatial network and spatial ontology matching

Let us consider the first elementary route $(p_s, action_1^{f_{d_0}}, p_1^{f_{d_0}})$ of the spatial network depicted in Figure 6: (1) p_s and $p_1^{f_{d_0}}$ are instances of the concept Regions, (2) $action_1^{f_{d_0}}$ is an instance of the concept OrientationRelations and (3) an instance of the HasRelation is created in order to link p_s and $p_1^{f_{d_0}}$, where p_s is the reference and $p_1^{f_{d_0}}$ the target. The result of the matching process gives a directed graph denoted *spatial semantic network*. Figure 8 illustrates the

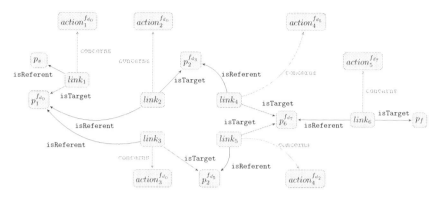

Fig. 8. Spatial semantic network

semantic network which corresponds to the spatial network described in Figure 6. The spatial semantic network combines static and dynamic views of the environment. The static component is given by the instantiation of the ontology, while the dynamic component is derived from the route descriptions derived from the verbal descriptions.

4   Experimental Evaluation

The evaluation of our modeling approach is conducted with a set of verbal descriptions experimented in a previous work [3].

(a) Map given to the orienteers   (b) A control point

Fig. 9. Experimental environment

Fourteen foot orienteering courses descriptions were recorded from path finding tasks where several orienteers were given the objective of reaching a series of control places in minimum time, all routes being set with same origins and destinations. The panelists were composed of eleven men and three women, aged between twenty to thirty five years old. The orienteers were asked to describe the route followed to another person unfamiliar with the environment and where the objective will be to follow the same route. The course took place in a natural environment of the French Naval Academy campus. Figure 9 shows an orienteering course map of the environment given to the orienteers and a view of a control point. From the routes recorded seven route descriptions have been selected to apply the fusion rules introduced in the previous section. It is worth noting that the experiments made show predominance of qualitative relations over quantitative relations, e.g., metrics. This has been observed in similar studies [4] and explains the fact that quantitative relations are not taken into account. The main features identified by the orienteers are labeled by their IOF codes: electric pole (517), hole (116), lamps (539), etc. All route descriptions are manually mapped to the most appropriate landmarks and spatial relations. The fusion process starts by creating a set of route descriptions where all routes modeled have same starting and ending locations (Fig. 10).

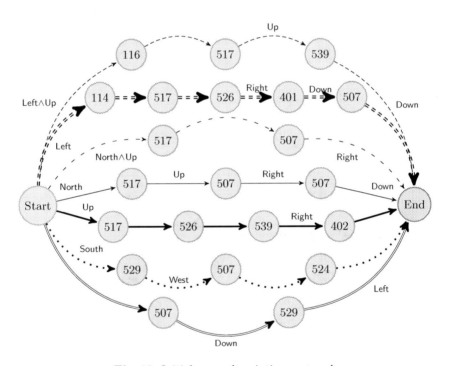

Fig. 10. Initial seven descriptions network

4.1   Formal Descriptions Fusion

f_{d_i} denotes the i-th description in the network presented in Figure 10, where i is valued between 0 and 6, f_{d_0} denotes the first route description from the top, while f_{d_6} denotes the last route description. In order to evaluate the similarity between nodes in different descriptions, let us consider the rules defined in Section 3.2. The type-distance threshold α is valued small enough to guarantee semantics similarity ($\alpha = 2$), while the action-distance β and the route-distance γ high enough to maximize similarity ($\beta = 0.75$ and $\gamma = 0.5$).

Similarity between Nodes 114 and 116: Rule 1 is applied in order to evaluate the similarity between the nodes labeled 114 and 116. The application of Rule 1 requires: (1) $computeTypeDist(114, 116) \leq \alpha$ is true and (2) $computeSim((Start, 114), (Start, 116)) \leq \beta$ is true, so:

(1) $computeTypeDist(114, 116) \leq \alpha$ is true. In fact, according to $\mathcal{T}$:
    $computeTypeDist(114, 116) = 2 \leq 2$;
(2) and $computeSim(Start, 114), (Start, 116)) \leq \beta$ is true. This is verified by the application of Rule 7 as:
    (a) $Similar_p(Start, Start)$ is true;
    (b) and $computeTypeDist(114, 116) \leq \alpha$ is also true;
    (c) and $computeActionDist(Left \wedge Up, Left) \leq \gamma$ is also true:
        i. first, $computeActionDist(Left, Left) = 0$;
        ii. then, $computeActionDist(Up, null) \leq \gamma$. In fact, we assume that the action distance $computeActionDist(a, null) \leq \gamma$ is true with $a \in O_r$.

Finally, the nodes 114 and 116 and be merged. The result of this process gives a node labeled by $114 \vee 116$ (Fig. 11).

Similarity between Nodes 517 in f_{d_2} and 517 in f_{d_3}: Rule 1 is applied in order to evaluate the similarity between the nodes labelled 517 in f_{d_2} and 517 in f_{d_3}. The application of Rule 1 requires: (1) $computeTypeDist(517_{f_{d_2}}, 517_{f_{d_3}}) \leq \alpha$ is true and (2) and $computeSim((Start, 517)_{f_{d_2}}, (Start, 517)_{f_{d_3}}) \leq \beta$ is true, so:

(1) $computeTypeDist(517_{f_{d_2}}, 517_{f_{d_3}}) \leq \alpha$ is true as 517 in $\mathcal{T}$ is denoted by one node;
(2) and $computeSim((Start, 517)_{f_{d_2}}, (Start, 517)_{f_{d_3}}) \leq \beta$. This is verified by the application of Rule 7 as:
    (a) $Similar_p(Start, Start)$ is true;
    (b) and $computeTypeDist(517_{f_{d_2}}, 517_{f_{d_3}}) \leq \alpha$ is also true;
    (c) and $computeActionDist((Start, 517)_{f_{d_2}}, (Start, 517)_{f_{d_3}}) \leq \gamma$ is also true.

Finally, the nodes labeled 517 in f_{d_2} and f_{d_3} can be merged. The result of this process gives a node labeled by 517 (Fig. 11).

Similarity between Nodes 401 and 402:   Rule 5 is applied in order to evaluate the similarity between the nodes labeled 401 and 402. The application of Rule 5 requires that:

1. there are a node $p_k^{f_{d_1}}$ in f_{d_1} and a node $p_l^{f_{d_4}}$ in f_{d_4}, where:
    (a) $Similar_p(p_k^{f_{d_1}}, p_l^{f_{d_4}})$ should be true;
    (b) and $computeSim((p_k^{f_{d_1}}, 526), (p_l^{f_{d_4}}, 539)) \leq \beta$ should be also true;
2. and $computeTypeDist(401, 402) \leq \alpha$ should be also true;
3. and $computeSim(526, 401), (539, 402) \leq \beta$ should be also true.

The evaluation of $computeSim(526, 401), (539, 402) \leq \beta$ depends on the type distance between nodes 526 and 539 and the type distance between nodes 401 and 402. According to $\mathcal{T}$, the type distance between nodes 526 and 539 is valued by 4. Therefore, $computeTypeDist(526, 539) \leq \alpha$ is false ($\alpha = 2$) then the result of Rule 2 gives that the nodes 526 and 539 are not similar. The result of Rule 4 application is that $computeSim((526, 401), (539, 402)) \leq \beta$ is false ($\beta = 0.75$), since $computeSim((526, 401), (539, 402)) = 1$. Therefore, the nodes labeled 401 and 402 cannot be merged (Fig. 11). After applying the rules described in the previous section a new network reduced by seven nodes is finally derived (Fig. 11). The respective roles of the landmarks, spatial entities and actions appear, as well as the semantics of the different routes result of the orienteering process.

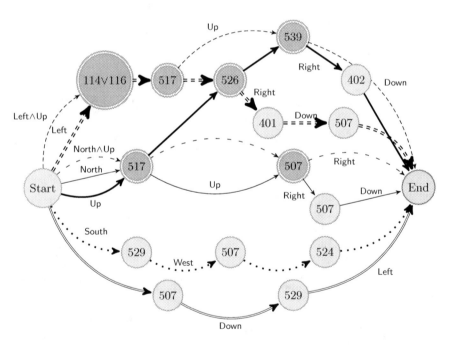

Fig. 11. Final result of the fusion process

4.2   Spatial Semantic Network Generation

The matching process between the nodes and edges of the spatial network depicted in Figure 11 and the considered ontology is applied. The result of this process gives the spatial semantic network described by Figure 12. The considered ontology provides inference mechanism that will be used to verify the consistency of the spatial semantic network, and discover additional knowledge. Each elementary route in the spatial network denotes an instance of the concept HasRelation. For instance, the elementary route (Start, North ∧ Up, 517) is described by the $link_3$, the instance of the concept HasRelation, where the node Start is the reference, the node 517 is the target and the node North ∧ Up is the spatial relation.

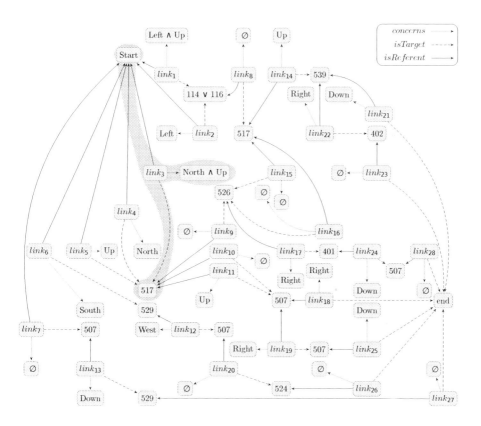

Fig. 12. Spatial semantic network

Overall the spatial semantic network result of the fusion process provides a graphic and global view of the environment used for our experimental evaluation. It appears that the different fusion rules identified can be successfully applied, and contribute to generate a synthetic view where all navigation routes are integrated in a common framework and reconciliated whenever possible. The spatial semantic network cannot be compared with the properties of conventional maps but its underlying structure gives a formal and qualitative representation of the environment.

An additional experimentation is now in progress. The objective is to start a new set of foot orienteering tasks where panelists will be given a spatial semantics network in input with a similar objective if reaching a destination. Preliminary results show that the spatial semantics network compares very well with the previous experiment. In particular, the global view of the environment and the respective roles of the landmarks and the route sequences identified help the orienteers to act in the environment.

5 Conclusion

Human beings have been long interacting in natural environments and attempting to communicate the spatial knowledge they are able to produce. Amongst many forms of expression used, it has been shown that verbal descriptions are particularly efficient to transmit navigation knowledge. The research developed in this paper proposes a semantic and rule-based approach whose objective is to model and reconcile several route descriptions derived from the perception of several humans navigating in a given environment. Verbal descriptions are analyzed using a graph representation where the navigation is decomposed into a sequences of actions and landmarks. Indeed, the mapping of the verbal descriptions towards appropriate landmarks and spatial relations might generate some ambiguities which are inherent to human language interpretation. Actions and landmarks are characterized by additional spatial semantics when available. These verbal descriptions are analyzed and processed by a set of rules whose objective is to fusion the route descriptions that contain similar knowledge. This permits to generate a spatial semantics network that offers a global view of the environment derived from the common knowledge that emerges from these route navigation and humans perception. The whole approach is illustrated by some experimental evaluations that show the potential of the approach.

The research developed so far opens several development perspectives. On the modeling side, the semantics that emerges from verbal descriptions can be explored and represented using additional semantics. Additional fusion rules are still to be explored. Moreover, we plan to consider route descriptions with additional constraints regarding starting and ending nodes. The consistency and inference rules required to enrich the spatial semantic network still need further implementations. On the computational side, the robustness of the approach and algorithm dependency on the initial conditions still require additional experimental evaluations. Finally, the potential of the spatial semantic network still needs to be compared with conventional map approaches in order to also evaluate its capability to provide a logical path towards cartographical representations, and how spatial semantic networks can be used in wayfinding processes.

References

1. Belouaer, L., Bouzid, M., Mouaddib, A.: Ontology based spatial planning for human-robot interaction. In: 17th International Symposium on Temporal Representation and Reasoning (TIME), pp. 103–110. IEEE (2010)

2. Brosset, D., Claramunt, C., Saux, É.: A location and action-based model for route descriptions. In: Fonseca, F., Rodríguez, M.A., Levashkin, S. (eds.) GeoS 2007. LNCS, vol. 4853, pp. 146–159. Springer, Heidelberg (2007)
3. Brosset, D.: Description d'itinéraire en milieu naturel: modèle intégré de description verbale et de représentation spatiale au sein des systèmes d'information géographique. Unpublished PhD report, Arts et Métiers Paritech (2008)
4. Brosset, D., Claramunt, C., Saux, E.: Wayfinding in natural and urban environments: a comparative study. Cartographica, Toronto University Press 43(1), 21–30 (2008)
5. Bunke, H., Shearer, K.: A graph distance metric based on the maximal common subgraph. Pattern Recognition Letters 19(3), 255–259 (1998)
6. Darken, R.P., Peterson, B.: Spatial orientation, wayfinding, and representation. In: Handbook of Virtual Environments, pp. 493–518 (2002)
7. Freksa, C., Moratz, R., Barkowsky, T.: Schematic maps for robot navigation. In: Habel, C., Brauer, W., Freksa, C., Wender, K.F. (eds.) Spatial Cognition 2000. LNCS (LNAI), vol. 1849, p. 100. Springer, Heidelberg (2000)
8. Habel, C., Kerzel, M., Lohmann, K.: Verbal assistance in tactile-map explorations: A case for visual representations and reasoning. In: Proceedings of AAAI Workshop on Visual Representations and Reasoning, pp. 34–41 (2010)
9. Mark, D., Freksa, C., Hirtle, S., Lloyd, R., Tversky, B.: Cognitive models of geographical space. International Journal of Geographical Information Science 13(8), 747–774 (1999)
10. Michon, P.-E., Denis, M.: When and why are visual landmarks used in giving directions? In: Montello, D.R. (ed.) COSIT 2001. LNCS, vol. 2205, pp. 292–305. Springer, Heidelberg (2001)
11. Nicholson, J., Kulyukin, V., Coster, D.: Shoptalk: Independent blind shopping through verbal route directions and barcode scans. The Open Rehabilitation Journal 2, 11–23 (2009)
12. Richter, K.F., Tomko, M., Winter, S.: A dialog-driven process of generating route directions. Computers, Environment and Urban Systems 32(3), 233–245 (2008)
13. Schuldes, S., Boland, K., Roth, M., Strube, M., Krömker, S., Frank, A.: Modeling spatial knowledge for generating verbal and visual route directions. In: König, A., Dengel, A., Hinkelmann, K., Kise, K., Howlett, R.J., Jain, L.C. (eds.) KES 2011, Part IV. LNCS, vol. 6884, pp. 366–377. Springer, Heidelberg (2011)
14. Thomas, K.E., Sripada, S., Noordzij, M.L.: Atlas. txt: exploring linguistic grounding techniques for communicating spatial information to blind users. Universal Access in the Information Society, 1–14 (2012)
15. Werner, S., Long, P.: Cognition meets le corbusier - cognitive principles of architectural design. In: Freksa, C., Brauer, W., Habel, C., Wender, K.F. (eds.) Spatial Cognition III. LNCS (LNAI), vol. 2685, pp. 112–126. Springer, Heidelberg (2003)
16. Westphal, M., Wölfl, S., Nebel, B., Renz, J.: On qualitative route descriptions: Representation and computational complexity. In: Proceedings of the Twenty-Second International Joint Conference on Artificial Intelligence, vol. Two, pp. 1120–1125. AAAI Press (2011)

A Computational Model for Reference Object Selection in Spatial Relations

Katrin Johannsen[1], Agnes Swadzba[2], Leon Ziegler[2],
Sven Wachsmuth[2], and Jan P. De Ruiter[1]

[1]Faculty of Linguistics and Literary Studies, Bielefeld University,
Universitätsstr. 25, 33615 Bielefeld, Germany
{katrin.johannsen,jan.deruiter}@uni-bielefeld.de
[2]Applied Informatics Group, Bielefeld University,
Universitätsstr. 25, 33615 Bielefeld, Germany
{aswadzba,lziegler,swachsmu}@techfak.uni-bielefeld.de

Abstract. Automatic generation of adequate spatial relations with regard to reference object selection is a nontrivial problem. In the present paper, we develop, implement and evaluate a computational model for reference object selection using empirically derived conceptual spatial strategies. The attribution of roles (reference object and located object) in object configurations was investigated to derive conceptual spatial strategies with regard to size and position of the objects. These strategies were implemented in a computational model to generate spatial descriptions of furniture configurations. To evaluate the automatically generated descriptions, we contrasted them with analog sentences with inverted roles of reference and located object and asked participants to choose the more adequate description out of the two. The spatial descriptions automatically generated were chosen significantly more often. This indicates that taking spatial conceptual strategies into account improves the adequacy of automatically generated spatial descriptions.

Keywords: Spatial strategies, reference object, spatial relation, localization, computational model, robot.

1   Introduction

Localizing an object in reference to another object is common in natural language. When a spatial relation between two objects is established, one object serves as *relatum* (reference object) and the other as *locatum* (located object). The position of the locatum is defined in relation to the relatum. Spatial descriptions thus require conceptual attributions of roles to objects. Consider the sentence "The chair is in front of the sofa". The chair is the locatum and its position is defined in relation to the sofa, which serves as relatum. Even though syntactically correct, the sentence with reversed thematic roles "The sofa is behind the chair" seems less adequate to describe the same spatial relation indicating that the roles are non-symmetric [1] and thus usually not interchangeable. This asymmetry should be taken into account for the automatic

T. Tenbrink et al. (Eds.): COSIT 2013, LNCS 8116, pp. 358–376, 2013.

generation of spatial descriptions in computational models. The generation of adequate relations is a challenging task. From the whole set of possible object pairs for which relations can be formulated, humans normally accept only a few of them as intuitive and matching the visual impression of a scene. Thus, the aim of the present paper is to empirically investigate regularities in the attribution of roles (relatum and locatum), implement these regularities in a computational model and evaluate the automatically generated spatial descriptions with regard attribution of roles.

2   Spatial Conceptual Strategies

The attribution of roles (relatum, locatum) is a conceptual process that underlies strategies of perceptual coding. The strategies can be deliberate, for instance, when we consciously choose to use one object as relatum and the other as locatum [2]. However, when there are multiple candidates, the choice underlies different influences which result from the properties of the objects and their spatial relation [3]. With regard to spatial relations, an object is preferably chosen as relatum when the locatum is positioned on an axis of the object's frame of reference [FOR, e.g. 4]. FOR are coordinate systems that define space [5] and allocate directions such as front, back, left and right to objects. Spatial strategies with regard to FOR selection in descriptions of complex object arrangements for small-scale settings have been investigated elsewhere [6] as have been influences resulting from functional relations between objects on FOR selection. It has been shown that functional relations between objects increase the choice of the object-centered FOR [7], however, the potential problem of relatum selection often remains. Ambiguities in reference frame interpretation have also been of interest in human-robot-communication [8, 9] but will not be considered in the present study.

Besides the influence of the spatial relation, perceptual and functional-interactive features of the objects impact the choice of relatum [3]. Subsuming different perceptual and conceptual dimensions, the salience of an object has been claimed to be decisive for relatum selection [10]. One of the perceptual features relates to the size of the objects. Out of two objects, the larger object is usually chosen as relatum and the smaller object as locatum [1, 2, 11]. Furthermore, movability has been described to influence the attribution of roles [1] as have been noun features such as animacy and countability [12]. We were interested in the attribution of roles (locatum and relatum) between two designated objects. Comprehensive work on how to computationally detect the most suitable relatum from a group of competing objects can be found elsewhere [e.g., 13, 14]. We focused on the question of how role attribution is influenced by the size and the position of objects in a room. Our aim was to deduce regularities for a computational model which generates verbal output with regard to spatial descriptions. As movability can only be assigned conceptually in static scenes, we focused on two dimensions that were computationally available: size of the objects and their position in a room.

As size is a decisive factor for the attribution of roles [1, 2, 15,], we assume that out of two objects, the larger object is chosen as relatum. Furthermore, Ehrich [11] has shown that, with regard to furniture, the position of an object in a room is critical for relatum selection. Participants tended to use objects positioned against a wall more often as relatum than free-standing objects. Thus, we follow the assumption that an object standing against a wall is chosen as relatum and a free-standing object as locatum [11,15]. However, the position of the objects was not controlled systematically, and the interaction between size and position was not taken into account in her analysis. As the two principles can be in opposition, e.g. when a smaller object is positioned against a wall and the bigger object is free-standing, we investigated which principle predominates. Furthermore, we tested whether size and position were mutually reinforcing.

2.1   Method

Participants. 250 students from Bielefeld University took part in the online study. For data analysis, 48 participants had to be excluded because of their different size ranking of the objects, a different native language or repeated participation. Thus, the data of 202 participants, ranging in age from 18 to 70 years (M=24.1 years, SD=5.8), were used for analysis.

Stimuli. Stimuli were pictures of object configurations created with indoor planning software (Sweet Home 3D) and German gapped sentences of the form "_____ steht _____." ("___ stands ____."). Thus, participants had to insert one of the two objects as locatum, a projective term (such as "neben", Engl. "next to") and the other object as relatum. The objects differed with regard to size and position and were presented according to a typical encounter, thus wall-standing objects were positioned with their back to the wall (facing the room).

As the size of the objects played a central role in our experiment, we conducted two pilot studies in which participants ranked the size of the objects. Six pieces of furniture were chosen (chair, armchair, chest of drawers, bookshelf, sofa and wardrobe). In the first pilot study, only 66% of the participants (N=20) agreed on the ordering with regard to the relative object size. Thus, we manipulated the size of the critical objects to achieve a more distinct difference. In the second pilot study, 100% of participants (N=10) agreed on size ranking.

Using the same objects as in the second pilot study, we constructed stimuli in 4 conditions. By using objects from the same semantic category, we controlled for animacy and countability [12]. It has been shown that an object is preferably chosen as relatum when the locatum is positioned on an axis of the object's FOR [4]. As we used two objects with intrinsic features and wanted to avoid such influences we positioned the objects so that they were standing on the same axis with regard to both objects' intrinsic FOR. Thus, we controlled for the objects' rotation with regard to the position on the axes which also helped keeping the number of trials within reasonable limits. However, limitations of the results may arise due to effects resulting from the orientation of the objects. In the *size* condition, the pictures displayed two different objects, varying in size but not in position. Both objects were standing next to each other on the left/right horizontal axis according to their typical encounter (facing the room). As we were only interested in effects of the size of objects, we did not include perspective changes. Each object was combined with each of the other objects, resulting in 15 stimulus pictures. In the *position* condition, stimuli depicted two identical objects of different colors (e.g. a blue and a brown chair). One object was standing with the back against the wall, the other object in the room. We positioned the objects on the intrinsic front/back axis and presented each configuration twice (12 stimuli altogether) either facing each other (except for three objects which would not have been identifiable seeing only the back) or turning the free-standing object with its back to the wall-standing object.

The *size versus position* condition consisted of pictures showing two different objects varying in size and position, thus opposing the first two hypotheses because the smaller object was positioned against the wall (facing the room) and the bigger object in the room. We included 5 stimuli in this condition. In the *size plus position* condition, objects again varied in size and position. However, in this condition, the bigger object was positioned against a wall (facing the room) and the smaller object was free-standing so that size and position supported the *size* and *position* condition (5 stimuli). In the latter two conditions (*size versus position* and *size plus position*), we used the front/back axis and turned the free-standing object with the backside to the wall-standing object. Fig. 1. Shows example stimuli for each condition.

In order to prevent effects resulting from left-right bias [e.g. 16, 17] stimuli pictures were presented in the original version and in a mirrored version.

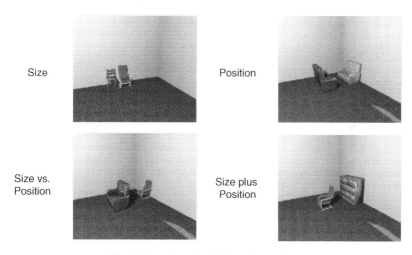

Fig. 1. Example stimuli for each condition

Procedure. Participants were recruited by email invitation, in which they received a link to the online study. First, instructions were given, stating that both pieces of furniture and a preposition should be inserted in the gapped sentences (in order to prevent participants from using the wall as relatum). Then, three examples were given using objects other than those in the study. Afterwards, the participants were shown the stimuli. The whole study comprised 37 pictures and lasted about 15 minutes. 50% of the participants completed the original study and 50% the study with mirrored images. At the end of the study, we asked participants to give a ranking on the size of the pieces of furniture. After that, participants could enter a prize draw for one of 10 prizes of 10 Euros.

2.2   Results

The descriptions were categorized with regard to the hypotheses in each condition. For example, in the *size* condition, answers that corresponded to our hypothesis and used the larger object as relatum were marked with a '1' and descriptions using the smaller object as relatum were marked with a '0'. The same procedure was carried out for the *position* and *size plus position* condition. In the *size versus position* condition, we distinguished between descriptions using the larger object and descriptions using the wall-standing object as relatum. Frequencies for the choices were tabulated (Table 1). As we focused on relatum selection and as the influence of FOR selection was minimized by considering object type and position of the objects with regard to FOR axes, we do not report results from FOR selection here.

We used a within-subjects design with regard to conditions. However, there was no theoretical motivation to address the differences between the conditions directly, and so we did not perform statistical procedures in this regard. Rather, we determined whether the choice of RO was above chance level in each condition using chi-squared tests. The tests indicated significant differences in all four conditions (all p < .001, *size* χ^2 (1, 3014) = 266.362; *position* χ^2 (1, 2292) = 675.19; *size vs. position* χ^2 (1, 1009) = 17.008, *size plus position* χ^2 (1,992) = 296.1).

Table 1. Percentages of relatum choice in each condition

Condition	Choice of object (%)	
Size	larger: 64.9	smaller: 35.1
Position	wall-standing: 77.1	free-standing : 22.9
Size plus Position	larger & wall-standing: 77.3	smaller & free-standing : 22.7
Size vs. Position	larger & free-standing: 43.5	smaller & wall-standing: 56.5

We formulated three principles with regard to relatum selection according to our results:

1. Out of two objects of different size, the larger object is preferred as relatum,
2. Out of two objects of different position, the wall-standing object is preferred as relatum,
3. If size and position are in opposition, the smaller, wall-standing object is preferred as relatum.

Using the empirical results, a computational model was developed which enables, e.g., a robot to generate a reliable set of spatial relations for an indoor scene which it currently perceives with a 3D sensor. This model is introduced in Section 4. As automatic recognition of furniture is a challenging problem, we have decided to evaluate the computational model on the NYU Dataset V2[1] [18], a database which provides data annotated indoor scenes as ground truth.

3   The NYU Depth Dataset V2

The NYU Depth Dataset V2 [18] is a recently assembled database consisting of depth and color data captured with a Kinect camera of 464 different indoor scenes in three different cities. A subset of the data is accompanied by dense multi-class labels across 26 scene classes. The labeling was acquired for each image using Amazon Mechanical

Fig. 2. This figure shows the 10 rooms chosen for the study on relatum selection. These rooms are part of the NYU Depth Dataset V2. On the left the RGB images of the Kinect camera are displayed and on the right the objects that were used for the generation of spatial descriptions.

---

[1] http://cs.nyu.edu/~silberman/datasets/nyu_depth_v2.html

Turk. Originally, this dataset served as ground truth for an approach which interprets major surfaces and objects in order to recover support relations between them, e.g., "picture on wall" or "cup on table", but this database is also ideally suited to test and evaluate our spatial relation generation. It contains many rooms of real apartments, where the furniture is already segmented. These controlled settings allow us to focus on the generation process without disturbance from segmentation errors. Nevertheless, we also work towards the automatic detection and segmentation of furniture and the incorporation of detection uncertainty into the understanding of spatial relations [9].

We chose 10 rooms from this database for our study. We focused on pieces of furniture as objects that could appear potentially in a reasonable spatial description of the scene. Tables, sofas, chairs, coffee tables, ottomans, bookshelves, cabinets, and walls were mainly chosen as starting set for the automatic spatial relation generation process. Fig. 2 shows the rooms and the objects we selected for spatial descriptions.

4   The Computational Model for Automatic Spatial Relation Generation

This section presents our computational approach to generate adequate spatial relations between spatial structures in a 3D scan of a scene. The empirical study presented in Section 2, provides specific principles with regard to relatum preferences. Using these principles, our algorithm for spatial description generation consists of three main steps:

1. the generation of spatial relations between all possible pairs of objects along the left/right and front/back axes within the relative FOR of the scene viewpoint,
2. the assignment of objects to walls they stand in front of, and
3. the elimination of object relations which violate the size, position, and spatial directness principles (Section 4.3).

The algorithm takes as input a 3D point cloud of a scene segmented into pieces of furniture and walls contained in the scene. The three steps are discussed in Section 4.1, 4.2, and 4.3, respectively.

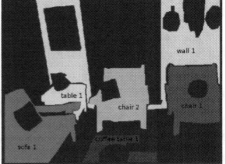

Fig. 3. This figure shows an example room with segmented spatial structures such as pieces of furniture and walls

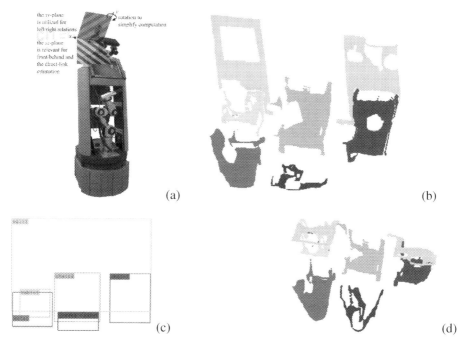

Fig. 4. This figure shows (a) the ego-centric FOR of our robot, (b) the 3D data of the selected spatial structures in our example room shown in Fig. 3, (c) the best bounding boxes of these structures projected in the *xy*-plane of the camera coordinate system, and (d) the projection of the 3D data in the *xz*-plane.

4.1 Generation of a Spatial Description

Given two objects in a scene, the relation of these two objects can always be described relative to the viewpoint of the speaker on the scene. Normally, objects are related on the *left/right* axis or on the *front/back* axis. Positions on the diagonal axis (e.g., left behind) are less preferred in descriptions [4]. Therefore, we also decided to focus on the automatic generation of the left/right and front/back relation.

In literature, few methods exist that generate relations on a fine grained level utilizing histograms of angles [19], histograms of forces between objects [20], acceptance volumes [21, 22], variations of acceptance areas [23, 24], or attentional vector-sum models [25]. Some of these approaches were utilized to equip robots with modules for understanding commands containing the primary directions "left", "right", "front", and "rear" of an object with respect to the robot's perspective. For example, Skubic et al. [26], utilized force histograms, which provide confidence about how well a position meets a given instruction like "go to the right of the object". Moratz et al. [27] equipped their robot with an ego-centered FOR by partitioning the environment along a reference direction into left-right and front-back. This reference direction is defined through a vector from the robot's center of mass to a relatum which could be the centroid of all perceived objects or a salient object. The approaches introduced work well for scenes with few compact objects, but will fail for elongated structures such as bookshelves and walls and in confined spaces where some pieces of furniture are only partially visible, which is typical for a shot from a real-world apartment. Schwering [28] introduced a similarity measure to ease human-computer communication about spatial configurations by enabling a comparison of natural language spatial relations with respect to their semantic similarity. In general, none of these approaches considered the problem of choosing the correct object as relatum. In the following, we present our approach to first generate all possible relations for furniture in confined indoor spaces, which is followed by an elimination of relations with inadequate relata (presented in Section 4.3).

We base our computation of left/right and front/behind directly on the x- and z-coordinates of the 3D point clouds. Using the coordinate system of the camera simulates an egocentric FOR of a robot as shown in Fig. 4. The xy-plane of the camera coordinate system serves as dimension where left/right relations are computed. The xz-plane is the dimension used to reason about front/behind relations and to calculate whether the space between two objects is empty. If necessary the computation is eased by rotating the 3D data around the y-axis such that a given reference direction becomes parallel to the z-axis (see Equation (2)).

First, for each spatial structure its x-interval is determined by computing the point with the minimum and the maximum x-value: $[x_{min}, x_{max}]$. This x-interval assembles the best bounding box for the y-values of the object points as displayed in Fig. 4 (c) with the same computation. Then for two objects obj1 and obj2 their x-intervals are compared. If they do not overlap:

$$\min(x^1_{max}, x^2_{max}) - \max(x^1_{min}, x^2_{min}) \leq 0 \tag{1}$$

the left/right relation is chosen: "obj1 left of obj2" and "obj2 right of obj1". For the objects chair1 and chair2 in Fig. 4 the relations "chair1 right of chair2" and "chair2 left of chair1" are generated as $x^{chair2}_{min} < x^{chair1}_{min}$. Due to noise we allow in our implementation a small overlap of maximum 5% of the union of both intervals, $[\min(x^1_{min}, x^2_{min}), \max(x^1_{max}, x^2_{max})]$.

If the x-intervals of the two objects overlap more than 5%, the front/back relation is chosen. It just needs to be determined which of the two objects is front of the other. If an object interval overlaps with a wall then this object will always be in front of the wall. In the other simple case all points of the one object have z-values that are smaller than the smallest z-value of the points from the other object. But due to noise and the influence of the viewing angle on the scene objects may overlap. Therefore, at least 30% of the points of an object must have a z-value smaller than the smallest z-value of the other object. For our example room shown in Fig. 4, 30 potential object relations are generated which are listed in Table 2.

Table 2. This table lists the 30 automatically generated potential relations between the selected spatial structures in the example room shown in Fig. 4.

chair1 right of *chair 2*	*chair2* right of *table1*	*table1* behind *sofa1*
chair2 left of *chair1*	*table1* left of *chair2*	*sofa1* in front of *table1*
chair1 right of *table1*	*chair2* in front of *wall1*	*table1* left of *coffee table1*
table1 left of *chair1*	*wall1* behind *chair2*	*coffee table1* right of *table1*
chair1 in front of *wall1*	*chair2* right of *sofa1*	*wall1* behind *sofa1*
wall1 behind *chair1*	*sofa1* left of *chair2*	*sofa1* in front of *wall1*
chair1 right of *sofa1*	*chair2* behind *coffee table1*	*wall1* behind *coffee table1*
sofa1 left of *chair1*	*coffee table1* in front of *chair2*	*coffee table1* in front of *wall1*
chair1 right of *coffee table1*	*table1* in front of *wall1*	*sofa1* left of *coffee table1*
coffee table1 left of *chair1*	*wall1* behind *table1*	*coffee table1* right of *sofa1*

4.2   Assignment of Objects to Walls

As shown in Section 2, the position of objects with regard to a wall influences whether an object is chosen as relatum or not. Therefore, it is necessary to find objects that stand at a wall and to encode this information in advance, as this information is important in order to be able to judge whether a given spatial relation is reliable.

Objects can only be assigned to a wall if they stand directly in front of it with no other piece of furniture between the object and the wall. Walls they stand beside cannot be chosen as walls influencing the position condition. At most, an object can only be assigned to one wall. If an object stands in front of several walls (which might be the case for an object standing near a corner) the wall is chosen where the distance of the object to the wall is least and only if this distance is less than 1.5 meters.

The wall assignment algorithm iterates through all walls contained in the scene and first rotates the wall and the remaining objects such that the normal vector $\vec{n} = (n_x, \ n_y, \ n_z)^{\mathrm{T}}$ of the wall is parallel to the z-axis of the coordinate system. For simplicity, only a rotation around the y-axis is performed. Perturbations of the scene's ground plane from the ideal xz-plane are neglected:

$$\alpha = \mathrm{sign}\,(n_x) \cdot (90^\circ - \sin^{-1}(n_z)),$$

$$R = \begin{bmatrix} \cos\alpha & 0 & -\sin\alpha \\ 0 & 1 & 0 \\ \sin\alpha & 0 & \cos\alpha \end{bmatrix} \tag{2}$$

The x-interval of each object can now be compared with the x-interval of the wall. If the object's interval overlaps at least 50% with the wall's interval:

$$\frac{\min\left(x^{obj}_{max}, \ x^{wall}_{max}\right) - \max\left(x^{obj}_{min}, \ x^{wall}_{min}\right)}{x^{obj}_{max} - x^{obj}_{min}} > 0.5 \tag{3}$$

it can be considered as being in front of the wall. All objects which fulfill the condition above are further checked if they stand directly in front of this wall or if other objects of the scene can be found between the object and the wall. Logically, this can be comprised as checking whether there is a spatial directness between an object and a wall [e.g., 29]. Here, we utilize the spatial directness implementation presented in Section 4.3. Finally, objects which were assigned to several walls $\{wall_i\}_{|i|>1}$ are revisited to select that wall as reference structure for the position condition which has the least distance to the object and the distance perpendicular to the wall being less than 1.5 meters:

$$d_{wall_i}^{obj} = \left| z_{med}^{obj} - z_{med}^{wall_i} \right|,$$

$$j = \underset{i}{\mathrm{argmin}} \left(d_{wall_i}^{obj} \right),$$

$$\text{select } wall_j \text{ if: } d_{wall_j}^{obj} < 1.5m \tag{4}$$

In the example room (Fig. 4), chair1, chair2, and table1 are assigned to wall1, which indicates that these objects stand in front of this wall.

4.3   Elimination of Unreliable Relations

The reduction of all potential relations between objects and walls in the scene follows principles that have been found to play a role when humans choose a relatum for a spatial relation. We have determined some of these principles in the empirical study described in Section 2 (Spatial Conceptual Strategies). Our results showed that not all objects will be chosen with the same probability as relatum. In particular, the size of the objects and their position with regard to a wall play an important role in choosing a relatum. These aspects are implemented as if-conditions for eliminating a relation in Principle 1, Principle 2, and Principle 3. Further, following previously reported results [e.g., 29], we extended them by a fourth principle (Principle 4) which determines that objects separated by other objects are not related to each other. Thus, two objects can only be related if a spatial directness between them can be established. The principles are explained in more detail as follows.

Principle 1. Walls always serve as relata. They will not be chosen as locatum. As a consequence, all relations where "wall" is the locatum are removed (as, e.g., "wall behind the table" or "wall left of chair").

Principle 2. The second principle deals with the position of locatum and relatum with respect to walls. Given the case that one of the two objects stands at a wall, the object at the wall will always be selected as relatum (as shown in Section 2.2). This means that if the locatum is assigned to a wall (as described in Section 4.2) while the relatum is standing free, this relation has to be removed. The reverse relation with the relatum assigned to a wall is kept independent of the size condition of Principle 3. In our example the relation "sofa1 in front of the table1" is kept while "table1 behind sofa1" is removed.

Principle 3. This principle deals with the influence of the size of the objects on the choice of relata. As humans normally select the larger object as relatum (cf. Section 2.2), relations have to be eliminated in which the smaller of two objects is the relatum (except for cases where Principle 2 applies). E.g., "sofa1 left of coffee table1" should be removed while "coffee table1 right of sofa1" is an appropriate relation. This principle applies if neither or both objects are assigned to a wall (which could be the same or a different one). Our implementation of the size estimation of an object consists of three steps: First, a plane is fitted into the 3D points assembling the object by minimizing the least-squares error. Second, the object points are projected onto this plane. And third, the best bounding box for the projected points is estimated by finding the directions of largest data variance in the plane using the Principal Component Analysis approach. The area of the best bounding box is then used as the value to estimate the size of an object.

Principle 4. The last step is to determine whether a locatum and relatum are adjacent (i.e. reveal spatial directness [cf. 29]), which means that no other object is standing between them. If an object is found between them, then this relation has to be removed. For example, the relations "chair1 right of table1" and "table1 left of chair1" should be removed because chair2 is standing between them (Fig. 3). The check for spatial directness is implemented in the following way: First, the two related objects together with all the remaining segmented objects in the scene are rotated around the y-axis such that a given reference direction is parallel to the z-axis of the global coordinate system. The reference direction is either the difference vector between the centroids of the two related objects or, in the case of a wall or bookshelf, their corresponding normal vector. The rotation estimation is given in Equation (2).

This coordinate transformation is necessary in order to determine the spatial area between the two objects for which the containment of other objects is calculated. A spatial directness can only be confirmed if no other object is contained in this area. The test area is constructed from two boundaries in the x-direction and the z-direction. For two arbitrary objects the left boundary $b_{left}: z = m_{left} \cdot x + t_{left}$ is estimated as a line through the point of object1 and object2, both with minimum x-value. The points from both objects with maximum x-value determine the right boundary $b_{right}: z = m_{right} \cdot x + t_{right}$. This boundary estimation is varied for widespread objects like walls or bookshelves where the other object is located in front of. Only a line through the point with minimum/maximum x-value of the non-wall or non-bookshelf object is estimated with an orientation parallel to the z-axis, $b_{left}: x_{min}^{obj}$ and $b_{right}: x_{max}^{obj}$. The two remaining boundaries of the testing area are estimated as lines parallel to the x-axis with an off-set in z-direction for each of the two objects. The off-sets are computed as follows. First, robust outer boundaries in z-direction are computed for each object by calculating the mean z-value and adding/subtracting 1.5-times the standard deviation of the points in z-direction to get the upper/lower object boundary. For clearly separated objects the lower boundary of the testing area would be determined by the upper boundary of the relatum z_{max}^{ref} while the lower boundary of the locatum z_{min}^{obj} gives the upper boundary of the testing area. But due to noise, the z-intervals of the two objects might overlap. Hence, we compute for robustness reasons the lower boundary b_{low}: $z_{\text{low}} = \max\bigl(\min(z_{\text{max}}^{\text{ref}}, z_{\text{min}}^{\text{obj}}), \ \min(z_{\text{mean}}^{\text{ref}}, z_{\text{mean}}^{\text{obj}})\bigr)$ of the test area by first finding the minimum of the two considered object boundaries and then choosing the maximum of the chosen minimum and the lower of the two object centroids. The upper boundary of the test area b_{up}: $z_{\text{up}} = \min\bigl(\max(z_{\text{max}}^{\text{ref}}, z_{\text{min}}^{\text{obj}}), \ \max(z_{\text{mean}}^{\text{ref}}, z_{\text{mean}}^{\text{obj}})\bigr)$ is the minimum of the upper object centroid and the maximum of the two considered object boundaries. Fig. 5 shows the boundaries of a testing area for two arbitrary objects and Fig. 6 depicts the area for relations where a wall is the relatum. All objects in the scene (except for walls which have been assigned to the currently related objects) are tested as to whether less than 5% of their object points fall into the testing area:

given object: $O = \{(x_i, \ y_i, \ z_i)^{\text{T}}\}_{i=1,\dots,n}$

$$\left\{ d_i^{\text{left}} = \text{sign}(m_{\text{left}}) \cdot (m_{\text{left}} \cdot x_i - z_i + t_{\text{left}}) \quad \text{or} \quad \begin{cases} 1 & : x_i > b_{\text{left}} = x_{\text{min}}^{\text{obj}} \\ 0 & :\text{else} \end{cases} \right\}$$

$$\left\{ d_i^{\text{right}} = \text{sign}(m_{\text{right}}) \cdot (m_{\text{right}} \cdot x_i - z_i + t_{\text{right}}) \ \text{or} \ \begin{cases} -1 & : x_i > b_{\text{right}} = x_{\text{max}}^{\text{obj}} \\ 0 & : \text{else} \end{cases} \right\}$$

$$t_{\text{in}} = \frac{\left|\{d_i^{\text{left}} > 0 \wedge d_i^{\text{right}} < 0 \wedge z_i > b_{\text{low}} = z_{\text{low}} \wedge z_i < b_{\text{up}} = z_{\text{up}}\}\right|}{n}$$

if: $t_{\text{in}} < 0.05 \ \rightarrow \ O \notin [b_{\text{left}}, b_{\text{right}}, b_{\text{up}}, b_{\text{low}}]$ (5)

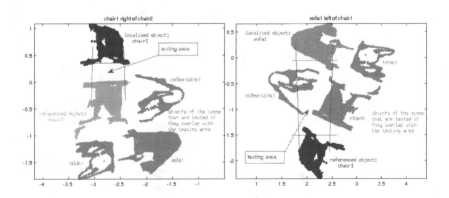

Fig. 5. This figure shows the testing area $[b_{\text{left}}, b_{\text{right}}, b_{\text{up}}, b_{\text{low}}]$ (red box) between two arbitrary objects where (left) no other scene object is contained in this area or (right) two objects partially overlap

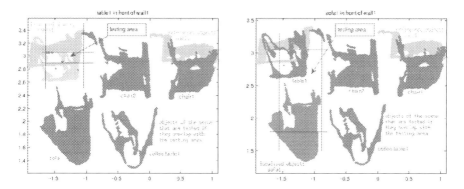

Fig. 6. This figure shows the testing area for relations where a *wall* is the relatum

The y-values of the object points are neglected in this test. If none of the objects has more than 5% within the testing area, a spatial directness between the related objects exists and the relation can be kept because all elimination principles are passed successfully.

Table 3. This is the final, automatically generated spatial description consisting of reliable relations which have passed the four elimination principles

chair1 right of *chair2*	*table1* left of *chair2*	*table1* in front of *wall1*
chair2 left of *chair1*	*chair2* in front of *wall1*	*table1* behind *sofa1*
chair1 in front of *wall1*	*sofa1* left of *chair2*	*coffee table1* right of *sofa1*
coffee table1 left of *chair1*	*coffee table1* in front of *chair2*	

5 Empirical Evaluation of Generated Descriptions

In order to evaluate the spatial descriptions generated by the computational model, we conducted an empirical study.

5.1 Method

Participants. 197 participants from Bielefeld University were recruited via email invitation ranging in age from 19 to 71 years (M=24.4, SD=6.1). All were native German speakers.

Stimuli. Stimuli consisted of the same 10 color photographs taken from the New York Depth Database [18] (Fig. 2) and sentences (in German) describing spatial relations between two of the objects in the picture. We chose several pieces of furniture in each picture and labeled them in order to ensure that participants mapped the objects mentioned in the spatial descriptions to the correct objects in the picture. Most of the pictures allowed testing of multiple conditions (as explained in Section 2), so we constructed 26 trials, of which 13 trials comprised spatial relations to test within the *size* condition, 8 trials within the *size plus position* condition and 5 trials within the *size versus position* condition. We had to exclude the *position* condition due to the lack of suitable stimuli. In order to investigate the implemented spatial directness principle, we included three trials for a qualitative analysis.

As stimulus sentences, we used the spatial description generated by the computational model (e.g. "The table is in front of the sofa") and contrasted it with an analog sentence with inversed thematic roles (e.g. "The sofa is behind the table"). However, for two of the object configurations chosen for testing, the model did not generate any spatial descriptions. We included those relations nevertheless to investigate which spatial description participants would have chosen, offering two sentences with inversed thematic roles. In the spatial directness trials, we used the generated description and contrasted it with a spatial description between one of the objects and the next non-adjacent object.

As the computational model did not consider spatial FOR, we adjusted the FOR in three stimuli, for which participants of a pilot study indicated that both sentences sounded inadequate due to the spatial FOR used. Keeping the thematic roles generated by the computational model we used the object-centered FOR for both sentences, in order to avoid interference with FOR preferences. Stimuli were randomized so that different conditions alternated between trials. Participants saw pictures that allowed testing of several conditions more than once but not subsequently. Sentences were randomly rotated so that the generated sentence appeared above or below the reversed sentence 50% of the time. Counterbalancing was achieved by presenting the study in a reversed order to half of the participants. Furthermore, in order to prevent a left-right bias resulting from reading directions [e.g. 16, 17], we produced two versions of the study: one with original pictures and another one with mirrored images.

Procedure. Participants received an email invitation to the online study. Having read the instructions, they completed three example trials with pictures not used in the study. Afterwards, participants completed 29 experimental trials. 50% of the participants saw the original version; 50% saw mirrored images. At the end of the study 5 participants had the chance to win 10 Euros.

5.2   Results

Trials testing the spatial directness principle and two trials for which no spatial descriptions were generated were excluded from quantitative analysis.

In order to test the distribution of answers, chi-squared tests were performed for each condition revealing significant preferences for the description generated by the computational model (all $p < .001$, *size* $\chi^2(1, N = 1970) = 589.89$; *size vs. position* $\chi^2(1, N = 591) = 81.15$; *size plus position* $\chi^2(1, N = 1576) = 370.37$). See Fig. 7 for percentages of the choices within conditions.

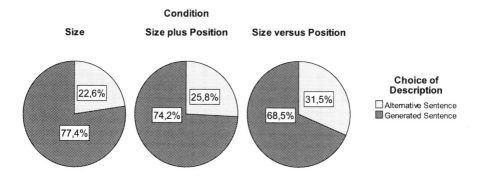

Fig. 7. Percentages for choice of spatial descriptions

In a second step, we analyzed the trials qualitatively. Even though the generated sentences conformed to the spatial conceptual principles, participants nevertheless preferred the inversed sentence in two of the trials. One of the trials was in the *size* condition (Fig. 2, picture 7: "The table is in front of the [smaller] sofa" was preferred to "The [smaller] sofa is behind the table") and one in the *size versus position* condition (Fig. 2, picture 9: "The armchair is behind the table" was preferred to "The table is in front of the armchair").

In three trials that were included to test the spatial directness principle we found a preference for the description generated by the computational model in two of the trials while in the third trial, participants chose the spatial relation that did not conform to the spatial directness principle.

The two trials for which no spatial descriptions were generated automatically revealed a high preference for the sentence which conformed to the spatial strategies (91.4% in the *size* condition and 80.2% in the *size plus position* condition).

6   Discussion

We implemented human conceptual spatial strategies in a computational model in order to automatically generate spatial descriptions of furniture configurations. Overall, the spatial descriptions generated were preferred with regard to relatum selection compared to analog descriptions with inversed thematic roles. This constitutes a promising step in improving verbal communication in human robot interactions. The principles may also bear the potential for a generalization to large-scale space and ontologically different objects.

However, there are a few points which we would like to discuss. In two of the trials, participants preferred the inversed sentence. One of the trials was in the *size* condition, the other one in the *size versus position* condition. With regard to the critical trial in the *size* condition, the two objects differed in size distinctively and were both free-standing yet participants nevertheless preferred the smaller object as relatum. We hypothesize that this choice might result from FOR preferences. It has been shown that functional relations lead to a preferred use of the intrinsic FOR [7], i.e. the FOR defined by the inherent orientation of the object. In the stimulus picture, the smaller object had an intrinsic orientation, and the object configuration revealed a functional relation (sofa and coffee table). This suggests that considering functional relations in spatial descriptions may help to improve automatically generated descriptions. This is a matter for further investigation.

The second case in which participants chose the inverse description was in the *size versus position* condition. The difference in preference was small (4.6%) which resembles the results in the first empirical study. The preference for the smaller but wall-standing object was statistically significant but less distinct than in the other conditions which might reflect the opposing effects of size and position. Another explanation might be that participants did not perceive the smaller object clearly as wall-standing and thus chose the bigger object as relatum (out of two free-standing objects). This fact points to a general problem of the wall assignment: what is the maximum distance between an object and the wall that still leads humans to perceive it as wall-standing? At the moment we assume that an object is considered as standing at the wall up to a distance of 1.5 m between the object and the wall. This is another matter for further investigation.

With regard to the three trials that were included to investigate the spatial directness principle, we found a general preference for descriptions between two directly adjacent objects. However, this should be interpreted with caution, as the stimuli were not designed for this hypothesis to be systematically tested and the number of stimuli is far from being representative. Furthermore, in one of the trials, participants preferred the object not directly adjacent to the locatum as relatum. In this case, it may be due to the fact that the alternative relatum was a lot larger than the one suggested by the model. This may have increased its salience [3] which is a decisive factor for the choice of relatum. The two trials in which no descriptions were generated for the chosen object configurations illustrate this problem. When defining objects in the picture to test the *size* condition, we chose – in our view – adjacent objects. However, in the 3D image, the object standing between the designated objects was considered as interfering with the spatial directness principle, thus the computational model did not generate a spatial relation. This discrepancy between human perception and computational modeling indicates that further information is required to specify which features and dimensions of an object are relevant.

We included the spatial directness principle [29] as a principle for defining which spatial relations between objects are relevant and which are not. The implemented constraints are helpful because they limit the total number of spatial relations between objects by eliminating irrelevant relations. However, whether humans consider the number of remaining spatial descriptions as sufficient is a matter for further investigation.

7   Conclusion

Using the results of our empirical study, we developed a computational model which considers conceptual spatial strategies as used by humans in relatum selection and generates spatial descriptions for object configurations. In a further empirical study, we contrasted some of the generated descriptions with analog sentences with inverted relatum-locatum roles. The results revealed a significant preference for the spatial descriptions automatically generated. In conclusion, the implementation of conceptual spatial strategies as used by humans in a computational model has led to significantly more adequate descriptions with regard to relatum selection. This indicates that taking spatial conceptual strategies into account improves the adequacy of automatically generated spatial descriptions and may improve human-robot communication.

Acknowledgments. This work was funded by the German Research Foundation (DFG) within the Collaborative Research Center 673 *Alignment in Communication*.

References

1. Talmy, L.: Figure and ground in complex sentences. In: Greenberg, J.H. (ed.) Universals of Human Language, pp. 625–649. Stanford University Press, Stanford (1975)
2. Clark, H.H., Chase, W.G.: Perceptual coding strategies in the formation and verification of descriptions. Memory and Cognition 1A, (101–111) (1974)
3. Miller, J.E., Carlson, L.A., Hill, P.L.: Selecting a Reference Object. Journal of Experimental Psychology: Learning, Memory, and Cognition 37(4), 840–850 (2011)
4. Logan, G.D., Sadler, D.D.: A Computational Analysis of the Apprehension of Spatial Relations. In: Bloom, P., Peterson, M., Nadel, L., Garrett, M. (eds.) Language and Space, pp. 493–529. MIT Press, Cambridge (1996)
5. Carlson, L.A.: Selecting a reference frame. Spatial Cognition and Computation 1(4), 365–379 (1999)
6. Tenbrink, T., Coventry, K., Andonova, E.: Spatial Strategies in the Description of Complex Configurations. Discourse Processes 48(4), 237–266 (2011)
7. Carlson-Radvansky, L.A., Radvansky, G.A.: The Influence of Functional Relations on Spatial Term Selection. Psychological Science 7(1), 56–60 (1996)
8. Liu, C., Walker, J., Chai, J.: Ambiguities in spatial language understanding in situated human robot dialogue. In: Dialogue with Robots: AAAI Fall Symposium, pp. 50–55 (2010)
9. Ziegler, L., Johannsen, K., Swadzba, A., de Ruiter, J., Wachsmuth, S.: Exploiting Spatial Descriptions in Visual Scene Analysis. Cognitive Processing 13, 369–374 (2012)
10. Tversky, B., Lee, P., Mainwaring, S.: Why do Speakers Mix Perspectives? Spatial Cognition and Computation 19(4), 399–412 (1999)
11. Ehrich, V.: Zur Linguistik und Psycholinguistik der sekundären Raumdeixis. In: Schweizer, H. (ed.) Sprache und Raum. Psychologische und linguistische Aspekte der Aneignung und Verarbeitung von Räumlichkeit, pp. 130–161. Stuttgart, Metzler (1985)
12. de Vega, M., Rodrigo, M.J., Ato, M., Dehn, D.M., Barquero, B.: How nouns and prepositions fit together. Discourse Processes 34, 117–143 (2002)
13. Barclay, M., Galton, A.: An Influence Model for Reference Object Selection in Spatially Locative Phrases. In: Freksa, C., Newcombe, N.S., Gärdenfors, P., Wölfl, S. (eds.) Spatial Cognition VI. LNCS (LNAI), vol. 5248, pp. 216–232. Springer, Heidelberg (2008)
14. Barclay, M.: Reference Object Choice in Spatial Language: Machine and Human Models. PhD thesis, University of Exeter (2010)
15. Shanon, B.: Room descriptions. Discourse Processes 7(3), 225–255 (1984)
16. Chatterjee, A., Maher, L.M., Heilman, K.M.: Spatial Characteristics of Thematic Role Representation. Neuropsychologia 33(5), 643–648 (1995)

17. Hartsuiker, R., Kolk, H.H.: Syntactic Facilitation in Agrammatic Sentence Production. Brain and Language 62, 221–254 (1998)
18. Silberman, N., Hoiem, D., Kohli, P., Fergus, R.: Indoor segmentation and support inference from RGBD images. In: Fitzgibbon, A., Lazebnik, S., Perona, P., Sato, Y., Schmid, C. (eds.) ECCV 2012, Part V. LNCS, vol. 7576, pp. 746–760. Springer, Heidelberg (2012)
19. Miyajima, K., Ralescu, A.: Spatial Organization in 2D Segmented Images: Representation and Recognition of Primitive Spatial Relations. Fuzzy Sets and Systems 65, 225–236 (1994)
20. Matsakis, P., Wendling, L.: A New Way to Represent the Relative Position between Areal Objects. Transactions on Pattern Analysis and Machine Intelligence 21(7), 634–643 (1999)
21. Socher, G., Sagerer, G., Perona, P.: Bayesian Reasoning on Qualitative Descriptions from Images and Speech. Image and Vision Computing 18(2), 155–172 (2000)
22. Vorwerg, C., Socher, G., Fuhr, T., Sagerer, G., Rickheit, G.: Projective Relations for 3D Space: Computational Model, Application, and Psychological Evaluation. In: Proceedings of the National Conference on Artificial Intelligence, Providence, Rhode Island, pp. 159–164. AAAI Press / The MIT Press (1997)
23. Moratz, R.: Intuitive linguistic joint object reference in human-robot interaction. In: Proceedings of the Twenty-first National Conference on Artificial Intelligence, Boston (2006)
24. Moratz, R., Tenbrink, T.: Spatial reference in linguistic human-robot interaction: Iterative, empirically supported development of a model of projective relations. Spatial Cognition and Computation 6(1), 63–107 (2006)
25. Regier, T., Carlson, L.A.: Grounding Spatial Language in Perception: An Empirical and Computational Investigation. Journal of Experimental Psychology: General 130(2), 273–298 (2001)
26. Skubic, M., Perzanowski, D., Blisard, S., Schultz, A., Adams, W., Bugajska, M., Brock, D.: Spatial Language for Human-Robot Dialogs. Transactions on Systems, Man, and Cybernetics 34(2), 154–167 (2004)
27. Moratz, R., Fischer, K., Tenbrink, T.: Cognitive Modeling of Spatial Reference for Human-Robot Interaction. Intl. Journal of Artificial Intelligence Tools 10(4), 589–611 (2001)
28. Schwering, A.: Evaluation of a semantic similarity measure for natural language spatial relations. In: Winter, S., Duckham, M., Kulik, L., Kuipers, B. (eds.) COSIT 2007. LNCS, vol. 4736, pp. 116–132. Springer, Heidelberg (2007)
29. Tenbrink, T., Ragni, M.: Linguistic principles for spatial relational reasoning. In: Stachniss, C., Schill, K., Uttal, D. (eds.) Spatial Cognition 2012. LNCS, vol. 7463, pp. 279–298. Springer, Heidelberg (2012)

Fundamental Cognitive Concepts
of Space (and Time): Using Cross-Linguistic, Crowdsourced Data to Cognitively Calibrate Modes of Overlap*

Alexander Klippel[1,], Jan Oliver Wallgrün[1], Jinlong Yang[1],
Jennifer S. Mason[1], Eun-Kyeong Kim[1], and David M. Mark[2]

[1] Department of Geography, GeoVISTA Center, The Pennsylvania State University
University Park, PA 16802, USA
{klippel,wallgrun,jinlong,jms1186,eun-kyeong.kim}@psu.edu
http://www.cognitiveGIScience.psu.edu
[2] Department of Geography, State University of New York at Buffalo
Buffalo, NY 14261-0023, USA
dmark@buffalo.edu

Abstract. This article makes several contributions to research on fundamental spatial and temporal concepts: First, we set out to render the notion of fundamental concepts of space and time more precise. Second, we introduce an efficient approach for collecting behavioral data combining crowdsourcing technology, efficient experimental software tools, and an effective and comprehensive analysis methodology. Third, we present behavioral studies that allow for identifying and calibrating potential candidates of fundamental spatial concepts from a cognitive perspective. Fourth, one prominent topic in the area of spatio-temporal cognition is the influence of language on how humans conceptualize their dynamic spatial environments. We used the aforementioned framework to collect data not only from English speaking participants but also from native Chinese and Korean speakers. Our application domain are the *modes of overlap* proposed by Galton [13]. We are able to show that the originally proposed spatial relations of the region connection calculus and intersection models are capturing cognitively fundamental distinctions that humans make with respect to modes of overlap. While finer distinctions are formally possible, they should not be considered fundamental conceptualizations in either Chinese, Korean, or English. The results show that our framework allows for efficiently answering questions about fundamental concepts of space, time, and space-time essential for theories of spatial information.

Keywords: Fundamental concepts of space and time, qualitative spatio-temporal reasoning, linguistic relativity, crowdsourcing.

* This research is funded through the National Science Foundations under grant number #0924534.

T. Tenbrink et al. (Eds.): COSIT 2013, LNCS 8116, pp. 377–396, 2013.

> *"To study concepts, cognitive [spatial] scientists must first identify some."* (Malt et al. 2011 [45])

1   Introduction

The Conference on Spatial Information Theory (COSIT) has a long tradition of addressing deeply theoretical questions of space and time, often from a cognitive perspective. In this spirit, we concentrate on the notion of concepts fundamental to the human understanding of space and time. We use the term fundamental loosely indicating concepts at the core of how humans are making sense of spatio-temporal information; alternative terms may be *intuitive, naive* [12,19], *common sense*, or *spatial conceptual primitives* [47]. We will use the acronym Func2st to indicate fundamental cognitive concepts of space, time, and space-time. Critical questions for cognitively oriented spatial scientists are how Func2st can be identified and defined, whether formally hypothesized Func2st have cognitive counterparts (i.e., are cognitively adequate [33]), how Func2st are related to language, and, importantly, whether Func2st in one language correspond to those in another language. This research relates to work on conceptual primitives [70,2], theories of image schemata [24,37,46], the spatial foundations of the human conceptual system [47], and the relation between concepts and words [22,71]. The importance of such research is manifold as it adds to the growing body of knowledge that addresses questions of linguistic relativity [17,71], it adds to the pursuit of defining the foundations of meaning (from a spatio-temporal perspective) [64], and results of this research potentially improve approaches for transforming data into knowledge, for instance when interpreting data from sensor networks as in [72]. The latter aspect is also important for cognitive scientists. An example is the continuous to discrete transformation essential for language production [16]. Continuously varying information needs to be translated into quasi-symbolic equivalence classes that allow for producing reliable behavior.

To answer the questions raised above, this article is structured as follows. First, we provide a background discussion on relevant topics: fundamental spatial and spatio-temporal concepts, crowdsourcing, and cross-linguistic differences. Second, we provide details on an experiment that uses Galton's [13] *modes of overlap* to demonstrate the overall approach we propose (comparing category construction behavior of English, Chinese, and Korean speakers) to shed light on questions surrounding Func2st. Third, we discuss results, present conclusions, and describe future research opportunities.

2   Background

2.1   Func2st and Their Behavioral Evaluation

The term *fundamental spatial concepts* is not well defined [73,56,37] (see also Table 1). However, we can use the notion of *invariance* to render it more precise—across disciplines. Researchers in many scientific fields from the cognitive to the spatial sciences have addressed the topic of invariance in the context of cognitive information processing as well as to formally distinguish fundamental concepts of of space and time. Worboys and Duckham [73], for example, utilize the concept of invariance as a framework for their chapter on Fundamental Spatial Concepts. Proposed first by Felix Klein (see Erlangen program [26]), geometries can be distinguished based on invariant properties under certain transformations. This approach allows for differentiating Euclidean geometry from set-based geometry as well as topology which can be seen as "rubber-sheet geometry" concerned with properties that remain invariant under topological transformations such as bending and stretching. Paralleling formal approaches, perceptual and cognitive invariants, which we also find to be associated with conceptual primitives (e.g., [46]), have long been of interest to the cognitive science community. Klix [32] stated that the human mind, in adapting to its environment, identifies invariant characteristics of information that form the basis for cognitive processes. Shaw [65] used the term transformational invariants to denote properties of objects and events that do not change from a group (set) theoretical perspective, stressing the importance of this concept for perception. Last, the classic work by Gibson [15] refers to temporally constant characteristics of environments as structural invariants. The basic work of these scholars is featured in many recent approaches utilizing the concept of invariance to explain, for example, perception, categorization, and event cognition [38,44,18,64,16]. The importance of identifying invariants of environmental information is prominently noted by Galton [14], who speaks of our ability to intersubjectively identify invariants of space and time which allow us to construct a shared understanding of our physical (and social) environments. Without the agreement that certain characteristics of spatial environments ground our meaningful understanding of spatial environments (e.g., [63]), the concept of a shared reality and our ability to communicate about this reality would not be possible. Many researchers have pointed to topology as the cognitively most important qualitative formal theory that allows for rendering the notion of invariance more precise (see [68,10,35,36,66,32]). While topology is unquestionably important for understanding cognition, there are other theories that identify invariants potentially relevant for humans' fundamental understanding of space and time.

Table 1 provides an overview of a first collection of what could be considered fundamental spatio-temporal concepts extracted from the relevant literature. Please note that this list does not claim to be comprehensive at this point. This list can and should be extended to capture concepts deemed fundamental for thinking spatially and temporally by researchers from both the cognitive and spatial sciences (see Section 5). The table also lists some cognitive evaluations that have been conducted to prove, disprove, or modify the cognitive adequacy of formal models attempting to better capture human understanding and processing of formally identified fundamental concepts.

Over the last years, members of the Human Factors in GIScience Lab[1] have established an efficient framework (see Figure 1) to conduct behavioral experiments with the goal to evaluate Func2st using category construction tasks [52].

---

[1] www.cognitiveGIScience.psu.edu

Table 1. Candidates for fundamental spatial, temporal, and spatio-temporal concepts

Spatio-temporal concepts	Short characterization	Cognitive calibration
Topology [60,11]	Different calculi that address the most basic properties of space, such as connectedness.	Static: [33,48] Dynamic: [27]
Modes of overlap [13]	Formal extension of the single overlap relation identified in topological calculi to capture more complex forms of overlap (see Figure 2).	Pre-study [69]
Perceptual topology (e.g., [8])	Real world relations may be perceived as being topological but formally they may not be: Example: a road enters a housing estate albeit the estate is constituted by individual houses.	No assessment
Allen's interval algebra (1D topology) [1]	Definition of possible relations between time intervals (events), six relations plus inverse relations and the identity of two intervals.	[43]
Renolen's basic types of change [61]	Seven basic types of change: creation, alteration, destruction, reincarnation, merging/annexation, splitting/deduction, reallocation.	No assessment
Direction calculi (e.g., [42])	Modeling of directions as sectors or projections, differentiating cardinal and relative directions.	[29]
Chorematic Modeling [5]	Theory established by a French geographer Roger Brunet detailing 26 basic models to characterize spatial relations and processes.	No assessment
Shape (e.g., [14])	Shape of objects including trajectories.	(e.g., [54])

In contrast to many linguistic-driven approaches (see Section 5.2 for a discussion) to understanding human concepts, this framework uses qualitative formal theories as a hypothesis for salient discontinuities of space and time (see also [48,33]). Given a formal model which categorizes relations over a fundamental spatial or temporal aspect into a finite (and typically small) number of equivalence classes, hereafter referred to as qualitative equivalence classes (QEC), a set of icons is generated consisting of an equal number of visual stimuli for each QEC. Human participants then group the icons (construct categories) in our own experimental software, CatScan, and the results can be analyzed using a comprehensive analysis methodology. Results from comparing the QEC-implied categories with the cognitive behavior shown in the experiment allows for drawing conclusions about the adequacy of the investigated formal model and may

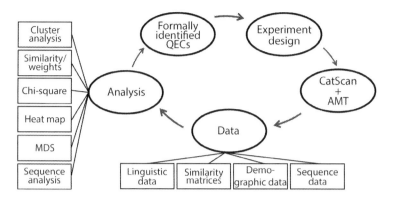

Fig. 1. Research framework (modified after [30])

also lead to refined or improved (calibrated) models that, in turn, can improve spatio-temporal data processing applications. Additionally, insights are gained of how space and time are cognitively processed. More details on the involved tools and methodology will be given in later sections of this paper.

2.2 Crowdsourcing

Crowdsourcing has the potential to change and substantially advance the way that knowledge about Func2st is elicited. Multiple crowdsourcing platforms (e.g., Survey Monkey, Lime Survey) have been developed over the last decade and are now central to scientific and especially behavioral research (e.g., [25]). Of the many options that exist, Amazon's Mechanical Turk (AMT) has attracted a lot of attention and has developed into a serious alternative to using laboratory experiments (e.g., [20,59,50]). One of the main advantages is the more diverse (in comparison to only college students) pool of participants that can be accessed [62]. Most commonly, simple tasks are outsourced to a network of registered users of AMT. These tasks are referred to as HITs (Human Intelligence Tasks) and AMT guarantees that the tasks are performed anonymously. "Turkers" receive payment directly through AMT. To control the quality of responses, Turkers will only receive reimbursement when they perform a task satisfactorily. Additionally, restrictions can be placed by allowing, for example, only Turkers above a certain HIT approval rate, obtained through successfully completing experiments. Several studies indicate that results are comparable to regular laboratory experiments (e.g., [55,59]). Some pre-tests that we conducted, however, suggest some cautions: In preparation for this article, we performed several (pre-)studies using AMT that allowed us to advise strategies to ensure the quality of crowdsourced data. It is essential to think of the experiments from the perspective of a Turker, that is, to maximize financial gain by finding the fastest yet correct way to perform a task. This leads to certain strategies that can be positive and/or negative. The bottom line is that Turkers' strategies have to be taken explicitly into account and channeled.

2.3   Cross-Linguistic Differences

While linguistic analysis as a window to cognition allows us to shed light on the question of underlying conceptual structures, the influence of language on the conceptualization of spatial relations (and on cognition in general) is a subject of ongoing debate [3,9,21,34,39,53].There are various perspectives on the role of language in cognition that range from assuming universal conceptual categories that are essentially uniform across languages [57,41] to assuming language dependent concepts [40] via recent approaches that propose a mapping between concepts and words in form of word-clusters [22]. A majority of researchers probably subscribe to the idea that language has some influence on cognition, specifically on the conceptualization of spatial relations, even though it is not completely deterministic. These theories are often discussed under the term *linguistic relativity* (e.g., [17,71]). The question, however, is how profound this influence is and how it is manifested. One of the major research approaches to addressing this debate is to compare different languages in their expression of spatial relations and how their expressions may in turn influence nonlinguistic thought (see, for example, [4] and a reply by January and Kako [23]).

Instead of starting with identified linguistic differences, our approach allows for using formally identified spatial and temporal invariants as a hypothesis of Func2st across different languages. While research has demonstrated potentially different influences of the three languages (English, Chinese, and Korean) that we chose on the conceptualization of space and time (see [4,51]), the level of detail with respect to formally identified invariants is often insufficient to derive exact hypotheses.

3   Methods

Revealing conceptual structures underlying humans understanding of space and time can be ideally addressed using category construction tasks [52], also referred to as conceptualization or free classification [58]. We have refined our custom-made software solution, CatScan, and extended it such that it is possible to run it in conjunction with Amazon Mechanical Turk. In the experiments, participants were presented with graphical stimuli that were created based on distinctions made by Galton's approach, exemplarily visualized in Figure 2 (details are given below). Participants created categories based on their own assessment of the stimuli. Participants' category construction behavior was subsequently compared against categories (modes of overlap) identified in Galton's approach. While we were able to show in previous experiments [27] that semantics has an influence on how people conceptualize spatial relations such as those identified by the region connection calculus (RCC) [60] and intersection models (IM) [11], we did not find any significant distinctions for the modes of overlap comparing a purely geometric scenario with the one involving an oil spill and protected habitat we used in this paper [69]. No sophisticated theory currently exists that predicts the influence of semantics on the conceptualization of qualitative spatial relations; we will return to this aspect in Section 6. The focus of this article is on whether someone's native language alters the conceptualization of spatial relations. Hence, we selected the oil spill plus protected habitat zone scenario as it does yield the same results as a purely geometric version.

Preliminary: Modes of Overlap. The formalism investigated in this paper is the *modes of overlap* model by Galton [13]. This approach has been developed in response to a somewhat indifferent or unspecific treatment of the concept of overlap in the most important qualitative spatial calculi dealing with topological relations: RCC and IM. In summary: on a coarse level, RCC and IM both distinguish the same eight spatial relations between two spatially extended entities. A subset of these relations is present in the table shown in Figure 2 (a, aa, c, d, dd, g; for extensive treatments see [7]). These topology-based distinctions of spatial relations have been extensively analyzed in behavioral studies (e.g., [27,33,49]). Galton's motivation for developing a more differentiated picture was that configurations such as g and q (see Figure 2) cannot readily be distinguished in RCC and IM. He therefore developed an overlap matrix that would capture different modes of overlap based on the number of connected components of regions with overlap properties: For two regions A and B, the 2 x 2 matrix

$$[A, B] = \left(\begin{smallmatrix} x & a \\ b & o \end{smallmatrix}\right)$$

contains the following information:

x is the number of connected components of the intersection $A \cap B$

a is the number of connected components of A without B ($A \setminus B$)

b is the number of connected components of B without A ($B \setminus A$)

o is the number of connected components of the outside area that belongs to neither A nor B ($(A \cup B)^c$)

For instance, the relation a in Figure 2 corresponds to the matrix $\left(\begin{smallmatrix} 0 & 1 \\ 1 & 1 \end{smallmatrix}\right)$ as there is no overlap between the circular object A and the candy-cane shaped object B, and $A \setminus B, B \setminus A, (A \cup B)^c$ each consist of a single connected region. In contrast, the matrix for h is $\left(\begin{smallmatrix} 1 & 1 \\ 1 & 2 \end{smallmatrix}\right)$ as all relevant regions consist of a single component except for the complement of $(A \cup B)$ (everything that does belong to neither A nor B), which has two connected components. Without going into any more detail (please see [13]), this approach allows for distinguishing 23 simple modes of overlap for regions without holes (see Figure 2). While the approach can distinguish an infinite number of increasingly complex topological relations, these 23 modes of overlap are the most basic cases where no number in the matrix gets higher than two. We use a slightly modified version (explained below) of this set of relations as a hypothesis to test whether humans naturally make similar distinctions regarding modes of overlap between two regions, a topic that has not been evaluated from a cognitive-behavioral perspective.

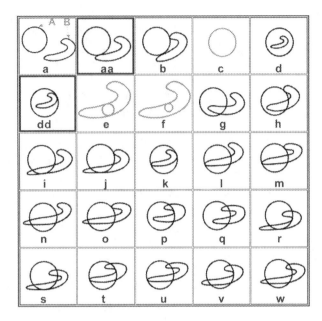

Fig. 2. Modes of overlap: 23 original modes of overlap relations between two spatially extended entities and modifications for the experimental setup used here. Two relations were added (aa, dd), three were omitted (c, e, f).

Participants. Sixty-eight participants were recruited through Amazon Mechanical Turk to participate in three experiments (i.e., English, Chinese, and Korean). Eight participants across all three language versions were excluded because they did not follow the instructions. Each language had 20 participants in the end. The English and Chinese version both included 8 females with an average age of 35.95 and 28.65, respectively. The Korean version had 10 female participants (average age: 30.25).

Materials. Based on the different modes of overlap defined by Galton [13] (see Figure 2), we created a set of icons showing different relations of an oil spill and a protected habitat zone. The icons show 22 different relations (see Figure 2) that resulted from making the following modifications to the original set of formally distinguished modes of overlap: First, we felt it is important to include a disconnected relation in addition to the relation where the two regions are only externally connected (see Figure 2, a and aa). Hence, the original relation a has been split into a for the disconnected case and aa for the externally connected case. These two situations are not distinguishable by Galton's overlap matrix but are distinguished in RCC and IM. Second, we wanted to include the distinction between tangential and non-tangential proper part also made in RCC and IM capturing whether or not the contained object in relation d touches the boundary of the containing object. As a result, the original relation d has been split into d for non-tangential proper part and dd for tangential proper part. Third, we omitted the following original relations as they would have required significant changes in the visualization: relation c, in which both entities have exactly the same spatial extension, relation e, the inverse of k, which would have required a reversal of the size differences between the two involved regions, and relation f, the inverse of d which we omitted for the same reason as e.

The full icon set was created by generating four random variations of an initial prototype consisting of a circular region and a region in the form of a candy-cane for each of the 22 relations, which allowed us to realize all relations without dramatic shape changes. The second, third, and fourth variations were created by randomly rotating the prototypical example at an angle between 0-90 degrees, 90-180 degrees, and 180-270 degrees, respectively; next, both regions were further scaled down by a random factor between 0.8-1.0. Due to the constraints imposed by some overlap relations, the size of the candy-cane region was restricted to be relatively large in some cases (e.g., i, j, and l) and relatively small in some others (e.g., d, dd, and k). For those overlap relations in which the candy-cane region can be scaled down by 50% without changing the relation, we did so for the first and second variation. In the instructions, however, we explicitly asked participants to ignore size. Considering that size is potentially a strong grouping criterion [28], this was the only way to ensure that certain modes of overlap are not singled out simply because they are larger or smaller. Finally, all icons were checked to ensure that the spatial relation in each variation is perceptually unambiguous as we are focusing on human concepts of space.

Procedure. We modified CatScan, our category construction assessment software, to enable compatibility with AMT. For this purpose, we created a stand-alone Java version that can be downloaded as a .jar file from a server and run locally on participants' computers. Recruitment and payment was organized through AMT. We used the available command line tools to automatically generate HIT descriptions with a running participant number. Participants read the general HIT instructions and entered the participant number into the interface. At the end of the experiment, CatScan generated a result file with this unique ID, which participants uploaded to AMT. We performed several pre-tests to tailor our experiment and instructions to the specifics of AMT and Turkers' perspectives: a) We only allowed Turkers with a HIT approval rating of at least 95% to participate in this experiment; b) We split the payment into two components: Turkers received $1.50 for their participation in the experiments and had the option to earn an additional $1.50 as a bonus. As we encountered problems recruiting sufficient participants for the Korean experiment, we incrementally increased the payment to $2.50 + $1 bonus and lowered the required qualitatication to 50% with mixed results (see discussion in Section 5.3). Participants were informed that they would only receive the bonus if they performed the category construction task thoroughly, label each category thoughtfully, and not use size as a grouping criterion. Each participant performed two tasks: a category construction task and a linguistic labeling task. All 88 overlap icons (22 × 4) were initially displayed on the left panel of the screen (Figure 3, top). Participants were required to create categories on the right panel of

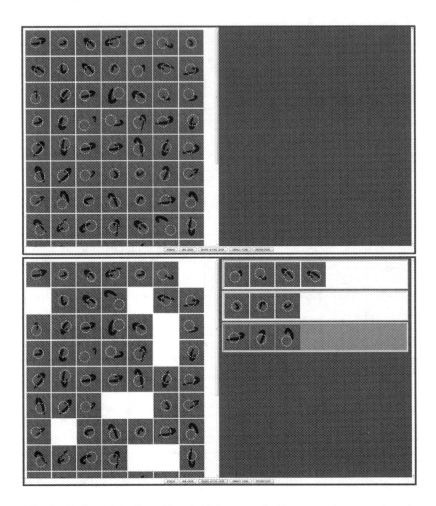

Fig. 3. CatScan interface. Top: initial screen; bottom: ongoing experiment.

the screen. After creating at least one empty category, they were able to drag icons from the left panel into a category on the right panel (Figure 3, bottom). They were explicitly advised that there was no right or wrong answer regarding the number of categories or which icons belong to which category (with the exception that they had to ignore size differences). They also had the opportunity to move icons between categories, move icons back to the left panel, or delete whole categories, in which case icons are placed back in the window on the left panel. The main category construction experiment was preceded by a warm up task (grouping animals) to acquaint participants with the software and category task. Participants performed a linguistic labeling task upon finishing the main category construction experiment. They were presented with the categories they created and provided two linguistic descriptions: a short name of no more than five words, and a longer description detailing their rationale(s) for placing icons into a particular category.

4   Analysis and Results

In our analysis, we will focus primarily on comparing the category construction (conceptualization) behavior of participants between the English, Chinese, and Korean experiments.

Our experimental software CatScan automatically recorded the number of groups created by each participant and the time (in seconds) the participant spent on the grouping task. On average, participants in the English experiment created 8.05 groups (SD 7.29) and spent 968.89 seconds on the grouping task (SD 824.43 seconds). In the Chinese experiment, participants created 7.80 groups (SD 6.81) on average and spent 777.97 seconds on the grouping task (SD 880.67 seconds). For the Korean experiment, participants created 5.30 groups (SD 2.45) on average and spent 617.30 seconds on the grouping task (SD 594.40 seconds). A one-way analysis of variance (ANOVA) comparing the number of categories created revealed that there are no statistically significant differences between English, Chinese, and Korean participants ($F(57,2) = 1.397$, $p = 0.260$). Similarly, there are no statistically significant differences in the time English, Chinese, and Korean participants spent on creating groups ($F(57,2) = 1.071$, $p = 0.350$).

The category construction behavior of each participant was recorded in an 88 * 88 sized individual similarity matrix (88 is the number of icons used in each experiment). In these matrices, the similarity of each pair of icons is binary-coded: A value of '1' indicates that a pair of icons was placed into the same group by that participant whereas a value of '0' indicates that a pair of icons was not placed into the same group. An overall similarity matrix (OSM) is obtained by summing over the individual similarity matrices of all 20 participants in each experiment. Therefore, the similarity value of a pair of icons in an OSM ranges from 0 (lowest similarity possible) to 20 (highest similarity possible).

Figure 4 synthesizes several aspects of the different analyses we performed. The images present a combination of heat maps and dendrograms (generated by Ward's clustering method) for the English, Chinese, and Korean experiments. The heat maps intuitively visualize the OSMs and, hence, the overall category construction behavior of all participants in each experiment. Red cells correspond to the highest similarity values (20) in the OSM, while white cells corresponds to the lowest similarity values (0) with the color varying between white and red on a continuous gradient. The dendrogram placed to the left side of each heat map was generated from a cluster analysis using Ward's method. The order of the icons (both in rows and columns) were rearranged such that icons that are similar to each other in the cluster analysis always neighbor with one another.

Overlap relations in all three experiments (English, Chinese, and Korean) form three general clusters, corresponding to the three larger red blocks found along the diagonal of the heat maps (see Figure 4). The results of both analysis methods are corroborated by visually comparing the top three clusters with the matching placement of the three larger blocks (hot spots) found on the heat maps. Upon analyzing the clustering structure across all three heat map/dendrogram combinations in more detail, the three major categories can be summarized as follows:

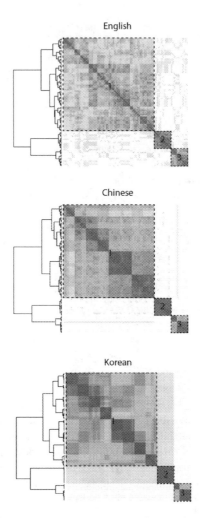

Fig. 4. Heat maps for the three experiments (English, Chinese, and Korean). Rows and columns are organized based on the results of Ward's clustering method as shown by the dendrograms. Numbers indicate a three cluster solution: 1. various overlap relations; 2. proper part relations; 3. non-overlapping relations.

- **Cluster 1** contains all g, h, i, j, l, m, n, o, p, q, r, s, t, u, v, and w icons. A commonality shared by all these icons (relations) is that the oil spill **overlaps** with the protected habitat zone.
- **Cluster 2** contains all d, dd, and k icons, in which the oil spill is found **inside** the protected habitat zones regardless of whether or not the boundary of the oil spill touches the boundary of the protected habitat zone.

  – **Cluster 3** contains all a, aa, and b icons, in which the oil spill is found **outside** the protected habitat zones regardless of whether or not the boundary of the oil spill touches the boundary of protected habitat zone.

To validate this interpretation, we followed a suggestion by Clatworthy and collaborators [6] and compared the structures of the dendrograms generated using Average Linkage and Complete Linkage in addition to the dendrograms generated using Ward's method. The cross-examination shows that the three-cluster structure is identical across all three dendrograms in all three experiments and that it is the only cluster structure that is identical. Additionally, we randomly sampled subsets of 10 participants (three times) and each time performed cluster analyses on these subsets using the three different clustering algorithms. The three-cluster structure again is a) consistent and b) the only consistent clustering solution. This strongly supports the validity of the three cluster solutions and also indicates that all three language experiments are comparable.

Despite the fact that the three cluster solution can be considered the natural category structure (as revealed by the different validation procedures), differences exist within the internal structure of the overlap category across the three languages. Intrigued by the differences, we took a closer look at participants' grouping behavior specifically for the overlap relations from Cluster 1. We extracted a sub-matrix containing only similarity values (i.e., 0 or 1) for Cluster 1 relations from each individual similarity matrix. Based on these sub-matrices, we constructed a between-participant similarity matrix (BSM) that encodes the similarity of grouping behavior for each pair of participants for each of the three experiments. In the BSM, the similarity value of each pair of participants is determined by the Hamming distance between the sub-matrices of these two participants. Hamming distance calculates the number of cells that differ across two matrices, thus effectively calculating the distance, or difference, between two participants' grouping behavior for Cluster 1 relations. Furthermore, a dendrogram was generated using Ward's method for each experiment, showing the similarity of participants based on their grouping behavior on Cluster 1 relations (see Figure 5 as an example from the Chinese experiment).

Guided by the results in Figure 5, we explored the groups created by participants and their linguistic description in KlipArt—a custom-made, freely available visual analysis tool [31]. In KlipArt, we were able to identify five major types of category construction strategies (as specified in Table 2) employed by participants in all three experiments.

Overall, more Chinese and Korean participants than English participants placed all overlap relations into one single group while more English participants grouped icons by the direction the oil spill faced than the Chinese and Korean participants. Korean participants grouped mainly based on the penetration of the oil spill in the protected habitat zone whereas the other two participant groups were more spread out among category construction strategies. For the final two grouping strategies, a similar number of participants (comparing between the Chinese, English, and Korean participants) were found to behave and group in the same way (i.e. grouping by the amount of the overlap: 20-25% of participants; grouping by the

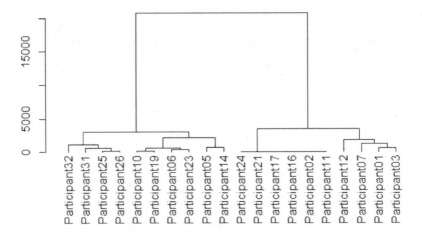

Fig. 5. This dendrogram shows the similarities of participants from the Chinese experiment based on their grouping behavior on Cluster 1 relations. The lower the distance at which two participants are merged with each other, the more similar they are.

Table 2. Category constructions strategies elicited from participants for the different languages

Grouping strategy	English experiment	Chinese experiment	Korean experiment
All overlap icons placed in one group	3 (15%)	6 (30%)	7 (35%)
Grouping by the amount of the overlap (large vs. small)	4 (20%)	4 (20%)	5 (25%)
Topological equivalence classes (according to modes of overlap relations)	2 (10%)	2 (10%)	0 (0%)
Whether the long or short tail of the oil spill penetrates the protected habitat zone or not	5 (25%)	6 (30%)	8 (40%)
Direction that the oil spill is facing	6 (30%)	2 (10%)	0 (0%)

topological equivalence classes implied by the modes of overlap relations: 0-10% of participants). This approach allows us to undergo finer-grained analyses that compare potential language related differences and grouping strategies. Furthermore, these findings allow us to identify new research questions to ascertain why there were grouping differences between the experimental groups. For instance, future research can analyze why English participants utilize direction more than participants in the other two language groups.

5   Conclusions

5.1   General Discussion

The results shed light on a number of essential details about humans' conceptualization of modes of overlap across different languages. Galton (1998) is making a convincing case that it would be desirable to adopt a finer level of granularity with respect to distinguishing modes of overlap rather than one overlap relation (as in RCC and IM). In contrast, our results indicate that this level of granularity is not common from a cognitive perspective in either of the three languages. The cross validation techniques applied to all three data sets reveal an extremely stable category structure distinguishing non-overlapping, overlapping, and proper part relations. This coarse and consistent level of granularity is remarkable for a number of reasons: First, we generally find that if spatial information is statically presented, distinctions made by qualitative calculi are reasonable predictors for category construction behaviors. This has been shown, for example, by Knauff and collaborators [33]. They revealed in their experiments using the eight relations identified by RCC/IM that participants did abide by formal distinctions. In contrast, in the results presented here, a very coarse perspective on spatial relations is adopted, that is, 22 different relations form only three spatial categories. Second, none of the languages shows a diversion from this clustering structure although previous research indicated that native Korean speakers might be inclined to distinguish different proper part relations [51]. Third, while finer distinctions are adopted by participants in all three experiments, they do not surface as a consistent pattern. Hence, they should not be considered as being fundamental (or salient discontinuities). This interpretation remains true even if participants are explicitly asked to create as many meaningful categories as possible [69].

5.2   Reflections on Research Framework

Our ultimate goal is the development of a research framework that allows for comprehensive behavioral research on fundamental concepts of space and time and space-time, Func2st. It is widely acknowledged (e.g., [32,35]) that equivalence classes identified by qualitative spatio-temporal calculi offer testable hypotheses that take the following general form: *Formally identified equivalence classes correspond to salient, quasi-symbolic equivalence that surfaces in reliable cognitive behavior.* In other words, discontinuities of space, time, and space-time relevant for cognitive information processing correspond to discrete distinctions introduced by qualitative spatio-temporal calculi. Our research framework adds an important aspect to the existing literature. In contrast to research that starts out analyzing language components such as spatial and temporal preposition, our work starts with eliciting conceptualizations of spatial relations. Given the often controversial discussion of the relation between language and cognitive concepts that ranges from assuming universal conceptual categories that are essentially uniform across languages [57,41,67] to assuming language dependent concepts [40] via recent approaches that propose a mapping between concepts and words in form of word-clusters [22], we consider it essential to add a non-linguistic approach that might be able to shed some light on the conceptual structure of human concepts and whether or not they are universal (even if no extensive linguistic theory exists).

5.3   Reflections on Crowdsourcing

Crowdsourcing is a powerful tool to scale up the behavioral calibration of formally (qualitatively) identified discontinuities of space and time. This approach is necessary as there is a large number of potential candidates for fundamental spatial concepts that require an efficient approach for validation.

Yet, there are some challenges in using the crowd that are important to keep in mind. When it comes to testing concepts using English speaking participants crowdsourcing is indeed a powerful and efficient tool for behavioral research. AMT, the platform we used for our experiments, allowed us to set very high standards with respect to Turker reliability and, nonetheless, data came in overnight (for English HITs). While we needed some experimentation with the exact phrasing of the experimental task and the reward structure, we have additional evidence in form of a pre-study, that the results obtained through AMT are comparable to regular lab studies. Our results are therefore in line with many current studies (e.g., [59]) that show that AMT is a reliable research tool. Additionally, and maybe even more importantly, using AMT can overcome the critical aspect that more lab studies rely almost exclusively on student participants. Several studies have shown that the population of Turkers is much more diverse than any undergraduate student body (e.g., [50]).

With respect to using AMT for cross cultural studies we have obtained mixed results. The Chinese version of our experiment allowed us to have similarly high standards as the English version. However, the data came in at a much slower pace and it took us roughly two weeks to collect data for 20 participants. We did not require participants to live in China but to be native language speakers of Chinese (compare [53]). The long time it took makes this approach difficult in case larger numbers are needed but it seems possible to use AMT for Chinese studies. Additionally, it may be possible to increase participant numbers by using higher payments and/or lower Turker ratings. One option we did not fully explore yet is advertising HITs outside of AMT.

The Korean version of the experiments was extremely slow such that we collected additional data through direct requests. We experimented with higher payments and lower Turker ratings but got mixed results. More Turkers signed up for our HITs but we also had to disregard data as we obtained invalid submissions. The most extreme case was a participant whose linguistic descriptions in the second part were, most likely, obtained through an online translation service such as Google Translate.

6   Outlook

We only provided a glimpse of how fundamental cognitive spatial, temporal, and spatio-temporal concepts, Func2st, can be evaluated in a crowdsourcing, cross-linguistic setting. We selected Galton's modes of overlap as a test case for our quest to deliver a comprehensive assessment of Func2st relating formal and human conceptualizations of space and time. However, the research framework we have developed is tailored to be efficient and effective such that a large scale assessment is indeed becoming feasible. We believe that every formalism that is proposed to enhance processes at the interface of humans and machines requires some validation. We have provided an overview of potential extensions of our work in Table 1. By no means do we claim that this list is complete and we invite others to comment and contribute to this ongoing research challenge.

References

1. Allen, J.F.: Maintaining knowledge about temporal intervals. Communications of the ACM 26(11), 832–843 (1983)
2. Aziz-Zadeh, L., Casasanto, D., Feldman, J., Saxe, R., Talmy, L.: Discovering the conceptual primitives. In: Love, B.C., McRae, K., Sloutsky, V.M. (eds.) Proceedings of the 30th Annual Conference of the Cognitive Science Society, pp. 27–28. Cognitive Science Society, Austin (2008)
3. Bloom, P., Peterson, M.P., Nadel, L., Garrett, M.F.: Language and space. MIT Press, Cambridge (1996)
4. Boroditsky, L.: Does language shape thought?: Mandarin and English speakers' conceptions of time. Cognitive Psychology 43, 1–22 (2001)
5. Brunet, R.: La carte, mode d'emploi. Fayard-Reclus, Paris (1987)
6. Clatworthy, J., Buick, D., Hankins, M., Weinman, J., Horne, R.: The use and reporting of cluster analysis in health psychology: A review. British Journal of Health Psychology 10, 329–358 (2005)
7. Cohn, A.G., Renz, J.: Qualitative spatial representation and reasoning. In: Harmelen, F.v., Lifschitz, V., Porter, B. (eds.) Handbook of Knowledge Representation. Foundations of artificial intelligence, pp. 551–596. Elsevier (2008)
8. Corcoran, P., Mooney, P., Bertolotto, M.: Spatial Relations Using High Level Concepts. ISPRS International Journal of Geo-Information 1(3), 333–350 (2012)
9. Crawford, L.E., Regier, T., Huttenlocher, J.: Linguistic and non- linguistic spatial categorization. Cognition 75(3), 209–235 (2000)
10. Egenhofer, M.J.: The family of conceptual neighborhood graphs for region-region relations. In: Fabrikant, S.I., Reichenbacher, T., van Kreveld, M., Schlieder, C. (eds.) GIScience 2010. LNCS, vol. 6292, pp. 42–55. Springer, Heidelberg (2010)
11. Egenhofer, M.J., Franzosa, R.D.: Point-set topological spatial relations. International Journal of Geographical Information Systems 5(2), 161–174 (1991)
12. Egenhofer, M.J., Mark, D.M.: Naive geography. In: Frank, A.U., Kuhn, W. (eds.) Spatial Information Theory, pp. 1–15. Springer, Berlin (1995)
13. Galton, A.: Modes of overlap. Journal of Visual Languages and Computing 9, 61–79 (1998)
14. Galton, A.: Qualitative spatial change. Spatial information systems. Oxford Univ. Press, Oxford (2000)

15. Gibson, J.: The ecological approach to visual perception. Houghton Mifflin (1979)
16. Goldstone, R.L., Hendrickson, A.T.: Categorical perception. Wiley Interdisciplinary Reviews: Cognitive Science 1(1), 69–78 (2010)
17. Gumperz, J.J., Levinson, S.C.: Rethinking linguistic relativity. Cambridge University Press, Cambridge (1996)
18. Harnad, S.: To cognize is to categorize: Cognition is categorization. In: Cohen, H., Lefebvre, C. (eds.) Handbook of Categorization in Cognitive Science. Elsevier, Amsterdam (2005)
19. Hayes, P.: The naive physics manifesto. In: Michie, D. (ed.) Expert Systems in the Microelectronic Age, pp. 242–270. Edinburgh University Press, Edinburgh (1978)
20. Heer, J., Bostock, M.: Crowdsourcing graphical perception: Using Mechanical Turk to assess visualization design. In: ACM CHI 2010 (2010)
21. Hermer-Vazquez, L., Moffet, A., Munkholm, P.: Language, space, and the development of cognitive flexibility in humans: the case of two spatial memory tasks. Cognition 79, 263–299 (2001)
22. Holmes, K.J.: Language as a Window into the Mind: The Case of Space. Ph.D. thesis, Emory University (2012)
23. January, D., Kako, E.: Re-evaluating evidence for linguistic relativity: Reply to Boroditsky (2001). Cognition 104, 417–426 (2007)
24. Johnson, M.: The body in the mind: The bodily basis of meaning, imagination, and reasoning. University of Chicago Press, Chicago (1987)
25. Kittur, A., Chi, E.H., Suh, B.: Crowdsourcing user studies with Mechanical Turk. In: CHI 2008: Proceeding of the Twenty-Sixth Annual SIGCHI Conference on Human Factors in Computing Systems, pp. 453–456. ACM (2008)
26. Klein, F.: Vergleichende Betrachtungen über neuere geometrische Forschungen ("A comparative review of recent researches in geometry"). Mathematische Annalen 43, 63–100 (1872)
27. Klippel, A.: Spatial information theory meets spatial thinking - Is topology the Rosetta Stone of spatio-temporal cognition? Annals of the Association of American Geographers 102(6), 1310–1328 (2012)
28. Klippel, A., Hardisty, F., Li, R.: Interpreting spatial patterns: An inquiry into formal and cognitive aspects of Tobler's first law of geography. Annals of the Association of American Geographers 101(5), 1011–1031 (2011)
29. Klippel, A., Montello, D.R.: Linguistic and nonlinguistic turn direction concepts. In: Winter, S., Duckham, M., Kulik, L., Kuipers, B. (eds.) COSIT 2007. LNCS, vol. 4736, pp. 354–372. Springer, Heidelberg (2007)
30. Klippel, A., Wallgrün, J.O., Yang, J., Li, R., Dylla, F.: Formally grounding spatio-temporal thinking. Cognitive Processing 13(suppl. 1), 209–214 (2012)
31. Klippel, A., Weaver, C., Robinson, A.C.: Analyzing cognitive conceptualizations using interactive visual environments. Cartography and Geographic Information Science 38(1), 52–68 (2011)
32. Klix, F.: Die Natur des Verstandes. Hogrefe, Göttingen (1992)
33. Knauff, M., Rauh, R., Renz, J.: A cognitive assessment of topological spatial relations: Results from an empirical investigation. In: Hirtle, S.C., Frank, A.U. (eds.) Spatial Information Theory: A Theoretical Basis for GIS, pp. 193–206. Springer, Berlin (1997)
34. Knauff, M., Ragni, M.: Cross-Cultural Preferences in Spatial Reasoning. Journal of Cognition and Culture 11(1), 1–21 (2011)
35. Kuhn, W.: An image-schematic account of spatial categories. In: Winter, S., Duckham, M., Kulik, L., Kuipers, B. (eds.) COSIT 2007. LNCS, vol. 4736, pp. 152–168. Springer, Heidelberg (2007)

36. Kurata, Y., Egenhofer, M.J.: Interpretation of behaviors from a viewpoint of topology. In: Gottfried, B., Aghajan, H. (eds.) Behaviour Monitoring and Interpretation, pp. 75–97. IOS Press, Amsterdam (2009)
37. Lakoff, G.: Women, fire and dangerous things. Chicago University Press (1987)
38. Lakoff, G.: The invariance hypothesis: is abstract reason based on image schemata? Cognitive Linguistics 1(1), 39–74 (1990)
39. Levinson, S.C., Kita, S., Haun, D.B.M., Rasch, B.H.: Returning the tables: language affects spatial reasoning. Cognition 84, 155–188 (2002)
40. Levinson, S.C., Sergio: 'Natural concepts' in the spatial topological domain—Adpositional meanings in crosslinguistic perspective: An exercise in semantic typology. Language 79(3), 485–516 (2003)
41. Li, P., Gleitman, L.: Turning the tables: Language and spatial reasoning. Cognition 83, 265–294 (2002)
42. Ligozat, G.: Qualitative spatial and temporal reasoning. ISTE, London (2012)
43. Lu, S., Harter, D., Graesser, A.C.: An empirical and computational investigation of perceiving and remembering event temporal relations. Cognitive Science 33, 345–373 (2009)
44. Maguire, M.J., Brumberg, J., Ennis, M., Shipley, T.F.: Similarities in object and event segmentation: A geometric approach to event path segmentation. Spatial Cognition and Computation (3), 254–279 (2011)
45. Malt, B.C., Ameel, E., Gennari, S., Imai, M., Saji, N., Majid, A.: Do words reveal concepts? In: Carlson, L.A., Hölscher, C., Shipley, T.F. (eds.) Proceedings of the 33rd Annual Conference of the Cognitive Science Society, pp. 519–525. Cognitive Science Society, Austin and TX (2011)
46. Mandler, J.M.: How to build a baby II. Conceptual primitives. Psychological Review 99(4), 587–604 (1992)
47. Mandler, J.M.: The spatial foundations of the conceptual system. Language and Cognition 2(1) (2010)
48. Mark, D.M., Egenhofer, M.J.: Calibrating the meanings of spatial predicates from natural language: Line-region relations. In: Waugh, T.C., Healey, R.G. (eds.) Advances in GIS Research, 6th International Symposium on Spatial Data Handling, Edinburgh, Scotland, UK, pp. 538–553 (1994)
49. Mark, D.M., Egenhofer, M.J.: Topology of prototypical spatial relations between lines and regions in English and Spanish. In: Proceedings, Auto Carto 12, Charlotte, North Carolina, pp. 245–254 (March 1995)
50. Mason, W.A., Suri, S.: Conducting behavioral research on amazon's mechanical turk. Behavior Research Methods, Instruments, & Computers 44(1), 1–23 (2011)
51. McDonough, L., Choi, S., Mandler, J.M.: Understanding spatial relations: Flexible infants, lexical adults. Cognitive Psychology 46(3), 229–259 (2003)
52. Medin, D.L., Wattenmaker, W.D., Hampson, S.E.: Family resemblance, conceptual cohesiveness, and category construction. Cognitive Psychology 19(2), 242–279 (1987)
53. Munnich, E., Landau, B., Dosher, B.A.: Spatial language and spatial representation: A cross-linguistic comparison. Cognition 81, 171–207 (2001)
54. Op Beeck, H.P.D., Wagemans, J., Vogels, R.: The representation of perceived shape similarity and its role for category learning in monkeys: A modeling study. Vision Research 48(4), 598–610 (2008)
55. Paolacci, G., Chandler, J., Ipeirotis, P.G.: Running experiments on amazon mechanical turk. Judgement and Decision Making 5(5), 411–419 (2010)
56. Peuquet, D.J.: Representations of space and time. Guilford Press, New York (2002)

57. Pinker, S.: The language instinct: How the mind creates language, 1st edn., New York. HarperPerennial ModernClassics (2007)
58. Pothos, E.M., Chater, N.: A simplicity principle in unsupervised human categorization. Cognitive Science 26(3), 303–343 (2002)
59. Rand, D.G.: The promise of Mechanical Turk: How online labor markets can help theorists run behavioral experiments: Evolution of Cooperation. Journal of Theoretical Biology 299(0), 172–179 (2012)
60. Randell, D.A., Cui, Z., Cohn, A.G.: A spatial logic based on regions and connections. In: Nebel, B., Rich, C., Swartout, W.R. (eds.) Proceedings of the 3rd International Conference on Knowledge Representation and Reasoning, pp. 165–176. Morgan Kaufmann, San Francisco (1992)
61. Renolen, A.: Modelling the Real World: Conceptual Modelling in Spatiotemporal Information System Design. Transactions in GIS 4(1), 23–42 (2000)
62. Ross, J., Irani, L., Silberman, M.S., Zaldivar, A., Tomlinson, B.: Who are the crowdworkers? Shifting demographics in mechanical turk. In: CHI 2012: Imagine all the People, pp. 2863–2872 (2012)
63. Scheider, S., Janowicz, K., Kuhn, W.: Grounding geographic categories in the meaningful environment. In: Hornsby, K.S., Claramunt, C., Denis, M., Ligozat, G. (eds.) COSIT 2009. LNCS, vol. 5756, pp. 69–87. Springer, Heidelberg (2009)
64. Scheider, S., Kuhn, W.: Affordance-based categorization of road network data using a grounded theory of channel networks. International Journal of Geographical Information Science 24(8), 1249–1267 (2010)
65. Shaw, R., McIntyre, M., Mace, W.: The role of symmetry in event perception. In: MacLeod, R.B., Pick, H.L. (eds.) Perception, pp. 276–310. Cornell University Press, Ithaca (1974)
66. Smith, B., Mark, D.M.: Geographical categories: An ontological investigation. International Journal of Geographical Information Science 15(7), 591–612 (2001)
67. Stock, K., Cialone, C.: Universality, Language-Variability and Individuality: Defining Linguistic Building Blocks for Spatial Relations. In: Egenhofer, M., Giudice, N., Moratz, R., Worboys, M. (eds.) COSIT 2011. LNCS, vol. 6899, pp. 391–412. Springer, Heidelberg (2011)
68. Talmy, L.: How language structures space. In: Pick, H.L., Acredolo, L.P. (eds.) Spatial Orientation: Theory, Research, and Application, New York, pp. 225–282 (1983)
69. Wallgrün, J.O., Yang, J., Klippel, A.: Investigating intuitive granularities of overlap relations. In: Proceedings of the 12th IEEE International Conference on Cognitive Informatics & Cognitive Computing, New York City, USA, Wiley-IEEE CS Press, Hoboken (2013)
70. Wierzbicka, A.: The search for universal semantic primitives. In: Pütz, M. (ed.) Thirty Years of Linguistic Evolution, Benjamins, pp. 215–242 (1992)
71. Wolff, P., Holmes, K.J.: Linguistic relativity. Wiley Interdisciplinary Reviews: Cognitive Science 2(3), 253–265 (2011)
72. Worboys, M., Duckham, M.: Monitoring qualitative spatiotemporal change for geosensor networks. International Journal of Geographical Information Science 20(10), 1087–1108 (2006)
73. Worboys, M., Duckham, M.: GIS: A Computing Perspective., 2nd edn. CRC Press, Boca Raton (2004)

Kinds of Full Physical Containment

Torsten Hahmann and Boyan Brodaric *

Department of Computer Science, University of Toronto, Toronto, ON, Canada
Geological Survey of Canada, Natural Resources Canada, Ottawa, ON, Canada

Abstract. Full physical containment is the relation in which one physical entity is completely inside another. It is central to the description of natural resources held in reservoirs above or below the surface. Previous ontological representations of containment are located in abstract space, incomplete, or insufficiently incorporate voids, so in this paper we develop a complete taxonomy for the full containment relation that is situated in physical space and integrates voids. The taxonomy is formalized in a mereotopological theory and specializes the DOLCE foundational ontology, thus advancing hydro ontology development.

Keywords: knowledge representation, geospatial ontology, spatial relations, hydro ontology, hydrogeology, physical containment, taxonomy, container schema, physical void, hole, material, immaterial, DOLCE.

1 Introduction

Containment is critical to the geosciences. It plays a foundational role in the description of both surface and subsurface resources, such as water, petroleum, and natural gas, where a nuanced notion is required to capture the subtle ways that something can be inside or surrounded by something else. For example, estimates of subsurface fluids or gases need to know how much underground space exists, how much of it is fillable and open to flow, and how much is actually filled or flowing. A container's spaces—its voids—assume particular significance in these scenarios, as it is their size and arrangement (together known as porosity [17]) that is used to determine storage and flow. This is particularly evident in models and simulations of subsurface resources, which inherently rely on a sophisticated containment schema that delineates three things, a container, its voids, and a containee, as first-class entities. However, these distinctions are largely absent in emerging domain representations, such as data standards for water and geology [1,21], which either ignore containment or omit voids from containment relations. Their omission is problematic given the expected role that such standards will play in delivering data from distributed databases, and sensor networks, to modeling and simulation tools. Theoretical work on qualitative topological relations [4,5,9,10,14,16] can help bridge this representation gap, but the relations are modeled in an abstract mathematical-geometric conception of space rather than in physical space, accounting for topological constraints but not for physical constraints. The relations also hold only between untyped regions, which are akin to mathematical objects, rather than between

---

* We gratefully acknowledge support from the Groundwater Program of Natural Resources Canada.

T. Tenbrink et al. (Eds.): COSIT 2013, LNCS 8116, pp. 397–417, 2013.

physically typed entities, such as voids or material bodies, thus the physical significance of the relation is not captured. In short, what is required is an ontological interpretation of topological relations in which abstract containment is considered physically.

In this paper, we begin such an interpretation motivated by various containment relations associated with both surface and subsurface water. An ultimate endpoint for this work is its incorporation into a hydro ontology. The paper thus aims to accomplish a specific piece of ontology engineering, and in doing so it makes the following original contributions: (1) it provides a taxonomy of the full containment relation, in which the relation and its components are interpreted physically (see Figure 4); (2) it specializes the DOLCE foundational ontology [19] to provide types for the components of the relation; and (3) it grounds these physical distinctions in a first-order theory of abstract space, which provides mereotopological and mereogeometrical relations.

The paper is organized as follows: Section 2 illustrates motivating examples; Section 3 discusses related work; Section 4 introduces some background material; Sections 5–7 describe the full containment relation, first generically and then partitioned into dependent and detachable containment; Section 8 provides some additional discussion and the paper concludes in Section 9 with a recap and future directions.

2   Containment Scenario

Various types of physical containment that motivate this work are illustrated in Figure 1. Included are the following relations:

$contains(\text{LB}, \text{SWB})$ $contains(\text{LB}, \text{Rock})$ $contains(\text{LB}, \text{Hole})$
$contains(\text{Hole}, \text{SWB})$ $contains(\text{Hole}, \text{Rock})$ $contains(\text{SWB}, \text{Rock})$
$contains(\text{AQ}, \text{GWB})$ $contains(\text{AQ}, \text{RM})$ $contains(\text{AQ}, \text{CT})$
$contains(\text{AQ}, \text{Gaps})$ $contains(\text{Gaps}, \text{GWB})$ $contains(\text{Gaps}, \text{CT})$
$contains(\text{GWB}, \text{CT})$

Described is a lake: a surface water body (SWB) located in a lakebed (LB) and thus also in its associated hole. Also described is a rock at the bottom of the lake, a groundwater body (GWB) in the gaps of an aquifer's (AQ) rock matter (RM), and a contaminant (CT) within the groundwater body. These examples show that full physical containment is primarily considered here, mainly for scoping reasons, but also because it is most frequently encountered in the water domain. Full physical containment involves complete immersion or enclosure of one entity within another, such as a rock completely immersed in the lake, or a body of water completely enclosed by the lakebed, but excludes partial containment such as a boat floating partially submerged on the lake. Despite this

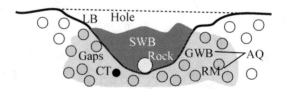

Fig. 1. Containment examples in surface and subsurface water. See text for details.

narrowing of scope, it is apparent that a physical interpretation of full containment can lead to useful specializations, e.g. being inside a hole is different from, but associated with, being surrounded by a material container. These and many other distinctions are elaborated in Sections 6 and 7.

3   Related Work

Containment relations have been considered in image schemata, qualitative spatial relations, formal top-level ontologies, and geoscience representations. In the work to date, however, containment relations are either detached from an associated physical ontology, the participation of voids is limited, or the taxonomy is incomplete.

Containment image schema is a cognitive-driven template for containment. Diversely studied in terms of its formalization and uses [18,24], it typically exhibits a rudimentary ontology consisting of the roles played by participants (container/containee), their potential behaviour (moving in/out), and states (inside/outside). However, this ontology lacks a key participant—the void afforded by a container and occupied by a containee— and also generally lacks a physical typing of participants, which could lead to erroneous interpretations. Also typically lacking are physical constraints, such as the existential dependence of a void on its host container.

Qualitative spatial relations have been studied extensively, most notably in the 9-intersection and RCC paradigms, which describe the possible topological relations that occur between abstract regions as well as their compositions and inferences [4,5,9,10]. Voids can implicitly participate in the defined abstract containment relations, inasmuch as a void can be represented as an abstract region, but there is no mechanism for typing the region as a void [12]. These relations are, then, subject to concerns similar to those expressed about image schema: a lack of physical typing and associated constraints.

Formal top-level ontologies such as DOLCE and BFO [13,19], include neither physical containment relations nor voids, but provide a superstructure from which they can be specialized. Indeed, DOLCE physical entities and their dependants have been extended to encompass containers and voids, respectively [15], laying the groundwork for a physical containment relation. The upper ontology SUMO [20] does include two basic containment relations, 'contains' and 'inside': while 'contains' is a partial containment relation that is further specialized into a proper and a complete version, the 'inside' relation lacks a clear definition and other kinds of containment are missing. Cyc [11], a large collection of commonsense knowledge, possesses a rich suite of physical containment relations that are, however, also incomplete: for example, a void cannot contain something, except implicitly via the very general 'inside' relation, which has no constraints on its domain and range. Conversely, voids can explicitly contain material entities in one of four containment relations in [8], but the remaining relations occur between abstract regions, again without physical typing and constraints. Similarly, holes (but not gaps) are fillable in [2], though other types of containment relations are missing.

Relevant geoscience representations for water or geology, such as ontologies or data transfer standards, e.g. [21,23], do not include containment relations nor voids. An exception is the Groundwater Markup language (GWML) [1], in which water reservoirs offer a limited notion of voids, and where a containment relation holds between a reservoir and the matter filling it, but GWML is not expressed formally and lacks a general approach to containment.

4 Background

This work follows in the footsteps of [15] and relies on a distinction between *physical space* and abstract spatial regions of purely topological/geometrical nature. For the former we specialize the DOLCE foundational ontology, and for the latter we reuse the first-order spatial ontology from [14,16], which is a multidimensional version of the first-order axiomatization of the Region Connection Calculus (RCC) [4]. This section reviews the relevant DOLCE categories of physical entities, the theory of abstract space, and the formalization of physical voids from [14,15]. We follow [14] in notation and axiom numbering. Free variables in our logical sentences are assumed to be implicitly universally quantified.

Physical endurants —real entities of primarily physical nature that populate *physical space* form the domain of interest for our study of physical containment. These physical entities are captured by the DOLCE category PED, as shown in Figure 2. They can be physical objects (e.g. rocks), POB, amounts of matter (e.g. clay), M, or physical features, F. Features are either relevant part features (e.g. a bump or an edge), RPF, which are constituted by a portion of the matter of their associated physical object, or dependent place features (e.g. a hole or a shadow), DPF, which are *immaterial*, i.e., not constituted of any matter. All other physical endurants are *material*, denoted by the predicate $mat(x)$. The predicate $DK_1(x, y)$, called *direct primary constitution*, denotes that x is the immediate matter constituting an object or relevant part y.

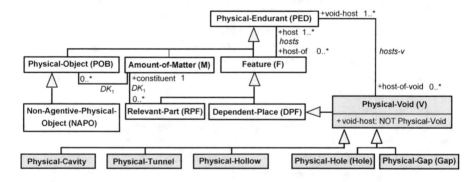

Fig. 2. The DOLCE category of physical endurants and its refinements, as UML diagram

(Mat-D) $mat(x) \leftrightarrow POB(x) \lor M(x) \lor RPF(x)$ (material endurant)

(DK₁-D) $DK_1(x, y) \to M(x) \land [POB(y) \lor RPF(y)] \land P(\mathrm{r}(x), \mathrm{r}(y))$
        (primary direct constitution of an object or relevant-part feature by matter)

Physical endurants are entities extended in physical space; following the ideas of [7,8] we assign them a location in abstract space using the $r(x)$ function. The resulting abstract regions, i.e., the entities satisfying $S(x) \leftrightarrow x = r(x)$, are disjoint from the set of physical endurants that populate physical space. This distinction between physical endurants and abstract regions allows us to consider abstract space as a mathematical-geometrical construct, which provides greater flexibility in all spatial operations. For example, many physical features of physical endurants, such as physical boundaries, are defined in terms of their regions.

Abstract regions of space are interpreted as topologically closed regions of abstract space obtainable by gluing together finite sets of *manifolds with boundaries* [14]. Their dimension can vary, e.g., both a 2D area and its 1D boundary are valid regions. We compare regions dimensionally using the predicates $<_{\dim}$, $=_{\dim}$, and $\prec_{\dim}$ that denote lower, equal, and next lower dimension, respectively. We assume that all physical endurants are located in a region of maximal dimension, that is, of codimension 0. For example, all physical endurants in a 3D space must also be 3D.

(MaxDim-D) $MaxDim(x) \leftrightarrow S(x) \wedge \neg \exists y[x <_{\dim} y]$ (regions of maximal dim.)
(S-A8) $PED(x) \to MaxDim(r(x))$ (*PEDs* are located in regions of maximal dim.)

Regions can be in various spatial relationships. These relationships are all founded on a single primitive spatial relation $Cont(x,y)$ that expresses 'x is a subregion (of equal or lower dimension) of y' and that is reflexive, antisymmetric, and transitive[1]. For mathematical simplicity we include an empty region of lowest dimension, denoted as $ZEX(x)$. Among other relations, we define contact C, parthood P, proper parthood PP, overlap PO, and superficial contact SC; those apply only to regions[2], but we can, for example, write $C(r(x), r(y))$ to state that the physical endurants x and y are in contact. Regions that overlap exactly the same set of regions are considered equivalent (PO-E1).

(C-D) $C(x,y) \leftrightarrow \exists z \, [Cont(z,x) \wedge Cont(z,y)]$ (contact: a shared entity exists)
(EP-D) $P(x,y) \leftrightarrow Cont(x,y) \wedge x =_{\dim} y$ (equi-dimensional parthood)
(EPP-D) $PP(x,y) \leftrightarrow P(x,y) \wedge \neg P(y,x)$ (equi-dimensional proper parthood)
(PO-D) $PO(x,y) \leftrightarrow \exists z \, [P(z,x) \wedge P(z,y)]$ (overlapping in a part)
(SC-D) $SC(x,y) \leftrightarrow C(x,y) \wedge \neg \exists z[Cont(z,x) \wedge P(z,y)] \wedge \neg \exists z[P(z,x) \wedge Cont(z,y)]$
     (superficial contact)
(PO-E1) $S(x) \wedge S(y) \to \big[\forall z[PO(x,z) \leftrightarrow PO(y,z)] \to x = y\big]$
     (*PO*-extensionality: two regions with the same *PO*-extension are equivalent)

The set of regions of maximal dimension is assumed mereologically closed, that is, for every pair of regions x and y with $MaxDim(x)$ and $MaxDim(y)$ and thus for every pair of regions corresponding to physical endurants, the intersection $x \cdot y$, the difference $x - y$, and the sum $x + y$ are defined (they may yield the zero region); for their precise definitions we refer to [14]. Moreover, we assume that a universal region S_u of maximal dimension exists with $MaxDim(S_u)$ and $\forall x[S(x) \wedge \neg ZEX(x) \to Cont(x, S_u)]$, so that all regions of maximal dimension have a complement $x' = S_u - x$. These mereological

---

[1] The relation $Cont(x,y)$ is equivalent to the relation $x \subseteq y$ used in [15].
[2] This deviates from the more relaxed use of the predicates in [15].

closure operations apply only to regions; we do not force the set of physical entities to be closed in the same way. The operations allow us to define strong contact, $C_S(x, y)$, as sharing a region of the next lower dimension. E.g., 3D bodies are in strong contact if they touch in a 2D surface, but not if they only touch in a curve segment or in points. This, in turn, lets us define (self-)connectedness and internal (self-)connectedness (also known as *strong self-connectedness*), meaning that the interior of $x+y$ is a single piece. The universal region S_u is assumed to be internally connected, that is, $ICon(S_u)$.

(C$_S$-D) $C_S(x, y) \leftrightarrow SC(x, y) \wedge x =_{\dim} y \wedge r(x) \cdot r(y) \prec_{\dim} x$ (strong contact)
(Con-D) $Con(x) \leftrightarrow \forall y[PP(y, x) \to C(y, r(x) - r(y))]$ (connected)
(ICon-D) $ICon(x) \leftrightarrow \forall y[PP(y, x) \to C_S(y, r(x) - r(y))]$ (internally connected)

We further use a primitive function $ch(x)$ to denote the convex hull region of x. It is needed to specify necessary spatial conditions for the existence of voids. See [14] for a more complete axiomatization of ch based on ideas from [4,8].

Voids , V, are special kinds of immaterial dependent place features, as they depend on some hosting, non-void, physical endurant. Voids include holes (following [2]) and gaps, which are differentiated according to whether their hosts are internally connected or not, respectively [15]. Any void must be located in the convex hull $ch(x)$ of a host x, but the region $ch(x) - r(x)$ need not be completely covered by voids. In other words, part of the convex hull of an endurant may be neither the location of a material part nor of a void of the endurant. For example, the space between the base and the bulb of a wine glass is typically not called a void. As identifying voids is thus somewhat arbitrary, *hosts-v* is a primitive, i.e., undefined relation. It can be refined in various ways [15], such as hosting a hole or a gap (depending on the host's internal connectedness); hosting a cavity, hollow, or tunnel (depending on the void's *opening*, see V-A12); or hosting an external or internal void (depending on the contact of the void's opening to the exterior or to other voids in the same host).

(V-A1′) $hosts\text{-}v(y, x) \to PED(y) \wedge \neg V(y) \wedge V(x) \wedge P(r(x), ch(y)) \wedge C_S(r(x), r(y)) \wedge$
     $PO(r(x), r(y))$ (hosting a void: relation between a void x and its physical host y)
(V-D) $V(x) \leftrightarrow \exists y[hosts\text{-}v(y, x)]$ (all voids are hosted)
(V-A11) $hosts\text{-}v(x, v) \to op(x, v) = r(v) \cdot (r(x) + r(v))'$
     (the opening of a void v is the part of its boundary that is not shared with its host)
(V-A12) $hosts\text{-}cav(x, y) \leftrightarrow hosts\text{-}v(x, y) \wedge op(x, y) \not\prec_{\dim} r(x)$ (hosting a cavity:
     hosting a void with an opening that is not of the dimension of its host's boundary)

We can ascribe voids to the level of granularity at which they occur [15]: a physical object directly hosts larger, macroscopic voids, while more minuscule voids are hosted by the object's constituent matter. The region encompassing all voids within an object's matter defines its *pore space*—it is the difference between the object's region and the region corresponding to its constituent matter. This two-level model of physical space can be extended to multiple levels to represent, for example, that a rock body constituted by grains of rock matter (first level) is also constituted by crystals (second level) and then molecules and atoms (third level). To accommodate multiple levels of granularity, we introduce a set of hosting relations expressing that a physical endurant hosts a void at the n-th level of granularity, written as $hosts\text{-}v_n(y, x)$ with n being a natural number.

$hosts\text{-}v(y, x)$ is then equivalent to $hosts\text{-}v_0(y, x)$. The relations $hosts\text{-}v_n$ will be properly defined and axiomitized in subsequent work. Here, the relation $hosts\text{-}v_{\mathrm{any}}(y, x)$ is used to express that y hosts the void x at *some* level of granularity—assuming a finite number of granularity levels. Then the entire void space of a physical endurant x, $voidspace_{\mathrm{all}}(x)$, is defined as the sum of the regions of all voids therein, independent of their level of granularity. This region must correspond to a void itself, which is intuitively hosted by the object's constituent matter and encompasses all the empty spaces within an object that can be filled by matter of the lowest granularity.

(V_{any}-D) $hosts\text{-}v_{\mathrm{any}}(y, x) \leftrightarrow \exists (n \geq 0)\big[hosts\text{-}v_n(y, x)\big]$

(hosting a void at any granularity level)

(V-A25) $PO(y, voidspace_{\mathrm{all}}(x)) \leftrightarrow \exists v[hosts\text{-}v_{\mathrm{any}}(x, v) \wedge PO(y, \mathrm{r}(v))]$

(the union of the voids of all levels of granularity define $voidspace_{\mathrm{all}}(x)$)

(V-A26) $mat(x) \wedge \neg ZEX(voidspace_{\mathrm{all}}(x)) \rightarrow \exists y, h[\mathrm{r}(y) = voidspace_{\mathrm{all}}(x) \wedge$
$hosts\text{-}v(h, y)]$ (some void is located in a material endurant's nonempty void space)

5 Generic Physical Containment

Three restrictions are imposed here on full physical containment relations. First, all participants must be physical endurants, i.e. physical objects, amounts of matter, or related features such as voids. Second, the full physical containment relation addresses spatial inclusion in physical space, eliminating other forms of non-physical or non-spatial containment, such as a file containing data, a book containing information, or my heart containing feelings. Third, only static containment relations are considered: change, motion, and other time-related ideas are beyond the scope of this paper. The predicate $fully\text{-}phys\text{-}contains(y, x)$ is used to denote this generic kind of full physical containment, which has the necessary condition that x's region is a subregion of y's convex hull. Any two physical endurants—material or immaterial—can participate in generic full physical containment. However, full containment in immaterial containers is still more restrictive: the containee x must not only be located within container y's convex hull, but within y's region. Figures 3(a) and (b) demonstrate the need for this restriction.

(FPCont-D) $fully\text{-}phys\text{-}contains(y, x) \leftrightarrow PED(x) \wedge PED(y) \wedge P(\mathrm{r}(x), \mathrm{ch}(y)) \wedge$
$\big[\neg mat(y) \rightarrow P(\mathrm{r}(x), \mathrm{r}(y))\big]$ (for x to be generically fully
physically contained in y, both x and y must be physical endurants and x's region
must be within y's convex hull and, if y is immaterial, within y's region)

Where feasible, we identify relationships to the containment relations presented in [8,5] and give explicit mappings (labelled **Mx**) to the relations from [5]. The relation $fully\text{-}phys\text{-}contains(y, x)$ resembles the union of Donnelly's generic and material-region containment predicates $CNT\text{-}IN_{\mathrm{g}}(x, y)$ and $CNT\text{-}IN_{\mathrm{mr}}(x, y)$ [8], as well as the $INSIDE(x, y)$ predicate defined in [5], but differs by its addition of physical typing.

(M1) $fully\text{-}phys\text{-}contains(y, x) \rightarrow INSIDE(\mathrm{r}(x), \mathrm{r}(y)) \wedge [CNT\text{-}IN_{\mathrm{g}}(x, y) \vee$
$CNT\text{-}IN_{\mathrm{mr}}(x, y)]$ (y generically physically contains x implies $INSIDE(x, y)$)

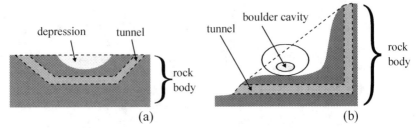

Fig. 3. Examples of a physical endurant spatially contained in the convex hull of an immaterial physical endurant, but not in its region. On the left, the depression is within the tunnel's convex hull, but not physically contained in the tunnel. On the right, the boulder's cavity is within the tunnel's and rock body's convex hulls, but is not physically contained in the tunnel or rock body.

For brevity, we subsequently drop the qualification "physical" with the understanding that *all* containment relations in this paper occur exclusively between two physical endurants. Two types of generic containment relations are distinguished first, using the physical dependence between participants as a discriminator: dependent containment and detachable containment, as illustrated by the initial division of full physical containment in the taxonomy in Figure 4. Intuitively, this division delineates whether the topological attachment between participants is necessary or accidental, respectively. Each of these two branches is subsequently delineated according to the (im)materiality of the container and containee, leading to two more horizontal levels of division in the taxonomy: in the first the container is either a material endurant or a void, and in the next the containee varies similarly. The last level in the taxonomy represents refinements of the previous distinctions; these are a partial selection from a greater number of possibilities, and they denote common uses that could eventually be further expanded. For example, containment involving material detachable containers is subdivided by the manner of enclosure, i.e. whether the container fully, partially, or incidentally encloses the containee. Containment with immaterial detachable containers is subdivided by the spatial positioning of the material containee, i.e. whether it splits or fills the void container; and containment in which the dependent container and containee are both im(material) is subdivided by parthood, i.e. whether the containee is a spatial part of the container. Other possible subdivisions are discussed where applicable within the relevant sections below. Note the taxonomy in Figure 4 is the product of a set of theorems, labelled **JPEDx**, which prove that the subrelations at each division are pairwise disjoint and jointly exhaustive with respect to the relation they refine.

While all kinds of physical endurants are valid participants in full physical containment relations, we restrict our study to voids as the only kind of immaterial physical endurant. In other words, voids are the only subtype of DOLCE's *DPF* category that are included. Other subtypes of *DPF*, such as shadows, might also participate in full physical containment relations, but they are beyond the scope of this paper.

6   Dependent Containment

Dependent containment is denoted as *dep-contains*(y, x), meaning that 'y physically contains x through some inherent physical dependency between x and y'. This

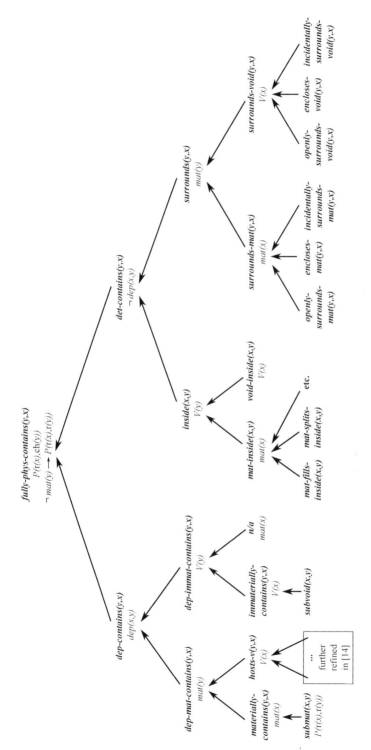

Fig. 4. The taxonomy of full physical containment relations

(a)   (b)   (c)   (d)   (e)

Fig. 5. Examples of two dependent physical endurants in a containment relation. The rock body (dark) (a) contains a submaterial part (light), (b) contains its constituent matter (light) as submaterial, (c) the rock body (dark and medium grey) materially contains a boulder (medium grey and white) without the latter being a submaterial, (d) hosts two holes: a depression and a cavity (both orange/medium grey), and (e) hosts a gap, a tunnel system (orange/medium grey).

(undirected) physical dependency is denoted by the symmetric primitive predicate $dep(x, y)$ meaning 'there is a dependency between the endurants x and y'. Dependent physical containment is further specialized in this section according to the types of physical endurants participating in the relation. Both containers and containees can be either material endurants or voids, resulting in four possible specializations. Figure 5 illustrates a selection of these dependent containment relations. It is noteworthy that while some specializations of $dep(x, y)$ might be strongly related to a form of ontological dependence, e.g. if x exists then y must exist [22], we interpret $dep(x, y)$ physically and topologically, rather than existentially, as seems more appropriate. The relationship to ontological dependence is left to future work. The four specializations of dependent containment are next described individually in Subsections 6.1–6.4.

(Dep-A1) $dep(x, y) \rightarrow PED(x) \land PED(y)$
　　　　　　　　　(dependence is a relation between physical endurants)
(Dep-A2) $dep(x, y) \rightarrow dep(y, x)$　　　　　(dependence is symmetric, i.e. undirected)
(DepCont-D) $dep\text{-}contains(y, x) \leftrightarrow fully\text{-}phys\text{-}contains(y, x) \land dep(y, x)$
　　　　(dependent containment is generic containment where x and y are dependent)

6.1　Material Containment

An obvious kind of dependent containment is *material containment*: a material endurant x that is physically contained in a second material endurant y, and whose region overlaps y's region, is materially contained in y. The most frequently encountered case occurs when x is located completely within y's region, which is known as x being a *submaterial* of y, as shown in Figures 5(a),(b), and (c). In this case, the dependency between x and y manifests itself in that all matter that constitutes x also constitutes y, at least in part, e.g. a particular rock formation within an aquifer, or the water in a bay and in the corresponding lake. The case of y materially containing x, but x not being a submaterial of y, is far less common but can occur, e.g., in a coral reef. The coral material—the containee—consists of dead as well as living corals. The dead corals are essentially a kind of rock matter, thus also part of the rock body that hosts and contains the reef, while the living corals are not part of the rock body.

(Dep-A3) $mat(x) \land mat(y) \rightarrow [dep(x, y) \leftrightarrow PO(\mathrm{r}(x), \mathrm{r}(y))]$
　　　　　　　　(material endurants are dependent iff they spatially overlap)

(MCont-D) $materially\text{-}contains(y, x) \leftrightarrow dep\text{-}contains(y, x) \wedge mat(x) \wedge mat(y)$
  (material containment is dependent containment between material endurants)
(SubMat-D) $submaterial(x, y) \leftrightarrow materially\text{-}contains(y, x) \wedge P(\mathrm{r}(x), \mathrm{r}(y))$
  (x is a submaterial of y iff y materially contains x and x is located in a subregion of y)
(SubMat-T1) $DK_1(x, y) \to submaterial(x, y)$
  (the constituent matter x of y is a submaterial of y)
(SubMat-T2) $P(\mathrm{r}(x), \mathrm{r}(y)) \wedge mat(x) \wedge mat(y) \to submaterial(x, y)$
  (a physical endurant whose region is located in part of another material endurant's region is a submaterial thereof)

The relation $submaterial(x, y)$ is the material version of Donnelly's *region contain-ment* [8], written as $CNT\text{-}IN_\mathrm{r}(x, y)$.

6.2   Hosting a Void

The second kind of dependent containment arises from *hosting a void*. Any void x hosted by a physical endurant y also depends on the host, because the void would not be present without the host. For this to be transitive over material containment, V-A27 specifies how voids are preserved when hosted by a material part [3].

(V-A27) $mat(y) \wedge mat(z) \wedge hosts\text{-}v(y, x) \wedge P(\mathrm{r}(y), \mathrm{r}(z)) \wedge \neg PO(\mathrm{r}(z), \mathrm{r}(x)) \to$
  $hosts\text{-}v(z, x)$ (any void x hosted by a material part y of z that is not even partially filled by z is also hosted by z)
(Dep-A4) $V(x) \wedge mat(y) \to [dep(y, x) \leftrightarrow hosts\text{-}v_{\mathrm{any}}(y, x)]$
  (a void and a material endurant are dependent iff they are in a hosts relation)
(DepCont-T1) $hosts\text{-}v_{\mathrm{any}}(y, x) \leftrightarrow dep\text{-}contains(y, x) \wedge V(x) \wedge mat(y)$
  (y hosts void x iff y is a material endurant that dependently contains void x)

6.3   Immaterial Containment

The third kind of dependent containment is *immaterial containment*. In order to define it, we first capture dependency between voids: two overlapping voids are dependent if their material hosts occupy overlapping regions. Then, a void that is dependent on another void, and that occupies a subregion of the other void, is also immaterially con-tained in the other void.

(Dep-A5) $V(x) \wedge V(y) \to \big[dep(x, y) \leftrightarrow PO(\mathrm{r}(x), \mathrm{r}(y)) \wedge \exists h_x, h_y[hosts\text{-}v(h_x, x) \wedge$
  $hosts\text{-}v(h_y, y) \wedge mat(h_x) \wedge mat(h_y) \wedge \big(P(\mathrm{r}(h_x), \mathrm{r}(h_y)) \vee P(\mathrm{r}(h_y), \mathrm{r}(h_x))\big)]\big]$
  (voids are dependent iff they overlap and have spatially nested material hosts)
(ImCont-D) $immaterially\text{-}contains(y, x) \leftrightarrow dep\text{-}contains(y, x) \wedge V(x) \wedge V(y)$
  (immaterial containment is dependent containment between voids)
(ImCont-T1) $immaterially\text{-}contains(y, x) \to P(\mathrm{r}(x), \mathrm{r}(y))$
  (immaterial containment requires that x is located in a subregion of y)

In immaterial containment, the dependency relation is more implicit because of the inclusion relation between their hosts' regions. Intuitively, if the larger host (or one of

---

[3] V-A27 is required in addition to the weaker condition previously imposed by V-A7 in [14,15].

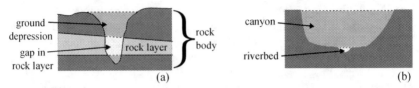

Fig. 6. Examples of two dependent voids in a containment relation. In (a) the gap hosted by the layer of rock is immaterially contained in the depression hosted by the rock body. In (b), the riverbed and canyon are both hosted by the rock body (voids do not need to be maximal), thus the riverbed is immaterially contained in the canyon.

its material parts) does not host the smaller void, the larger void would not exist in its present form. A general form of immaterial containment is illustrated in Figure 6(a). A special case is illustrated in Figure 6(b), where two voids have the same host; then we call the contained void a *subvoid* of the container void.

(SubVoid-D) $subvoid(x, y) \leftrightarrow immaterially\text{-}contains(y, x) \wedge \exists h[hosts\text{-}v(h, y) \wedge$
    $hosts\text{-}v(h, x)]$
        (x is a subvoid of y iff y immaterially contains x and they have a common host)

6.4   Can a Material Endurant Be Dependently Contained in a Void?

So far, we have considered three types of dependent containment between material or void endurants. The remaining combination involves a void as container and a dependent material endurant as containee. It is difficult to imagine this case unless the container void is the entire space of interest, e.g., physical space. However, that does not fit our framework requiring voids to be hosted by some physical endurant, because space as such is not physically hosted. Hence voids and physical space are distinct notions, with the latter being a container for all empirical entities, including voids. Physical space should then not be confused with related notions of voids, e.g. the term "outer space" is more precisely described as the space "that separates the planets, stars, and galaxies" [6, p. 132], which is in fact a gap (a void) hosted by celestial bodies. In essence, void containers cannot dependently contain material endurants. We can prove this in our formalization with the help of Dep-A4 and DepCont-D.

(Dep-T1) $dep\text{-}contains(y, x) \wedge mat(x) \rightarrow \neg V(y)$
                (a material entity cannot be dependently contained in a void)

The inclusion of physical space as a participant in containment relations remains a potential future task.

6.5   Classification of Dependent Containment

Because voids cannot dependently contain material endurants, the remaining three relations are exhaustive subrelations of $dep\text{-}contains(y, x)$. Their typing immediately entails they are disjoint relations.

(JEPD1) $[mat(x) \vee V(x)] \wedge [mat(y) \vee V(y)] \wedge dep\text{-}contains(y, x) \rightarrow$
        $[materially\text{-}contains(y, x) \vee immaterially\text{-}contains(y, x) \vee hosts\text{-}v(y, x)]$

(JEPD2) $\neg materially\text{-}contains(y,x) \lor \neg immaterially\text{-}contains(y,x)$
(JEPD3) $\neg materially\text{-}contains(y,x) \lor \neg hosts\text{-}v(y,x)$
(JEPD4) $\neg immaterially\text{-}contains(y,x) \lor \neg hosts\text{-}v(y,x)$

To complete the dependent containment taxonomy, we consider notions of containment in which the container type is fixed, but the containee can vary as material endurant or void: dependent material containment, $dep\text{-}mat\text{-}contains$, has a material container, and dependent immaterial containment, $dep\text{-}immat\text{-}contains$, has an immaterial container. Dependent material containment is thus specialized by the material containment and the hosting relations, and dependent immaterial containment has immaterial containment as its only feasible specialization, so that their extensions are equivalent.

(DepMCont-D) $dep\text{-}mat\text{-}contains(y,x) \leftrightarrow dep\text{-}contains(y,x) \land mat(y)$
(dependent material containment)
(DepImCont-D) $dep\text{-}immat\text{-}contains(y,x) \leftrightarrow dep\text{-}contains(y,x) \land V(y)$
(dependent immaterial containment)
(JEPD5) $dep\text{-}mat\text{-}contains(y,x) \leftrightarrow materially\text{-}contains(y,x) \lor hosts\text{-}v(y,x)$
(JEPD6) $dep\text{-}immat\text{-}contains(y,x) \leftrightarrow immaterially\text{-}contains(y,x)$
(JEPD7) $\neg dep\text{-}mat\text{-}contains(y,x) \lor \neg dep\text{-}immat\text{-}contains(y,x)$

7 Detachable Containment

Detachable containment holds between physical endurants that are not physically dependent. It occurs when the containee is physically contained strictly due to a non-necessary spatial arrangement. The term 'detachable' emphasizes that the two physical endurants are independent and, in principle, separable[4].

(DetCont-D) $det\text{-}contains(y,x) \leftrightarrow fully\text{-}phys\text{-}contains(y,x) \land \neg dep(y,x)$
(detachable containment is generic containment between independent endurants)

By the definitions DepCont-D and DetCont-D, detachable and dependent containment are subrelations of generic physical containment; now we can prove that they are jointly exhaustive, pairwise disjoint (JEPD) subrelations.

(JEPD8) $fully\text{-}phys\text{-}contains(y,x) \leftrightarrow dep\text{-}contains(y,x) \lor det\text{-}contains(y,x)$
(JEPD9) $\neg dep\text{-}contains(y,x) \lor \neg det\text{-}contains(y,x)$

To achieve our objective of classifying and formalizing the various kinds of *detachable containment*, we study its specializations that arise from the four combinations of material endurants and voids as container and containee.

7.1 An Endurant Inside a Void

Arguably, the most foundational form of detachable containment involves something being inside a void. This relation is denoted by the predicate $inside(x,y)$ understood

---

[4] Two physical endurants in a detachable containment relation may be inseparable for reasons other than a physical dependence, for example, because they are interlocked, glued or otherwise fused together, or because one endurant fully encloses the other.

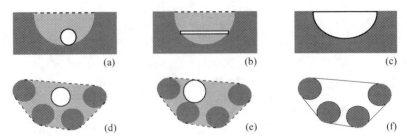

Fig. 7. A material endurant (white) being inside a void (orange/medium grey) and being surrounded by another material endurant (dark). The void is a hole in the top row, and a gap in the bottom row. The white containee may split (center column) or fill (right column) the void.

as 'the physical endurant x is spatially located within the void y'. Being located inside a void requires that the containee's region $r(x)$ is completely within the region of the void container, that is, $P(r(x), r(y))$, and not just within the container's convex hull. Beware that this relation switches the positions of the two parameters: x is inside the void y, that is, $inside(x, y)$, means y detachably contains x, that is, $det\text{-}contains(y, x)$.

(INSIDE-D) $inside(x, y) \leftrightarrow det\text{-}contains(y, x) \wedge V(y)$
$\qquad\qquad\qquad$ ($inside(x, y)$ is detachable containment in a void container)
(INSIDE-T1) $inside(x, y) \rightarrow P(r(x), r(y))$
$\qquad\qquad\qquad$ ($inside(x, y)$ requires x to be located within the region of y)

Next, the $inside(x, y)$ relation is further specialized by the containee's type.

A Material Endurant Inside a Void. If a material endurant is inside a void, we use the relation $mat\text{-}inside(x, y)$. This relation is equivalent to the only relation discussed in [8] that explicitly involves immaterial entities, namely *material-region containment*, $CNT\text{-}IN_{mr}(x, y)$.

(MINSIDE-D) $mat\text{-}inside(x, y) \leftrightarrow inside(x, y) \wedge mat(x)$
$\qquad\qquad\qquad$ ($mat\text{-}inside(x, y)$ denotes that the material endurant x is inside the void y)

More fine-grained specializations of $mat\text{-}inside(x, y)$ can be derived according to (1) the kind of container void, and (2) the location of the containee within the container void. The first choice includes distinctions based on (1a) the void's internal connectedness [3] (whether it is a simple or complex void), (1b) the host's internal connectedness (whether the void is a hole or a gap), (1c) the void's opening (whether the void is a cavity, a hollow, or a tunnel), and (1d) the void's connection to other voids within the same host (whether the void is an internal or external void). As these choices simply mirror the classification of voids in our earlier work, we invite the reader to consult [15] for the relevant axioms necessary to expand the taxonomy accordingly.

The second choice includes distinctions based on the containee's connection to (2a.I) the container's host, (2a.II) the outside, and (2a.III) other voids in the container's host. These three distinctions involve the container void's host and are thus not definable using the void alone. But they can be coupled with further distinctions that fall within the

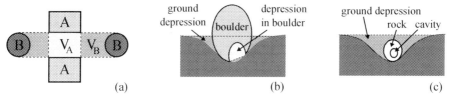

Fig. 8. Examples of a void detachably containing another void

second choice, namely whether the containee (2b.I) splits the void into disconnected parts, (2b.II) completely fills the void, or (2b.III) is "stuck" in the void. Except for "being stuck" in a void, which is an intricately shape-based relation, these relations are definable using our underlying mereotopological theory. Figure 7 gives examples of splitting and filling a void. *mat-fills-inside* is equivalent to the intersection of SUMO's [20] relations 'completelyFills' and 'properlyFills'.

(MSINSIDE-D) $mat\text{-}splits\text{-}inside(x, y) \leftrightarrow mat\text{-}inside(x, y) \wedge PP(\mathrm{r}(x), \mathrm{r}(y)) \wedge$
   $ICon(\mathrm{r}(y)) \wedge \neg ICon(\mathrm{r}(y) - \mathrm{r}(x))$ (a material containee x splits a void
   y iff it is located in a proper subregion of y, the void y is internally connected, and
   the part of y that is not occupied by the containee is not internally connected)
(MFINSIDE-D) $mat\text{-}fills\text{-}inside(x, y) \leftrightarrow mat\text{-}inside(x, y) \wedge \mathrm{r}(x) = \mathrm{r}(y)$
          (a containee fills a void if it is material and its region saturates the entire void)
(JEPD10) $\neg mat\text{-}splits\text{-}inside(x, y) \vee \neg mat\text{-}fills\text{-}inside(x, y)$

A Void Inside Another Void. Voids can be inside other voids without their hosts being in a containment relation. We capture this relation between two independent voids using the predicate $void\text{-}inside(x, y)$.

(VINSIDE-D) $void\text{-}inside(x, y) \leftrightarrow inside(x, y) \wedge V(x)$
          ($void\text{-}inside(x, y)$ is the relation of a void x inside another void y)

Figure 8(a) and (b) illustrate this relation. As the example (a) demonstrates, it is not required that the containee's host (A) spatially overlaps the container's convex hull $(\mathrm{r}(V_B) + \mathrm{r}(B))$. However, this is often the case, as shown in example (b). A special case of $void\text{-}inside(x, y)$ occurs when the convex hull of the containee's host is entirely contained in the container void, as demonstrated by Figure 8(c). Then *every* void within that particular host z is inside the void y.

(VINSIDE-T1) $P(\mathrm{ch}(z), \mathrm{r}(y)) \wedge mat(z) \wedge V(y) \to \forall v[hosts\text{-}v_{\mathrm{any}}(z, v) \to$
   $void\text{-}inside(v, y)]$ (if the convex hull of a material endurant z is completely
   contained in the void y, then every void v hosted by z is inside y as well)

Classification of *inside*. The two relations *mat-inside* and *void-inside* are JEPD subrelations of *inside*.

(JEPD11) $inside(x, y) \leftrightarrow mat\text{-}inside(x, y) \vee void\text{-}inside(x, y)$
(JEPD12) $\neg mat\text{-}inside(x, y) \vee \neg void\text{-}inside(x, y)$

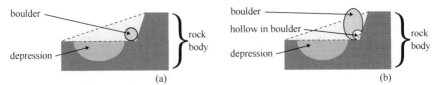

Fig. 9. Example of a (a) material endurant (a boulder) or a (b) void (a hollow in a boulder) located within the convex hull of a material endurant (the rock body) without being located within any of its voids (the depression). Hence, the boulder in (a) and the hollow in (b) are only incidentally surrounded by the rock body.

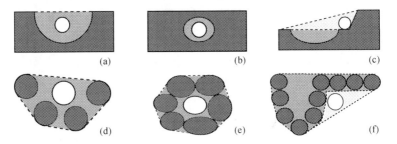

Fig. 10. A material endurant (dark) surrounding another material endurant (white) with the surrounding container's void (orange/medium grey) being either a hole (top row) or a gap (bottom row). From left to right: openly surrounds, encloses, and incidentally surrounds.

7.2   A Material Endurant Surrounding Another Endurant

When a material endurant y detachably contains another endurant x, we say 'y surrounds x' denoted by the predicate $surrounds(y, x)$.

(SUR-D) $surrounds(y, x) \leftrightarrow det\text{-}contains(y, x) \wedge mat(y)$
   ($surrounds(y, x)$ is detachable containment with a material container)

This relation is equivalent to *surround containment*, $CNT\text{-}IN_s(x, y)$, from [8]. It can be further refined according to the (im)materiality of the containee, as discussed in the remainder of this section.

A Material Endurant Surrounding Another Material Endurant. A material endurant detachably contained in another material endurant is *materially surrounded* by the latter. We denote this relation using the predicate $surrounds\text{-}mat(y, x)$, which is read as 'the material endurant y partially or fully surrounds the material endurant x'.

(MSUR-D) $surrounds\text{-}mat(y, x) \leftrightarrow surrounds(y, x) \wedge mat(x)$
   (material endurant y surrounds a material endurant x)

$surrounds\text{-}mat(y, x)$ does not rule out the case in which x's region is within y's convex hull, but outside any void (or set of voids) hosted by y. For example, the boulder in Figure 9(a) is surrounded by the rock body, yet is not contained in the rock body.

This can occur when the convex hull of a material container has spaces that are neither material nor voids. There is no principled way to identify voids amongst candidate spaces at this time [3], hence their identification is somewhat arbitrary. When another material endurant is located in such a non-void space, the relation is called *incidentally materially surrounds*, written as $incidentally\text{-}surrounds\text{-}mat(y, x)$.

(IMSUR-D) $incidentally\text{-}surrounds\text{-}mat(y, x) \leftrightarrow surrounds\text{-}mat(y, x) \land$
$\neg P(r(x), voidspace_{all}(y))$ (*y* incidentally materially surrounds *x* iff *y* materially surrounds *x*, but *x*'s region is not within *y*'s entire void space)

A special case of the material surrounds relation is *fully materially surrounds*, written as $encloses\text{-}mat(y, x)$. In this case, the containee must be located within some [3] cavity of the container. Examples of a container fully surrounding a containee are water in a closed bottle, or water in the subterranean cavity of a rock body.

(MENCL-D) $encloses\text{-}mat(y, x) \leftrightarrow surrounds\text{-}mat(y, x) \land \exists v[hosts\text{-}cav_{any}(y, v) \land$
$P(r(x), r(v))]$ (a material container *y* encloses a material containee *x*
iff it hosts a cavity wherein *x* is located)

The relation $encloses\text{-}mat(y, x)$ is a physical version of being *topologically inside*, $TOP\text{-}INSIDE(x, y)$, as defined in [5], for detachable material endurants.

(M2) $encloses\text{-}mat(y, x) \leftrightarrow TOP\text{-}INSIDE(r(x), r(y)) \land mat(y) \land mat(x) \land$
$\neg dep(x, y)$ (*encloses-mat* is the *topologically inside* relation for detachable material endurants)

It is obvious that *incidentally-surrounds-mat* and *encloses-mat* are disjoint relations. When the containee is neither incidentally surrounded nor enclosed by its material container, the container *openly materially surrounds* the containee, denoted as *openly-surrounds-mat(y, x)*. Notice that $openly\text{-}surrounds\text{-}mat(y, x)$ is agnostic about whether the containee can exit the container—we consider physical accessibility to be an associated but different relation. The three relations *openly-surrounds-mat*, *incidentally-surrounds-mat*, and *encloses-mat* form a set of JEPD subrelations of *surrounds-mat*. Examples for each are given in Figure 10.

(OMSUR-D) $openly\text{-}surrounds\text{-}mat(y, x) \leftrightarrow surrounds\text{-}mat(y, x) \land \neg encloses\text{-}$
$mat(y, x) \land \neg incidentally\text{-}surrounds\text{-}mat(y, x)$ (openly materially surrounds)
(JEPD13) $\neg incidentally\text{-}surrounds\text{-}mat(y, x) \lor \neg encloses\text{-}mat(y, x)$
(JEPD14) $surrounds\text{-}mat(y, x) \leftrightarrow openly\text{-}surrounds\text{-}mat(y, x) \lor$
$encloses\text{-}mat(y, x) \lor incidentally\text{-}surrounds\text{-}mat(y, x)$

Together, *openly-surrounds-mat* and *incidentally-surrounds-mat* are the physical version of being *geometrically inside*, $GEO\text{-}INSIDE(x, y)$, from [5], for detachable material endurants.

(M3) $openly\text{-}surrounds\text{-}mat(y, x) \lor incidentally\text{-}surrounds\text{-}mat(y, x) \leftrightarrow GEO\text{-}$
$INSIDE(r(x), r(y)) \land mat(y) \land mat(x) \land \neg dep(x, y)$
(*encloses-mat* and *incidentally-surrounds-mat* together are the geometrically inside relation for detachable material endurants)

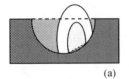  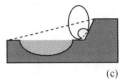

(a)   (b)   (c)

Fig. 11. Examples of a material endurant (dark) (a) openly surrounding, (b) enclosing, and (c) incidentally surrounding an independent void (light grey) hosted by the white object. The case for a hole (orange/medium grey) is depicted, but works equally for a gap.

A Material Endurant Surrounding a Void. If a material endurant y surrounds a void x that is independent of y, we write $surrounds\text{-}void(y, x)$ and say y *void-surrounds* x.

(VSUR-D) $surrounds\text{-}void(y, x) \leftrightarrow surrounds(y, x) \wedge V(x)$

(a material endurant y surrounds a void x)

Again, this kind of surrounds relation may be incidental, in which case we say y *incidentally void-surrounds* x and write $incidentally\text{-}surrounds\text{-}void(y, x)$. The enclosed and open analogues can also be defined: y detachably *void-encloses*, that is, fully surrounds the void x, if x is located in a cavity of y. When the void containee is neither incidentally surrounded nor enclosed by its material container, then it is *openly void-surrounded* by the container. Examples are given in Figure 11. These three subrelations of *surrounds-void* are the physical versions of *topologically* and *geometrically inside* involving a material container and a void, and they are JEPD subrelations.

(IVSUR-D) $incidentally\text{-}surrounds\text{-}void(y, x) \leftrightarrow surrounds\text{-}void(y, x) \wedge$
$\neg P(\mathrm{r}(x), \mathrm{voidspace}_{\mathrm{all}}(y))$ (y incidentally void-surrounds the void x
            iff y void-surrounds x but y's void space does not spatially contain x)
(VENCL-D) $encloses\text{-}void(y, x) \leftrightarrow surrounds\text{-}void(y, x) \wedge$
$\exists v[hosts\text{-}cav_{\mathrm{any}}(y, v) \wedge P(\mathrm{r}(x), \mathrm{r}(v))]$
            (a material container y encloses a void x iff it hosts a cavity wherein x is located)
(OVSUR-D) $openly\text{-}surrounds\text{-}void(y, x) \leftrightarrow surrounds\text{-}void(y, x) \wedge \neg encloses\text{-}$
$void(y, x) \wedge \neg incidentally\text{-}surrounds\text{-}void(y, x)$ (openly void-surrounds)
(JEPD15) $\neg incidentally\text{-}surrounds\text{-}void(y, x) \vee \neg encloses\text{-}void(y, x)$
(JEPD16) $surrounds\text{-}void(y, x) \leftrightarrow openly\text{-}surrounds\text{-}void(y, x) \vee$
$incidentally\text{-}surrounds\text{-}void(y, x) \vee encloses\text{-}void(y, x)$

Classification of *surrounds*. It is easy to see that the relations $surrounds\text{-}mat(y, x)$ and $surrounds\text{-}void(y, x)$ are JEPD subrelations of $surrounds(y, x)$. Both relations are further specialized into three JEPD subrelations as already discussed.

(JEPD17) $surrounds(y, x) \leftrightarrow surrounds\text{-}mat(y, x) \vee surrounds\text{-}void(y, x)$
(JEPD18) $\neg surrounds\text{-}mat(y, x) \vee \neg surrounds\text{-}void(y, x)$

The following three subrelations of *surrounds* are introduced for convenience only.

(ISUR-D) $incidentally\text{-}surrounds(y, x) \leftrightarrow$
$incidentally\text{-}surrounds\text{-}void(y, x) \vee incidentally\text{-}surrounds\text{-}mat(y, x)$
(ENCL-D) $encloses(y, x) \leftrightarrow encloses\text{-}void(y, x) \vee encloses\text{-}mat(y, x)$
(OSUR-D) $openly\text{-}surrounds(y, x) \leftrightarrow$
$openly\text{-}surrounds\text{-}void(y, x) \vee openly\text{-}surrounds\text{-}mat(y, x)$

7.3   Classification of Detachable Containment

If we only consider material endurants and voids as possible participants, then *inside* and *surrounds* are the only kinds of detachable containment, that is, they are JEPD subrelations of *det-contains*.

(JEPD19) $[mat(x) \lor V(x)] \land [mat(y) \lor V(y)]$
    $\to [det\text{-}contains(y, x) \leftrightarrow inside(x, y) \lor surrounds(y, x)]$
(JEPD20) $\neg inside(x, y) \lor \neg surrounds(y, x)$

8   Discussion

The various types of containment described above lead to several implications. First, material constitution is a special case of full physical containment.

(DetCont-T1) $DK_1(x, y) \to materially\text{-}contains(y, x)$

Second, the surrounds and inside relations are somewhat reciprocal: the two main kinds of the surrounds relation, namely *openly-surrounds(y, x)* and *encloses(y, x)*, can always be traced back to the relation of x being inside y's entire void space, voidspace$_{all}(y)$, which must exist according to V-A26. This accounts for the case where the surrounded entity x is distributed across voids at multiple levels of granularity within the container y. For example, an amount of water surrounded by a rock body can be located partly in the rock body's macroscopic voids and partly in the rock matter's microscopic voids. In the other direction, *inside(x, y)* entails that any host of y openly surrounds or encloses x.

(DetCont-T2) $openly\text{-}surrounds(y, x) \lor encloses(y, x) \to \exists z[inside(x, z) \land$
    $P(\mathrm{r}(z), voidspace_{all}(y))]$ (y openly surrounding or enclosing x requires x to be inside some void located in y's void space)
(DetCont-T3) $inside(x, y) \to$
    $\forall h [hosts\text{-}v_{any}(h, y) \to openly\text{-}surrounds(h, x) \lor encloses(h, x)]$
    (x being inside void y requires any host of y to openly surround or to enclose x)

Third, the most interesting relations, *mat-inside* and *surrounds-mat*, might be refined further if we take not only the relative location of the two participating endurants into account, but also the location of the indirectly involved host or void. DetCont-T2 and DetCont-T3 demonstrate that such a third, indirect, participant must exist. Consider the example of a material containee inside a void hosted by some material endurant, e.g. a rock in a hole hosted by a lakebed. In this case, it is possible to distinguish whether the containee is (non)tangentially inside the void or (non)tangentially surrounded by the host. Similar to the definition of (non)tangential parthood in [4], this can be expressed in a multidimensional setting using a definable relation of tangential containment, $TCont$, that specializes the spatial inclusion relation $Cont$ [14]. We can then express more sophisticated relations, such as whether the rock protrudes from the lake. A protruding rock is not only tangentially contained in the water body but is also in contact, by means of partial overlap or tangential contact, with the lake's exterior—the space where neither the lakebed nor the water body nor the water body's voids are located.

Lastly, all relations in our motivating example (Figure 1) can now be much more specifically expressed using the various kinds of full physical containment:

$openly\text{-}surrounds\text{-}mat(\text{LB}, \text{SWB})$ $openly\text{-}surrounds\text{-}mat(\text{LB}, \text{Rock})$
$hosts\text{-}v(\text{LB}, \text{Hole})$ $mat\text{-}inside(\text{SWB}, \text{Hole})$
$mat\text{-}inside(\text{Rock}, \text{Hole})$ $openly\text{-}surrounds\text{-}mat(\text{SWB}, \text{Rock})$
$materially\text{-}contains(\text{AQ}, \text{GWB})$ $materially\text{-}contains(\text{AQ}, \text{RM})$
$encloses\text{-}mat(\text{AQ}, \text{CT})$ $hosts\text{-}v_{\text{any}}(\text{AQ}, \text{Gaps})$
$mat\text{-}inside(\text{Gaps}, \text{GWB})$ $mat\text{-}inside(\text{Gaps}, \text{CT})$
$encloses\text{-}mat(\text{GWB}, \text{CT})$

By accounting for physical constraints in the definitions of the various containment relations we further ensured that the rock is not physically contained in the gaps, i.e. that $\neg fully\text{-}phys\text{-}contains(\text{Gaps}, \text{Rock})$, because while the rock's region is fully inside the gap's convex hull, it is not inside the gap's region.

We further have two examples of the incidental surrounds relation:

$incidentally\text{-}surrounds\text{-}mat(\text{AQ}, \text{SWB})$ $incidentally\text{-}surrounds\text{-}mat(\text{AQ}, \text{Rock})$

Some of the detachable containment relations, notably a void being inside another void, or a void being surrounded by a material endurant, are included here merely for completeness. They are not required to express the various containment relations in our motivating example, and will likely not play a prominent role in practical settings.

9   Conclusion

Full physical containment—the notion that one physical entity is completely inside or surrounded by another—plays a central role in describing many natural resources, especially water. To date, ontological representations of the full physical containment relation are limited to abstract space, incomplete, or they insufficiently incorporate voids. In this paper we argue that a thorough interpretation of this relation must accommodate both voids and material entities as containers and containees, and must account for the physical differences between voids and material entities. From this we develop a taxonomy, summarized in Figure 4, in which such containment relations are first differentiated according to the dependency between container and containee, and then according to their (im)materiality. This results in a delineation of full physical containment that is grounded in physical space and more comprehensive than prior efforts. The taxonomical distinctions are expressed using a formal multidimensional mereotopological theory, inspired by RCC, and are integrated into the DOLCE foundational ontology, as another step towards a rigorous hydro ontology. Potential future research directions include extensions to partial containment, and to containment across various levels of physical granularity, e.g. to better examine the relation between a containee held in the gaps of some matter (such as a contaminant), and the physical object constituted by the matter. Work is also underway to integrate the notion of physical space—distinct from voids—as a container for all physical entities.

Acknowledgements. We are grateful for the anonymous reviewers' thoughtful and meticulous comments, which helped improve the final version of the paper.

References

1. Boisvert, E., Brodaric, B.: GroundWater Markup Language (GWML) – enabling groundwater data interoperability in spatial data infrastructures. J. Hydroinformatics 14(1), 93–107 (2012)
2. Casati, R., Varzi, A.C.: Holes and other Superficialities. MIT Press (1994)
3. Casati, R., Varzi, A.C.: Parts and Places. MIT Press (1999)
4. Cohn, A.G., Bennett, B., Gooday, J.M., Gotts, N.M.: Qualitative spatial representation and reasoning with the Region Connection Calculus. GeoInformatica 1, 275–316 (1997)
5. Cohn, A.G., Randell, D.A., Cui, Z.: Taxonomies of logically defined qualitative spatial relations. Int. J. Hum.-Comput. St. 43(5-6), 831–846 (1995)
6. Dainton, B.: Space and Time. McGill-Queens Press (2001)
7. Donnelly, M.: Layered mereotopology. In: Int. Joint Conf. on Artif. Intell. (IJCAI 2003), pp. 1269–1274 (2003)
8. Donnelly, M.: Containment relations in anatomical ontologies. In: Symp. of the Amer. Medical Inform. Assoc. (AMIA 2005), pp. 206–210 (2005)
9. Egenhofer, M.J., Clementini, E., Di Felice, P.: Topological relations between regions with holes. Int. J. Geogr. Inf. Sci. 8(2), 129–144 (1994)
10. Egenhofer, M.J., Vasardani, M.: Spatial reasoning with a hole. In: Winter, S., Duckham, M., Kulik, L., Kuipers, B. (eds.) COSIT 2007. LNCS, vol. 4736, pp. 303–320. Springer, Heidelberg (2007)
11. Foxvog, D.: Cyc. In: Poli, R., Healy, M., Kameas, A. (eds.) Theory and Applications of Ontology: Computer Applications, pp. 259–278. Springer (2010)
12. Grenon, P.: Tucking RCC in Cyc's ontological bed. In: Int. Joint Conf. on Artif. Intell. (IJCAI 2003), pp. 894–899 (2003)
13. Grenon, P., Smith, B.: SNAP and SPAN: towards dynamic spatial ontology. J. Spat. Cogn. Comput. 4(1), 69–104 (2004)
14. Hahmann, T.: A Reconciliation of Logical Representations of Space: from Multidimensional Mereotopology to Geometry. PhD thesis, Univ. of Toronto, Dept. of Comp. Science (2013)
15. Hahmann, T., Brodaric, B.: The void in hydro ontology. In: Conf. on Formal Ontology in Inf. Systems (FOIS 2012), pp. 45–58. IOS Press (2012)
16. Hahmann, T., Grüninger, M.: A naïve theory of dimension for qualitative spatial relations. In: Symp. on Logical Formalizations of Commonsense Reasoning. AAAI Press (2011)
17. Hook, J.R.: An introduction to porosity. Petrophysics 44(3) (2003)
18. Kuhn, W.: An image-schematic account of spatial categories. In: Winter, S., Duckham, M., Kulik, L., Kuipers, B. (eds.) COSIT 2007. LNCS, vol. 4736, pp. 152–168. Springer, Heidelberg (2007)
19. Masolo, C., Borgo, S., Gangemi, A., Guarino, N., Oltramari, A.: WonderWeb Deliverable D18 – Ontology Library (final report). Technical report, ISTC-CNR, Trento (2003)
20. Niles, I., Pease, A.: Towards a standard upper ontology. In: Conf. on Formal Ontology in Inf. Systems (FOIS 2001), pp. 2–9. IOS Press (2001)
21. Sen, M., Duffy, T.: GeoSciML: development of a generic geoscience markup language. Comput. Geosci. 31(9), 1095–1103 (2005)
22. Simons, P.: Parts - A Study in Ontology. Clarendon Press (1987)
23. Tripathi, A., Babaie, H.: Developing a modular hydrogeology ontology by extending the SWEET upper-level ontologies. Comput. Geosci. 34(9), 1022–1033 (2008)
24. Walton, L., Worboys, M.: An algebraic approach to image schemas for geographic space. In: Hornsby, K.S., Claramunt, C., Denis, M., Ligozat, G. (eds.) COSIT 2009. LNCS, vol. 5756, pp. 357–370. Springer, Heidelberg (2009)

A Vocabulary of Topological and Containment Relations for a Practical Biological Ontology

Brandon Bennett[1], Vinay Chaudhri[2], and Nikhil Dinesh[2]

[1] University of Leeds, Leeds, UK
B.Bennett@leeds.ac.uk
[2] SRI International, Menlo Park, CA, USA
vinay.chaudhri@sri.com, dinesh@ai.sri.com

Abstract. We describe the development of a formal language for expressing qualitative spatial knowledge. The language is intended as a practical tool for knowledge representation and has been designed with the particular aim of encoding the qualitative spatial information found in an introductory college level biology textbook. We have taken a corpus-driven approach in which we first identify the requirements by analysing the sentences in the book, design the vocabulary, and then check its adequacy by applying it to model the sentences containing spatial knowledge. Our technical solution extends the well-known *Region Connection Calculus* with predicates for referring to surfaces, cavities and different forms of containment. We illustrate the application of this vocabulary in encoding sample sentences from the book, and give empirical results regarding the correspondence between the defined relations of our formal theory and the actual usage of vocabulary pertaining to 'surrounding', 'enclosure' and 'containment' in the textbook.

1 Introduction

Our goal is to represent the knowledge contained in a biology textbook (Reece et al. 2011) in a digital knowledge base that can provide intelligent functionality to support teachers and students in their exploration and understanding of biology. The diverse and multi-faceted nature of biological information make this an extremely challenging problem. In order to make progress we have taken the approach of identifying significant sub-domains of conceptual vocabulary, which can be modelled separately and then combined into a more general representational framework. Such sub-domains may correspond to particular topics within biology or to certain distinctive categories of more generic vocabulary. Among the generic categories, that of *spatial* concepts and relations play an extremely important role both in descriptions of biological structures and in explaining the processes that they undergo.

In this paper we develop a *Spatial Ontology for Knowledge Engineering* (SpOKE), which provides a vocabulary of predicates and relations that can be used by a variety of reasoning systems. Our specific technical solution extends the prior work on the *Region Connection Calculus* (RCC) (Randell et al. 1992) by adding concepts such as cavities, surfaces and different forms of containment.

Our design approach is corpus-driven in that the selection of formal vocabulary included in SpOKE is guided by the sentences in the textbook that contain spatial knowledge. We use the corpus both to identify the most commonly used spatial concepts

T. Tenbrink et al. (Eds.): COSIT 2013, LNCS 8116, pp. 418–437, 2013.

and relations and also to elicit the semantics of these terms by examining the range of spatial configurations to which they are applied. Some examples of the sentences we want to represent are the following:

(1)   The stroma *surrounds* a third *compartment*, the thylakoid *space*, delineated by the thylakoid membrane.

(2)   Integral proteins are *embedded* in the lipid bilayer; peripheral proteins are *attached* to the *surfaces*.

(3)   Parasites that live *within* their host, such as tapeworms and malarial parasites, are called endoparasites; parasites that feed on the *external surface* of a host, such as mosquitoes and aphids, are called ectoparasites.

In the preceding examples, the words with spatial content are shown in italics. For example, the use of *surrounds* in example (1) and *embedded* in example (3) requires representing the notions of containment. The word *space* in example (1) and *surface* in examples (2) and (3) requires representing cavities and surfaces. In Section 3, we discuss the frequency of occurrences of such words in more detail to provide an empirical justification for our focus on notions of containment, cavities, surfaces, and boundaries. And in Section 5 we shall present statistical information regarding the correspondence of our formal relations to particular usage of natural language terminology found in our sentence corpus.

Use of a precisely defined formal vocabulary for representing spatial knowledge has the advantage of clarifying the meaning of text by resolving ambiguities, and thus, substantially improving the understanding and communication of that knowledge. An ambiguity free representation also enables automated reasoning which can be used for answering questions.

As an initial set of requirements, we are interested in the questions of the following form:

Q1 Description: What is the structure of a stroma?
Q2 Comparison: What is the structural difference between a peripheral protein and an integral protein?
Q3 Relationship: What is the structural relationship between an ectoparasite and its host?

Here, the notion of structure encompasses both the mereological and spatial notions of parthood and other relations associated with spatial arrangement of parts. A complete answer to each of these questions will obviously also include non-spatial information. A preliminary investigation has recently shown that such capabilities can have a significant positive impact on students learning from a biology textbook [1].

In the next section we give an overview of previous research relating to our aims and methodology. Following that, we identify the requirements for adequately capturing the spatial knowledge contained in our textbook. We then introduce the vocabulary of SpOKE and explain how it can be used to formalise the spatial content of a wide variety of sentences taken from our corpus. We conclude with a summary of our contribution and discussion of directions for future work.

---

[1] http://www.aaaivideos.org/2012/inquire_intelligent_textbook/

2   Background and Related Work

Early work on representing spatial information for AI systems was conducted by Davis (1990), who used a rich formal language to describe a wide range of spatial situations. The *Region Connection Calculus* (RCC) (Randell et al. 1992) presents a much more constrained formal language for representing and reasoning about topological relationships. A system based on a similar set of topological relations had been proposed by Egenhofer and Franzosa (1991). Despite the limitation of topological relations, the theory has been found to be well suited to representing and reasoning with a wide variety of commonsense spatial information and has been widely used as a core theory for *Qualitative Spatial Reasoning* (QSR). Many authors working on qualitative spatial reasoning have argued that topological relations are of particular importance in the way that humans understand and communicate spatial information and the significance of such relations is supported by certain empirical cognitive tests, such as those conducted by Knauff et al. (1997).[2]

The basic RCC framework has been modified and extended in various ways (e.g. Cohn (1995), Bennett (2001a)) demonstrating it can be used to describe a huge variety of spatial properties and relations. An extension of the theory goes beyond pure topology by introducing a *convex hull* operator, which enables description of certain shape-related spatial properties and configurations.

From its inception, the field of QSR (and Knowledge Representation more generally) has been inspired and informed by naturalistic ways of thinking and talking about spatial information. On the one hand researchers have looked to natural language as a source or guide for the construction of formalised vocabularies (Mark and Egenhofer 1994, Shariff et al. 1998, Shihong et al. 2006). On the other hand designers of interfaces have been interested in finding natural language descriptions of formal relations that are used within spatial information systems (Clementini et al. 1993). However, establishing a correspondence between formal concepts and natural language expressions is by no means easy. Riedemann (2005) conducted a range of experiments that revealed a rather poor match between the names used in certain GIS query interfaces and typical human usage and understanding of vocabulary describing spatial relationships. Other areas in which the semantics of natural language expressions is difficult to relate to formal theories have been studied by a number of authors. For example, Bennett (2001b), Agarwal (2004), Montello et al. (2003), Bennett and Agarwal (2007). The problems arise partly from the generality of natural language concepts, which can typically be applied to a wide variety of specific situations, and partly from the pervasive presence of vagueness and ambiguity in natural language expressions. The combination of both generality and ambiguity is particularly hard to tease apart.

In the past few years, investigation of the complex relationship between the means of expression used in natural languages and the formal symbolism used spatial ontologies has become a major topic, with a variety of different approaches being pursued. Recent work in this area includes that of Bateman et al. (2010), Tenbrink and Kuhn (2011), Stock and Cialone (2011) and Davis (2012).

---

[2] However, authors have subsequently argued that topological aspects of spatial information are not always the most salient (Klippel et al. 2013).

Early work on developing formal vocabularies (i.e. ontologies) for biology and medicine tended to focus on constructing large concept hierarchies to capture semantic dependencies among the huge vocabulary of terms used in the biological and medical literature (e.g. Rector et al. (1997)). The specific meaning of particular terms was not directly addressed. However, it came to be recognised that lack of clarity regarding the meaning of terms (especially relational terms) can easily lead to incoherence and incorrect inference (Schulz et al. 2006, Schulz and Jansen 2009). Subsequently, a number researchers have worked on developing more secure foundations for biological ontology by rigorously defining vocabulary based on an axiomatised theory of key primitives. Because of their generality and significance, spatial relations have received particular attention. Work in this area includes that of Smith et al. (2005), Donnelly et al. (2006) and Bittner (2009). The work of Rosse et al. (2003) on the development of a *Foundational Model of Anatomy* (FMA) should also be mentioned. The FMA incorporates both a conceptual taxonomy and a physical partonomy, although it does not attempt to capture complex spatial relationships.

3   Requirements Specification

A formal vocabulary of spatial concepts and relations for a textbook needs to meet the following, often conflicting, desiderata:

(a) **Coverage**: The vocabulary should provide good coverage of the spatial content in the textbook.
(b) **Clarity**: The vocabulary should have a clear semantics.
(c) **Accessibility**: The language should lend itself to use by teachers and students who are not expert in formal semantics.

Much of the conflict between these desiderata stems from the variety and ambiguity of natural language descriptions. To achieve clarity and coverage we need to distinguish and make precise many alternative senses of common natural language vocabulary. But this results in a highly complex artificial vocabulary, which is inaccessible to those unaccustomed to the rigidity of formal representations. In this section, we narrow our focus to the specification of *coverage* — i.e. the proportion of information that can be represented out of the total body of spatial information that could be represented. We will return to the other desiderata in later sections.

The coverage of *all* spatial information is difficult to quantify because the variety of possible spatial information is not well defined. To address this difficulty, we consider a simpler requirement – *coverage of frequent spatial information*. We use lexical resources to identify frequently occurring spatial terms in the textbook and then provide ontological primitives to model these terms. Our corpus is taken from the general biology text book *Campbell Biology* (Reece et al. 2011). It consists of 30346 sentences, which are all the sentences occurring within the main body of the text book. To identify the sentences in the book containing spatial knowledge, we used two resources: a general purpose vocabulary of spatial terms (in short, SCorpus) (Stock et al. n.d.) and Wordnet (Miller 1990, Fellbaum 1998).

We searched the Campbell Biology textbook for occurrence of spatial verbs (spatial verbs (e.g., *encloses, attaches*), location and movement verbs (e.g., *appears, crouches*) from SCorpus,and identified 4263 sentences that use a spatial verb that we could use for further analysis. Amongst these sentences the following verbs occurred most frequently: *includes* (708 times), *contains* 399 times, *passes* (215 times), *attach* (177 times), *surrounds* (168 times) and *encloses* (56 times). We omit rest of the data for brevity. Motivated by this data, it was obvious that SpOKE needed to address the spatial notions related to inclusion, containment, surrounding and enclosing. Examples (1) and (2) considered earlier are illustrative of such information. A similar search for spatial nouns from SCorpus was too noisy to be useful because some of the spatial nouns such as *on* have too many senses, and SCorpus omits some key spatial terms such as *cavities* and *pores*. Therefore, we leveraged Wordnet to identify the spatial nouns.

The sense 1 of "location" in Wordnet has the meaning of "a point or extent in space" and we found it to be the most general sense which had a predominantly spatial interpretation. Sense 1 of "location" has 43 hyponyms, which occur 2442 times in Campbell, giving it a hyponym average of 56.8 (2442/43).[3] We then ranked each of the hyponyms of "location" by their hyponym average and selected the one with the highest average. The highest ranked hyponyms of "location" are senses 1 and 3 of "region" and sense 2 of "space". We then explored the hyponyms of these terms. The hyponyms of the sense 1 of "region" are "outside", "inside", "extremity" and "layer". The hyponyms of the sense 2 of "region" are "district", "area" and "domain". The hyponyms of "space" are "compartment", "angle", and "cavity". By searching these hyponyms in the textbook, we identified that the following were the most frequent: *outside, inside, extremity, layer, compartment* and *cavity*. These terms create the requirements to support incorporation of the concepts of "boundary" (which is a hyponym of extremity), "surface" (which is a hyponym of boundary) and "cavity" into the ontology.

4 Design

In the current work we confine our attention to purely spatial entities and relations, and consider only those properties of biological entities that are determined by their spatial extensions. A more comprehensive semantics would need to specify more fully the relation between material entities and space (which is particularly important in defining processes and moving entities). We aim to provide this in future developments of our ontology. Our design is based on the foundation of basic RCC-8 relations which we extend by adding vocabulary for cavities, surfaces, and different forms of containment.

4.1 Basic Topological Relations

Our core vocabulary of spatial relations is a set of topological relations adapted from the well-known RCC-8 relations of the *Region Connection Calculus* (Randell et al. 1992). To assess the accessibility of the basic RCC-8 relationships, we presented them visually

---

[3] We compute hyponym average as the ratio of number of times that word occurs in Campbell Biology divided by the number of hyponyms. We use averages rather than absolute counts to include words that have fewer hyponyms in Wordnet.

to the biologists and asked them to describe them in English. The biologists had significant difficulty in distinguishing between 'tangential' and 'non-tangential' part by any commonly used concise natural language expression. Therefore, we chose to combine the 'tangential' and 'non-tangential' parthood into a single relation, and likewise for the inverses of these relations. This gave us a core vocabulary of six relations. (Although the tangential/non-tangential part distinction is not made by the basic topological relation set, it can still be made within our language by describing the relationship between the surfaces of entities, as explained in Section 4.2.)

Another difference from RCC terminology is that our parthood relations correspond to what would be 'proper part' in RCC — i.e. is-spatial-part-of(x, y) excludes the case where x and y have equal spatial extent. This is because our experience in training biologists to use our representation suggests that the idea that an entity is a part of itself causes confusion. Thus our relations are as follows:

SpOKE	RCC equivalent
spatially-equal(x, y)	$EQ(x, y)$
has-spatial-part(x, y)	$TPPi(x, y) \vee NTPPi(x, y)$
is-spatial-part-of(x, y)	$TPP(x, y) \vee NTPP(x, y)$
partial-overlap(x, y)	$PO(x, y)$
abut(x, y)	$EC(x, y)$
apart(x, y)	$DC(x, y)$

The correspondence with the RCC-8 relations achieves our clarity requirement, since these relations have a precise semantics, which (notwithstanding certain subtleties — see e.g. Cohn and Varzi (2003)) may be defined in terms of basic set theory. Each of the SpOKE topological relations has a clear natural language reading. Moreover, like RCC-8, our set is pairwise disjoint and exhaustive, meaning that each (ordered) pair of objects (or spatial regions) stands in exactly one of our 6 relations.

The least natural of our relations is 'abut', which is rare in English, and does not occur at all in our biology textbook. Nevertheless, the corpus contains a large number of hyponyms, which do occur frequently.[4] These can be divided into those involving physical attachment and those where there is mere contact. In describing an actual situation one would normally want to distinguish between these cases (which would explain the rarity of 'abut'). In practice, our biologists did not seem to have any problem with the term 'abut' and were able to use it in the sense we intended. Nevertheless, in future development of our ontology it is likely that we will require that knowledge be encoded using more specific relations such as is-attached-to and touches, which are not purely spatial in meaning and thus beyond the scope of the current formalism.

Though, our aim has been to make our formal relations as accessible as possible, we do use a small number of relations in the development of our theory that do not correspond to ordinary English expressions. It is intended that knowledge encoders without expertise in formal representations should not be directly exposed to these terms, though they may occur in formal definitions of the vocabulary that the encoders will use.

---

[4] We identified 22 common hyponyms of 'abut' (e.g. 'touch', 'connect', 'join'), which occur in a total of 497 sentences (around 3.7% of all the sentences that contain spatial vocabulary).

4.2   Additional Entities and Primitive Relations

Although the 6 basic topological relations cover all possible situations, they do not differentiate between the many ways in which these relations may occur. One could simply introduce a range of more specific relations corresponding to significant sub-cases specified in terms of more particular topological and geometric properties. Although this could be made to work, the resulting system would be rather open-ended, and care would have to be taken to ensure that each new sub-case were correctly and unambiguously specified. SpOKE is designed according to a somewhat different approach. We first introduce additional entities corresponding to surfaces and cavities and a limited set of additional primitive relations, with a clear semantics, describing their relationship to 3-dimensional (3D) regions. These auxiliary entities and relations enable us to define refinements of the basic topological relations in a well-structured and systematic way.

Surfaces. Many descriptions of biological structures and processes refer to surfaces. Typically, these are surfaces of some 3D material entity or distinguished 3D region within such an entity. In general a 3D region has exactly one outer surface and may also have one or more inner surfaces (walls of closed cavities within the entity). We provide the following primitive predicates relating a 3D entity r to a 2D surface s:

-   has-outer-surface(r, s)
-   has-inner-surface(r, s)

Figure 1 illustrates a cross section of some material entity, which has three surfaces: one outer surface and two inner surfaces. The situation satisfies the formulae: has-outer-surface(x, s_1), has-inner-surface(x, s_2), has-inner-surface(x, s_3).

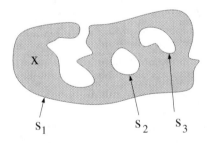

Fig. 1. Outer and Inner Surfaces, illustrated by cross-section of a 3D entity, x

Since every biological entity will have a unique outer surface, it will often be convenient to use a functional notation the-outer-surface-of(x), which avoids the need to use existential quantification to refer to this surface.

In addition to referring to the entirety of a surface, we may want to refer to only part of the surface. We may generalise the has-spatial-part and is-spatial-part-of relations so that they can apply to a pair of surfaces (as well as to a pair of 3-dimensional extents). This enables us to distinguish the tangential and non-tangential parts RCC. E.g. for tangential proper part (TPP) we have:

$$\mathsf{TPP}(x,y) \equiv_{def} \text{is-spatial-part-of}(x,y) \wedge$$
$$\exists s_1 s_2 s'[\text{has-surface}(x,s_1) \wedge \text{has-surface}(y,s_2)$$
$$\wedge \text{is-spatial-part-of}(s',s_1) \wedge \text{is-spatial-part-of}(s',s_2)]$$

where $\text{has-surface}(x,s) \equiv_{def} \text{has-outer-surface}(x,s) \vee \text{has-inner-surface}(x,s)$. Using this definition, it is easy to define its inverse relation $\mathsf{TPPi}(x,y) \equiv_{def} \mathsf{TPP}(y,x)$. Non-tangential (proper) parthood can then be defined by $\mathsf{NTPP}(x,y) \equiv_{def} \text{is-spatial-part-of}(x,y) \wedge \neg\mathsf{TPP}(x,y)$, and its inverse $\mathsf{NTPPi}(x,y) \equiv_{def} \mathsf{NTPP}(y,x)$.

More generally we can apply a wide variety of topological relations to pairs of surfaces. If we consider two surfaces that are both parts of the same larger surface, then the same set of 6 relations that was specified for 3D spatial extents will also apply to 2D spatial extents. However, if we consider surfaces that may be arbitrarily embedded with 3D space then a wider range of possibilities will arise. Furthermore, if we consider the spatial relationship between a surface and a 3D spatial region, yet more relationships occur. Thus, by combining our entity-surface relations with simple topological relationships a highly expressive vocabulary can be defined. In the current work we shall concentrate on the representational requirements of our corpus of biological knowledge and use our core topological and entity-surface relations to define a subset of key relationships that arise in this domain.

Cavities. The significance of holes and cavities in spatial information is widely recognised and has received considerable attention from the point of view of theoretical ontology (e.g. Casati and Varzi (1994)). But the ways in which such entities relate to natural language descriptions has received far less attention. Cavities are extremely common in biological entities and are referred to very often (either directly or implicitly) in descriptions of biological structures and processes.

Two fundamental kinds of cavity can be distinguished.

– Closed-cavity — a closed cavity is one whose **Cavity-wall** completely surrounds the cavity. Nothing can pass in or out of the cavity, except by crossing the cavity wall.
– Open-cavity — an open cavity is one that is only partially enclosed by its **Cavity-wall**. An open cavity will have one or more **Cavity-mouths**.

Out of 148 occurrences of the words 'cavity' and 'cavities' in our corpus, in 75 cases the cavity was open and in 63 it was closed. In the other cases the word was either used to refer generically to cavities (3) or the cavity (e.g. the mouth cavity) could change between being open or closed (2 cases). In a few cases (5) it was unclear whether the cavity referred to was open or closed.[5]

---

[5] The distinction is sometimes complicated by an issue of scale: if one considers small scale structure, a seemingly closed cavity may well have tiny openings. The appropriate granularity may vary according *context* — e.g. what other entities are being considered? Can they pass in and out of the cavity? Our current approach does not tackle such issues.

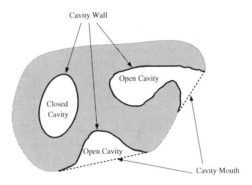

Fig. 2. Open and Closed Cavities (cross-section)

A typical example of a reference to a closed cavity is: ... *enclose the embryo in a fluid-filled amniotic* **cavity**; and a reference to an open cavity is made in: ... *the northern flying squirrel ... sleeps in a nest* **cavity** *in a hollow tree* .... There are also many references to specific body cavities, most of which have at least one opening (e.g. nasal and oral cavities).

We assume that every cavity must be a cavity of some material entity. Hence, the basic way in which cavities will be introduced into a spatial description is by relations associating an entity with a cavity. Since there is a fundamental difference between open and closed cavities, and this difference will normally be clear-cut in any given situation, it will be advantageous to build this distinction into the relations used to introduce cavities. Hence the following primitive relations are provided:[6]

- has-closed-cavity(x, c)
- has-open-cavity(x, c)

Like the topological relations, our semantics for the meaning of these formal relations is grounded in the RCC theory of spatial regions, which provides definitions for *convex-hull* (Randell et al. 1992) and for *well-connected* (Gotts 1994) region (a region such that all its interior points can be connected by a line that is fully interior to the region). Using these concepts, a *cavity* of a spatial region r is a maximal well-connected region r that is spatially included in the *convex-hull* of r but does not overlap with r. Closed cavities are those that are non-tangential proper parts of the convex hull, while open cavities are tangential proper parts.[7]

Since our formal spatial vocabulary relates to the purely spatial characteristics of a situation, it does not distinguish whether the entities involved are solid or liquid (indeed, biological information often involves complex tissues incorporating both fluid and solid

---

[6] It may also be useful to be able to assert that an entity or region is convex. This can be defined as meaning that the entity or region does not have any cavities.

[7] Some further technical issues must be addressed to handle special cases, but these are not of particular relevance to the current work.

materials). Thus, if a solid is surrounded by a liquid entity, it will be regarded as filling a cavity within the liquid entity. Unfortunately, this does not fit well with the normal usage of the word 'cavity' and it seems there is no established terminology in ordinary English to refer to the hole within a fluid within which a solid object is present.

According to our intended meaning of 'cavity', if an entity has a cavity then the contents of the cavity cannot be parts of the entity itself. However, certain cases arise where one would like to say that something is a cavity within another entity but would also like to say that its contents are part of that entity. For instance the inner region of many bones, where bone marrow is stored is known as the medullary cavity. However, in so far as a bone is considered to be an organ incorporating a quantity marrow as a part, the medullary cavity is *not* a true cavity of the bone. Rather it is a cavity of the osseous tissue (i.e. hard material) of the bone. This distinction is important because, whether or not a region and its contents are spatial parts of a larger entity is very significant in determining inferences about spatial structure.

Cavities have fundamental importance in many biological structures, affecting the function and processes carried out by organs and other biological entities ranging from digestive cavities to protein molecules. Biological descriptions contain a wide variety of vocabulary relating to cavities, including spatial relation verbs such as: 'fill', 'fit', 'surround', 'enclose', 'contain'; and others with dynamic or non-spatial aspects, such as: 'expand', 'contract', 'inflate', 'protect', 'isolate', 'conceal'.

4.3   Surrounding, Enclosure and Containment

In this section, we will concentrate on the relational verbs 'surround', 'enclose' and 'contain'. These words are relatively frequent, with a total of 673 of our corpus containing at least one of these words (or an inflected form); and, according to semantic analysis of these sentences by the authors, approximately 97% of these clearly express spatial content. This corresponds to around 17% of the total number of those sentences in Campbell that express spatial content, according to our sample analysis.

English has a variety of verbs and prepositions that are used to describe notions relating to containment of one form or another. These include: 'is in', 'is inside', 'is within', 'contains', 'is contained in/within', 'encloses', 'is enclosed by/within', 'surrounds', 'is surrounded by'. Also there are many different spatial configurations that intuitively constitute some kind of containment relationship but differ in the particular way that this occurs. Although each English phrase has certain typical usages and connotations there is a large overlap in the situations to which the different phrases are applied so that it seems unlikely that one can establish an uncontroversial mapping from phrases to spatial situations.

In this section we consider a number of formal definitions of precise spatial relationships that capture spatial conditions relevant to notions of surrounding, enclosure and containment.

is-in-closed/open-cavity-of.   The notion of an object being fully within a cavity is both easy to understand and frequently salient to spatial descriptions. Moreover, it is very easy to define in terms of our fundamental topological and cavity relations. Since

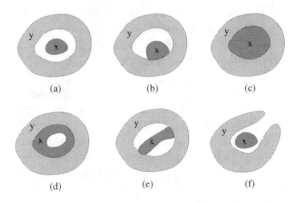

Fig. 3. Some representable relations involving a cavity

it is often relevant to biological functionality, we define separate relations for closed and open cavities:

$$\text{is-in-closed-cavity-of}(x, y) \equiv_{def}$$
$$\exists c[\text{has-closed-cavity}(y, c) \wedge \text{is-spatial-part-of}(x, c)]$$

$$\text{is-in-open-cavity-of}(x, y) \equiv_{def}$$
$$\exists c[\text{has-open-cavity}(y, c) \wedge \text{is-spatial-part-of}(x, c)]$$

In Figure 3, configurations (a)–(e) are all special cases of the relation is-in-closed-cavity-of(x, y), whereas (f) is a case of is-in-closed-cavity-of(x, y). Each of the sub-cases (a)–(e), and many more, can easily be distinguished using the primitive relations already presented to add further conditions. For instance, in case (a) we also have apart(x, y), and in case (b) we have abut(x, y).

There are many sentences in our corpus where the is-in-closed-cavity relation captures the spatial aspect of the meaning. For example: *The seed coat **encloses** the embryo ...; The stroma **surrounds** a third compartment, the thylakoid space, delineated by the thylakoid membrane.*

Cases (c) and (d) satisfy the condition that the outer surface of x is equal to the inner surface of y. In such cases, we say that x is *bounded* by y and define:

$$\text{is-bounded-by}(x, y) \equiv_{def}$$
$$\text{has-inner-surface}(y, \text{the-outer-surface-of}(x))$$

This means that y closely surrounds x so that the whole outside of x is in contact with an inner surface of y. A paradigm and very frequent case of this is where some material entity x is surrounded by some quantity of fluid y. Out of 86 cases of "surrounding" used in this sense, 52 were surrounding by fluid (e.g. *Most cells are **surrounded** by water*), 16 were surrounding by some kind of tissue sac or encasement and 18 referred to the surrounding environment in general terms without saying exactly what it is.

Figure 3 case (f) shows a case where x is within an open cavity of y. Many possible variations of is-in-open-cavity-of can be represented in SpOKE, but there is not space here to go into details. A distinction that we do consider in our evaluation is that an entity within an open cavity may either abut or be apart from the containing entity. We can also easily represent cases where an entity is only partially within an open cavity. In our analysis of word senses we also distinguish the relation is-bounded-by-2D to describe cases where a region of a surface is surrounded along its boundary by another surface region. This can apply to geographic areas or regions on a skin or membrane.

is-surrounding-part-of. A rather different sense in which one entity may be said to surround or enclose another is where the first is part of the other and extends around exterior. For example, *All cells are **enclosed** by a membrane that regulates the passage of materials between the cell and its surroundings.* Here, the cell membrane is a part of the cell and thus may be described as 'surrounding part'. Formally, we define:

$$\text{is-surrounding-part-of}(x, y) \equiv_{def} \text{is-spatial-part-of}(x, y)$$
$$\wedge \text{the-outer-surface-of}(x) = \text{the-outer-surface-of}(y)$$

has-spatial-part. We found that in many corpus sentences the verb 'contain' (and sometimes 'enclose') was used to refer to a situation satisfying our basic topological relation has-spatial-part. For example: *All cells **contain** chromosomes, carrying genes in the form of DNA.*; *The fruit protects the **enclosed** seeds.* In many cases where an entity x is described as 'containing' or 'enclosing' another entity y, and where the relation has-spatial-part(x, y) is an appropriate interpretation , it was found that entity y is actually a 'non-tangential proper part' of x (i.e. NTPP(y, x) or equivalently NTPPi(x, y) holds). As we have seen, this relation can also easily be captured in SpOKE.

Aggregate and Network Surrounding. A fairly common use of the verb 'surround' is to describe the location of an aggregate of entities in relation to another entity. E.g. *water molecules **surround** the individual sodium and chloride ions*; *the egg cell is **surrounded** by other cells.* In most cases an entity is surrounded in three dimensions; however there are also cases where the surrounding is relevant to a 2-dimensional plane or surface that is salient to the situation: e.g. ... *ciliated tentacles **surround** the mouth.*.

To provide a precise semantics for aggregate surrounding relations, we make use of the RCC *convex-hull* operator (Randell et al. 1992).[8] Although the surrounded entity is separate from each member of the surrounding aggregate, it is a (non-tangential) spatial part of the convex-hull of the aggregate.

Certain examples led us to introduce a relation network-surrounds for cases where something is surrounded by a network-like structure. For example: *The peri-tubular capillaries, which **surround** the proximal and distal tubules.* This is similar to the aggregate case, in that there are a lot of gaps, although the network's structure is connected. The convex-hull can also be used to provide the semantics for this relation.

---

[8] The α-volume of Edelsbrunner et al. (1983) can also be used to give a more refined notion of the space occupied by an aggregate.

5   Evaluation

Our development of the SpOKE vocabulary was continually informed by consideration of example sentences of our corpus, so that the choice of formal concepts used to describe notions of containment was tailored to accounting for the meanings of actual natural language expressions. Although, there is a large diversity of terminology relating to surrounding, enclosing and containment, and also a wide range of different spatial situations that are described by such terms, we eventually reached a point where it seemed that our formalism could capture most of the spatial information relating to this sub-domain that was being conveyed by our examples. In order to more thoroughly test the coverage of our formalism, we then undertook an exhaustive analysis of the occurrences of 'surround', 'enclose', 'contain' their cognates within the *Campbell Biology* text (Reece et al. 2011). As well as considering whether we could represent the spatial aspect of meanings of these sentences this analysis also enabled us to determine the frequency of the different senses of 'surround', 'enclose' and 'contain' employed in actual biological descriptions, in relation to the formal relations that were defined in Section 4.3. This evaluation involves a combination of statistical and semantic analysis, which we believe is an original and promising approach to using a natural corpus to inform the development of logic-based formal representation.

Specifically, we examined all 169 sentences including the words 'surround', 'surrounds', 'surrounded' and 'surrounding' that occurred in the text book. Similar sets were considered for cognates of 'enclose' (70 sentences) and 'contain' (434 sentences). In each case we considered which if any of the formally defined relations in our vocabulary corresponds to sense expressed by word as it occurs in the particular sentence. Of course this involves human judgement, and in some cases the relation corresponding to the intended meaning is not completely clear. In several cases we had to clarify our understanding of the situation being described by discussion with biology experts. But we found that, as long as we could determine an unambiguous interpretation of the particular spatial situation begin described, the correspondence with one or other of our formally defined relations was uncontroversial. Only around 2% of situations described using these vocabulary terms were deemed impossible to reliably classify.

In the following subsections we survey our results by giving examples from the corpus corresponding to each of our formal relations. We shall see that all the natural language terms are ambiguous, although some senses are much more common than others. Here, we are not concerned with distinguishing the polarity of the relationship involved (i.e. whether it should be modelled as $R(x, y)$ or $R(y, x)$). This varies according to the particular grammar employed (e.g. 'surrounds' *vs* 'is surrounded by') but is generally easy to determine. The intended sense of these terms seems to be largely independent of such grammatical details. However, the intended sense is highly dependent on assumed background information regarding the objects being described.

5.1   Senses of 'Surround'

The most common sense that we found for 'surround', is that corresponding to our formal relation **is-bounded-by**, where the outer surface of the surrounded entity coincides with an inner surface of the surrounding entity. Examples include: *Most cells are **surrounded** by water*; *The fruit tissues **surrounding** the seed are omitted in this diagram.* This sense accounted for just over half (50.9%) of all uses of the verb 'surround'. In a further 9.5% or cases, 'surround' was applied to a distinguished part of a 2D surface bounded by some other distinctive part of the surface: *Areas of private and public land **surrounding** reserves will have to contribute to the conservation of bio-diversity.* But 'surround' can also be used with a wide variety of other senses, as we now illustrate:

is-surrounding-part-of (0.6%): *The ovule has developed into a pine seed, which consists of an embryo , its food supply , and a **surrounding** seed coat derived from the integuments of the parent tree.* (Here, we take the seed coat as being part of the seed, which seems to be the most obvious interpretation.)

is-in-open-cavity-of (13%): *The pituitary gland, located at the base of the brain and **surrounded** by bone, consists of the posterior pituitary and the anterior pituitary. An exoskeleton is a hard case that **surrounds** the soft parts of an animal.* (We assume there will be some openings in the exoskeleton.) The is-in-open-cavity-of relation may be refined to distinguish whether the surrounded entity abuts or is apart from the surrounding cavity wall.

is-in-closed-cavity-of (3.6%): *The original cell replicates its chromosome, and one copy becomes **surrounded** by a durable wall.*

aggregated-around-3D (11.2%): *Such compounds dissolve when water molecules **surround** each of the solute molecules.*

aggregated-around-2D (1.2%): *The lophophore is a horseshoe-shaped or circular fold of the body wall bearing ciliated tentacles that **surround** the mouth.*

network-surrounds (4.1%): *Neurons of the central nervous system **surround** a continuous system of fluid-filled cavities.*

We also found 'surrounding' being used in a temporal senses (0.6%): *Whereas human females may be receptive to sexual activity throughout their cycles, most mammals will copulate only during the period **surrounding** ovulation*); and in metaphorical senses (2.4%): *Here, we take a closer look at a few of the controversies **surrounding** plant biotechnology in agriculture.*

5.2 Senses of 'enclose'

In terms of our formal vocabulary, the most common specific relation corresponding to 'enclose' and its cognates was found to be is-surrounding-part-of (32.9% of uses): *All cells are **enclosed** by a membrane ...*; of is-partially-surrounding-part-of (4.3%): *The shoot of a grass seedling is **enclosed** in a sheath called the coleoptile ....*

However, 'enclose' is often used when referring to to an 'enclosed' entity that is (wholly or partly) within a cavity. For example, *A virus is little more than DNA **enclosed** by a protective coat of protein.* and *The seed coat **encloses** the embryo, along with a food supply stocked in either the cotyledons or the endosperm.* Typically, such enclosure would be within a closed cavity (34.2%). These cases were divided equally into cases where we have is-bounded-by and those where we have is-in-closed-cavity but not is-bounded-by. We also found some cases (18.5%) involving an open cavity: *Their segmentally arranged vertebrae **enclose** the dorsal hollow nerve cord.*

'Enclose' is also occasionally (5.7%) used in the sense of NTPPi: *The fruit protects the enclosed seeds.* Since, seeds are normally considered parts of the fruit, the fruit both encloses the seeds and has them as a (non-tangential) part.

5.3 Senses of 'contain'

'Contain' turned out to occur very frequently in our corpus. It was present in 1.43% of all sentence in the text book, and in 11.53% of those sentences that contained any of our set of designated 'spatial' words. By far the most common usage was found to be to convey the sense has-spatial-part, as in *All cells **contain** chromosomes, carrying genes in the form of DNA.*

If an entity is described as being contained within a cavity, then in SpOKE we would have the relation has-spatial-part holding between the cavity and the contained item. The spatial extent of the cavity is the entire space bounded by the entity within which the cavity occurs. Thus if another entity is within the cavity, its spatial extent will be part of the spatial extent of the cavity. Thus, a piece of food within the stomach cavity would be considered as a spatial part of the cavity[9] Sentences, using 'contain' in relation to a cavity are surprisingly rare in our corpus (we suspect the corpus is atypical in this respect), but one example is: *The vestibular and tympanic canals **contain** a fluid called perilymph.*

In all, we found that the formal relation has-spatial-part could be appropriately applied to representing the situation described by 82.9% of all the occurrences of 'contain'. However, it was notable that some particular, more specific types of part relation were frequently associated with the uses of 'contain' in the our biology text corpus. In 22.2% of all its uses, 'contain' was applied to a molecular structure (e.g. *Most organic compounds **contain** hydrogen atoms.*); and in a further 22.2% of occurrences, the containment relation concerned ingredients of some compound or heterogeneous substance: *The spray **contains** irritating chemicals*). *Seawater, for instance, **contains** a great variety of dissolved ions, as do living cells.* The high frequencies of these types of usage of 'contain' are likely to be particular to our biology corpus, rather than typical of the everyday use of the word.

There is also a small but clear cut range of uses where 'contain' is used to describe one entity being within a cavity of another, where the cavity itself is not referred to: For example, *The DNA samples are placed in vials **containing** a fluid that emits flashes of light whenever ...* We speculate that in a more general context, this usage of 'contain' would be much more common than it is in the biology corpus (sentences such as *This box **contains** a hat,* seem to be very typical uses of 'contain'.) There are several cases where a cavity is described as containing a fluid: *The inner membrane encloses a second compartment, **containing** a fluid called stroma.* In this case the portion of fluid completely fills the compartment so the two would be (approximately) spatially equal.

Finally, there were a small number of cases (2.6%) where 'contain' was used in relation to a place or location rather than a physical containing entity. For example: *... a particularly rich fossil site ... **containing** a diversity of algae and animals, ...*

---

[9] Though of course the food would not be considered to be a *mereological* part of the stomach cavity. But that is another issue beyond the scope of the current paper.

5.4   Statistical Tabulation of Sense Frequencies

The results are summarised in Table 1 (below). We see that each of the verbs has a different distribution of senses, having one or two predominant meanings as well as less common uses conveying other senses. These results not only indicate the coverage of the SpOKE ontology over these kinds of natural language expression but also give insight into the degree of ambiguity in language. Moreover, they are very helpful in designing the wording used in guidelines for knowledge encoders, since they enable us to point out possible confusions between different interpretations of English vocabulary in relation to our formal terminology.

Table 1. 'surround', 'enclose' and 'contain': frequency of use of different word senses

Formal Relation	Word occurrences		
	Surround	Enclose	Contain
frequency in all sentences	0.56%	0.23%	1.43%
frequency in spatial sentences	4.49%	1.86%	11.53%
is-partially-surrounding-part	–	4.3%	–
is-surrounding-part-of	0.6%	32.9%	–
is-in-closed-cavity-of (inv)[10]	3.6%	17.1%	5.1%
is-bounded-by (inv)	50.9%	17.1%	2.6%
is-in-open-cavity-of (inv) + apart	4.7%	7.1%	0.9%
is-in-open-cavity-of (inv) + abut	8.9%	11.4%	1.7%
has-spatial-part (NTPPi)	–	5.7%	38.5%
has-spatial-part (molecular)	–	–	22.2%
has-spatial-part (ingredient)	–	–	22.2%
is-spatially-equal	–	–	0.9%
is-bounded-by-2D (inv)	9.5%	–	–
aggregated-around-2D	1.2%	–	–
aggregated-around-3D	11.2%	–	–
network-surrounds	4.1%	1.4%	–
is location of (inv)	–	–	2.6%
temporal	0.6%	–	–
metaphorical	2.4%	–	2.6%
hard to classify	1.2%	2.9%	1.7%

It will be seen that some of the formal relations in the table are followed by '(inv)'. This means that the formal relation acts in the inverse direction relative to the usual ordering of grammatical subject and object in the active use of the verb. For example, the sense is-bounded-by is noted as 'inv' because a sentence of the form 'x *surrounds* y' would (in its most common sense) be interpreted as is-bounded-by(y, x). (In these cases we could have specified the inverse of the formal relation but we find the inverses of these relations somewhat less intuitive to understand.)

---

[10] Note that the figures in this row exclude cases where the is-bounded-by relation holds even though is-bounded-by is a special case of is-in-closed-cavity-of. In other words the figures here are for cases represented by is-in-closed-cavity-of$(y, x) \land \neg$is-bounded-by(y, x).

The analysis of word senses used to produce Table 1 was done using our own judgement in considering the particular meaning of given sentences in the corpus. In some cases it was not immediately clear what was the actual situation being described, since we do not have expert knowledge of anatomy. In such cases we referred to the surrounding text and in some cases to other biology resources to find a clearer or more explicit account. We also looked at labelled diagrams which occur very frequently in *Campbell Biology* (Reece et al. 2011). Once one has a clear picture of the spatial situation that is being described, the choice of relation is in the great majority of situations clear cut. This is in part because, during the development of our system of relations, in cases where we encountered examples of situations that could not be described we augmented our relation set by defining additional relations. After several iterations of this process, we believe that our relation set is now sufficiently comprehensive and discriminating to capture a large proportion of the spatial information in our corpus. With regard to sentences containing cognates of 'surround', 'enclose' and 'contain' our representation captures relationship referred to by these words in over 98% of the cases we have examined. A total of 356 sentences were individually analysed, including all those containing 'surround' (169) or 'enclose' (70) and a large fraction (117/434) of the sentences containing 'contain'.

The fact that in nearly all cases we could confidently assign a particular sense to a given word occurrence may be due to the nature of the corpus. Most biological information is describing very specific kinds of spatial configuration, exemplified by actual biological entities. If we were to use a corpus made up of, say, fictional stories, we would expect to find many sentences where the intended meaning of spatial vocabulary is unclear, and there is no way to establish the situation that is being described (it is not even a real situation). Nevertheless, in the biology corpus there there are examples when the appropriate choice requires some discretion. For instance, we classified the case *..., most mollusks are **enclosed** in calcareous shells secreted by the mantle ...* as is-surrounding-part-of, but one could argue that the shell only partially surrounds the mollusk, since there will be small gaps. As mentioned above, the choice may depend on the *granularity* of the description, which will in turn depend on the context within which the spatial structure is being described.

6   Summary and Conclusions

As developed so far, SpOKE has only partially met its original goals of coverage, clarity and accessibility. Since we have focused our attention on the most frequently occurring spatial information in the textbook, we believe SpOKE provides good coverage of spatial knowledge, and adequately meets this goal. Because the core of SpOKE is based on well-known RCC-8 vocabulary, it partially satisfies the clarity requirement. More work needs to be done to fully satisfy the clarity requirement for surfaces and cavities. Although the notion of surface is well-defined from a mathematical point of view, an exact definition relating to physical entities has not been given. Various issues relating to granularity arise, which cause complications that are likely to be significant in the biological domain, because it involves entities on a wide range of different scales (e.g. an egg shell might be considered as forming a closed cavity fully surrounding the yolk and albumen of the egg — i.e. is-in-closed-cavity-of(*yolk*, *shell*) — but the shell is permeable to gas molecules such as oxygen, which is required for a bird embryo to survive). Although, it is not difficult to give a precise definition of cavity based on convexity (and hence on affine geometry), it is not clear that the resulting mathematical definitions would correspond well with the informal notion of cavity used in natural language (and in particular in biology texts). Finally, since we have not yet tested the vocabulary with biologists, we do not have sufficient data to claim that it will be accessible to them.

A further limitation of our work so far is that we have given relatively little consideration to computational properties. On the one hand we may argue that, since we are using standard first-order logic notation, reasoning can be done *via* generic reasoning systems. On the other hand, we may say that since we take topological relations based on RCC-8 as fundamental, we can make use of special purpose topological reasoning algorithms (Renz and Nebel 1999). Nevertheless, our use of surfaces and cavities as basic entities, will certainly give rise to particular characteristics of the problem of reasoning with SpOKE representations, which need further attention. Given the expressive power of our formal language, we would expect it to be intractable in a general sense. However, we believe that it may give good support for localised modes of reasoning using limited inference chains, and also for certain kinds global information completion and inconsistency checking (e.g. by using compositional reasoning).

In this paper, we introduced SpOKE, a *Spatial Ontology for Knowledge Engineering*. This is intended to provide a general purpose tool for symbolic representation of spatial information. But our particular goal was to develop a formal vocabulary suitable for modelling the spatial knowledge found in a college-level biology textbook. We used a novel methodology for developing a QSR representation, driven by the analysis of a large corpus of natural language sentences. For each of the formal spatial relations specified in SpOKE, we gave examples of specific sentences in the textbook that motivate that relation. We hope that others will find our methodology useful and further develop the idea of closely coupling the development of formal theories with the corpus-based evaluation techniques. We believe that this approach can be particularly useful in constructing theories that both have a clear semantics and can deal with the subtleties, ambiguities and enormous complexity found in natural language descriptions of spatial properties and configurations.

References

Agarwal, P.: Contested Nature of place: Knowledge Mapping for Resolving Ontological Distinctions Between Geographical Concepts. In: Egenhofer, M., Freksa, C., Miller, H.J. (eds.) GIScience 2004. LNCS, vol. 3234, pp. 1–21. Springer, Heidelberg (2004)

Bateman, J.A., Hois, J., Ross, R.J., Tenbrink, T.: A linguistic ontology of space for natural language processing. Artif. Intell. 174(14), 1027–1071 (2010)

Bennett, B.: A categorical axiomatisation of region-based geometry. Fundamenta Informaticae 46(1-2), 145–158 (2001)

Bennett, B.: What is a forest? on the vagueness of certain geographic concepts. Topoi 20(2), 189–201 (2001)

Bennett, B., Agarwal, P.: Semantic categories underlying the meaning of 'place'. In: Winter, S., Duckham, M., Kulik, L., Kuipers, B. (eds.) COSIT 2007. LNCS, vol. 4736, pp. 78–95. Springer, Heidelberg (2007)

Bittner, T.: Logical properties of foundational mereogeometrical relations in bio-ontologies. Applied Ontology 4(2), 109–138 (2009)

Casati, R., Varzi, A.: Holes and Other Superficialities. MIT Press, Cambridge (1994)

Clementini, E., Di Felice, P., van Oosterom, P.: A small set of formal topological relationships suitable for end-user interaction. In: Abel, D.J., Ooi, B.-C. (eds.) SSD 1993. LNCS, vol. 692, pp. 277–295. Springer, Heidelberg (1993)

Cohn, A.: A hierarchcial representation of qualitative shape based on connection and convexity. In: Kuhn, W., Frank, A.U. (eds.) COSIT 1995. LNCS, vol. 988, pp. 311–326. Springer, Heidelberg (1995)

Cohn, A.G., Mark, D.M. (eds.): COSIT 2005. LNCS, vol. 3693. Springer, Heidelberg (2005)

Cohn, A.G., Varzi, A.C.: Mereotopological connection. Journal of Philosophical Logic 32(4), 357–390 (2003)

Davis, E.: Representations of Commonsense Knowledge. Morgan Kaufmann, San Mateo (1990)

Davis, E.: Qualitative spatial reasoning in interpreting text and narrative, Spatial Cognition and Computation (2012) (to appear)

Donnelly, M., Bittner, T., Rosse, C.: A formal theory for spatial representation and reasoning in biomedical ontologies. Artificial Intelligence in Medicine 36(1), 1–27 (2006)

Edelsbrunner, H., Kirkpatrick, D.G., Seidel, R.: On the shape of a set of points in the plane. IEEE Transactions on Information Theory 29(4), 551–558 (1983)

Egenhofer, M., Franzosa, R.: Point-set topological spatial relations. International Journal of Geographic Information Systems 5(2), 161–174 (1991)

Egenhofer, M., Giudice, N., Moratz, R., Worboys, M. (eds.): COSIT 2011. LNCS, vol. 6899. Springer, Heidelberg (2011)

Fellbaum, C.: WordNet: an electronic lexical database. Bradford Books (1998)

Gotts, N.M.: How far can we 'C'? defining a 'doughnut' using connection alone. In: Doyle, J., Sandewall, E., Torasso, P. (eds.) Principles of Knowledge Representation and Reasoning: Proceedings of the 4th International Conference (KR 1994). Morgan Kaufmann (1994)

Klippel, A., Li, R., Yang, J., Hardisty, F., Xu, S.: The egenhofer cohn hypothesis or, topological relativity? In: Raubal, M., Mark, D.M., Frank, A.U. (eds.) Cognitive and Linguistic Aspects of Geographic Space. Lecture Notes in Geoinformation and Cartography, pp. 195–215. Springer, Heidelberg (2013)

Knauff, M., Rauh, R., Renz, J.: A cognitive assessment of topological spatial relations: Results from an empirical investigation. In: Frank, A.U. (ed.) COSIT 1997. LNCS, vol. 1329, pp. 193–206. Springer, Heidelberg (1997)

Mark, D., Egenhofer, M.: Calibrating the meanings of spatial predicates form natural language: Line-region relations. In: Proceedings of the Sixth International Symposium on Spatial Data Handling (SDH), pp. 538–553. Edinborough (1994)

Miller, G.: Wordnet: A lexical database for english. Communications of the ACM 38(11), 39–41 (1990)

Montello, D.R., Goodchild, M.F., Gottsegen, J., Fohl, P.: Where's downtown?: Behavioral methods for determining referents of vague spatial queries. Spatial Cognition & Computation 3(2-3), 185–204 (2003)

Randell, D.A., Cui, Z., Cohn, A.G.: A spatial logic based on regions and connection. In: Proc. 3rd Int. Conf. on Knowledge Representation and Reasoning, pp. 165–176. Morgan Kaufmann, San Mateo (1992)

Rector, A.L., Bechhofer, S., Goble, C.A., Horrocks, I., Nowlan, W.A., Solomon, W.D.: The grail concept modelling language for medical terminology. Artificial Intelligence in Medicine 9(2), 139–171 (1997)

Reece, J.B., Urry, L.A., Cain, M.L., Wasserman, S.A., Minorsky, P.V., Jackson, R.B.: Campbell Biology, 9th edn. Pearson (2011)

Renz, J., Nebel, B.: On the complexity of qualitative spatial reasoning: A maximal tractable fragment of the Region Connection Calculus. Artificial Intelligence 108(1-2), 69–123 (1999)

Riedemann, C.: Matching names and definitions of topological operators. In: Cohn, A.G., Mark, D.M. (eds.) COSIT 2005. LNCS, vol. 3693, pp. 165–181. Springer, Heidelberg (2005)

Rosse, C., Mejino Jr., J.L., et al.: A reference ontology for biomedical informatics: the foundational model of anatomy. Journal of Biomedical Informatics 36(6), 478–500 (2003)

Schulz, S., Jansen, L.: Molecular interactions: On the ambiguity of ordinary statements in biomedical literature. Applied Ontology 4(1), 21–34 (2009)

Schulz, S., Kumar, A., Bittner, T.: Biomedical ontologies: What part-of is and isn't. Journal of Biomedical Informatics 39(3), 350–361 (2006)

Shariff, A., Egenhofer, M., Mark, D.: Natural language spatial relations between linear and areal objects: the topology and metric of english-language terms. International Journal of Geographical Information Systems 12(3), 215–245 (1998)

Shihong, D., Qiao, W., Qiming, Q.: Definitions of natural-language spatial relations: Combining topology and directions. Geospatial Information Science 9(1), 55–64 (2006)

Smith, B., Mejino Jr., J.L.V., Schulz, S., Kumar, A., Rosse, C.: Anatomical information science. In: Cohn, A.G., Mark, D.M. (eds.) COSIT 2005. LNCS, vol. 3693, pp. 149–164. Springer, Heidelberg (2005)

Stock, K., Cialone, C.: Universality, language-variability and individuality: Defining linguistic building blocks for spatial relations. In: Egenhofer, M., Giudice, N., Moratz, R., Worboys, M. (eds.) COSIT 2011. LNCS, vol. 6899, pp. 391–412. Springer, Heidelberg (2011)

Stock, K., Pasley, R.C., Gardner, Z., Brindley, P., Moreley, J., Cialone, C.: Creating a corpus of geospatial language. Draft available on web at,
http://www.nottingham.ac.uk/~lgzwww/contacts/
staffPages/kristinstock/documents/CorpusPaperv1.4forweb.pdf

Tenbrink, T., Kuhn, W.: A model of spatial reference frames in language. In: Egenhofer, M., Giudice, N., Moratz, R., Worboys, M. (eds.) COSIT 2011. LNCS, vol. 6899, pp. 371–390. Springer, Heidelberg (2011)

A Geo-ontology Design Pattern
for Semantic Trajectories

Yingjie Hu[1], Krzysztof Janowicz[1], David Carral[2], Simon Scheider[3],
Werner Kuhn[3], Gary Berg-Cross[4], Pascal Hitzler[2], Mike Dean[5],
and Dave Kolas[5]

[1] Department of Geography, University of California Santa Barbara, USA
{yingjiehu,jano}@geog.ucsb.edu
[2] Kno.e.sis Center, Wright State University, USA
{carral.2,pascal.hitzler}@wright.edu
[3] Institute for Geoinformatics University of Münster, Germany
{simon.scheider,kuhn}@uni-muenster.de
[4] Spatial Ontology Community of Practice (SOCOP), USA
gbergcross@gmail.com
[5] Raytheon BBN Technologies, USA
{mdean,dkolas}@bbn.com

Abstract. Trajectory data have been used in a variety of studies, including human behavior analysis, transportation management, and wildlife tracking. While each study area introduces a different perspective, they share the need to integrate positioning data with domain-specific information. Semantic annotations are necessary to improve discovery, reuse, and integration of trajectory data from different sources. Consequently, it would be beneficial if the common structure encountered in trajectory data could be annotated based on a shared vocabulary, abstracting from domain-specific aspects. Ontology design patterns are an increasingly popular approach to define such flexible and self-contained building blocks of annotations. They appear more suitable for the annotation of interdisciplinary, multi-thematic, and multi-perspective data than the use of foundational and domain ontologies alone. In this paper, we introduce such an ontology design pattern for semantic trajectories. It was developed as a community effort across multiple disciplines and in a data-driven fashion. We discuss the formalization of the pattern using the Web Ontology Language (OWL) and apply the pattern to two different scenarios, personal travel and wildlife monitoring.

1 Introduction

The term *trajectory* is used in many different contexts. It can be defined as a path through space on which a moving object travels over time. For example, the path of a projectile can be described by a mathematical model which returns the idealized position of the projectile at each point in time. In other cases, such as studying animal movement, trajectories are defined by a sparse set of temporally-indexed positions or "fixes", while the exact path between these fixes is unknown and has to be estimated, e.g., by using Brownian Bridges [23].

T. Tenbrink et al. (Eds.): COSIT 2013, LNCS 8116, pp. 438–456, 2013.

In some of these cases, the fixes have no specific meaning and are purely an artifact of the used positioning technology, restrictions imposed by energy requirements, area coverage, and so forth. In other cases, the fixes denote important activities and decision points, and researchers may be interested in labeling and classifying them. We will refer to the latter cases as *semantic trajectories* [1]. An example of such semantic trajectories occurs in location-based social networks (LBSN), where the fixes are user check-ins to places and the labels are the names and types of these places [39,28]. The user's location between check-ins is unknown. The distinction between semantic trajectories and other fixes is not always crisp. For instance, the OCEARCH's Global Shark Tracker[1] can only record *pings* of tagged sharks if they surface for a certain amount of time. One could argue that these fixes do not carry any semantics and just reflect technological limitations of the used positioning technology. However, they reveal some important information, namely the event of surfacing and, thus, can be meaningfully labeled. Summing up, with the fast development of location-enabled mobile devices, it has become technically and economically feasible to record a large number of (semantic) trajectories generated by vehicles, animals, humans, and other moving objects (e.g., from the Internet of Things). While GPS has been widely used to detect the outdoor locations of moving objects, WiFi[11,10], RFID[31], and other sensor-tracking techniques have been employed to extend the geo-locating capability to indoor environments [20,32].

There are multiple ways to publish trajectory data in order to make it accessible to others. During the last few years, Linked Data [4] has become one of the methods of choice. It opens up data silos by providing globally unique identifiers for physical objects and information entities, links between them, and semantic annotations to foster discovery, retrieval, and integration. The semantic annotations are realized using shared vocabularies. In a highly heterogeneous and dynamic environment, such as the Web, arriving at commonly agreed and stable domain ontologies is a difficult task and progress has been slow over the last years. Foundational ontologies, such as DOLCE [16], have been usefully applied as a common ground for geo-ontologies [7]. In a Linked Data context, however, foundational ontologies tend to be too abstract and introduce a hardly comprehensible set of ontological commitments difficult to handle for laypersons. Ontology design patterns [14] have emerged as more flexible, reusable, manageable, and self-contained building blocks that help to model reoccurring tasks and provide common ground for more complex ontologies. To reach a higher degree of formalization and further improve interoperability, these patterns can be combined and ultimately aligned with foundational ontologies that act as glue between patterns. An increasing number of geo-ontology design patterns has been developed as joint community effort by domain experts and ontology engineers during so-called Geo-Vocabulary Camps (GeoVocamps) [9,8].

In this paper, we propose an ontology design pattern for semantic trajectories and demonstrate its applicability. While trajectory ontologies have been

---

[1] http://sharks-ocearch.verite.com/

developed before [37,34], they were confined to specific application areas and were not optimized for querying Linked Data, e.g., via the GeoSPARQL query language [2]. The proposed pattern is developed with two major goals. First, it should be directly applicable to a variety of trajectory datasets and, thus, reduce the initial hurdle for scientists to publish Linked Data [3]. Secondly, it should be easily extensible, e.g., by aligning to or matching with existing trajectory ontologies, foundational ontologies, or other domain specific vocabularies.

The remainder of this paper is structured as follows. First, we introduce some background materials and related work supporting the understanding of the proposed ontology design pattern. Section 3 introduces the conceptual foundation for the pattern. Next, in section 4, we discuss the formalization of the pattern using the Web Ontology Language (OWL). In section 5, we demonstrate how to annotate two trajectory datasets using the proposed pattern, in order to evaluate its applicability. We conclude by summarizing our results and pointing out directions for future work.

2   Background and Related Work

In this section we introduce related research and background materials relevant for the presented geo-ontology design pattern.

2.1   Semantic Trajectories

A trajectory consists of a series of spatiotemporal points generated by the moving object. These points are often represented as $\{x_i, y_i, t_i\}$ (with x_i, y_i denoting a position in the 2D geographic plane, and t_i representing a time point) or $\{x_i, y_i, z_i, t_i\}$ (with z_i denoting the elevation information) if the trajectory should be analyzed in a 3D space. While such spatiotemporal points support an exploration of the mobility pattern of a moving object [13], many applications require an understanding of additional information to interpret the trajectories. For example, a traffic analysis based on car trajectories may not be able to derive meaningful results without incorporating information about road networks. Similarly, studies on bird migration patterns may require an understanding of the features of the particular bird species (e.g., their body sizes, food sources, and competitors) as well as information about the weather conditions during their flight.

Semantic trajectories fill this gap by associating the spatiotemporal points and segments with geographic and domain knowledge, as well as other related information [5,1,33]. These semantically enriched trajectories facilitate the discovery of new knowledge, which otherwise may not be easily found. For example, human trajectories are best understood when the positional fixes can be labeled with activities performed at these places and the places are associated with semantic categories such as *restaurant* or *grocery store*.

2.2   Ontology Design Patterns

Ontology design patterns are derived from the common conceptual patterns that emerge in different domains when solving different tasks. A good example (given by Gangemi) is the *participation pattern*, which can be observed in enterprise models, software management, fishery, and many other domains [14,17,29,15]. Ontology design patterns capture the common conceptualization among knowledge engineers and domain experts, and can serve as building blocks or strategies for the design of future (more complex) ontologies. These flexible and self-contained building blocks appear to fit the needs to model knowledge for domain applications where more complicated and abstract ontologies may be difficult to apply. Lately, ontology design patterns have become popular in the geospatial semantics community [12,9,8]. A series of so-called Geo-Vocabulary Camps (GeoVoCamps[2]) have been held to promote the joint development, documentation, and testing of geo-patterns. These camps try to bring ontology engineers and domain experts together for two to three days to discuss and implement pattern ideas. Usually, patterns are conceptually developed during the camp and implemented and tested later. A follow-up camp (potentially with different participants and also embarking on new patterns) evaluates and refines the results. This paper is the result of such a community process.

There are two major types of ontology design patterns: logical patterns and content patterns, though other types have also been discussed in the literature [9,14]. Logical patterns deal with issues arising from the formal semantics of a knowledge representation languages, and therefore are independent from application domains. Content patterns often focus on domain knowledge and are used to model recurrent domain facts. The ontology design pattern proposed in this work is a content pattern addressing the design of classes and properties found commonly in semantic trajectories across application domains.

2.3   Semantic Trajectory Ontologies

As an ontology design pattern reflects a common conceptualization of domain experts with respect to a modeling problem, it is worthwhile to review the existing semantic trajectory ontologies in order to ensure consistency. A conceptual view on trajectories has been proposed by Spaccapietra et al. [34] who decompose a trajectory into a series of moves and stops. This stop-move conceptualization has been applied in several other trajectory studies, and the stops and moves are often coupled with corresponding geographic information to help interpret them [1,18,30]. Transportation networks are an important type of geographic information which is often utilized to make sense of the trajectories [35,27]. Other geo-data, such as those on Points Of Interest (POI), weather, land use, vegetation, and habitats, have also been employed to improve the understanding of trajectories [34,6,19,38]. While geographic information is the key contributor and commonality, domain knowledge is included in trajectories and their ontologies to help understand domain facts [37].

---

[2] http://vocamp.org/wiki/Main_Page

3   Conceptual Foundation and Motivation for the Pattern

Creating an ontology design pattern requires a generic use case (GUC), general enough to capture the recurring issues in cross-domain projects [14]. *Competency questions* are often utilized to discover and refine the GUC in a particular domain. A competency question is a typical query that a domain expert might want to submit to a knowledge base to complete a particular task [17]. A good ontology design pattern should define all and only the conceptualizations that are necessary to answer the competency questions shared by domain experts.

We conceptualize and motivate the trajectory design pattern using competency questions. For readability, we will use particular examples, e.g., related to animal movement, without restricting the pattern to those application areas. With the spatiotemporal information of the points contained in the trajectories, we can answer queries such as these:

Question 1. *"Show the birds which stop at x and y"*

Question 2. *"Show the birds which move at a ground speed of 0.4 m/s"*

While spatiotemporal points may only provide a basic level of movement understanding, geographic information, such as on points of interests, allows queries like:

Question 3. *"Show the trajectories which cross national parks"*

Domain knowledge is another important information source, enabling queries like:

Question 4. *"Show the trajectories of the birds which are less than one year old".*

In addition, information about the data creator (such as the location-tracking device) is necessary to answer related queries such as:

Question 5. *"Show the trajectories captured by Gamin GPS"* or *"show the trajectories generated by iPhone users".*

In order to answer these questions, an ontology design pattern needs to distinguish a number of relations. We introduce these abstract relations in Table 1, before formalizing them in Section 4. First, in order to query trajectories by spatiotemporal positions (Question 1), we need to segment the trajectory through fixes, which, in turn, require spatial and temporal reference systems (*hasSegment, startsFrom, endsAt, hasLocation, atTime*). Second, in order to answer questions about movement properties (Question 2), we need attributes for fixes and segments (*hasAttribute*). Third, in order to describe the geography of a trajectory (Question 3), it needs to be related to relevant geographic features. In the simplest case, this can be done by relating fixes to geographic features. Fourth, in order to identify and categorize a moving object (such as a bird, Question 4), we need to relate it to segments of the trajectory (*isTraversedBy*).

Table 1. Basic relations needed to answer the competency questions

Name	Type	Explanation
hasSegment	*SemanticTrajectory* × *Segment*	A segment of a trajectory
startsFrom	*Segment* × *Fix*	The from fix of a segment
endsAt	*Segment* × *Fix*	The to fix of a segment
· *isTraversedBy*	*Segment* × *MovingObject*	A moving object traversing a segment
hasLocation	*Fix* × *Position*	The spatial position of a fix
atTime	*Fix* × *TemporalThing*	The temporal position of a fix
hasAttribute	*Segment* ⊔ *Fix* × *Attribute*	An attribute of a segment or a fix
hasCreator	*Fix* × *Source*	The creator of a fix

Fifth, in order to query for properties of the trajectory data creator (Question 5), we need a relation between fixes and their source (*hasCreator*).

Figure 1 illustrates the creation of a semantic trajectory by integrating relevant knowledge with a person's daily trajectory. From top to bottom, the trajectory is enriched with a variety of data and acquires the capability to answer more advanced queries. This example demonstrates a general use case which is designed to support as many kinds of queries as possible. In a particular application, it may not be necessary to include some information (e.g., about the data creator). The pattern is designed such that information can be added at different levels and resolutions, e.g., by sub-typing. This idea will be discussed in more detail in the following section.

4   OWL Formalization

In this section, we present our geo-ontology design pattern, based on the previously described conceptual foundations. A schematic view of the pattern is shown in Figure 2. In the following paragraphs, we respectively discuss the classes and properties within the pattern and formally encode them using Web Ontology Language (OWL). We make use of Description Logics (DL) [22] notation, as we believe this improves the readability and understandability of the axioms presented. To encode our pattern, we make use of the logic fragment DLPE as defined in [8], which allows for tractable reasoning. Note that tractable reasoning is important for producing an efficient implementation of the pattern.

Fix. A fix is defined as a spatiotemporal point $\{x_i, y_i, t_i\}$ indicating the position of a moving object at an instant of time. It can be captured by a location measurement device (such as a GPS), but can also involve other types of points, such as check-ins on location-based social networks (LBSN) or centroids of regions passed by a moving object, but not automatically recorded by a device. Fixes are the atoms of the presented ontology design pattern: they not only capture the spatiotemporal information of a trajectory, but also link the

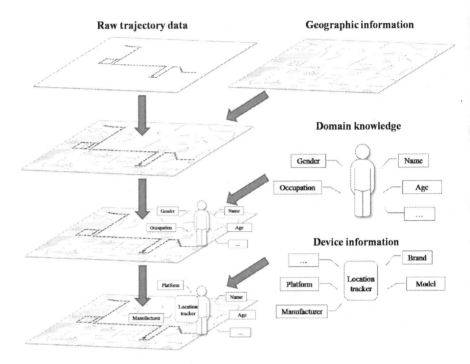

Fig. 1. An example semantic trajectory of an individual's daily activities

segments and provide information on attributes and metadata. By Axiom 1, a fix is enforced to have a timestamp and a position and to belong to a trajectory.

$$Fix \sqsubseteq \exists atTime.OWL\text{-}Time\text{:}Temporal\ Thing \sqcap \exists hasLocation.Position$$
$$\sqcap \exists hasFix^-.SemanticTrajectory \tag{1}$$

The number of fixes in a trajectory depends on the requirements of the particular application. Resolution can be as coarse as containing only the important trajectory points (e.g., check-ins on LBSN), but can also be as fine as including points recorded according to a sampling rate of the location-tracking device. This scale-neutral design makes the pattern flexible, allowing users to model trajectories at different scales. Real-world examples of fixes include a stop of a migration flock in a wet land area, an intersection a vehicle has passed, or a restaurant visited.

Segment. A segment is defined by a starting fix $\{x_i,y_i,t_i\}$ and an ending fix $\{x_j,y_j,t_j\}$. $t_i < t_j$, $\{x_i,y_i\}$ is not necessarily different from $\{x_j,y_j\}$, as the moving object may stay at a the same position for a time period. An encoded formalization of a segment is given by Axioms 2–5. Axiom 2 enforces that every segment is connected to some fixes through the properties *startsFrom* and *endsAt*. Axioms 3 and 4 enforce that every segment is connected to at most two fixes, as

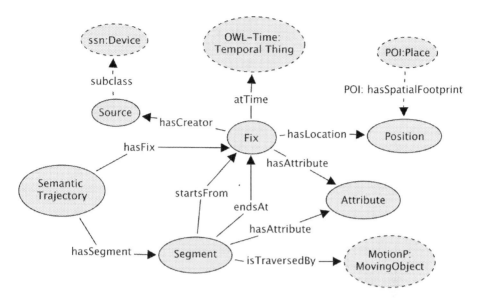

Fig. 2. Schematical description of the pattern

with these two axioms we declare both *startsFrom* and *endsAt* to be functional. Every segment is related to a trajectory as enforced by Axiom 5.

$$Segment \sqsubseteq \exists startsFrom.Fix \sqcap \exists endsAt.Fix \qquad (2)$$

$$\top \sqsubseteq \leq 1\,startsFrom.\top \qquad (3)$$

$$\top \sqsubseteq \leq 1\,endsAt.\top \qquad (4)$$

$$Segment \sqsubseteq \exists hasSegment^-.SemanticTrajectory \qquad (5)$$

As a segment is determined by its corresponding *startsFrom* and *endsAt* fixes, it inherits scalability from the fixes. A segment can, thus, be a route connecting two cities in a coarse-scale application, but also a line linking two spatiotemporal points on the same road in an application at a finer scale.

OWL-Time:Temporal Thing. We reuse *OWL-Time* to express the temporal information associated with a fix. As shown in Figure 2, the relation of *atTime* links a fix with an instance of the class *OWL-Time:Temporal Thing*. *OWL-Time* is part of the W3C Semantic Web Activity and has been used in many applications before, e.g., [36]. It can express rich temporal information using relations and classes such as *owl:before* and *owl:timeZone*. Embedding *OWL-Time* in the semantic-trajectory ontology design pattern not only captures the temporal relations among fixes, but also makes the pattern more reusable for those familiar with *OWL-Time*.

Position and Point-of-Interest (POI). As a fix is a spatiotemporal point, it contains a *position*. A *position* is defined as a coordinate tuple $\{x_i, y_i\}$ in a 2D plane, or $\{x_i, y_i, z_i\}$ in 3D space. It acts as an *interface* to integrate geographic information into the ontology design pattern. The concept of *interface* is well known from object-oriented programming language, as an enabler for a class to acquire additional functions. Existing POI ontologies can be integrated with the trajectory pattern to include geographic data. In Figure 2, we show an example of integrating the POI ontology developed in GeoVoCampSB2012[3] (the classes and relations from the POI ontology are shown by dotted lines). A *POI* can be any geographic feature that the user is interested in (e.g., a gas station or a tourist attraction), and can be represented by various vector geometries (e.g., polygons, polylines, and points).

Ordering Fixes within a Trajectory. We automatize the creation of properties *hasNext*, *hasSuccessor*, *hasPrevious*, and *hasPredecessor* making use of DL Axioms 6–10. These properties link fixes in the appropriate order within a given trajectory. The property *hasNext* is automatically instantiated between the two fixes related to the same segment by Axiom 6.[4] Note that the functionality of roles *startsFrom* and *endsAt* prevents the creation of incorrect instances of the *hasNext* property. With Axiom 7 we define *hasNext* to be a subrole of *hasSuccessor*, which is enforced to be transitive due to Axiom 8. Properties *hasPrevious* and *hasPredecesor* are defined to be inverses of *hasNext* and *hasSuccessor* with Axioms 9 and 10.

$$startsFrom^- \circ endsAt \sqsubseteq hasNext \tag{6}$$

$$hasNext \sqsubseteq hasSuccessor \tag{7}$$

$$hasSuccessor \circ hasSuccessor \sqsubseteq hasSuccessor \tag{8}$$

$$hasNext^- \sqsubseteq hasPrevious \tag{9}$$

$$hasSuccessor^- \sqsubseteq hasPredecesor \tag{10}$$

As previously stated, we can enforce an ordering among the fixes within a trajectory, something that may be useful to query data within an application. Also, we can easily verify that the time restrictions for a set of fixes are consistent with respect to the timestamps, now that these are related by the *hasNext* and *hasSuccessor* properties.

StartingFix, EndingFix, and Stop. The concepts of *StartingFix*, *EndingFix*, and *Stop* are important for the queries on trajectory data [37,34]. These concepts are not explicitly defined in the ontology design pattern; instead they are derived from the fixes and segments. The *StartingFix* has the earliest timestamp and links to only one segment through the property of *startsFrom^-*. Similarly the *EndingFix* is the one which has the latest timestamp and which links to only one

---

[3] http://geog.ucsb.edu/~jano/POIpattern.eps

[4] Axiom 6 is a role chain. Due to this axiom, *hasNext(a, b)* is entailed if *startsFrom^- (a, c)* and *endsAt(c, b)* are the case for any individual *c*.

segment through to^-. A stop is a segment whose length (the Euclidean distance between the *startsFrom* fix and the *endsAt* fix) is shorter than a threshold defined by the user, and the time difference between the *startsFrom* fix and the *endsAt* fix indicates the duration of the stop.

We can automatically detect all fixes that are the start and end of a trajectory, but this comes at the price of loosing tractability. In any case, we include the necessary axioms to automatize this classification of fixes and leave it to the user to choose whether to utilize these axioms to the pattern. Fixes where a trajectory starts or ends are appropriately placed in classes *StartingFix* and *EndingFix* with Axioms 11 and 12. We also extend the starting and ending classification to segments and automatically classify these into *StartingSegment* and *EndingSegment* with Axioms 13 and 14.

$$Fix \sqcap \neg \exists endsAt.Segment \sqsubseteq StartingFix \qquad (11)$$

$$Fix \sqcap \neg \exists startsFrom.Segment \sqsubseteq EndingFix \qquad (12)$$

$$Segment \sqcap \exists startsFrom.StartingFix \sqsubseteq StartingSegment \qquad (13)$$

$$Segment \sqcap \exists endsAt.EndingFix \sqsubseteq EndingSegment \qquad (14)$$

Attribute and hasAttribute. The class *Attribute* and the corresponding relation *hasAttribute* have been defined as the generic class and relation to connect fixes and segments to their attribute values, such as the speed at a particular fix or the bearing of a segment. Users of the pattern can either remain on this level or define their own subclasses and subroles, e.g., hasSpeed.Speed, based on the requirements of the particular applications. This strategy is a well-established practice, and has been used in many applications and patterns [36]. Both are key for the development of a successful and reusable pattern, i.e., sub-typing them gives the pattern the required flexibility without introducing domain knowledge. *Attribute* and *hasAttribute* can also be used to store the pre-calculated spatial distance or time duration of a segment so that such values do not need to be dynamically calculated for each query.

Source. The *Source* class captures the knowledge about the device or the subject which has collected the trajectory data. Potential device information may include the device's manufacturer, produced year, accuracy in terms of location and time, product model, and so forth. Such information has important meaning since even for the same moving object in the same trajectory, different devices or subjects may generate different degrees of uncertain data. Similar to the *Position* class, this class also serves as an interface that allows the ontology design pattern to acquire additional information to support more complex queries. To give a concrete example, Figure 2 shows the integration of the W3C SSN-XG semantic sensor network ontology developed in [12].[5]

---

[5] The fact that the W3C SSN-XG ontology was developed around an ontology design pattern as its skeleton is further evidence for the effectiveness of patterns.

isTraversedBy. This relation links a *Segment* with the corresponding moving object. The *MotionP:MovingObject* class is borrowed from the *Motion Pattern* developed in a previous GeoVoCamp. It can be used as a hook for the integration of domain knowledge about the moving object, such as the name of a person, the species of a bird, the manufacturer of a car, and many other types of information that are necessary to answer the user's queries. Users can also utilize other ontologies, such as FOAF (which is used to model information about people and the relations with their friends) or the bird ontology ONKI[6], to capture related knowledge.

Semantic Trajectory. This class serves as the access point for the ontology design pattern. A semantic trajectory conveys fixes, segments, and related knowledge into a meaningful path connecting the origin and destination. We encode some features over individuals in class *SemanticTrajectory* with Axioms 15–17. Axiom 15 enforces that every trajectory is linked to at least one segment through the *hasSegment* property. Axioms 15 and 17 automatize the *hasFix* relationship from every trajectory to every related segment within this trajectory.

$$SemanticTrajectory \sqsubseteq \exists hasSegment.Segment \qquad (15)$$

$$hasSegment \circ startsFrom \sqsubseteq hasFix \qquad (16)$$

$$hasSegment \circ endsAt \sqsubseteq hasFix \qquad (17)$$

Domain and Ranges and Class Disjointness. We declare all classes defined for the pattern to be disjoint (not shown here for lack of space and to improve readability). This is not only considered to be a good practice while modeling with OWL, as it allows for further inference, but also a necessary condition for the pattern to be expressed in DLPE.

We also recommend the definition of domains and ranges for existing classes, as these axioms are useful in order to complete missing information in some scenarios. We include Axioms 18–21 as an example, to show how to enforce some of these restrictions.

$$\exists hasSegment.Segment \sqsubseteq SemanticTrajectory \qquad (18)$$

$$\exists hasSegment^-.SemanticTrajectory \sqsubseteq Segment \qquad (19)$$

$$\exists hasFix.Segment \sqsubseteq SemanticTrajectory \qquad (20)$$

$$\exists hasFix^-.SemanticTrajectory \sqsubseteq Fix \qquad (21)$$

Note that we do **not** include strict domain and range declarations such as $\exists hasSegment.\top \sqsubseteq SemanticTrajectory$. Defining strict domain and ranges over the properties in an ontology have proven to reduce interoperability instead of

---

[6] http://onki.fi/en/browser/overview/avio

fostering it. Defining domains and ranges over existing classes is less intrusive, and we believe it will be more useful in practice. It is easy to see how these axioms enforcing domain and range could be extended to the rest of classes and relationships presented in Figure 2.

Since pair class disjointness is enforced across all classes presented in the ontology, the pattern satisfies all conditions and can be expressed within the DL fragment DLPE as described in [8]. As previously commented, this is only the case if we have that Axioms 11–14 are not part of the pattern. These axioms were depicted in this section as we believe they may become useful in some particular situation. Nonetheless, we do not recommend to include them a priori since the addition of these axioms makes the reasoning process exponential with respect to the ontology size.

In summary, the geo-ontology design pattern uses *fixes* and *segments* to capture the trajectory data, and defines a number of interfaces to integrate related geographic information, domain knowledge, and device data.

5   Applications to Trajectory Data

A successful ontology design pattern should have the usability that allows it to be applied to a wide range of datasets, solving problems of discovery and integration. It should not be too specific, nor introduce particular application perspectives. In this section, we use our semantic trajectory pattern to annotate datasets of two kinds: trajectories generated by human travelers and by animals. We also show how existing ontologies (such as a POI ontology) can be combined with the design pattern to capture related knowledge.

5.1   Modeling Human Trajectories

Human trajectories have been investigated by psychologists, anthropologists, geographers, and traffic planners to understand human behavior. In recent years, trajectory data from individuals have also been used to improve personal information management by providing information which is related to the user's current activities [24]. In the following paragraphs, we apply our ontology design pattern to an individual's trajectory data recorded by a handheld GPS receiver. During the trip, the user switched the transportation mode from walking to driving a car, so that the moving object changed between different segments of the trajectory. Graphic notations are employed to visualize the integration of our ontology design pattern with existing ontologies.

Figure 3 shows part of Mike's trajectory annotation, using the proposed design pattern, for his trip to the GeoVoCamp Dayton 2012, integrating location data, GPS positions, personal data, vehicle information, and so forth. We extracted two representative segments and four fixes from the entire dataset to illustrate the application. The two segments are *traversedBy* the person and his car respectively, and the information about the moving objects is included.

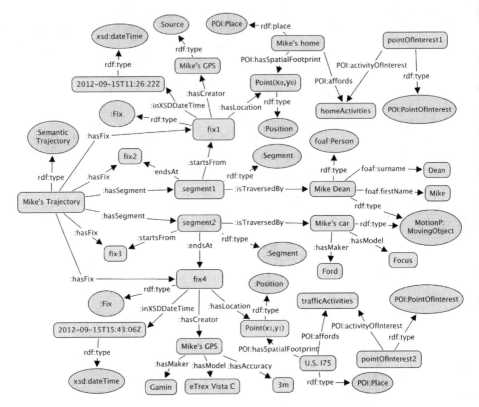

Fig. 3. Graphic notation for part of a person's trajectory annotation (blue rectangles represent entities and orange circles represent classes)

This integration enables queries such as "show the segments traversed by Mike without his car". While each segment has two fixes, we only display the relations of one fix in the figure, as the others have similar relations. A *fix* is linked to a timestamp represented by *xsd:dateTime* and a position[7]. To demonstrate the trajectory pattern's capability of integration with exiting ontologies, we combine it with the POI pattern developed in a previous GeoVoCamp through the POI interface. A *fix* is also linked with a timestamp, and a *source* (the Gamin GPS in this case). Information about the GPS device (such as the maker and mode) is also integrated into the example. We formally encode this ontology using the N3 syntax. Fragments of the code are shown in Table 2.

5.2   Application to Wildlife Monitoring

Here, we apply the design pattern to wildlife monitoring, using the trajectory data for a bird as example. The dataset is from the MoveBank, an online database

---

[7] For privacy considerations, we replace the real coordinates with x_i and y_i

Table 2. Part of the code for the individual trajectory using N3

:mikesTrajectory	a	:SemanticTrajectory;
	:hasSegment	:segment1, :segment2, ...;
	:hasFix	:fix1, :fix2, :fix3, :fix4, ...;
:mike	a	foaf:Person;
:mikesCar	a	MotionP:MovingObject;
:mikesGPS	a	:Source;
:mikesHome	a	POI:Place;
	POI:hasSpatialFootprint	:pos1;
:segment1	a	:Segment;
	:startsFrom	:fix1;
	:endsAt	:fix2;
	:isTraversedBy	:fordFocus;
:fix1	a	:Fix;
	:hasCreator	:mikesGPS;
	:inXSDDataTime	:2012-09-15T11:26:22Z;
	:hasLocation	:pos1;
:pos1	a	:Position;
	:geo:astWKT	Point(x_0,y_0);

providing animal track data openly to researchers[8]. The moving object is a toucan (a type of ramphastos sulfuratus), and the trajectory data contains the information about its positions, timestamps, temperatures of the environment, speeds, accuracy, directions, as well as some tracking device information, such as battery voltage [25,26]. Figure 4 shows the semantic annotation for part of the toucan's trajectory.

For reason of space limitation and readability, only one segment and two fixes are shown in the figure; the full data can be stored in any RDF triple store. The segment is *traversedBy* the toucan, and the application-specific information of this bird, such as its taxonomic name and local identifier, is also included in this example. Each *fix* has its corresponding time and position information. Unlike the human example, each *fix* of the bird's trajectory has several additional attributes, such as the temperature, its speed, and heading direction, which describe the status of the toucan and the environment at that *fix*. These attributes are expressed by sub-typing the *Attribute* class and the *hasAttribute* relation in the ontology design pattern. Such attributes enable queries such as "show the

---

[8] https://www.movebank.org/

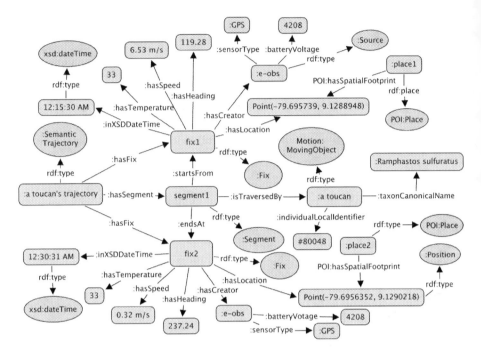

Fig. 4. Graphical notation for part of the annotation of the toucan's trajectory (blue rectangles for entities and orange circles for classes).

fixes where the toucan is moving at a speed higher than 6 m/s". The fixes are generated by the *e-obs*, a tracking device whose *sensorType* and *batteryVotage* are also captured by this trajectory ontology.

6   Conclusions

In this paper we presented a geo-ontology design pattern for semantic trajectories and demonstrated its applicability. The pattern resulted from a joint effort of domain experts and knowledge engineers. It can be used to semantically annotate trajectory data from a range of different domains such as navigation and wildlife monitoring. The major advantages of using the proposed pattern (also in comparison to existing work) are:

– **Expressiveness.** The design pattern can express a trajectory's spatiotemporal properties, geographic knowledge, domain knowledge, as well as relations among them. The pattern's formalization goes far beyond the typical surface semantics that reduces ontologies to mere subsumption hierarchies. Instead, it uses the expressive power of OWL to support a wide range of inferences.

This makes the pattern suitable for semantic annotation (of Linked Data), reasoning, and support for retrieval of scientific data (e.g., in semantics-enabled cyber-infrastructures).

- **Simplicity**. Only a minimal number of classes and relations are defined, which makes the design pattern easy to understand, reuse, and extend. The pattern can be used as a skeleton for more complex ontologies by sub-typing. This is in line with the approaches taken previously by other researchers, e.g. in the Simple Event Model (SEM) [21]. We do not restrict the domains and ranges of the used relations on a global level, in order to avoid unintended inferences. Misunderstandings of the formal semantics underlying such restrictions have been identified as a common source of errors (especially made by those new to ontology engineering).

- **Flexibility**. The provided interfaces (generic classes such as *Source*) allow the user to integrate related knowledge according to the specific needs of the application (users can also leave interfaces open and use the pattern directly without sub-typing). The pattern can model not only trajectories that have already been recorded, but also hypothesized or planned trajectories (e.g., in the context of navigation).

- **Scalability**. Depending on the required granularity of a particular application, the ontology design pattern can model trajectories at different scales. For instance, the physical movement path can be resolved to any degree based on the sample interval for fixes.

In a broader context, the pattern also contributes to a data-driven geo-ontology engineering paradigm. The success of a pattern and the methodology as such can only be evaluated over the years based on its usage *in the wild*. Nonetheless, the ability to develop such patterns (see also [9,8]) in a community process, agree on their ontological commitments, implement them in OWL, document them using real data from different domains, and publish the results, provides insights into the significant potential of the use of patterns, and their appeal to domain scientists, e.g., in a setting such as NSF's EarthCube [3].

While the presented work also demonstrates how to combine different patterns, so far they have mostly been developed independently and without an overarching structure. In fact, and in contrast to other domains, there is no common platform for geo-ontology design patterns, documentations, best practice, examples, and so forth that would significantly lower the initial hurdle for domain scientists interested in semantically annotating their data. This will be one of the main goals for the next years of geospatial semantics research. To address this challenge, we recently propose *Descartes-Core* as a community-wide collection of vocabularies, (geo-)ontology design patterns, best practice guides, examples, software, and services, with the aim to foster semantic interoperability between different sources without restricting semantic heterogeneity.

Finally, we plan to develop an optional alignment layer between the trajectory pattern and the DOLCE foundational ontology in a similar way as done for the W3C SSN-XG ontology before [12]. We expect that this will further foster interoperability and reuse of the pattern.

Acknowledgements. This work is a collaborative outcome of the GeoVoCam-pDayton2012[9]. Some of the authors credit funding from European Commission (ICT-FP7-249120 ENVISION project), as well as the German Research Foundation (Research fellowship grant DFG SCHE 1796/1-1). Authors from Wright State University acknowledge funding from by the National Science Foundation under award 1017225 III: Small: TROn – Tractable Reasoning with Ontologies. Mike Dean, Gary Berg-Cross, and Dave Kolas acknowledge funding from the NSF grant 0955816, INTEROP–Spatial Ontology Community of Practice.

References

1. Alvares, L.O., Bogorny, V., Kuijpers, B., De Macedo, J.A.F., Moelans, B., Vaisman, A.: A model for enriching trajectories with semantic geographical information. In: Samet, H., Shahabi, C., Schneider, M. (eds.) Proceedings of the 15th ACM International Symposium on Geographic Information Systems, ACM-GIS 2007, Seattle, Washington, USA, November 7-9. ACM Press (2007)

2. Battle, R., Kolas, D.: Enabling the geospatial Semantic Web with Parliament and GeoSPARQL. Semantic Web 3(4), 355–370 (2012)

3. Berg-Cross, G., Cruz, I., Dean, M., Finin, T., Gahegan, M., Hitzler, P., Hua, H., Janowicz, K., Li, N., Murphy, P., Nordgren, B., Obrst, L., Schildhauer, M., Sheth, A., Sinha, K., Thessen, A., Wiegand, N., Zaslavsky, I.: Semantics and Ontologies for EarthCube. In: Proceedings of the 2012 Workshop on GIScience in the Big Data Age, In Conjunction with the Seventh International Conference on Geographic Information Science 2012 (GIScience 2012), Columbus, Ohio, USA, September 18 (2012)

4. Bizer, C., Heath, T., Berners-Lee, T.: Linked Data – The Story So Far. International Journal on Semantic Web and Information Systems 5(3), 1–22 (2009)

5. Bogorny, V., Kuijpers, B., Alvares, L.O.: ST-DMQL: A semantic trajectory data mining query language. International Journal of Geographical Information Science 23(10), 1245–1276 (2009)

6. Brakatsoulas, S., Pfoser, D., Tryfona, N.: Modeling, storing, and mining moving object databases. In: 8th International Database Engineering and Applications Symposium (IDEAS 2004), Coimbra, Portugal, July 7-9, pp. 68–77. IEEE Computer Society (2004)

7. Brodaric, B., Probst, F.: Enabling cross-disciplinary e-science by integrating geoscience ontologies with Dolce. IEEE Intelligent Systems 24(1), 66–77 (2009)

8. Carral, D., Scheider, S., Janowicz, K., Vardeman, C., Krisnadhi, A.A., Hitzler, P.: An ontology design pattern for cartographic map scaling. In: Cimiano, P., Corcho, O., Presutti, V., Hollink, L., Rudolph, S. (eds.) ESWC 2013. LNCS, vol. 7882, pp. 76–93. Springer, Heidelberg (2013)

9. Carral, D., Janowicz, K., Hitzler, P.: A logical geo-ontology design pattern for quantifying over types. In: Cruz, I.F., Knoblock, C., Kröger, P., Tanin, E., Widmayer, P. (eds.) SIGSPATIAL 2012 International Conference on Advances in Geographic Information Systems (Formerly Known as GIS), SIGSPATIAL 2012, Redondo Beach, CA, USA, November 7-9, pp. 239–248. ACM (2012)

---

[9] http://vocamp.org/wiki/GeoVoCampDayton2012

10. Chan, L.-W., Chiang, J.-R., Chen, Y.-C., Ke, C.-N., Hsu, J., Chu, H.-H.: Collaborative localization: Enhancing WiFi-based position estimation with neighborhood links in clusters. In: Fishkin, K.P., Schiele, B., Nixon, P., Quigley, A. (eds.) PERVASIVE 2006. LNCS, vol. 3968, pp. 50–66. Springer, Heidelberg (2006)

11. Chiou, Y., Wang, C., Yeh, S., Su, M.: Design of an adaptive positioning system based on WiFi radio signals. Computer Communications 32(7), 1245–1254 (2009)

12. Compton, M., Barnaghi, P.M., Bermudez, L., Garcia-Castro, R., Corcho, Ó., Cox, S., Graybeal, J., Hauswirth, M., Henson, C.A., Herzog, A., Huang, V.A., Janowicz, K., Kelsey, W.D., Phuoc, D.L., Lefort, L., Leggieri, M., Neuhaus, H., Nikolov, A., Page, K.R., Passant, A., Sheth, A.P., Taylor, K.: The SSN ontology of the W3C semantic sensor network incubator group. Journal on Web Semantics 17, 25–32 (2012)

13. Dodge, S., Weibel, R., Lautenschütz, A.K.: Towards a taxonomy of movement patterns. Information Visualization 7(3), 240–252 (2008)

14. Gangemi, A.: Ontology design patterns for semantic web content. In: Gil, Y., Motta, E., Benjamins, V.R., Musen, M.A. (eds.) ISWC 2005. LNCS, vol. 3729, pp. 262–276. Springer, Heidelberg (2005)

15. Gangemi, A., Fisseha, F., Keizer, J., Lehmann, J., Liang, A., Pettman, I., Sini, M., Taconet, M.: A core ontology of fishery and its use in the fishery ontology service project. In: First International Workshop on Core Ontologies, EKAW Conference. CEUR-WS, vol. 118 (2004)

16. Gangemi, A., Guarino, N., Masolo, C., Oltramari, A., Schneider, L.: Sweetening ontologies with DOLCE. In: Gómez-Pérez, A., Benjamins, V.R. (eds.) EKAW 2002. LNCS (LNAI), vol. 2473, pp. 166–181. Springer, Heidelberg (2002)

17. Grüninger, M., Fox, M.S.: The role of competency questions in enterprise engineering. In: Proceedings of the IFIP WG5, vol. 7, pp. 212–221 (1994)

18. Güting, R.H., Böhlen, M.H., Erwig, M., Jensen, C.S., Lorentzos, N.A., Schneider, M., Vazirgiannis, M.: A foundation for representing and querying moving objects. ACM Transactions on Database Systems (TODS) 25(1), 1–42 (2000)

19. Güting, R., De Almeida, V., Ding, Z.: Modeling and querying moving objects in networks. The VLDB Journal 15(2), 165–190 (2006)

20. Gwon, Y., Jain, R., Kawahara, T.: Robust indoor location estimation of stationary and mobile users. In: Proceedings IEEE INFOCOM 2004, the 23rd Annual Joint Conference of the IEEE Computer and Communications Societies, Hong Kong, China, March 7-11, pp. 1032–1043. IEEE (2004)

21. van Hage, W.R., Malaise, V., Segers, R.H., Hollink, L., Schreiber, G.: Design and use of the Simple Event Model (SEM). Journal on Web Semantics 9(2), 128–136 (2011)

22. Hitzler, P., Krötzsch, M., Rudolph, S.: Foundations of Semantic Web Technologies. CRC Press (2010)

23. Horne, J.S., Garton, E.O., Krone, S.M., Lewis, J.S.: Analyzing animal movements using Brownian bridges. Ecology 88, 2354–2363 (2007)

24. Hu, Y., Janowicz, K.: Improving personal information management by integrating activities in the physical world with the semantic desktop. In: Cruz, I.F., Knoblock, C., Kröger, P., Tanin, E., Widmayer, P. (eds.) SIGSPATIAL 2012 International Conference on Advances in Geographic Information Systems (formerly known as GIS), SIGSPATIAL 2012, Redondo Beach, CA, USA, November 7-9, pp. 578–581. ACM (2012)

25. Kays, R., Jansen, P.A., Knecht, E.M., Vohwinkel, R., Wikelski, M.: The effect of feeding time on dispersal of virola seeds by toucans determined from gps tracking and accelerometers. Acta Oecologica 37(6), 625–631 (2011)

26. Kays, R., Jansen, P.A., Knecht, E.M., Vohwinkel, R., Wikelski, M.: Data from: The effect of feeding time on dispersal of virola seeds by toucans determined from gps tracking and accelerometers. Movebank Data Repository (2012)

27. Li, X., Claramunt, C., Ray, C., Lin, H.: A semantic-based approach to the representation of network-constrained trajectory data. In: Riedl, A., Kainz, W., Elmes, G.A. (eds.) Progress in Spatial Data Handling, pp. 451–464. Springer (2006)

28. McKenzie, G., Adams, B., Janowicz, K.: A thematic approach to user similarity built on geosocial check-ins. In: Proceedings of the 2013 AGILE Conference (to appear, 2013)

29. Mika, P., Oberle, D., Gangemi, A., Sabou, M.: Foundations for service ontologies: aligning OWL-S to Dolce. In: Proceedings of the 13th World Wide Web Conference, pp. 563–572. ACM (2004)

30. Mouza, C., Rigaux, P.: Mobility patterns. GeoInformatica 9(4), 297–319 (2005)

31. Ni, L., Liu, Y., Lau, Y., Patil, A.: LANDMARC: indoor location sensing using active RFID. Wireless Networks 10(6), 701–710 (2004)

32. Priyantha, N.: The cricket indoor location system. Ph.D. thesis, Massachusetts Institute of Technology (2005)

33. Schmid, F., Richter, K.-F., Laube, P.: Semantic trajectory compression. In: Mamoulis, N., Seidl, T., Pedersen, T.B., Torp, K., Assent, I. (eds.) SSTD 2009. LNCS, vol. 5644, pp. 411–416. Springer, Heidelberg (2009)

34. Spaccapietra, S., Parent, C., Damiani, M., De Macedo, J., Porto, F., Vangenot, C.: A conceptual view on trajectories. Data and Knowledge Engineering 65(1), 126–146 (2008)

35. Vazirgiannis, M., Wolfson, O.: A spatiotemporal model and language for moving objects on road networks. In: Jensen, C.S., Schneider, M., Seeger, B., Tsotras, V.J. (eds.) SSTD 2001. LNCS, vol. 2121, pp. 20–35. Springer, Heidelberg (2001)

36. Willems, N., van Hage, W.R., de Vries, G., Janssens, J.H.M., Malaise, V.: An integrated approach for visual analysis of a multisource moving objects knowledge base. International Journal of Geographical Information Science 24(10), 1543–1558 (2010)

37. Yan, Z., Macedo, J., Parent, C., Spaccapietra, S.: Trajectory ontologies and queries. Transactions in GIS 12, 75–91 (2008)

38. Yan, Z.: Towards semantic trajectory data analysis: A conceptual and computational approach. In: Rigaux, P., Senellart, P. (eds.) Proceedings of the VLDB 2009 PhD Workshop. Co-located with the 35th International Conference on Very Large Data Bases (VLDB 2009), Lyon, France, August 24. VLDB Endowment (2009)

39. Ying, J.J.C., Lu, E.H.C., Lee, W.C., Weng, T.C., Tseng, V.S.: Mining user similarity from semantic trajectories. In: Proceedings of the 2nd ACM SIGSPATIAL International Workshop on Location Based Social Networks, pp. 19–26. ACM (2010)

RCC and the Theory of Simple Regions in $\mathbb{R}^2$

Stefano Borgo

Laboratory for Applied Ontology, ISTC-CNR and KRDB FUB, Italy

Abstract. The theory of mereology and its topological extensions, called mereotopologies, are point-free approaches that allow to model information while avoiding several puzzling assumptions typical of set-theoretical systems. Although points can be introduced in a mereology or mereotopology, the idea is that one does so only when their existence is clearly motivated. The Region Connection Calculus, RCC, is a mereotopological system that assumes upfront the existence of points: points must be accepted for the system to have the desired semantics. This is awkward from the mereological viewpoint and is considered unsatisfactory from the cognitive and the philosophical perspectives on which mereology rests.

We prove that, in dimension two and with the standard semantics, a theory equivalent to RCC can be given without any reference to points at the semantic level also. The theory, based on mereology, uses the topological primitive 'being a simple region' (aka 'being strongly self-connected').

1 Introduction

Qualitative space representation (QSR) is a fruitful interdisciplinary topic spanning areas like knowledge representation and reasoning (KR), artificial intelligence (AI), cognitive science, logic, linguistics and philosophy. Representation formalisms in QSR have received increasing attention from the 90s when a series of theories, e.g. [19,2,8,7,4], have been coherently motivated, developed, compared and applied, see [9,17,24,5] for reviews.

Over the years, the Region Connection Calculus (RCC) [19] has become one of the most applied approaches to model relationships among regions in space. RCC is a mereotopological theory: it takes only extended regions as individuals and can express their topological relationships but no metrical or geometrical information. Technically, RCC is the theory of the binary relation C (connection) where $C(x, y)$ holds for x, y, interpreted as regular sets in the given space (tipically $\mathbb{R}^2$ or $\mathbb{R}^3$), when their closures share at least one point.

While RCC is syntactically a region-based theory, its formal dependence on points at the semantic level has always been an issue of dissatisfaction from at least the cognitive and the philosophical viewpoints. N. M. Gotts, one of the developers of RCC, wrote: "Using an interpretation expressed in terms of point-sets might seem inconsistent with the spirit underlying the RCC approach" [11, pg.5]. While the commitment to extended regions is considered philosophically and cognitively motivated, the direct syntactic or semantic commitment to points is generally seen as a choice in need of justifications. This criticism does not aim at

T. Tenbrink et al. (Eds.): COSIT 2013, LNCS 8116, pp. 457–474, 2013.

ruling out the point-based relation of connection; nonetheless it is doubtful that it is the best primitive for mereotopology. After all, even in RCC-driven cognitive experiments [20] the results show that other forms of connections inevitably show up as cognitively relevant. A general comparison of forms of connections at the cognitive level is however lacking.

The goal of this paper is to show that RCC (more precisely, RCC8 with universe, binary sum and the other operators) is expressively equivalent to a mereotopological theory in which the notion of point has no role whatsoever, including at the semantic level, and that is coherent with respect to the usual philosophical and cognitive arguments. Note that point-free semantics for RCC have already been provided [22,21]. The approach proposed by Stell and Worboys is based on Boolean connection algebras, i.e., it applies structures based on lattice theory. From these results, it follows that the semantics of RCC can be given independently from the notion of point making the system philosophically more coherent. This result answers the problem raised by Gotts but, unfortunately, at the cost of loosing the cognitive appeal of the theory. Indeed, the structures based on (different versions of) Boolean algebras do not seem cognitively relevant when modeling spatial relationships.

This paper tackles the same problem while taking in a different direction. Instead of changing the RCC semantics in oder to get rid of points, it studies a different theory which is motivated by the same cognitive arguments given for RCC, is independent of points at any level, and is expressively equivalent to RCC. The results we obtain are promising but not yet general enough: our proofs apply to mereotopological structures in $\mathbb{R}^2$ and are not conclusive for other important domains, primarily $\mathbb{R}^3$. The results, even if restricted to dimension 2, are however quite relevant for Euclidean logics [16] and seem related to recent works on the complexity of mereotopologies [13,12].

The theory we propose exploits two primitives: the binary relation of parthood (P) and the unary relation of simple region (SR). In particular, we show that these, in the discussed structures, are sufficient to define the connection relation C of RCC. Since it is known that C suffices to define both P and SR [10], the relative equivalence of these theories follows. The proof is divided in two steps. First, we show that C can be defined from P and SR in the mereological substructures described in [17] where regions are finitely decomposable. This part is based on a classification and discussion of the relevant cases. Second, we use topological considerations to generalize the result to the structure of open (equivalently, closed) regular regions in $\mathbb{R}^2$. The overall proof is based on standard logical and topological arguments. All the definitions are explicitly stated.

The fact that in the important models of RCC our theory is equivalent to RCC, leads to claim that RCC can be restated as a truly point-free theory. This, in our view, strengthens the foundations of that system. Another consequence of our result, which we do not investigate here, is the possibility to use SR as a grip for the assessment of the relationship between mereology and (mereo-)topology. Since C is expressively sufficient to model mereology in homogeneous domains, the use of P as a primitive in these mereotopologies has always been seen as a concession to the philosophically and foundationally inclined people without any practical or technical motivation. The independence of P and SR in our theory suggests that a notion of "cognitive topology" can be developed within mereology as the study of point-free topology with parthood.

The paper is structured as follows. Section 2 gives the basic backgrounds on RCC. Section 3 discusses the notion of simple region. Section 4 introduces the mereological structures studied in this paper. Section 5 defines C from P and SR in the chosen structures. Section 6 generalizes the result to the structure of open regular regions in $\mathbb{R}^2$. Section 7 adds some observations on the generality of the proofs and points to future research.

2 Basic Mereology and the Region Connection Calculus (RCC)

We begin by introducing some definitions based on the binary parthood predicate P [17,24].

(D1) $PP(x,y) \overset{\text{def}}{=} P(x,y) \wedge \neg P(y,x)$ $\hspace{3cm}$ [x is a proper part of y]

(D2) $O(x,y) \overset{\text{def}}{=} \exists z\, [P(z,x) \wedge P(z,y)]$ $\hspace{3cm}$ [x and y overlap]

(D3) $DR(x,y) \overset{\text{def}}{=} \neg O(x,y)$ $\hspace{3cm}$ [x and y are disjoint]

(D4) $PO(x,y) \overset{\text{def}}{=} O(x,y) \wedge \neg P(x,y) \wedge \neg P(y,x)$ $\hspace{1.5cm}$ [x and y properly overlap]

A mereology is a first-oder logical theory with relation P as the only primitive element. We do not present an actual axiomatization of the system since our discussion will remain at the semantic level but remind the reader that P is a partial order (in particular, extensional), that any region has a proper subregion (there are no atoms), and that the functional operators of sum, product (on overlapping regions) and complement (except for the universe itself) are defined [4].

The semantics of mereology is given by the standard first-order semantics where variables are interpreted in a domain $\mathcal{R}$ of non-empty regular open sets of $\mathbb{R}^n$ (equivalently, regular closed sets) with the following clause for the primitive relation P:

$$\langle \mathcal{R}, I \rangle \models P(x,y) \quad if \quad x^I \subseteq y^I.$$

As anticipated, the RCC system [19] is a mereotopology that takes the connection relation C as the only primitive. In RCC the parthood relation is defined by

(D5) $P(x,y) \overset{\text{def}}{=} \forall z\, [C(z,x) \to C(z,y)]$ $\hspace{3cm}$ [RCC parthood]

which is correctly interpreted at the semantic level because of the RCC axiomatization. Thus, the mereological definitions (D1)-(D4) hold in RCC with their intended meaning. In RCC we also have sum, product and complement, with the restrictions seen above for mereology, plus the universal region and the following

(D6) $DC(x,y) \stackrel{\text{def}}{=} \neg C(x,y)$ [x and y are disconnected]

(D7) $EC(x,y) \stackrel{\text{def}}{=} C(x,y) \wedge \neg O(x,y)$ [x and y are externally connected]

(D8) $TPP(x,y) \stackrel{\text{def}}{=} PP(x,y) \wedge \exists z\, [EC(z,x) \wedge EC(z,y)]$ [x is a tangential
proper part of y]

(D9) $NTPP(x,y) \stackrel{\text{def}}{=} PP(x,y) \wedge \neg \exists z\, [EC(z,x) \wedge EC(z,y)]$ [x is a nontangential
proper part of y]

The semantics of RCC is also based on the standard first-order semantics
where variables are interpreted in a domain $\mathcal{R}$ of non-empty regular open sets of
$\mathbb{R}^n$ (equivalently, regular closed sets) with the following clause for the primitive
relation C:

$$\langle \mathcal{R}, I \rangle \models C(x,y) \quad if \quad [x^I] \cap [y^I] \neq \emptyset.$$

Here $[\cdot]$ is the topological closure on sets.

As before, the specific axiomatization of RCC is not relevant in our work
and we report here just the basic axioms that enforce C to be reflexive and
symmetric, and the whole system to be extensional and atomless. The interested
reader can find the whole axiomatization in [19] and a different version in [21].

(A1) $\forall x\, [C(x,x)]$ [reflexivity]

(A2) $\forall x,y\, [C(x,y) \rightarrow C(y,x)]$ [symmetry]

(A3) $\forall x,y\, [\forall z\, [C(z,x) \leftrightarrow C(z,y)] \rightarrow x = y]$ [extensionality]

(A4) $\forall x \exists y\, [NTPP(y,x)]$ [nontangential proper part existence]

3 Simple Regions

The cognitive interest for the simple region predicate is clear: everyday things
relevant at the human (mesoscopic) level tend to occupy regions that are "everywhere thick". Everywhere thick means that a small enough object can move
to any place within the region without going through the region boundary. This
intuition may correspond to different situations in the real world depending on
ontological and epistemical views of notions like boundary and moving. However,
the interest for everywhere thick regions, in contrast to regions connected at a
single (boundary) point, is obvious.

When limiting the above intuition to the domain of topology and mereotopology, the interest boils down to the distinction between regions that are self-
connected (they may be composed of just point-connected pairs of subregions),
and those that are strongly self-connected (simple regions as described above).
From now on, we will use the term 'region' to mean an element of some suitable
domain of interpretation, in particular a non-empty regular region. Without loss
of generality, we also assume that regions are open.

Topologically speaking a self-connected region x is a region that cannot be
divided in two subregions whose closures share no point. This notion, formally
written $PntC(x)$, is dubbed *point-connection* or *self-connection* in mereotopology
and is definable from C [19]:

(D10) $PntC(x) \overset{\text{def}}{=} \forall y, z \ [x = y + z \to C(y, z)]$ [RCC self-connection]

The predicate of strong self-connection or simple region, formally SR, has been used in [2,3] but is known in the literature under different names like $ICON$ (interior connection) [10] (later modified in [1]); SSC (strong connection) [4]; $SCON$ (strongly self connection) [6]; SR (simple region) [14]; FSC (firmly self-connection) [24]; and C_b (strong connection) [23].

Here is the definition of SR in RCC [1]:

(D11) $SR_C(x) \overset{\text{def}}{=} \forall y \exists z \ [NTPP(y, x) \to (P(y, z) \wedge NTPP(z, x) \wedge PntC(z))]$ [RCC simple region]

From these definitions, it follows that enhancing mereology with the unary predicate SR gives a system within the realm of mereotopology.

4   Mereotopological Structures

In this section we work within the mereotopological structures $ROQ(\mathbb{R}^2)$, $ROP(\mathbb{R}^2)$ and $ROS(\mathbb{R}^2)$, subsystems of the mereotopology of the regular open sets in $\mathbb{R}^2$, $RO(\mathbb{R}^2)$. These restricted structures have been discussed in [17, Sect. 2.3] and, as Pratt-Hartmann argues, seem to correspond to a region-based model of space much better than the whole of regular open sets. The main argument is that among the regular open sets of $\mathbb{R}^2$ (and $\mathbb{R}^n$ in general) there are various pathological sets whose existence does not seem cognitively justified nor needed in qualitative knowledge representation [18].

The following definitions, where we always assume $n > 0$, are adapted from [17, Sect. 2.3]. For background topological notions see any modern textbook, e.g. [15].

Definition 1. *Let u be a subset of some topological space X. We say that u is regular open (in X) if u is equal to the interior of its closure. We denote the set of regular open subsets of X by $RO(X)$.*

Definition 2. *A set $u \subseteq \mathbb{R}^n$ is said to be semi-algebraic if for some m there exists a formula $\phi(x_1, \ldots, x_n, y_1, \ldots, y_m)$ in the first order language with arithmetic signature $\langle \leq, +, \cdot, 0, 1 \rangle$ (interpreted over $\mathbb{R}^n$ in the usual way) and m real numbers $b_1, \ldots, b_m$ such that*

$$u = \{(a_1, \ldots, a_n) \in \mathbb{R}^n \mid \mathbb{R}^n \models \phi(a_1, \ldots, a_n, b_1, \ldots, b_m)\}.$$

Definition 3. *We denote the set of regular open, semi-algebraic sets in $\mathbb{R}^n$ by $ROS(\mathbb{R}^n)$.*

Definition 4. *A basic polytope in $\mathbb{R}^n$ is the product, in $RO(\mathbb{R}^n)$, of finitely many half-spaces. A polytope in $\mathbb{R}^n$ is the sum, in $RO(\mathbb{R}^n)$, of any finite set of basic polytopes. We denote the set of polytopes in $\mathbb{R}^n$ by $ROP(\mathbb{R}^n)$.*

Definition 5. *A basic rational polytope in $\mathbb{R}^n$ is the product, in $RO(\mathbb{R}^n)$, of finitely many rational half-spaces. A rational polytope in $\mathbb{R}^n$ is the sum, in $RO(\mathbb{R}^n)$, of any finite set of basic rational polytopes. We denote the set of rational polytopes in $\mathbb{R}^n$ by $ROQ(\mathbb{R}^n)$.*

Lemma 1. *For $n > 0$, $ROQ(\mathbb{R}^n) \subseteq ROP(\mathbb{R}^n) \subseteq ROS(\mathbb{R}^n) \subseteq RO(\mathbb{R}^n)$ [17, Sect. 2.3]*

Definition 6. *Let X be a topological space. A mereotopology over X is a Boolean sub-algebra M of $RO(X)$ such that, if o is an open subset of X and $p \in o$, there exists $r \in M$ such that $p \in r \subseteq o$. We refer to the elements of M as regions. If M is a mereotopology such that any component of a region in M is also a region in M, then we say that M respects components.*

Lemma 2. *For $n > 0$, the structures $ROS(\mathbb{R}^n), ROP(\mathbb{R}^n)$ and $ROQ(\mathbb{R}^n)$ are finitely decomposable mereotopologies and respect components [17, Lemma 2.12, 13,16].*

5   Connection in $ROQ(\mathbb{R}^2)$, $ROP(\mathbb{R}^2)$ and $ROS(\mathbb{R}^2)$

In this part we provide a definition of the binary relation C that uses only the predicates P, SR and derived notions. We begin with the characterization of regions which are strong-connected:

(D12) $SC(a, b) \overset{\text{def}}{=} \exists x, y\, [P(x, a) \wedge P(y, b) \wedge SR(x+y)]$ [strong-connected regions]

We also define external strong-connection as a relation holding when two regions are strong-connected and do not overlap.

(D13) $ESC(a, b) \overset{\text{def}}{=} SC(a, b) \wedge \neg O(a, b)$ [externally strong-connected regions]

The notion of *full-simple region* corresponds to predicate $WCON$ in [1] and characterizes the simple regions whose complement is also a simple region. First, let us define predicate U for universal the region

(D14) $U(a) \overset{\text{def}}{=} \forall b\, [P(b, a)]$ [universe]

Using $*$ for the complement operator, we can now define full-simple regions as follows

(D15) $SR_{full}(a) \overset{\text{def}}{=} \neg U(a) \wedge SR(a) \wedge SR(a^*)$ [full-simple region]

In $\mathbb{R}^2$ a disk is SR_{full}. Also any half-plane satisfies SR_{full} while a stripe does not. If a regions satisfies SR_{full}, then the complement satisfies it also, for instance the region $\mathbb{R}^2$ minus a disk is SR_{full}.

The next definition, *finger (strong) connectivity two* or *f2.connectivity* for short, is adapted from [10] where it is introduced using the standard point-connection relation. The definition, as we use it, isolates the annuli where two touching holes count as one and holes touching the external boundary are ignored. Formally, the definition holds for a simple region that can be partitioned in four simple subregions, but not more than four, such that two of these are non strong-connected, the other two are also non strong-connected but each of the first pair is strong-connected to each of the second pair, see Figure 1. Note that we define *f2.connectivity* only to characterize annuli in dimension 2 so our definition covers only a case of the definition in [10]. (We use "Σ" for the sum of the indexed regions.)

(D16) $FC_2(a) \stackrel{\text{def}}{=} SR(a) \wedge \exists u_1, u_2, u_3, u_4 \, [a = \Sigma_i u_i \wedge \bigwedge_i SR_{full}(u_i) \wedge \neg SC(u_1, u_2) \wedge$
$ESC(u_1, u_4) \wedge ESC(u_1, u_3) \wedge ESC(u_2, u_4) \wedge ESC(u_2, u_3) \wedge \neg SC(u_3, u_4)] \wedge$
$\neg \exists u_1, u_2, u_3, u_4, u_5 \, [a = \Sigma_i u_i \wedge \bigwedge_i SR(u_i) \wedge \neg SC(u_1, u_2) \wedge ESC(u_1, u_3) \wedge$
$ESC(u_1, u_4) \wedge ESC(u_1, u_5) \wedge ESC(u_2, u_3) \wedge ESC(u_2, u_4) \wedge ESC(u_2, u_5) \wedge$
$\neg SC(u_3, u_4) \wedge \neg SC(u_3, u_5) \wedge \neg SC(u_4, u_5)]$
[finger strong-connectivity two]

A key notion in the following work captures configurations of four simple regions among which those formed by the subregions in definition (D16). Since this configuration is a kind of partitioning of a part of space that reminds a cake splitting, we dub it *cake partition* or *cake 4-partition*, formally $CkCut_4$.

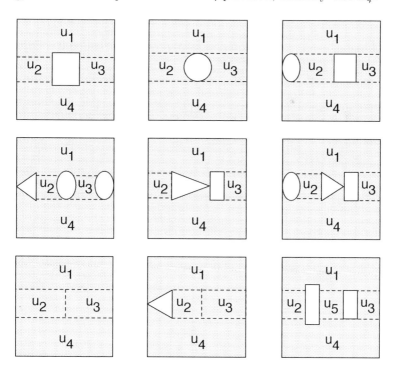

Fig. 1. Regions in the top two rows satisfy finger strong-connectivity (D16), regions in the bottom row do not. Possible subdivisions are shown.

A cake partition is formed by four non-overlapping full-simple regions, say z_1, z_2, z_3, z_4 (in this order) with the following properties: the sum $z_1 + z_2$ and the sum $z_3 + z_4$ are not simple regions while the sum of any other combination of two or more z_i regions is simple.

(D17) $CkCut_4(z_1, z_2, z_3, z_4) \stackrel{\text{def}}{=} \bigwedge_i SR_{full}(z_i) \wedge \neg SC(z_1, z_2) \wedge \neg SC(z_3, z_4) \wedge$
$\qquad ESC(z_1, z_3) \wedge ESC(z_1, z_4) \wedge ESC(z_2, z_3) \wedge ESC(z_2, z_4)$
$\hfill$ [cake 4-partition]

Definition (D17) forces regions z_i to satisfy only instances of the schema in Figure 2 where black squares stand for disjoint simple regions and solid edges indicate that the corresponding simple regions are strong-connected. Instead, dotted lines indicate that the joined simple regions must not be strong-connected: they can be point-connected or completely disconnected. The schema is also designed to convey some local information: when two dotted lines meet the same square but are in areas separated by solid lines, then the region represented by the square and the regions on the other sides of the dotted lines cannot all share a common point. For instance, z_1 and z_2 can be point-connected in three ways, but the point(s) of connection they share along the horizontal dotted line must be distinct from the point(s) of connection they share along the curved dotted lines. This constraint is enforced by the fact that z_1 is strong-connected to both z_3 and z_4. No further information on region location, arrangement and connection should be inferred from the schema of Figure 2. In particular, there is no assumption that these regions cover the whole space, are comparable in size, etc.

In Figure 3 we presents relevant instances of the schema of Figure 2. Some combinations are not shown for different reasons. For example, we do not show the analogous cases where regions z_1 and z_2 have switched positions, similarly for z_3 with z_4. More importantly, here the dotted lines on the outside of the schema are all instantiated by completely disconnected regions; it follows that in none of the shown cases a z_i region is surrounded by a z_j region. Some of these more general examples are depicted in Figure 4 without aiming at covering all cases; and further examples can be inferred from Figure 5.

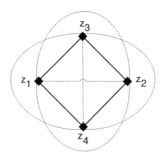

Fig. 2. Schema for cake 4-partition (D17)

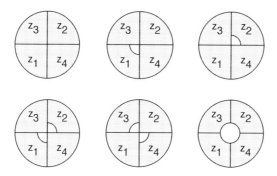

Fig. 3. Instances of cake 4-partitions, see definition (D17)

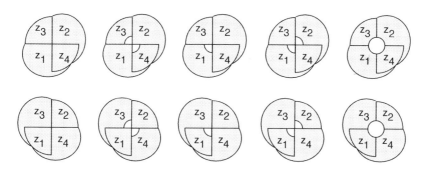

Fig. 4. Instances of cake 4-partitions, see definition (D17).

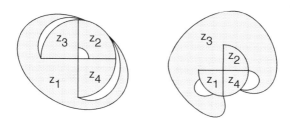

Fig. 5. Cases of multiple point-connections in cake 4-partitions (D17). Left: z_1, z_2 have two "outside" points of connection and are "inside" completely disconnected. Right: z_3, z_4 have three "outside" points of connection and have one "inside" point of connection.

Finally, for simplicity of the presentation we require each z_i to be a full-simple region (SR_{full}). This is not necessary since possible holes cannot be strong-connected to other z_i regions. The SR_{full}-constraint can be weakened in a SR constraint for the z_i regions.

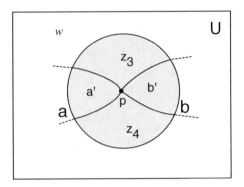

Fig. 6. A choice of a', b', z_3, z_4 for definition (D18). U is the universal region

We now exploit the notions of cake 4-partition and of maximal simple region to define the connection relation C in dimension two. Our first result applies only to the discussed substructures of $RO(\mathbb{R}^2)$ and we write $C_{\mathbb{P}^2}$ for our SR-definition of point-connection limited to these structures. We will deal with the general case in next section. In the formula below, the superscript * is the complement operator, thus a^* is the region that complements a with respect to the universe.

(D18) $C_{\mathbb{P}^2}(a, b) \overset{\text{def}}{=} SC(a, b) \ \lor \exists a', b', z_3, z_4, w, w' \ [P(a', a) \land P(b', b) \land$
$$SR_{full}(a') \land SR_{full}(b') \land CkCut_4(a', b', z_3, z_4) \land$$
$$w = (a' + b')^* \land FC_2(w) \land w' = (a' + b' + z_3 + z_4)^* \land SR(w')]$$
$$[\text{(restricted) } SR \text{ point-connection}]$$

Note that the definition is just existential and is not designed for closed spaces, i.e. for spaces that include their own boundary points like $\mathbb{R}_\infty^2$.

Recall that SR_C is defined within the RCC system via definition (D11).

Theorem 1. *In* $ROQ(\mathbb{R}^2), ROP(\mathbb{R}^2)$ *and* $ROS(\mathbb{R}^2)$, *let* P *be interpreted as the subset relation and assume* $SR(a)$ *holds if and only if* $SR_C(a)$ *holds for* RCC, *then:*
$$C(a, b) \text{ holds if and only if } C_{\mathbb{P}^2}(a, b) \text{ holds.}$$

Proof. (Left to right.) Assume a, b are two regions connected in dimension 2, i.e., $C(a, b)$ holds. We show that $C_{\mathbb{P}^2}(a, b)$ holds as well. If a, b are strong-connected we are done. Assume they are only point-connected and let p be one of their isolated points of connection (if none is, then by regularity a, b must be strong-connected.) Since regions in $ROQ(\mathbb{R}^2), ROP(\mathbb{R}^2)$ and $ROS(\mathbb{R}^2)$ have the finite decomposability property, Proposition 2, there exists a simple region part of a and a simple region part of b that have both p as boundary point. Also, these regions are in $ROQ(\mathbb{R}^2), ROP(\mathbb{R}^2)$ and $ROS(\mathbb{R}^2)$, respectively, because these spaces respect components. Clearly, we can choose a finite region, without holes, which is a subregion of each (these regions have finitely many holes, if any). Let a' be such a region. Similarly, let b' be a finite simple region for b with p as boundary point and without holes. Since a', b' are not strong-connected, choose regions z_3, z_4 to be simple, finite without holes and positioned as in Figure 6. Note that this is possible by considering a suitable neighbor of p in the chosen space with regions that (locally) complement $a' + b'$. Then, a', b', z_3, z_4 satisfy $CkCut_4$ plus all the mereological and the topological constraints in definition (D18). Finally, note that in this construction it is irrelevant whether a and b are finite or else.

(Right to left) By contraposition. Let a, b be two regions not connected in dimension 2, i.e., $C(a, b)$ fails. We show that $C_{\mathbb{P}^2}(a, b)$ does not hold as well. Clearly, $SC(a, b)$ fails, thus we need to concentrate on the second disjunct of (D18). Since a, b are finitely decomposable and the second disjunct looks at the existence of full-simple regions only, without loss of generality we can assume a, b are themselves full-simple regions. Figure 7 lists the cases we need to consider. Here a, b are depicted in gray color and are not strong-connected, i.e., they are disconnected or at most point-connected. (We do not need to label the regions in the figure because the definition is symmetrical.) Note that not all topological variations are listed, yet the topological variations not listed are irrelevant. For instance, it is irrelevant whether the regions in A.3) share a piece of the boundary (as on the left) or just one point (as on the right) with the boundary of the universe since what we consider is the "simple region" status of their complement. Analogously for B.3). The same argument applies to the annulus in cases A.5) and B.5). It is also irrelevant whether there are multiple connection points or lines since the definition considers only existential conditions of (sub-) regions of a and b with disk or annulus-like complements. If a region has multiple point-connections, we just need to consider a subregion with a single point-connection. These observations allow us to focus on the simplified cases in the figure. More precisely, we need to show that none of the cases in column B satisfies the second disjunct of (D18).

We should find a', b', respectively full-simple parts of a and b, such that their sum has a simple complement which satisfies FC_2. This is clearly impossible for case B.1) since a', b' force the complement to have disconnected holes. We can find suitable a', b' in case B.2) provided one region shares part of the boundary of the universe, e.g. taking $a' = a$. In this case, the complement of $a' + b'$, say w, is FC_2, see Figure 8 (left) where we take also $b' = b$. Note however, that these a', b' make impossible to find disconnected regions z_3, z_4 so that condition $CkCut_4$ is satisfied and the complement is simple, see Figure 8 (right). The reason is a direct conflict between constraints in planar graph theory and our request of having a simple region as complement: given four nodes a', b', z_3, z_4, if two are connected to all the others and cannot connect themselves, there are two distinct areas to block, one internal the graph and one external (Figure 2). Since the complement is a simple region, it can be located in either part but not in both. We conclude that the second disjunct of definition (D18) must fail. (Figure 9 shows why the disjunct holds when the regions are point-connected. The remaining cases in column A are solved similarly).

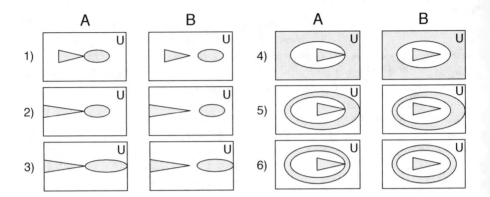

Fig. 7. Instances of relevant cases for definition (D18). Regions a, b are the point-connected (column A) or disconnected (column B) simple regions in gray. As before, U is the universal region.

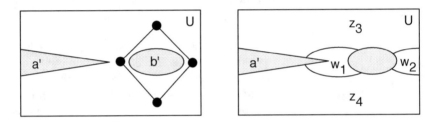

Fig. 8. Choice of a', b' for case B.2) of Figure 7

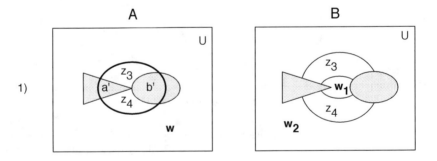

Fig. 9. Choice of a', b' for case A.1) of Figure 7 and the impossibility of finding z_3, z_4 such that $CkCut_4$ is satisfied in case B.1)

In case B.3) at least one of a' and b' must not share boundary with the universe otherwise the complement would not satisfy FC_2. Then, this case reduces to B.1) or B.2). Case B.4) is analogous to B.2) with the cone region completely covering the boundary of the universe. Case B.5) also reduces to B.1) or B.2): one needs to take a section of the annulus region to ensure the complement of $a' + b'$ is a simple region. If the section includes a point in common with the universe boundary, then we are in case B.2). Otherwise we are in case B.1). Finally, in case B.6) one again needs to take a section of the annulus region and the case reduces to B.1).

6   Connection in $RO(\mathbb{R}^2)$

In the previous section we have seen that strong-connection is expressively equivalent to point-connection in the structures proposed in [17] as *the* standard mereotopological structures. The proof takes advantage of a common property of these structures, namely finite decomposability. Nonetheless, the *intended* interpretation of most planar mereotopological theories, among which RCC, is the full range of open regions in $\mathbb{R}^2$, that is, $RO(\mathbb{R}^2)$. We now show that the equivalence holds also in this structure.

First, note that $C_{\mathbb{P}^2}$ immediately extends to any pair of simple regions in $RO(\mathbb{R}^2)$. That is, it holds even for simple regions of $\mathbb{R}^2$ which are not in $ROS(\mathbb{R}^2)$. This follows from the fact that to verify $C_{\mathbb{P}^2}(a, b)$ for a, b simple regions in $RO(\mathbb{R}^2) \setminus ROS(\mathbb{R}^2)$ it suffices to consider simple subregions which have as boundary point the same point shared by a and b. It follows that we do not need to introduce any restriction on the form of simple regions in $RO(\mathbb{R}^2)$ to apply $C_{\mathbb{P}^2}$.

We now introduce a new clause that takes into account pairs of regions in $RO(\mathbb{R}^2)$ at least one of which is not simple and not in $ROS(\mathbb{R}^2)$. Positively put, we extend the previous definition to cover the case of a regular open region connected to another via a point which is not on the boundary of any of its simple subregions. See [17, Sect. 2] for a discussion of these special regions of $RO(\mathbb{R}^2)$ and Figures 10 and 11 (left) for two examples.

First of all, recall that all definitions in Section 5, with the only exception of $C_{\mathbb{P}^2}$, hold in any mereotopological structure, in particular $RO(\mathbb{R}^2)$.

We need just one another auxiliary notion, that of maximal simple region within a given region or, in other terms, of maximal simple component.

(D19) $MxIn_{SR}(a, b) \overset{\text{def}}{=} P(a, b) \wedge SR(a) \wedge \forall x \ [(P(a, x) \wedge P(x, b) \wedge SR(x)) \to a = x]$
[maximal simple component]

Here is the general definition of point-connection in $\mathbb{R}^2$ that uses only P and SR as primitives.

(D20) $C_{\mathbb{R}^2}(a, b) \overset{\text{def}}{=} C_{\mathbb{P}^2}(a, b) \ \vee$
$\exists w \ [\exists u[SR_{full}(u) \wedge P(u, w) \wedge O(u, w \cdot a) \wedge \forall z \ [(MxIn_{SR}(z, a) \wedge O(z, w)) \to O(z, u)]] \wedge$
$\exists v \ [SR_{full}(v) \wedge P(v, w) \wedge O(v, w \cdot b) \wedge \forall z \ [(MxIn_{SR}(z, b) \wedge O(z, w)) \to O(z, v)]] \wedge$
$\forall s, t \ [(SR_{full}(s) \wedge P(s, w) \wedge \forall x \ [(MxIn_{SR}(x, a) \wedge O(x, w)) \to O(x, s)] \wedge$
$SR_{full}(t) \wedge P(t, w) \wedge \forall x \ [(MxIn_{SR}(x, b) \wedge O(x, w)) \to O(x, t)]) \to$
$(C_{\mathbb{P}^2}(s, t) \vee C_{\mathbb{P}^2}(s, b) \vee C_{\mathbb{P}^2}(a, t))]]$
[SR point-connection]

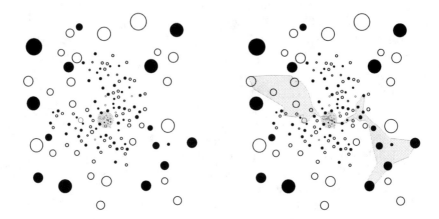

Fig. 10. The scattered region formed by the black disks and the scattered region formed by the white disks are point-connected at the center of the shaded area via infinite converging chains of disks (left). The region shaded on the right is suitable for w in (D20).

The core of the definition is the quantification on w with the constraints on s, t. Informally, the definition says that two arbitrary open regular regions in $\mathbb{R}^2$, say a and b, not satisfying $C_{\mathbb{P}^2}$, are point-connected provided one can find a region w such that any full-simple region that overlaps all a's subregions in w is connected to b (condition $C_{\mathbb{P}^2}(s, b)$) or to any full-simple region in w that overlaps all b's subregions in w (condition $C_{\mathbb{P}^2}(s, t)$). Of course, there is the symmetric condition on b (condition $C_{\mathbb{P}^2}(a, t)$). The universal quantification on s, t is guaranteed to be non-vacuous by the initial existential quantifications on u and v. Figures 10 and 11 (right) show suitable regions w in two cases.

Theorem 2. *In $RO(\mathbb{R}^2)$ let P be interpreted as the subset relation and assume $SR(a)$ holds if and only if $SR_C(a)$ holds for RCC, then:*

$$C(a, b) \text{ holds if and only if } C_{\mathbb{R}^2}(a, b) \text{ holds.}$$

Proof. (Left to right.) If a, b are strong-connected or, in alternative, point-connected because of some of their simple subregions, there is nothing to prove. Assume that (1) a, b are point-connected and that (2) they have only point(s) of connection which do not belong to any of their simple subregions or boundaries of these, i.e. $C_{\mathbb{P}^2}(a, b)$ fails. Let p be one of these points of connection. We now proceed by induction, the goal is to isolate a region w which satisfies the second disjunct of (D20) for the given regions a, b.

Let U_1 be the open ball centered at p and with radius $1/2$. Since p is a (non-internal) accumulation point for both a and b, there exist points $r_1 \in U_1 \cap a$ and $r_1' \in U_1 \cap b$ with $r_1 \neq r_1'$ (since $a \cap b = \emptyset$). Since $U_1 \cap a$ is open, let a_1 be a disk centered in r_1 and contained in $U_1 \cap a$. Analogously fix b_1 for r_1' in $U_1 \cap b$. By construction, p belongs to neither a_1 and b_1 nor their boundaries.

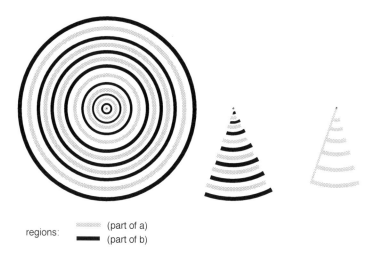

regions: ░░░░░ (part of a)
           ▬▬▬ (part of b)

Fig. 11. The region formed by the gray annuli and the region formed by the black annuli have the (common) center of the annuli as point of connection (left). The quasi sector-shaped region (center) is suitable as w in (D20) for a the sum of the gray annuli and b the sum of black annuli. The SR_{full} region (right) is covered by the quantification on s.

Fix now an open ball $U_2 \subseteq U_1$ with center p and radius $\leq 1/2^2$ such that $U_2 \cap (a_1 \cup b_1) = \emptyset$. As before, there exist points $r_2 \in U_2 \cap a$ and $r'_2 \in U_2 \cap b$. As before, it follows that $r_2 \neq r'_2$. Let a_2 be a disk centered in r_2 and contained in $U_2 \cap a$. Analogously fix b_2 for r'_2 in $U_2 \cap b$. By construction, p belongs to neither a_2 and b_2 nor their boundaries.

Assume now we have isolated a_{n-1} and b_{n-1} in a and b (resp.ly). Note that these do not contain p nor have p on their boundaries. We show how to isolate a_n, b_n. Fix an open ball $U_n \subseteq U_{n-1}$ with center p and radius $\leq 1/2^n$ such that $U_n \cap (a_{n-1} \cup b_{n-1}) = \emptyset$. Since p is accumulation point for a and for b, there exist points $r_n \in U_n \cap a$ and $r'_n \in U_n \cap b$ (clearly, $r_n \neq r'_n$). Let a_n be a disk centered in r_n and contained in $U_n \cap a$. Analogously fix b_n for r'_n in $U_n \cap b$. By construction, p belongs to neither a_n and b_n nor their boundaries.

Let $a' = \Sigma_i a_i$ and $b' = \Sigma_i b_i$. Since the points r_i (the centers of a_i for each i) by construction converge to p, p is a (non-internal) accumulation point for a'. Note that the same result holds for any sequence of points q_i provided $q_i \in a_i$ for each i. It follows that, given an open set c, if $a_i \cap c \neq \emptyset$ for all i, then p is an accumulation point for c as well. Analogously for b'.

We now build a regular open set $\hat{a}$ with $\hat{a} \cap a_i \neq \emptyset$ for each i which, by construction, has p as accumulation point. (The same method applies to obtain $\hat{b}$ over the sequence of the b_i's.) Recall that each a_i is an open disk belonging to the annulus $U_n \setminus U_{n+1}$. In particular the a_i's are all disjoint and we can connect them in the order of construction $a_1, a_2, a_3 \ldots$ with stripe-like regions without dissecting the space. In particular, note that this ensures we do not circumscribe any of the b_i's regions (which, in turn, can be connected similarly). This ensures that there exists a full-simple region $\hat{a}$ which overlaps all a_i and does not overlap any b_i. Similarly for $\hat{b}$. (Note that $\hat{a}$ and b may very well overlap and so $\hat{b}$ and a). Clearly, both $\hat{a}$ and $\hat{b}$ are regular open sets in the domain, satisfy the simple region property and have p as accumulation point. Thus, $C_{\mathbb{P}^2}(\hat{a}, \hat{b})$.

Now let $w = \hat{a} \cup \hat{b}$. By construction, any simple region s in w that overlaps all SR subregions of a in w (among which $\hat{a}$) has p as accumulation point. Similarly for analogous simple regions t relatively to b (among which $\hat{b}$). We conclude that for all these regions s, t, relation $C_{\mathbb{P}^2}(s, t)$ holds. This proves that, with the given assumptions, in $\mathbb{R}^2$ the regions required by (D20) exist, thus $C_{\mathbb{R}^2}(a, b)$ holds as well.

We left out the case where the point of connection p is on the boundary of a simple subregion of, say, a but not of b (or vice versa). Clearly, we can repeat the previous construction on a only. Let $w = \hat{a} + b^*$ where b^* is a simple region in b which has p as accumulation point. We conclude that $C_{\mathbb{P}^2}(s, b^*)$, thus $C_{\mathbb{R}^2}(a, b)$.

(Right to left) We only need to show that the second disjunct implies that the regions are point-connected. Suppose not. Then for any point p which is accumulation for, say, a there is a neighbor U of p with $U \cap b = \emptyset$. Fix any suitable region w and let t satisfy $SR_{full}(t) \wedge P(t, w) \wedge \forall x \, [(MxIn_{SR}(x, b) \wedge O(x, w)) \rightarrow O(x, t)]$ so that t is full-simple, part of w and for any maximal part of b which overlaps w then t overlaps it. (We know from the existential condition on v that such region t exists and that it overlaps b.) Let U' be a neighbor of p overlapping $U \cap t$ and let t' be a SR_{full} region in $t \setminus U'$. Note that such U' and t' exist because w is SR and U' can be taken arbitrarily small. Then, t' satisfies the antecedent whenever t does since $t' \cap b = t \cap b$. Yet, $C_{\mathbb{P}^2}(s, t')$ fails because on the simple regions s, t' the relation $C_{\mathbb{P}^2}$ is equivalent to C and, by construction, the latter fails for s, t'.

7   Conclusion and Future Work

We have shown that in the relevant mereotopological structures of $\mathbb{R}^2$ with the standard interpretation of the primitives, RCC is equivalent to the theory based on the parthood relation and the simple region predicate. This latter theory is better suited for spatial representation from the cognitive and philosophical perspectives since it does not commit to points neither at the syntactic nor at the semantical level.

The result relies on some auxiliary definitions, SR_{full}, which are independent of the structures one considers provided the axioms of closure extensional mereology (CEM, see [4]) are satisfied. More interestingly, theorem 1 makes use of two restrictions: (a) the regions are finitely decomposable and respect complement; (b) the space is restricted to dimension two. Note that theorem 1 does not use any metric property of the reals. Apparently, theorem 2 generalizes restriction (a) at the cost of using the metric properties of $\mathbb{R}^2$. This is not really true. The proof of theorem 2 is just easier to present using the metric properties of the reals. Since we introduce theorem 2 as an extension of a theorem in a subdomain of $\mathbb{R}^2$ and since the intended interpretation of the target theory, RCC, has this domain, we did not see the point of complicating the proof. Nonetheless, the proof of theorem 2 can be given in purely topological terms. The "right to left" part is already stated in topological terms. The lengthy construction in the "left to right" part ensures that there is a sequence of points in region a converging to the given point p. Any numerable subsequence must converge to p as well and since regions are open, we can find distinct SR subregions of each point (only the topological separability property is required, a property already enforced by RCC itself) to build the a_i's and the a' regions in the proof. Similarly for b. Note also that theorem 2 does not use any specific property of the overall space (not even connectedness is required) but the combination of theorems 1 and 2, as used here to reach the general result, requires that the target mereotopological structure has a suitable subset of finitely decomposable (and respecting component) regions.

To generalize the result to spaces of higher dimension, it suffices to extend the first theorem to $ROS(\mathbb{R}^n)$. However, the proof of theorem 1 in this paper is not extendible to other dimensions without relevant changes. From the proof, theorem 2 already holds in any dimension.

Beside studying how to achieve this generalization, we also plan to develop a direct axiomatization of the theory of P and SR, and to study its relationship with point-free topology.

Acknowledgements. The author thanks the reviewers for pointing to relevant references and for the stimulating questions, and A. G. Cohn for his comments on an earlier draft of this work.

References

1. Bennett, B.: Carving up space: Steps towards construction of an absolutely complete theory of spatial regions. In: Alferes, J.J., Pereira, L.M., Orlowska, E. (eds.) JELIA 1996. LNCS, vol. 1126, pp. 337–353. Springer, Heidelberg (1996)
2. Borgo, S., Guarino, N., Masolo, C.: A pointless theory of space based on strong connection and congruence. In: Carlucci Aiello, L., Doyle, J., Shapiro, S.C. (eds.) International Conference on Principles of Knowledge Representation and Reasoning (KR 1996), Boston, MA, pp. 220–229. Morgan Kaufmann (1996)
3. Borgo, S., Guarino, N., Masolo, C.: Qualitative spatial modelling based on parthood, strong connection and congruence. Technical Report Internal Report 03/97, LADSEB-CNR, March 1997 (1997)
4. Casati, R., Varzi, A.C.: Parts and Places. The Structure of Spatial Representation. MIT Press, Cambridge, MA (1999)
5. Cohn, A.G., Renz, J.: Qualitative spatial representation and reasoning. In: van Harmelen, F., et al. (eds.) Handbook of Knowledge Representation, pp. 551–596. Elsevier (2007)
6. Dugat, V., Gambarotto, P., Larvor, Y.: Qualitative theory of shape and structure. Progress in Artificial Intelligence, 850–850 (1999)
7. Eschenbach, C.: A predication calculus for qualitative spatial representations. In: Freksa, C., Mark, D.M. (eds.) COSIT 1999. LNCS, vol. 1661, pp. 157–172. Springer, Heidelberg (1999)
8. Galton, A.: Towards a qualitative theory of movement. In: Kuhn, W., Frank, A.U. (eds.) COSIT 1995. LNCS, vol. 988, pp. 377–396. Springer, Heidelberg (1995)

9. Galton, A.: Spatial and temporal knowledge representation. Earth Science Informatics 2(3), 169–187 (2009)
10. Gotts, N.M.: How far can we 'C'? defining a 'doughnut' using connection alone. In: Proceedings of the 4th International Conference on Principles of Knowledge Representation and Reasoning (KR 1994). Morgan Kaufmann (1994)
11. Gotts, N.M.: Formalizing commonsense topology: The inch calculus. In: Kautz, H., Selman, B. (eds.) International Symposium on Artificial Intelligence and Mathematics (AI/MATH 1996), pp. 72–75 (1996)
12. Kontchakov, R., Nenov, Y., Pratt-Hartmann, I., Zakharyaschev, M.: On the decidability of connectedness constraints in 2D and 3D euclidean spaces. In: Proceedings of the Twenty-Second International Joint Conference on Artificial Intelligence, IJCAI 2011, vol. Two, pp. 957–962. AAAI Press (2011)
13. Kontchakov, R., Pratt-Hartmann, I., Zakharyaschev, M.: Interpreting Topological Logics over Euclidean Space. In: Twelfth International Conference on Principles of Knowledge Representation and Reasoning, KR 2010 (2010)
14. Muller, P.: Topological Spatio-Temporal Reasoning and Representation. Computational Intelligence 18(3), 420–450 (2002)
15. Munkres, J.R.: Topology, 2nd edn. Prentice Hall (2000)
16. Nenov, Y., Pratt-Hartmann, I.: On the computability of region-based euclidean logics. In: Dawar, A., Veith, H. (eds.) CSL 2010. LNCS, vol. 6247, pp. 439–453. Springer, Heidelberg (2010)
17. Pratt-Hartmann, I.: First-order mereotopology. In: Aiello, M., Pratt-Hartmann, I., van Benthem, J. (eds.) Handbook of Spatial Logics, pp. 13–97. Springer (2007)
18. Pratt-Hartmann, I., Schoop, D.: Elementary polyhedral mereotopology. Journal of Philosophical Logic 31, 469–498 (2002)
19. Randell, D.A., Cui, Z., Cohn, A.G.: A spatial logic based on regions and connections. In: Nebel, B., Rich, C., Swartout, W. (eds.) International Conference on Principles of Knowledge Representation and Reasoning (KR 1992), pp. 165–176. Morgan Kaufmann (1992)
20. Renz, J., Rauh, R., Knauff, M.: Towards cognitive adequacy of topological spatial relations. In: Habel, C., Brauer, W., Freksa, C., Wender, K.F. (eds.) Spatial Cognition 2000. LNCS (LNAI), vol. 1849, pp. 184–197. Springer, Heidelberg (2000)
21. Stell, J.G.: Boolean connection algebras: a new approach to the region-connection calculus. Artificial Intelligence 122(1), 111–136 (2000)
22. Stell, J.G., Worboys, M.F.: The algebraic structure of sets of regions. In: Spatial Information Theory A Theoretical Basis for GIS, pp. 163–174. Springer (1997)
23. Thompson, R.J., Oosterom, P.: Connectivity in the regular polytope representation. GeoInformatica 15(2), 223–246 (2009)
24. Varzi, A.C.: Spatial reasoning and ontology: Parts, wholes, and locations. In: Aiello, M., Pratt-Hartmann, I., van Benthem, J. (eds.) Handbook of Spatial Logics, pp. 945–1038. Springer (2007)

The Logic of NEAR and FAR

Heshan Du, Natasha Alechina, Kristin Stock, and Michael Jackson

University of Nottingham

Abstract. We propose a new qualitative spatial logic based on metric (distance) relations between spatial objects. We provide a sound and complete axiomatisation of the logic with respect to metric models. The logic is intended for use in checking consistency of matching geospatial individuals from different data sets, where some data sets may be imprecise (e.g. crowd-sourced data).

1 Introduction

The work on the spatial logic presented in this paper is motivated by our work on integrating geospatial data from crowd-sourced and authoritative sources. We match geospatial ontologies which contain both terminological definitions of concepts such as Shop, Bank, University etc. and geospatial individuals (concrete objects such as Victoria Centre in Nottingham) with associated geometry information [1,2]. We generate candidate matchings between concepts in different ontologies and between individuals, and then check the result for consistency using the description logic reasoner Pellet [3]. A match between two concepts in two ontologies is expressed as an equivalence statement of the form $OSGB : Bank \equiv OSM : Bank$ (which means that $Bank$ in the Buildings and Places ontology of the Ordnance Survey of Great Britain (OSGB) [4] has the same meaning as $Bank$ in the ontology generated from the OpenStreetMap (OSM) dataset [5][1]). A match between two individuals is expressed as

$$OSGB : 1000002309000257 = OSM : 116824670$$

(which means that an object with ID 1000002309000257 from the OSGB ontology is the same as the object with ID 116824670 in the OSM ontology). In addition, to help verify the matchings, we generate disjointness statements which are not part of the original ontologies, but should intuitively hold there, for example

$$OSGB : Bank \sqsubseteq \neg OSGB : Clinic$$

which says that in the OSGB ontology concepts $Bank$ and $Clinic$ are disjoint.

At the moment, we check consistency of matchings only with respect to definitions of concepts and statements of disjointness of concepts. For example,

---

[1] OSM does not have a standard ontology, but maintains a collection of commonly used tags for main map features. An OSM feature ontology is generated automatically from the existing classification of main features.

T. Tenbrink et al. (Eds.): COSIT 2013, LNCS 8116, pp. 475–494, 2013.

the following is a typical example of a minimal inconsistent set found by Pellet which would be given to a human expert to decide whether to retract the match between the two *Bank* concepts (1), or the match between the two objects a and b (2), or the disjointness statement (3). The remaining sentences state that a is a *Clinic* in OSGB and b is a *Bank* according to OSM (hence by (1) also according to OSGB).

Example 1.

$$OSGB : Bank \equiv OSM : Bank \tag{1}$$

$$OSGB : a = OSM : b \tag{2}$$

$$OSGB : Bank \sqsubseteq \neg OSGB : Clinic \tag{3}$$

$$OSGB : a \in OSGB : Clinic \tag{4}$$

$$OSM : b \in OSM : Bank \tag{5}$$

We use spatial information about individuals (their respective geometries) to generate a candidate matching, but we do not use spatial information about their relations to other matched individuals to check consistency of the resulting matching. It would increase the reliability of matching and utilise available information more fully if we could use for example the fact that individual a from ontology 1 is externally connected to individual b in the same ontology. If b is matched to b' from ontology 2, and a is matched to a', then a' should also be externally connected to b'. In general, we could have required that topological relations are preserved by the matching, and if two individuals a and b are in relation R, then their candidate matches a' and b' should also be in R, otherwise there is a contradiction and we should retract one of the matches ($a = a'$ or $b = b'$). However, this does not work given the quality of geometry information in crowd-sourced data, such as the OSM data set. Consider the following example. Objects representing the Prezzo Ristorante in OSGB and in OSM are correctly matched, and so are objects corresponding to the Blue Bell Inn. However, in the OSGB ontology the geometries of Prezzo Ristorante and Blue Bell Inn are correctly represented as disconnected (DC), whilst in OSM they are represented as externally connected (EC) [2].

Example 2.

$$OSGB : 1000002309000257 = OSM : 116824670^3 \tag{6}$$

$$OSGB : 1000002308429988 = OSM : 116824687^4 \tag{7}$$

$$(OSGB : 1000002309000257, OSGB : 1000002308429988) \in DC \tag{8}$$

$$(OSM : 116824670, OSM : 116824687) \in EC \tag{9}$$

---

[2] In RCC8, DC and EC are disjoint.
[3] Prezzo Ristorante, Restaurant
[4] Blue Bell Inn, Pub

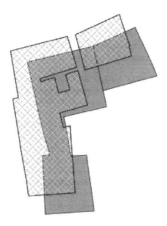

Fig. 1. In OSM (dark), Prezzo Ristorante is externally connected to Blue Bell Inn, whilst in OSGB (light), they are disconnected, but very near. The smaller polygons represent Prezzo Ristorante.

We have discovered empirically that locations and shapes of buildings in Nottingham are shifted and distorted in the OSM data set by as much as 20 meters. This means that we have no guarantee that even very loose pure topological constraints are preserved (e.g. two objects which are disconnected in one data set may be not disconnected in another data set). On the other hand, it seems a reasonable hypothesis that objects shown to be hundreds of meters apart in one data set will be disconnected in another one. Capturing this intuition however involves some metric information, namely distances, and not only pure topological relations. This is what we do in this paper: introduce a very simple logic for capturing the notions of 'far enough' (definitely disconnected even given the margin of error for this data set) and 'near' (could intersect given the margin of error). We call these relations FAR and NEAR for brevity, although we should say that we are not trying to capture the colloquial meaning of 'far' and 'near', rather try to reason qualitatively about distance with respect to some fixed margin of error in positioning. We also assume for simplicity that the error is the same for the whole OSM data set, in particular it is the same for different types and sizes of objects, and that it is completely random (the actual geometry of a building can be of any size or shape within a 'buffer' determined by the margin of error).

2   Related Work

The integration of data from crowd-sourced (e.g. OSM) and authoritative (e.g. OSGB) sources offers several new opportunities, such as lowering the cost, capturing richer user-based information, and providing data which is more up-to-date and complete [6]. However, whilst government collected data typically have a specified accuracy, quality standards and validation procedures, crowd-sourced data does not. Accuracy can vary widely, and the data may include systematic errors (for example, due to incorrectly calibrated data collection methods), gross errors due to mistakes or the insertion of deliberately spurious data, and a range of other types of errors. Furthermore, classification systems (for example, school, playing field) used by crowd-sourced sites are often informal, unlike the ontology-based systems used by some authoritative data sources.

Many previous approaches to the goal of semantically aligning data sources have taken the approach of using known ontologies that are aligned either manually [7] or using automated or semi-automated alignment approaches [8,9]. Due to limited space, we recommend Euzenat and Shvaiko [10] for general ontology matching methods, and Cali et al [11] for probabilistic mappings. Alignment approaches make use of lexical (for example, class names), structural (parent/child relations) and other information (for example, properties). Recent work has considered the use of topological spatial relations between geometries to augment semantic matching [12,13]. However, for the purposes of integration of sources with potentially poor accuracy like crowd-sourced geographic information, strict topological relations are unsuitable, and relations that incorporate geometry boundary uncertainty are required.

Clementini and De Felice [14] identify three broad methods for dealing with boundary uncertainty with respect to topology: fuzzy, probabilistic and extensions of exact models. Fuzzy approaches model the vagueness involved in the assertion of spatial relations (attaching degrees of truth to topological relations) and include applications of the 9 intersection [15] model (such as [16,17,18,19,20]) and the region connection calculus (RCC)[21] (such as [22,23,24]).

Probabilistic approaches model likelihood, and applications to spatial relations include the work of Winter [25], in which the problem of imprecision in the location of the objects resulting in uncertain spatial relations is addressed, and Montello et al [26], in which frequency is used as a measure to determine confidence intervals in defining a fuzzy region ('downtown').

In contrast to fuzzy and probabilistic approaches, tolerance relations have been studied using rough sets, i.e. the lower and the upper approximations of the original sets, based on rough set theory [27]. Clementini and De Felice [14] present the notion of approximate topological relations, extending the existing 9 intersection model [15] with broad geometry boundaries. Broad boundaries identify the maximum and minimum extent of a boundary for a spatial object. They define 44 9 intersections representing the possible relations between two regions with broad boundaries. Buffers are one application of this work, and thus approximate topological relations provide some foundations for our work. Another related approach is Cohn and Gotts' egg-yolk theory [28]. They define relations between regions with indeterminate boundaries (conceptualised as eggs with a yolk and a white, the latter being the indeterminate area), and also address differences in boundary crispness.

Other extensions to the exact model include the use of three valued logic based on extending RCC to cater for indeterminate regions using Lukasiewicz algebras [29]. Also related is Blakemore's work [30], which does not use logic, but identifies a set of 7 values resulting from an analysis of digitising error: including possibly in; unassignable, ambiguous; etc. Some work has also been done on the use of supervaluation semantics to define vagueness [31], but this focused on vagueness in attribute definitions of geographic features rather than uncertainty. In addition, a large number of logical treatments of non-topological aspects of space have been developed, including Zimmerman and Freksa [32]; Dutta [33] and Frank [34] among others, along with work combining topological and metric operators (such as [35,36]).

Our work differs from this previous work on topology in that we do not adopt fuzzy or probabilistic approaches. We define a simple logic that can be used for comparisons between pairs of objects to determine how to treat them in an integrated database. Like the approximate topological relations and egg-yolk theories, we use a buffer to deal with uncertainty. However, unlike these approaches, we do not define a region of certainty inside the uncertain region. For our purposes, the entire region, including the region itself and the buffer (the yolk and the white) is uncertain. This is because in crowd-sourced data, errors that result in complete displacement of objects are common. Secondly, unlike previous buffer-based approaches, we explore in more detail the relations between the proximity of the buffer zones, and define a logic to determine the nearness and farness of pairs of uncertain regions for the purpose of determining whether they represent the same object.

Previous work on nearness has explored the factors that define nearness, including Euclidean distance; relative distance between the objects and a reference point; the size of the area being assessed; attractiveness; reachability and density of surrounding objects [37,38,39,40]. Attempts to mathematically model nearness have included fuzzy approaches ([41,42]) and the definition of nearness relations in terms of a ratio between the area of a buffer to the area of the region [36]. Worboys [43] compares three different methods for representing nearness: broad boundaries [14]; fuzzy boundaries and nearness relations and four valued logic (near, far, both and neither). Frank [34] defines a qualitative distance function, ranging through five values: very close, close, neutral, far and very far, including some properties and rules relating to the function, which has some similarities to our work, as does Hernandez et al [38], who combines distance and orientation.

Our work differs from the previous work on nearness in that we define a logic of near and far for the purposes of semantic matching of objects from different data sources, and do not attempt to model human notions of nearness. We define exact values for the calculation of whether two objects are to be considered near or far, based on an error function σ, thus providing a strict mathematical definition. While this makes our approach unlikely to be suitable for the simulation of human notions of nearness, it provides a useful tool for geographic data comparison and integration.

3 Logic

In this section, we present the logic we use to represent distance constraints in the presence of a possible measurement error. We fix this error to be a positive constant number σ. We assume that σ can be determined empirically by comparing two geospatial ontologies representing the same objects, and finding the largest 'distortion' which exists between any pair of objects from the two ontologies. As we mentioned before, in the case of OSGB and OSM ontologies for Nottingham the empirically determined σ is 20 m.

3.1 Syntax

The syntax of our logic contains binary predicates $NEAR$, FAR and BEQ ('Buffered EQual'). BEQ is used to generate initial matches between objects based on their location: two points a and b are considered potentially the same if they are within σ distance of each other [5]. Below we present a simple version of the logic which assumes that each object has a single point characterising it. For polygon geometries, buffered equals means that a is included in the σ buffer of b, this is, for every point p in a, there exists a point q in b such that p and q are within σ distance, and vice versa. Two objects are $NEAR$, if they possibly contain some common point(s), given the margin of error σ in representation, this is, they are within $2 * \sigma$ distance of each other. Two objects are FAR, if after shifting each towards the other by σ, they are still not $NEAR$, this is, their distance is larger than $4 * \sigma$.

- Individual names (terms): $a, b, c...$;
- Predicate letters (binary): $BEQ, NEAR, FAR$;
- Logical connectives: $\neg, \wedge, \vee, \rightarrow$.

The predicate letters applied to terms yield the atomic formulas, for example, $BEQ(a, b)$. The well-formed formulas (wffs) are defined as follows:

- Every atomic formula is a wff;
- If α and β are wffs, then $\neg\alpha, \alpha \wedge \beta, \alpha \vee \beta, \alpha \rightarrow \beta$ are wffs.

The logic we are proposing is essentially propositional, since we only allow constants as arguments of predicates. The reason we do not introduce quantification over points is to keep the complexity of reasoning in the logic low and its use as part of an automated ontology matching tool practicable. (It is well known that first order logic even with one binary predicate is undecidable.)

3.2 Semantics

We interpret the logic over models which are based on a metric space (similar to other spatial logics, such as [45] and [46]).

---

[5] Similar to the definition of tolerance space by Poston [44].

Definition 1 (Metric Space). *A metric space is a pair* (Δ, d), *where* Δ *is a set and* d *is a metric on* Δ, *i.e., a function* $d : \Delta \times \Delta \longrightarrow \mathbb{R}$ *such that for any* $x, y, z \in \Delta$, *the following holds:*

1. $d(x, y) = 0$ *iff* $x = y$
2. $d(x, y) = d(y, x)$
3. $d(x, z) \leq d(x, y) + d(y, z)$

Definition 2 (Metric Model). *A metric model* M *is a tuple* (Δ, d, I, σ), *where* (Δ, d) *is a metric space,* I *is an interpretation function which maps each constant to an element of* Δ, *and* $\sigma \in \mathbb{R}$ *is the margin of error,* $\sigma > 0$. *The notion of* $M \models \phi$ *(* ϕ *is true in model* M *) is defined as follows:*

$M \models BEQ(a, b)$ *iff* $d(I(a), I(b)) \in [0, \sigma]$
$M \models NEAR(a, b)$ *iff* $d(I(a), I(b)) \in [0, 2 * \sigma]$
$M \models FAR(a, b)$ *iff* $d(I(a), I(b)) \in (4 * \sigma, +\infty)$
$M \models \neg\alpha$ *iff* $M \not\models \alpha$
$M \models \alpha \wedge \beta$ *iff* $M \models \alpha$ *and* $M \models \beta$
$M \models \alpha \vee \beta$ *iff* $M \models \alpha$ *or* $M \models \beta$
$M \models \alpha \rightarrow \beta$ *iff* $M \not\models \alpha$ *or* $M \models \beta$

The notions of validity and satisfiability in metric models are standard. A formula is satisfiable if it is true in some metric model. A formula ϕ is valid ($\models \phi$) if it is true in all metric models (hence if its negation is not satisfiable).

3.3 Axioms

The following calculus LNF (which stands for "Logic of NEAR and FAR") will be shown to characterise metric models:

Axiom 0 All tautologies of classical propositional logic
Axiom 1 $BEQ(a, a)$;
Axiom 2 $BEQ(a, b) \rightarrow BEQ(b, a)$;
Axiom 3 $NEAR(a, b) \rightarrow NEAR(b, a)$;
Axiom 4 $FAR(a, b) \rightarrow FAR(b, a)$;
Axiom 5 $BEQ(a, b) \wedge BEQ(b, c) \rightarrow NEAR(c, a)$;
Axiom 6 $BEQ(a, b) \wedge NEAR(b, c) \wedge BEQ(c, d) \rightarrow \neg FAR(d, a)$;
Axiom 7 $NEAR(a, b) \wedge NEAR(b, c) \rightarrow \neg FAR(c, a)$;
MP Modus ponens: $\phi, \phi \rightarrow \psi \vdash \psi$

The notion of derivability $\Gamma \vdash \phi$ in LNF is standard. A formula ϕ is LNF-derivable if $\vdash \phi$. A set Γ is (LNF) inconsistent if for some formula ϕ it derives both ϕ and $\neg\phi$.

We have the following derivable formulas (which we will refer to as facts in the completeness proof):

Fact 8 $BEQ(a, b) \wedge BEQ(b, c) \wedge NEAR(c, d) \rightarrow \neg FAR(d, a)$;
Fact 9 $BEQ(a, b) \rightarrow NEAR(a, b)$;

Fact 10 $FAR(a, b) \rightarrow \neg NEAR(a, b)$;

Fact 11 $BEQ(a, b) \wedge FAR(b, c) \rightarrow \neg NEAR(c, a)$;

Fact 12 $BEQ(a, b) \rightarrow \neg FAR(a, b)$;

Fact 13 $BEQ(a, b) \wedge BEQ(b, c) \rightarrow \neg FAR(c, a)$;

Fact 14 $BEQ(a, b) \wedge BEQ(b, c) \wedge BEQ(c, d) \rightarrow \neg FAR(d, a)$;

Fact 15 $BEQ(a, b) \wedge BEQ(b, c) \wedge BEQ(c, d) \wedge BEQ(d, e) \rightarrow \neg FAR(e, a)$

3.4   Soundness and Completeness

In this section we prove that LNF is sound and complete for metric models, namely that

$$\vdash \phi \Leftrightarrow \models \phi$$

(every derivable formula is valid and every valid formula can be derived.)

Theorem 1. *Every LNF derivable formula is valid:*

$$\vdash \phi \Rightarrow \models \phi$$

Proof. The proof is by an easy induction on the length of the derivation of ϕ. Axioms 1-7 are valid (by the truth definition of BEQ, $NEAR$ and FAR) and modus ponens preserves validity.

In the rest of this section, we prove completeness:

$$\models \phi \Rightarrow \vdash \phi$$

We will actually prove that given a finite consistent set of formulas, we can build a satisfying model for it. This shows that $\nvdash \phi \Rightarrow \nmodels \phi$ and by contraposition we get completeness.

Given a consistent set of formulas Σ, we try to construct a metric model for a maximal consistent set Σ^+ containing Σ. Firstly, we show that, there is a corresponding set of distance constraints of Σ^+, represented as $D(\Sigma^+)$. Then, we define path-consistency in this context, and show that, if $D(\Sigma^+)$ is path-consistent, then there is a metric space (Δ, d) such that all constraints in $D(\Sigma^+)$ are satisfied by d. At last, we show that, $D(\Sigma^+)$ is path-consistent. Therefore, there is a metric space that can be extended to a metric model satisfying Σ^+, thus, Σ.

Firstly, we will show how to transform a consistent set of formulas to a set of distance constraints.

Definition 3 (MCS). *A set of formulas Γ is maximal consistent if Γ is consistent, and any set of formulas properly containing Γ is inconsistent. If Γ is a maximal consistent set of formulas, then we call it a MCS.*

The following are standard properties of $MCSs$.

Proposition 1 (Properties of MCSs). *If Γ is a MCS, then,*

- *Γ is closed under modus ponens: if ϕ, $\phi \rightarrow \psi \in \Gamma$, then $\psi \in \Gamma$;*

- *if ϕ is derivable, then $\phi \in \Gamma$;*
- *for all formulas ϕ: $\phi \in \Gamma$ or $\neg\phi \in \Gamma$;*
- *for all formulas ϕ, ψ: $\phi \wedge \psi \in \Gamma$ iff $\phi \in \Gamma$ and $\psi \in \Gamma$;*
- *for all formulas ϕ, ψ: $\phi \vee \psi \in \Gamma$ iff $\phi \in \Gamma$ or $\psi \in \Gamma$.*

Lemma 1 (Lindenbaum's Lemma). *If Σ is a consistent set of formulas, then there is a MCS Σ^+ such that $\Sigma \subseteq \Sigma^+$.*

We assume that Σ is a finite consistent set of formulas. Then, Σ^+, a MCS containing Σ, contains the same set of constants as Σ.

Lemma 2. *If Σ^+ be a MCS, then, for any pair of individual names a, b occurring in Σ, exactly one of the following cases holds:*

- $BEQ(a, b) \in \Sigma^+$;
- $\neg BEQ(a, b) \wedge NEAR(a, b) \in \Sigma^+$;
- $\neg NEAR(a, b) \wedge \neg FAR(a, b) \in \Sigma^+$;
- $FAR(a, b) \in \Sigma^+$.

Proof. For any pair of individual names a, b occurring in Σ, we have:

$$\vdash (B \wedge N \wedge F) \vee (B \wedge N \wedge \neg F) \vee (B \wedge \neg N \wedge F) \vee (B \wedge \neg N \wedge \neg F) \vee (\neg B \wedge N \wedge F) \vee (\neg B \wedge N \wedge \neg F) \vee (\neg B \wedge \neg N \wedge F) \vee (\neg B \wedge \neg N \wedge \neg F)$$

where B, N, F stand for $BEQ(a, b), NEAR(a, b), FAR(a, b)$ respectively. By Fact 10, Fact 9 and Fact 10, Fact 12, Fact 9, Fact 10, Fact 10, Fact 9 and Fact 10, and Fact 9 respectively, we have

$$\vdash \perp \vee B \vee \perp \vee \perp \vee \perp \vee (\neg B \wedge N) \vee F \vee (\neg N \wedge \neg F)$$

this is,

$$\vdash B \vee (\neg B \wedge N) \vee (\neg N \wedge \neg F) \vee F$$

QED.

Definition 4. *An interval h is non-negative, if $h \subseteq [0, +\infty)$.*

Definition 5. *A distance constraint tells in what range the distance between two points falls, it is represented as $d(a, b) \in g$, where a, b are constants representing points, g is a non-negative interval. g is called the distance range for a, b.*

Lemma 3. *If Σ^+ be a MCS, then, for any pair of individual names a, b occurring in Σ, $d(a, b) = d(b, a)$.*

Proof. Follows from Axiom 2, 3, 4 ($BEQ, NEAR, FAR$ are all symmetric).

Definition 6. *Let Σ^+ be a MCS. Given a fixed positive σ, a corresponding set of distance constraints $D^\sigma(\Sigma^+)$ is constructed as follows. Initially, $D^\sigma(\Sigma^+) = \{\}$. For every individual name a in A, if $BEQ(a, a) \in \Sigma^+$, then we add $d(a, a) \in \{0\}$ to $D^\sigma(\Sigma^+)$. For every pair of different individual names a, b involved in Σ^+, if*

- $BEQ(a, b) \in \Sigma^+$: add $d(a, b) = d(b, a) \in [0, \sigma]$ to $D^\sigma(\Sigma^+)$;
- $\neg BEQ(a, b) \wedge NEAR(a, b) \in \Sigma^+$: add $d(a, b) = d(b, a) \in (\sigma, 2*\sigma]$ to $D^\sigma(\Sigma^+)$;
- $\neg NEAR(a, b) \wedge \neg FAR(a, b) \in \Sigma^+$: add $d(a, b) = d(b, a) \in (2*\sigma, 4*\sigma]$ to $D^\sigma(\Sigma^+)$;
- $FAR(a, b) \in \Sigma^+$: add $d(a, b) = d(b, a) \in (4*\sigma, +\infty)$ to $D^\sigma(\Sigma^+)$.

For readability, we will omit σ in $D^\sigma(\Sigma^+)$.

Lemma 4. *Let Σ^+ be a MCS. Then in $D(\Sigma^+)$, for every pair of constants, there is only one distance range for them.*

Proof. Follows from Lemma 2, 3 and Definition 6.

In this paper, a set of distance constraints refers to a set where for every pair of constants involved, there is only one distance range for them.

Definition 7 (Composition). *Given d_1, d_2 are non-negative numbers, the composition $\{d_1\} \circ \{d_2\} = [|d_1 - d_2|, d_1 + d_2]$ [6]. Given g_1, g_2 are non-negative intervals, their composition is an interval which is the union of all $\{d_1\} \circ \{d_2\}$, where $d_1 \in g_1, d_2 \in g_2$, this is,*

$$g_1 \circ g_2 = \bigcup\nolimits_{d_1 \in g_1, d_2 \in g_2} \{d_1\} \circ \{d_2\}$$

Due to limited space, we will not provide the proofs for some of the lemmas below.

Lemma 5. *Given g_1, g_2 are non-negative non-empty intervals, if $d_3 \in g_1 \circ g_2$, then there exist $d_1 \in g_1$, $d_2 \in g_2$ such that $d_3 \in [|d_1 - d_2|, d_1 + d_2]$.*

Lemma 6 (Calculation of Composition). *Given $(m, n), (s, t)$, $\{l\}, \{r\}, g_1, g_2$ are non-negative non-empty intervals, h_1, h_2, h are non-negative intervals, based on Definition 7, their composition is calculated as follows:*

1. $\{l\} \circ \{r\} = [l - r, l + r]$, *if* $l \geq r$;
2. $\{l\} \circ (s, t) = (s - l, t + l)$, *if* $s \geq l$;
3. $\{l\} \circ (s, t) = [0, t + l)$, *if* $l \in (s, t)$;
4. $\{l\} \circ (s, t) = (l - t, t + l)$, *if* $t \leq l$;
5. $\{l\} \circ (s, +\infty) = (s - l, +\infty)$, *if* $s \geq l$;
6. $\{l\} \circ (s, +\infty) = [0, +\infty)$, *if* $s < l$;
7. $(m, n) \circ (s, t) = (s - n, t + n)$, *if* $s \geq n$;
8. $(m, n) \circ (s, t) = [0, t + n)$, *if* $(m, n) \cap (s, t) \neq \emptyset$;
9. $(m, n) \circ (s, +\infty) = (s - n, +\infty)$, *if* $s \geq n$;
10. $(m, n) \circ (s, +\infty) = [0, +\infty)$, *if* $s < n$;
11. $(m, +\infty) \circ (s, +\infty) = [0, +\infty)$;
12. $\emptyset \circ \emptyset = \emptyset$;
13. $g_1 \circ \emptyset = \emptyset$;
14. $g_1 \circ g_2 \neq \emptyset$;
15. $h_1 \circ h_2 = h_2 \circ h_1$;

---

[6] Based on $d(x, z) \leq d(x, y) + d(y, z)$ (Property 3 of Definition 1).

16. $(h_1 \cup h_2) \circ h = (h_1 \circ h) \cup (h_2 \circ h)$;
17. $(h_1 \cap h_2) \circ h = (h_1 \circ h) \cap (h_2 \circ h)$, if $(h_1 \cap h_2) \neq \emptyset$;
18. $h_1 \circ h_2 \circ h = h_1 \circ (h_2 \circ h)$.

Given an interval h of the form (l, u), $[l, u)$, $(l, u]$ or $[l, u]$, let us call l the lower bound of h, represented as $lower(h)$, and u the upper bound of h, represented as $upper(h)$.

Lemma 7. *Given g, h are non-negative non-empty intervals, the following properties hold:*

- $lower(g) \geq 0$;
- $upper(g) \geq lower(g)$;
- $upper(g \circ h) = upper(g) + upper(h)$
- $lower(g \circ h) \leq max(lower(g), lower(h))$

Definition 8 (Path Consistency). *Given a set of distance constraints D, for every pair of constants a, b, the range of their distance is strengthened by enforcing path-consistency as follows until a fixed point is reached:*

$$\forall c : g(a, b) \leftarrow g(a, b) \cap (g(a, c) \circ g(c, b))$$

where c is a constant different from a, b, $g(a, b)$ denotes the distance range for a, b. If at the fixed point, for every pair of constants a, b, there exists a valid value for their distance, this is, $g(a, b) \neq \emptyset$, then D is path-consistent.

In the following, we will show there is a metric space satisfying $D(\Sigma^+)$. By Definition 6, an interval h contained in $D(\Sigma^+)$ is one of $\{0\}$ ($[0, 0]$), $[0, \sigma]$, $(\sigma, 2 * \sigma]$, $(2 * \sigma, 4 * \sigma]$, $(4 * \sigma, +\infty)$, which are defined as identity or primitive intervals.

Definition 9. *A non-negative non-empty continuous interval h is an identity interval, if $h = \{0\}$. h is primitive, if h is one of $[0, \sigma]$, $(\sigma, 2 * \sigma]$, $(2 * \sigma, 4 * \sigma]$, $(4 * \sigma, +\infty)$. h is composite, if it can be composed using primitive intervals. h is definable, if it is primitive or composite.*

Lemma 8. *For any identity interval or definable interval h, $h \circ \{0\} = h$.*

Lemma 9. *If an interval h is definable and $h \neq \emptyset$, the following properties hold:*

- $lower(h) = n * \sigma, n \in \{0, 1, 2, 3, 4\}$;
- $upper(h) = m * \sigma, m = 1, 2, 3, ...$;
- $lower(h) = 4 * \sigma$, iff $h = (4 * \sigma, +\infty)$;
- $upper(h) = \sigma$, iff $h = [0, \sigma]$

Let us call a set of distance constraints which involve only right-closed intervals (that is, of the form $(m, n]$ or $[m, n]$) or right-infinite intervals (with ∞ upper bound) *right-closed*.

Lemma 10. *If D is a set of distance constraints involving only identity or primitive intervals, D is path-consistent and D^f is a fixed point of enforcing path consistency on D, then D^f is right-closed.*

Proof. The set of constraints involving only identity or primitive intervals, $\{0\}$, $[0, \sigma]$, $(\sigma, 2 * \sigma]$, $(2 * \sigma, 4 * \sigma]$, $(4 * \sigma, \infty)$ is clearly right-closed. Enforcing path consistency produces new intervals by two operations, composition and intersection. By inspection of composition rules (Lemma 6), we see that composing two intervals which are right-closed or right-infinite gives a right-closed or a right-infinite interval. The same obviously applies to intersection. So if we start with a right-closed set, then enforcing path consistency will always produce a right-closed set. QED.

Then, we show how to construct a metric space from D^f, a fixed point.

Lemma 11. *Given g_1, g_2, g_3 are non-negative non-empty right-closed intervals, if*

 - $g_1 \subseteq g_2 \circ g_3$
 - $g_2 \subseteq g_1 \circ g_3$
 - $g_3 \subseteq g_1 \circ g_2$

then,

 - $upper(g_1) \leq upper(g_2) + upper(g_3)$;
 - $upper(g_2) \leq upper(g_1) + upper(g_3)$;
 - $upper(g_3) \leq upper(g_1) + upper(g_2)$.

Proof. Suppose $g_1 \subseteq g_2 \circ g_3$. Since $upper(g_1) \in g_1$, $upper(g_1) \in g_2 \circ g_3$. By Lemma 5, there exists $d_2 \in g_2$, $d_3 \in g_3$, such that $upper(g_1) \leq d_2 + d_3$. Since $d_2 \leq upper(g_2)$, $d_3 \leq upper(g_3)$, $upper(g_1) \leq upper(g_2) + upper(g_3)$. Similarly, $upper(g_2) \leq upper(g_1) + upper(g_3)$, $upper(g_3) \leq upper(g_1) + upper(g_2)$. QED.

Lemma 12. *Let g_1, g_2, g_3 be non-negative non-empty intervals, g_1 is right-infinite, and*

 - $g_1 \subseteq g_2 \circ g_3$
 - $g_2 \subseteq g_1 \circ g_3$
 - $g_3 \subseteq g_1 \circ g_2$

then g_2 or g_3 is right-infinite.

Proof. Suppose g_1 is right-infinite. Since $g_1 \subseteq g_2 \circ g_3$, $g_2 \circ g_3$ is right-infinite. By Definition 7 and Lemma 6, g_2 or g_3 is right-infinite. QED.

Lemma 13. *If D is a set of distance constraints involving only identity or primitive intervals over a set of m ($m > 0$) constants, D is path-consistent, D^f is a fixed point of enforcing path consistency on D, D_s is obtained from D^f by replacing every right-infinite interval with $\{5 * \sigma * m\}$, every right-closed interval h with $\{upper(h)\}$, then D_s is path-consistent.*

Proof. Suppose D is path-consistent. By Lemma 10, D^f is right-closed. By Definition 8, for every interval h in D^f, $h \neq \emptyset$. To prove D_s is path-consistent, we only need to show that for any three distance ranges, $\{n_{ab}\}, \{n_{bc}\}, \{n_{ac}\}$ in D_s over three constants a, b, c, we have

1. $n_{ab} \leq n_{bc} + n_{ac}$;
2. $n_{bc} \leq n_{ab} + n_{ac}$;
3. $n_{ac} \leq n_{ab} + n_{bc}$.

Let h_{ab}, h_{bc}, h_{ac} denote the corresponding distance ranges of $\{n_{ab}\}, \{n_{bc}\}, \{n_{ac}\}$ respectively in D^f, by Definition 8, we have

- $h_{ab} \subseteq h_{bc} \circ h_{ac}$;
- $h_{bc} \subseteq h_{ab} \circ h_{ac}$;
- $h_{ac} \subseteq h_{ab} \circ h_{bc}$;

- if all h_i ($i = ab, bc, ac$) are right-closed, then, $n_i = upper(h_i)$. By Lemma 11, $1 - 3$ hold.
- else, not all of them are right-closed, by Lemma 12, at least two of them are right-infinite.
  - if all of them are right-infinite, then $n_i = 5 * \sigma * m$. Since $5 * \sigma * m \leq 5 * \sigma * m + 5 * \sigma * m$, $1 - 3$ hold.
  - else, only one of them is right-closed. Let h_{ab} be right-closed. Then, $n_{ab} = upper(h_{ab})$, $n_{bc} = 5 * \sigma * m$, $n_{ac} = 5 * \sigma * m$. In D, there are m constants, and for every pair of constants a, b, their distance range r is primitive or $\{0\}$. If r is right-closed, then $upper(r) \leq 4 * \sigma$. By Lemma 6, for every composite interval c, if c is right-closed, then $0 < upper(c) < 4 * \sigma * m$. Since intersection does not generate new upper bounds, for every right-closed interval h in D^f, $0 < upper(h) < 4 * \sigma * m$. Thus, $0 < upper(h_{ab}) < 4*\sigma*m < 5*\sigma*m$. Since $upper(h_{ab}) < 5*\sigma*m+5*\sigma*m$ and $5 * \sigma * m < 5 * \sigma * m + upper(h_{ab})$, $1 - 3$ hold. QED.

Lemma 14. *Let Σ^+ be a MCS, if its corresponding set of distance constraints $D(\Sigma^+)$ is path-consistent, then there is a metric space (Δ, d) such that all constraints in $D(\Sigma^+)$ are satisfied by d.*

Proof. Suppose $D(\Sigma^+)$ is path-consistent. By Definition 6 and Definition 9, $D(\Sigma^+)$ is a set of distance constraints involving only identity or primitive intervals. Let D^f be a fixed point of enforcing path consistency on $D(\Sigma^+)$, D_s is obtained from D^f by replacing every right-infinite interval with $\{5 * \sigma * m\}$, every right-closed interval h with $\{upper(h)\}$. Let Δ be the set of constants occurring in Σ^+. By Definition 6, for constants x, y, if $x = y$, then $d(x, y) = 0$. By Lemma 13, if $x \neq y$, $d(x, y) \geq \sigma > 0$. Thus, we have $d(x, y) = 0$ iff $x = y$. By Definition 6, for any pair of constants x, y, $d(x, y) = d(y, x)$ holds. By Lemma 13, D_s is path-consistent. Thus, $d(x, z) \leq d(x, y) + d(y, z)$ holds. By Definition 1, there is a metric space (Δ, d) such that all constraints in $D(\Sigma^+)$ are satisfied by d. QED.

We will prove that $D(\Sigma^+)$ is path-consistent by case analysis. Given an upper bound or a lower bound of a definable interval h, Lemma 15 - 20 show all the possibilities of h. Due to limited space, we will not show the proofs for them here.

Lemma 15. *If an interval h is definable, $upper(h) = 2 * \sigma$, then $h = (\sigma, 2 * \sigma]$ or $h = [0, \sigma] \circ [0, \sigma]$.*

Lemma 16. *If an interval h is definable, $upper(h) = 3 * \sigma$, then $h = [0, \sigma] \circ (\sigma, 2 * \sigma]$ or $h = [0, \sigma] \circ [0, \sigma] \circ [0, \sigma]$.*

Lemma 17. *If an interval h is definable, $upper(h) = 4 * \sigma$, then $h = (2 * \sigma, 4 * \sigma]$ or $h = (\sigma, 2 * \sigma] \circ (\sigma, 2 * \sigma]$ or $h = [0, \sigma] \circ [0, \sigma] \circ [0, \sigma] \circ [0, \sigma]$ or h is composed from two $[0, \sigma]$ and one $(\sigma, 2 * \sigma]$.*

Lemma 18. *If an interval h is definable, $lower(h) = 3 * \sigma$, then $h = [0, \sigma] \circ (4 * \sigma, +\infty)$.*

Lemma 19. *If an interval h is definable, $lower(h) = 2 * \sigma$, then $h = (2 * \sigma, 4 * \sigma]$ or $h = (\sigma, 2 * \sigma] \circ (4 * \sigma, +\infty)$ or h is composed using two $[0, \sigma]$ and one $(4 * \sigma, +\infty)$.*

Lemma 20. *If an interval h is definable, $lower(h) = \sigma$, then $h = (\sigma, 2 * \sigma]$ or $h = [0, \sigma] \circ (2 * \sigma, 4 * \sigma]$ or h is composed using three $[0, \sigma]$ and one $(4 * \sigma, +\infty)$ or h is composed using one $[0, \sigma]$, one $(\sigma, 2 * \sigma]$ and one $(4 * \sigma, +\infty)$.*

Lemma 21. *If an interval h is primitive or $\{0\}$, then h cannot be composed using primitive intervals.*

Definition 10 (Cycle). *Given a set of formulas B of the language, there exists a cycle, if $R_0(a_0, a_1), R_1(a_1, a_2), ..., R_n(a_n, a_0)$ is derivable from B, where $R_i \in \{BEQ, NEAR, FAR, \neg BEQ, \neg NEAR, \neg FAR\}$.*

Lemma 22. *Let Σ^+ be a MCS, then its corresponding set of distance constraints $D(\Sigma^+)$ is path-consistent.*

Proof. Suppose $D(\Sigma^+)$ is not path-consistent. Then $\emptyset$ is obtained when strengthening intervals. Let's look at how the first $\emptyset$ can be obtained. It must be obtained using the strengthening operator as follows:

$$g_1 \cap (g_2 \circ g_3) = \emptyset$$

and each $g_i \neq \emptyset$. g_i may be an identity or primitive interval, or it can be written as $x_i \cap (y_i \circ z_i)$, where each of x_i, y_i, z_i may not be an identity or primitive interval also. In the latter case, $g_i = x_i \cap (y_i \circ z_i) \neq \emptyset$. Thus, x_i, y_i, z_i are all not empty, etc. Given $s_1 \cap s_2 \neq \emptyset$, we use $(s_1 \cap s_2) \circ s_3 = (s_1 \circ s_3) \cap (s_2 \circ s_3)$ (Rule 17 in Lemma 6), where s_i is a non-negative interval, repeatedly until every interval is an identity or primitive interval. The final form is the intersection of identity or definable intervals, as follows:

$$h_1 \cap ... \cap h_n = \emptyset, n > 1$$

where h_i, $(1 \leq i \leq n)$, is either $\{0\}$ or definable and $h_i \neq \emptyset$. Thus there exist two intervals h_i, h_j, $(1 \leq i \leq n, 1 \leq j \leq n, i \neq j)$, such that $h_i \cap h_j = \emptyset$, h_i, h_j cannot both be primitive or $\{0\}$ (since there is only one identity or primitive interval for each pair of constants (Lemma 4) and Lemma 21). There are only the following cases:

- $upper(h_i) = 0$, $lower(h_j) \in \{\sigma, 2 * \sigma, 3 * \sigma, 4 * \sigma\}$;
- $upper(h_i) = \sigma$, $lower(h_j) \in \{\sigma, 2 * \sigma, 3 * \sigma, 4 * \sigma\}$;
- $upper(h_i) = 2 * \sigma$, $lower(h_j) \in \{2 * \sigma, 3 * \sigma, 4 * \sigma\}$;
- $upper(h_i) = 3 * \sigma$, $lower(h_j) \in \{3 * \sigma, 4 * \sigma\}$
- $upper(h_i) = 4 * \sigma$, $lower(h_j) = 4 * \sigma$

We will check whether $\perp$ can be derived in every case using axioms (or derivable facts). From Lemma 15 - 20, we know all the ways a definable interval, whose upper or lower bound is given, can be composed from primitive intervals. We also know from Lemma 21 that primitive intervals cannot be obtained by composition, hence they come from the original set of constraints $D(\Sigma^+)$ and for each of them there is a corresponding formula in Σ^+ (see Definition 6). By Axiom 2-4, $BEQ, NEAR, FAR$ are symmetric.

- $upper(h_i) = 0$: by Definition 9, $h_i = [0, 0]$, then it comes from $BEQ(a, a)$ for some individual name a. By Axiom 1, $BEQ(a, a)$ is valid.
  - $lower(h_j) = \sigma$: by Lemma 20, h_j has the following cases:
    * $h_j = (\sigma, 2 * \sigma]$: invalid case, h_i, h_j cannot both be primitive or $\{0\}$.
    * $h_j = [0, \sigma] \circ (2 * \sigma, 4 * \sigma]$:
      $h_j = h_1 \circ h_2$, $h_1 = [0, \sigma]$, so it comes from $BEQ(a, b)$. $h_2 = (2 * \sigma, 4 * \sigma]$, so it comes from $\neg NEAR(b, a)$ and $\neg FAR(b, a)$, where b is an individual name. Invalid case.
    * h_j is composed using three $[0, \sigma]$ and one $(4 * \sigma, +\infty)$:
      $BEQ(a, a), BEQ(a, b), BEQ(b, c), BEQ(c, d), FAR(d, a)$ form a cycle. By Fact 14, $BEQ(a, b), BEQ(b, c), BEQ(c, d) \rightarrow \neg FAR(d, a)$, and $\neg FAR(d, a) \wedge FAR(d, a) \rightarrow \perp$.
    * h_j is composed using one $[0, \sigma]$, one $(\sigma, 2 * \sigma]$ and one $(4 * \sigma, +\infty)$ one BEQ, one $NEAR$ and one FAR, using Fact 11.
  - $lower(h_j) = 2 * \sigma$: by Lemma 19, h_j has the following cases:
    * $h_j = (2 * \sigma, 4 * \sigma]$: invalid case
    * $h_j = (\sigma, 2 * \sigma] \circ (4 * \sigma, +\infty)$: invalid case.
    * h_j is composed using two $[0, \sigma]$ and one $(4 * \sigma, +\infty)$:
      two BEQ, one FAR, using Fact 13.
  - $lower(h_j) = 3 * \sigma$: by Lemma 18, $h_j = [0, \sigma] \circ (4 * \sigma, +\infty)$. Invalid case.
  - $lower(h_j) = 4 * \sigma$: by Lemma 9, $h_j = (4 * \sigma, +\infty)$. Invalid case.
- $upper(h_i) = \sigma$: by Lemma 9, $h_i = [0, \sigma]$, so we have one BEQ in the cycle.
  - $lower(h_j) = \sigma$: by Lemma 20, h_j has the following cases:
    * $h_j = (\sigma, 2 * \sigma]$: invalid case.
    * $h_j = [0, \sigma] \circ (2 * \sigma, 4 * \sigma]$: one BEQ, one $\neg NEAR$, using Axiom 5.
    * h_j is composed using three $[0, \sigma]$ and one $(4 * \sigma, +\infty)$:
      three BEQ and one FAR in the cycle, using Fact 15.
    * h_j is composed using one $[0, \sigma]$, one $(\sigma, 2 * \sigma]$ and one $(4 * \sigma, +\infty)$ one BEQ, one $NEAR$ and one FAR, using Axiom 6, Fact 8.
  - $lower(h_j) = 2 * \sigma$: by Lemma 19, h_j has the following cases:
    * $h_j = (2 * \sigma, 4 * \sigma]$: invalid case
    * $h_j = (\sigma, 2 * \sigma] \circ (4 * \sigma, +\infty)$:
      one $NEAR$, one FAR in the cycle, using Fact 11.

       * h_j is composed using two $[0, \sigma]$ and one $(4 * \sigma, +\infty)$:
         two BEQ, one FAR in the cycle, using Fact 14.
- $lower(h_j) = 3 * \sigma$: by Lemma 18, $h_j = [0, \sigma] \circ (4 * \sigma, +\infty)$.
   one BEQ, one FAR, using Fact 13.
- $lower(h_j) = 4 * \sigma$: by Lemma 9, $h_j = (4 * \sigma, +\infty)$. Invalid case.

— $upper(h_i) = 2 * \sigma$: by Lemma 15, h_i has the following cases:
- $h_i = (\sigma, 2 * \sigma]$: one $NEAR$
     * $lower(h_j) = 2 * \sigma$: by Lemma 19, h_j has the following cases:
       · $h_j = (2 * \sigma, 4 * \sigma]$: invalid case.
       · $h_j = (\sigma, 2 * \sigma] \circ (4 * \sigma, +\infty)$:
         one $NEAR$, one FAR in the cycle, using Axiom 7.
       · h_j is composed using two $[0, \sigma]$ and one $(4 * \sigma, +\infty)$:
         two BEQ, one FAR in the cycle, using Axiom 6, Fact 8.
     * $lower(h_j) = 3 * \sigma$: by Lemma 18, $h_j = [0, \sigma] \circ (4 * \sigma, +\infty)$.
       one BEQ, one FAR, using Fact 11.
     * $lower(h_j) = 4 * \sigma$: by Lemma 9, $h_j = (4 * \sigma, +\infty)$. Invalid case.
- $h_i = [0, \sigma] \circ [0, \sigma]$: two BEQ
     * $lower(h_j) = 2 * \sigma$: by Lemma 19, h_j has the following cases:
       · $h_j = (2 * \sigma, 4 * \sigma]$: one $\neg NEAR$, using Axiom 5.
       · $h_j = (\sigma, 2 * \sigma] \circ (4 * \sigma, +\infty)$:
         one $NEAR$, one FAR in the cycle, using Axiom 6, Fact 8.
       · h_j is composed using two $[0, \sigma]$ and one $(4 * \sigma, +\infty)$:
         two BEQ, one FAR in the cycle, using Fact 15.
     * $lower(h_j) = 3 * \sigma$: by Lemma 18, $h_j = [0, \sigma] \circ (4 * \sigma, +\infty)$.
       one BEQ, one FAR, using Fact 14.
     * $lower(h_j) = 4 * \sigma$: by Lemma 9, $h_j = (4 * \sigma, +\infty)$.
       one FAR, using Fact 13.

— $upper(h_i) = 3 * \sigma$: by Lemma 16, h_i has the following cases:
- $h_i = [0, \sigma] \circ (\sigma, 2 * \sigma]$: one BEQ, one $NEAR$
     * $lower(h_j) = 3 * \sigma$: by Lemma 18, $h_j = [0, \sigma] \circ (4 * \sigma, +\infty)$.
       one BEQ, one FAR, using Axiom 6, Fact 8.
     * $lower(h_j) = 4 * \sigma$: by Lemma 9, $h_j = (4 * \sigma, +\infty)$.
       one FAR, using Fact 11.
- $h_i = [0, \sigma] \circ [0, \sigma] \circ [0, \sigma]$: three BEQ
     * $lower(h_j) = 3 * \sigma$: by Lemma 18, $h_j = [0, \sigma] \circ (4 * \sigma, +\infty)$.
       one BEQ, one FAR, using Fact 15.
     * $lower(h_j) = 4 * \sigma$: by Lemma 9, $h_j = (4 * \sigma, +\infty)$.
       one FAR, using Fact 14.

— $lower(h_j) = 4 * \sigma$: by Lemma 9, $h_j = (4 * \sigma, +\infty)$, one FAR in cycle.
- $upper(h_i) = 4 * \sigma$: by Lemma 17, h_i has the following cases:
     * $h_i = (2 * \sigma, 4 * \sigma]$: invalid case;
     * $h_i = (\sigma, 2 * \sigma] \circ (\sigma, 2 * \sigma]$: two $NEAR$ in cycle, using Axiom 7.
     * $h_i = [0, \sigma] \circ [0, \sigma] \circ [0, \sigma] \circ [0, \sigma]$:
       four BEQ in cycle, using Fact 15.
     * h_i is composed from two $[0, \sigma]$ and one $(\sigma, 2 * \sigma]$:
       two BEQ, one $NEAR$ in cycle, using Axiom 6, Fact 8.

In each case, $\perp$ is derivable using the corresponding axiom, this contradicts the assumption that Σ^+ is consistent. Therefore, $D(\Sigma^+)$ is path-consistent. QED.

Theorem 2. *If a finite set of formulas Σ is LNF-consistent, there exists a metric model satisfying it.*

Proof. Given Σ, by Lemma 1, we can construct a MCS Σ^+ containing it. If Σ is LNF-consistent, so is Σ^+, and hence by Lemma 22 and Lemma 14 there is a metric space (Δ, d) such that all constraints in $D(\Sigma^+)$ are satisfied by d. Extend this to a model M by setting $I(a) = a$ for every a occurring in Σ. By the definition of $D(\Sigma^+)$ (Definition 6) and by properties of maximal consistent sets (Proposition 1), for every $\phi \in \Sigma^+$,

$$\phi \in \Sigma^+ \;\Leftrightarrow\; M \models \phi$$

hence since $\Sigma \subseteq \Sigma^+$, M satisfies all formulas in Σ.

4   Preliminary Experimental Results

In this section, we sketch how the Logic of NEAR and FAR can be used to debug matchings between different geospatial ontologies. At the moment, we do not have a complete implementation of all LNF rules in a tableau theorem prover. Preliminary experiments reported below use standard OWL-DL reasoning in Pellet together with the statement about disjointness of NEAR and FAR roles.

In our experiments, we used the two data sets mentioned in the Introduction: OSGB and OSM ontologies representing respectively 713 and 253 objects in Nottingham[7]. We investigated how many incorrect matchings are found by using the logic of NEAR and FAR to check consistency. To do this, we generated initial matchings between objects by using purely lexical matching of their labels. Then given $\sigma = 20m$ and initial matchings, we generated correct $NEAR$ and FAR facts for all matched objects based on their location. That is, for every object which has a match in the other data set, we generated NEAR and FAR relations to every other object which also has a match to some object in the other data set. For example, if we have two matchings, $OSGB : a = OSM : b$ and $OSGB : c = OSM : d$, we generate NEAR and FAR relations for (a, b), (a, c), (a, d), (b, c), ..., and also $NEAR(a, a)$, $NEAR(b, b)$, $NEAR(c, c)$ and $NEAR(d, d)$. Then we used Pellet to detect inconsistencies in the resulting set of statements (the two data sets, the matchings between individuals, and the $NEAR$ and FAR relations). Intuitively, each inconsistency contains a wrong matching between two individuals. The experiments are performed on an Intel Dual Core 2.00 GHz, 3.00 GB RAM personal computer from command line. Times are in seconds, averaged over 5 runs.

Below is an example of the kind of inconsistencies arising between NEAR and FAR relations for wrong matchings.

---

[7] The data used in experiments is available online
http://www.cs.nott.ac.uk/~hxd/GeoMap.html

Table 1. Detecting errors in lexical matchings between OSGB and OSM instances using the logic of NEAR and FAR

number of matches	108
number of matched objects	133
number of $NEAR, FAR$ facts	16907
time (seconds)	47
number of errors found	72

Example 3. The object a in $OSGB$ and the object b in OSM are lexically matched (10), since both have the same label, for example 'Prezzo Ristorante'. However, there exists an object c, for example 'Blue Bell Inn', in $OSGB$ such that a is far from it (11), whilst b is close to it (12).

$$OSGB : a = OSM : b \qquad (10)$$

$$(OSGB : c, OSGB : a) \in FAR \qquad (11)$$

$$(OSGB : c, OSM : b) \in NEAR \qquad (12)$$

From (10) and (12), we have $(OSGB : c, OSGB : a) \in NEAR$, which contradicts (11), since $NEAR$ and FAR are disjoint. Therefore, (10) is incorrect.

5 Conclusion

We presented a logic LNF which formalises concepts of being 'definitely near' and 'definitely far' with respect to a metric distance and a fixed margin of error. We give a sound and complete axiomatisation for this logic. In our future work, we plan to extend the logic from single points to polygon geometries, and provide decision procedures for the logics. The concepts formalised in the logic are motivated by using metric constraints to find errors in matches between geospatial objects from different ontologies. We present preliminary experimental results which illustrate the use of the logic.

References

1. Du, H., Alechina, N., Jackson, M., Hart, G.: Matching geospatial ontologies. In: Proceedings of the ISWC Workshop Ontology Matching, OM 2012 (2012)
2. Du, H., Alechina, N., Jackson, M., Hart, G.: Matching formal and informal geospatial ontologies. In: Geographic Information Science at the Heart of Europe. Lecture Notes in Geoinformation and Cartography. Springer (2013)
3. Sirin, E., Parsia, B., Grau, B.C., Kalyanpur, A., Katz, Y.: Pellet: A Practical OWL-DL Reasoner. Web Semantics: Science, Services and Agents on the World Wide Web 5, 51–53 (2007)
4. Hart, G., Dolbear, C., Kovacs, K., Guy, A.: Ordnance Survey Ontologies (2008), http://www.ordnancesurvey.co.uk/oswebsite/ontology
5. OpenStreetMap: The Free Wiki World Map (2012), http://www.openstreetmap.org

6. Jackson, M.J., Rahemtulla, H., Morley, J.: The Synergistic Use of Authenticated and Crowd-Sourced Data for Emergency Response. In: International Workshop on Validation of Geo-Information Products for Crisis Management (2010)
7. Stock, K., Cialone, C.: An approach to the management of multiple aligned multi-lingual ontologies for a geospatial earth observation system. In: Claramunt, C., Levashkin, S., Bertolotto, M. (eds.) GeoS 2011. LNCS, vol. 6631, pp. 52–69. Springer, Heidelberg (2011)
8. Rodríguez, M.A., Egenhofer, M.J.: Determining semantic similarity among entity classes from different ontologies. IEEE Transactions on Knowledge and Data Engineering 15(2), 442–456 (2003)
9. Cruz, I.F., Sunna, W.: Structural Alignment Methods with Applications to Geospatial Ontologies. Transactions in GIS 12(6), 683–711 (2008)
10. Euzenat, J., Shvaiko, P.: Ontology Matching. Springer (2007)
11. Calì, A., Lukasiewicz, T., Predoiu, L., Stuckenschmidt, H.: A Framework for Representing Ontology Mappings under Probabilities and Inconsistency. In: The Third ISWC Workshop on Uncertainty Reasoning for the Semantic Web (2007)
12. Fonseca, R.L., Llano, E.G.: Automatic representation of geographical data from a semantic point of view through a new ontology and classification techniques. Transactions in GIS 15(1), 61–85 (2011)
13. Klien, E., Lutz, M.: The role of spatial relations in automating the semantic annotation of geodata. In: Cohn, A.G., Mark, D.M. (eds.) COSIT 2005. LNCS, vol. 3693, pp. 133–148. Springer, Heidelberg (2005)
14. Clementini, E., Di Felice, P.: Approximate topological relations. International Journal of Approximate Reasoning 16(2), 173–204 (1997)
15. Egenhofer, M.J., Herring, J.: Categorizing binary topological relations between regions, lines, and points in geographic databases. Technical report, Department of Surveying Engineering, University of Maine (1991)
16. Zhan, F.B.: Approximate analysis of binary topological relations between geographic regions with indeterminate boundaries. Soft Computing 2(2), 28–34 (1998)
17. Schneider, M.: A design of topological predicates for complex crisp and fuzzy regions. In: Kunii, H.S., Jajodia, S., Sølvberg, A. (eds.) ER 2001. LNCS, vol. 2224, pp. 103–116. Springer, Heidelberg (2001)
18. Tang, X., Kainz, W.: Analysis of topological relations between fuzzy regions in a general fuzzy topological space. In: Proceedings of the Symposium on Geospatial Theory, Processing and Applications (2002)
19. Liu, K., Shi, W.: Computing the fuzzy topological relations of spatial objects based on induced fuzzy topology. International Journal of Geographical Information Science 20(8), 857–883 (2006)
20. Du, S., Qin, Q., Wang, Q., LI, B.: Fuzzy description of topological relations I: A unified fuzzy 9-intersection model. In: Wang, L., Chen, K., S. Ong, Y. (eds.) ICNC 2005. LNCS, vol. 3612, pp. 1261–1273. Springer, Heidelberg (2005)
21. Cohn, A.G., Bennett, B., Gooday, J., Gotts, N.M.: Qualitative spatial representation and reasoning with the region connection calculus. GeoInformatica 1(3), 275–316 (1997)
22. Guesgen, H.W., Albrecht, J.: Imprecise reasoning in geographic information systems. Fuzzy Sets and Systems 113(1), 121–131 (2000)
23. Li, Y., Li, S.: A fuzzy sets theoretic approach to approximate spatial reasoning. IEEE Transactions on Fuzzy Systems 12(6), 745–754 (2004)
24. Schockaert, S., Cock, M.D., Cornelis, C., Kerre, E.E.: Fuzzy region connection calculus: Representing vague topological information. International Journal of Approximate Reasoning 48(1), 314–331 (2008)

25. Winter, S.: Uncertain topological relations between imprecise regions. International Journal of Geographical Information Science 14(5), 411–430 (2000)
26. Montello, D., Goodchild, M., Gottsegen, J., Fohl, P.: Where's downtown?: Behavioral methods for determining referents of vague spatial queries. Spatial Cognition and Computation 3, 185–204 (2003)
27. Pawlak, Z., Polkowski, L., Skowron, A.: Rough Set Theory. In: Wiley Encyclopedia of Computer Science and Engineering (2008)
28. Cohn, A.G., Gotts, N.M.: The 'Egg-Yolk' Representation of Regions with Indeterminate Boundaries. In: GISDATA Specialist Meeting on Geographical Objects with Undetermined Boundaries, pp. 171–187 (1996)
29. Roy, A.J., Stell, J.G.: Spatial Relations between Indeterminate Regions. International Journal of Approximate Reasoning 27(3), 205–234 (2001)
30. Blakemore, M.: Generalization and error in spatial databases. Cartographica 21, 131–139 (1984)
31. Bennett, B.: What is a forest? on the vagueness of certain geographic concepts. Topoi 20(2), 189–201 (2001)
32. Zimmermann, K., Freksa, C.: Qualitative spatial reasoning using orientation, distance, and path knowledge. Applied Intelligence 6(1), 49–58 (1996)
33. Dutta, S.: Qualitative spatial reasoning: A semi-quantitative approach using fuzzy logic. In: Buchmann, A.P., Smith, T.R., Wang, Y.-F., Günther, O. (eds.) SSD 1989. LNCS, vol. 409, pp. 345–364. Springer, Heidelberg (1990)
34. Frank, A.U.: Qualitative spatial reasoning about distances and directions in geographic space. Journal of Visual Languages and Computing 3(4), 343–371 (1992)
35. Gerevini, A., Renz, J.: Combining Topological and Size information for Spatial Reasoning. Artificial Intelligence 137(1-2), 1–42 (2002)
36. Shariff, A.R., Egenhofer, M., Mark, D.: Natural-Language Spatial Relations Between Linear and Areal Objects: The Topology and Metric of English-Language Terms. International Journal of Geographical Information Science 12(3) (1998)
37. Gahegan, M.: Proximity operators for qualitative spatial reasoning. In: Kuhn, W., Frank, A.U. (eds.) COSIT 1995. LNCS, vol. 988, pp. 31–44. Springer, Heidelberg (1995)
38. Hernández, D., Clementini, E., Felice, P.D.: Qualitative distances. In: Kuhn, W., Frank, A.U. (eds.) COSIT 1995. LNCS, vol. 988, pp. 45–57. Springer, Heidelberg (1995)
39. Dolbear, C., Hart, G., Goodwin, J.: From theory to query: Using ontologies to make explicit imprecise spatial relationships for database querying. Poster presented at Conference on Spatial Information Theory, COSIT (2007)
40. Fisher, P., Orf, T.: An investigation of the meaning of near and close on a university campus. Computers, Environment and Urban Systems 15, 23–35 (1991)
41. Robinson, V.B.: Individual and multipersonal fuzzy spatial relations acquired using human-machine interaction. Fuzzy Sets and Systems 113(1), 133–145 (2000)
42. Schockaert, S., Cock, M.D., Cornelis, C., Kerre, E.E.: Fuzzy region connection calculus: An interpretation based on closeness. International Journal of Approximate Reasoning 48(1), 332–347 (2008)
43. Worboys, M.F.: Nearness relations in environmental space. International Journal of Geographical Information Science 15(7), 633–651 (2001)
44. Poston, T.: Fuzzy geometry. PhD thesis, University of Warwick (1971)
45. Wolter, F., Zakharyaschev, M.: Reasoning about distances. In: Gottlob, G., Walsh, T. (eds.) Proceedings of the Eighteenth International Joint Conference on Artificial Intelligence (IJCAI 2003), pp. 1275–1282. Morgan Kaufmann (2003)
46. Aiello, M., Pratt-Hartmann, I.E., van Benthem, J.F.: Handbook of Spatial Logics. Springer-Verlag New York, Inc., Secaucus (2007)

The Topology of Spatial Scenes in $\mathbb{R}^2$

Joshua A. Lewis, Matthew P. Dube, and Max J. Egenhofer

School of Computing and Information Science, University of Maine
5711 Boardman Hall, Orono, ME 04469-5711, USA
{joshua.lewis,matthew.dube}@umit.maine.edu, max@spatial.maine.edu

Abstract. Spatial scenes are abstractions of some geographic reality, focusing on the spatial objects identified and their spatial relations. Such qualitative models of space enable spatial querying, computational comparisons for similarity, and the generation of verbal descriptions. A specific strength of spatial scenes is that they offer a focus on particular types of spatial relations. While past approaches to representing spatial scenes, by recording exhaustively all binary spatial relations, capture accurately how pairs of objects are related to each other, they may fail to distinguish certain spatial properties that are enabled by an ensemble of objects. This paper overcomes such limitations by introducing a model that considers (1) the topology of potentially complexly structured spatial objects, (2) modeling applicable relations by their boundary contacts, and (3) considering exterior partitions and exterior relations. Such qualitative scene descriptions have all ingredients to generate topologically correct graphical renderings or verbal scene descriptions.

Keywords: Spatial scenes, topological relations, compound objects, ensembles of spatial regions.

1 Introduction

Spatial scenes are abstractions of some geographic reality, focusing on the spatial objects identified and their spatial relations [3]. Objects may also be associated with non-spatial properties [31], yielding comprehensive qualitative representations of some mini-world. Such qualitative models of space enable spatial querying, computational comparisons for similarity, and the generation of verbal descriptions. A specific strength of spatial scenes is that they offer a focus on particular types of spatial relations so that users' analyses may put explicit emphasis on properties that may be of particular importance for the task at hand. For instance, matching a sketch, represented as a spatial scene [17], with the closest configurations in a spatial database may ignore metric relations embodied in the sketch, as sketches are often not to scale, while emphasizing relations about relative directions among the objects drawn. This paper focuses on the *topology of spatial scenes* that are embedded in $\mathbb{R}^2$.

Fig. 1 depicts two simple scenes consisting of an ensemble of *spatial regions*. A spatial region is closed, homogenously 2-dimensional without separations or holes [14]. A scene representation may be thought of as a directed graph, in which each node represents a region and each directed edge, connecting two distinct nodes, captures the spatial relations between the two regions.

T. Tenbrink et al. (Eds.): COSIT 2013, LNCS 8116, pp. 495–515, 2013.

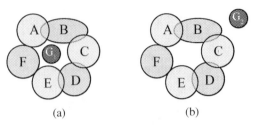

(a)   (b)

Fig. 1. A graphical depiction of a spatial scene in $\mathbb{R}^2$ for which the mere set of binary topological relations is insufficient to distinguish between (a) object G_1 being surrounded by the union of A through F and (b) object G_2 being outside the union

Representations of spatial scenes typically resort to capturing exhaustively all binary spatial relations, yielding for a scene of n objects and m types of spatial relations $m*(n^2-n)/2$ relations. In some cases this set of qualitative scene relations may be further reduced by eliminating redundancies that are implied by compositions [18,36] as well as dependencies across relations [20,29,38]. While this approach captures accurately how each object is related to each other object in a spatial scene, it may fail to distinguish certain spatial properties that are enabled by the *ensemble* of objects. The two configurations shown in Fig. 1 share the same pairs of topological relations as modeled by the 4-intersection [14,22] or RCC-8 [34]. While direction relations, such as G_1 is south of B, but G_2 is north-east of B, would contribute to distinguishing the two configurations, on a purely topological basis the set of binary topological relations would suggest they are topologically equivalent [17,21].

The mere pairwise recording of binary relations may have another shortcoming when multiple objects interact with the same object. The distinction between different types of *meet, overlap, covers,* and *coveredBy* relations between two regions have been addressed extensively [13,21,26], but these methods may fall short of capturing topological equivalence if multiple, possibly complex boundary-boundary interactions occur. Fig. 2 shows for two configurations how the breakdown of such a complex multi-object relation into all pairs of binary relations is insufficient to distinguish some critical topological information enabled only by the ensemble of the objects.

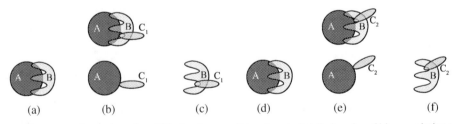

(a)   (b)   (c)   (d)   (e)   (f)

Fig. 2. Two configurations that differ by the way C touches A, but their pairs of binary relations show, however, pairwise the same topological relations: (a) and (d) with the same type of *overlap,* (b) and (e) with a simple *meet,* and (c) and (f) with the same type of *overlap*

This paper develops a formal model that addresses an arbitrarily complex topology of region-based spatial scenes, embedded in $\mathbb{R}^2$. After a summary of related work (Section 2), Section 3 develops a model for complexly structured objects. Section 4 accounts for intersection ambiguities among multiple objects (Fig. 2) by modeling boundary contacts, while Section 5 addresses relations in different exterior partitions. Section 6 demonstrates how both boundary contacts and exterior partitions may need to be used simultaneously for capturing the topology of complex spatial scenes. The paper concludes in Section 7 with a summary and a discussion of future work.

2   Topological Relations

Existing methods for modeling spatial relations have been comprehensively compiled in several survey articles [10,11,25]. While early approaches [24] addressed spatial relations in an integrated fashion, the use of tailored methods for different types—topological, direction, and metric—has prevailed during the last decade. Current models for *topological* relations fall primarily into two major categories: (1) those based on connection [34] and (2) those based on intersection [14,15]. For two simple regions, both models yield the same topological relations (Fig. 3).

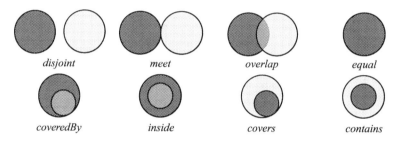

Fig. 3. The eight topological relations between two simple regions in $\mathbb{R}^2$ resulting from both the 9-intersection and the region-connection calculus (using the 9-intersection terminology)

Their extensions to capture more details—up to topological equivalence [21]—are particularly relevant for relations with boundary intersections, such as *overlap* [26], *meet*, and *covers/coveredBy* [13]. Some alternative models for spatial relations based on interval relations [26], direction [30,39], and 2-D strings [5], have been primarily applied to image retrieval. While RCC as a first-order theory deals with sets of regions, including possible holes, one could assume that this approach would lead directly to capturing the topology of a scene. The idiosyncrasies of the connection-base relations of single regions with multiple parts, however, yield undesired effects, as it captures topologically distinct configurations by the same relation (Fig. 4).

Fig. 4. Two scenes with a set of regions, both of which RCC classifies as *partially overlapping* (PO)

2.1   Models of Complex Spatial Features

Complex regions mainly refer to regions with separations of the interior (multiple components) and of the exterior (holes). Worboys and Bofakos [44] introduced the concept of a *generic area* based on a labeled-tree representation. Areal objects with separations, holes, and islands nested to any finite level are modeled by describing each level with a set of nodes. These nodes are topologically equivalent to the unit disk, pairwise disjoint or with a finite intersection, and may spatially contain their child nodes. The root node represents the entire region object. Clementini *et al.* [8] define *composite regions* as closed, non-empty, 2-dimensional subsets of $\mathbb{R}^2$. Each component is a simple region (with no holes). The components' interiors are pairwise disjoint and their boundaries are either pairwise disjoint or intersect at a finite set of points. A *complex region,* however, has components that may be either simple regions or regions with holes [7].

The OpenGIS Consortium (OGC) has informally described geometric features, called *simple features*, in the OGC Abstract Specification [32] and in the Geography Markup Language GML [33] an XML encoding for the transport and storage of geographic information. Among these features, *MultiPolygons* refer to complex regions that may have multiple components and holes. A complex region [37] comprises one or several regular sets, called *faces* that may have holes and other faces as islands in the holes. A hole within a face can at most touch the boundary of the face or of another hole at a single point. Each face is atomic and cannot be further decomposed. A complex region can comprise multiple faces, which can either be *disjoint, meet* at one or many single boundary points, or lie *inside* of the hole of another face and possibly share one or more single boundary points with the hole.

Ontological aspects of parts [12,42] and holes [4] have been studied extensively, with a specific emphasis on part-whole relations as well as mereology—the theory of parthood and relations—and mereotopology [41]—a first-order theory that blends mereological and topological concepts. These reflections and formalizations have been linked mostly to the region-connection calculus [34].

2.2   Models for Topological Relations with Holes or Separations

Among the many variations of models for topological relations are six previous approaches that address explicitly some aspects of a qualitative spatial-relation system for holed or separated regions. We briefly summarize their key features and analyze what aspects of holes and separations they cover, including their existing support for similarity and composition reasoning.

- The region-connection calculus RCC-8 [34] specifies eight relations that also apply to regions that may have holes and/or separations. This elegant yet coarse model suffers from a representation shortcoming, since it does not distinguish between significantly different topological configurations as it makes no distinction for holed regions whether an interaction takes place from the outside or the inside. Since the RCC-8 composition table [35] applies to an all-encompassing region, the particularities of holes and parts that constrain inferences at times are not accounted for, so that only an upper bound for the inference possibilities is available at times, which renders some of these composition inferences imprecise, because impossible derivations are included in the set of inferred possibilities.

- The use of the vanilla 9-intersection for complex regions [37] yields 33 relations, some of which expose the same issues as RCC-8. It further considers only holes that are fully contained, but none that are on the fringe (i.e., tangential). No neighborhood graphs or compositions have been derived.
- The Topological-Relations-for-Composite-Regions (TRCR) model [7,8] captures primarily relations of regions with separations, but indicates TRCR's extension to holed regions. TRCE uses a coarse underlying relation model derived from disjunctions of 9-intersection relations. But these disjunctions are not jointly exhaustive and pairwise disjoint, so that, for instance, equal must be derived from A *in* B and B *in* A. Therefore, the set of relations has no explicit identity relation. Here also no neighborhood graphs or compositions have been derived.
- Egenhofer *et al.* [19] treat topological relations of holed regions exhaustively, allowing for tangential holes, holes that fill a region completely, and multiple holes, but at the expense of a tedious enumeration of all details. While they address the transition to hole-free regions through examples, no generic method for determining a conceptual neighborhood graph or the composition of relations is given.
- Vasardani's model for relations with holed regions is highly expressive, identifying 23 relations between a hole-free to a holed region [23] and 152 between two single-holed regions, together with neighborhood graphs, compositions, and a summary model that applies to multi-holed regions [43]. It, however, deals only with fully contained holes and does not address regions with separations.
- Tryfona and Egenhofer [40] treat relations under the transition from regions with separations to generalized regions by filling in the gaps between parts with a (not necessarily convex) hull. These separations exclude touching components and have no account for holes.

3   Relation-Based Compound Spatial Objects

In order to capture complexly structured spatial objects, we use the *relation-based compound spatial object model*, which creates compound objects from geometric combinations—unions and set differences—of basic spatial object types that fulfill explicitly specified constraints about topological relations [16]. Although the basic object types may be spatial regions [14], simple lines or points [15], within the scope of this paper only compound spatial objects made up of regions are considered.

The topological-relation constraints between object parts are critical to construct the intended topology of the objects. For two regions R_1 and R_2, with the constraint R_1 *contains* R_2, the closure of the set difference $\overline{R_1 \backslash R_2}$ forms a (closed) region with a single hole (Fig. 5a). On the other hand, if R_3 *disjoint* R_4 then $R_3 \cup R_4$ forms a (closed) separated region (Fig. 5b); the construct $\overline{(R_5 \backslash R_6)} \cup R_7$, with the constraint that R_5 *contains* R_6 and R_5 is *disjoint* from R_7, yields a holed region with a separation in the region's exterior (Fig. 5c); while the construct $\overline{(R_8 \backslash R_9)} \cup R_{10}$, with the constraints R_8 *contains* R_9 and R_9 *contains* R_{10}, provides a holed region with a separation in the region's hole (Fig. 5d).

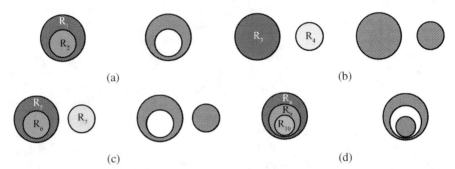

Fig. 5. Construction of compound spatial objects: (a) a holed region from $\overline{R_1\backslash R_2}$ with R_1 *contains* R_2; (b) a separated region from $R_3 \cup R_4$ with R_3 *disjoint* R_4; (c) a holed region with an exterior separation from $\overline{(R_5\backslash R_6)}\cup R_7$ with R_5 contains R_6 and R_7 disjoint R_5; and (d) a holed region with a separation in the hole from $\overline{(R_8\backslash R_9)}\cup R_{10}$ with R_8 *contains* R_9 and R_9 *contains* R_{10}

The advantage of such constructions of compound spatial objects is that the objects' topology enables immediately a specification of the topological relation involving compound spatial objects. There is no need to continue the search for more and more complicated spatial relations for complex objects (e.g., relations between a simple region and a 2-holed region), as the relation-based compound-object model lends itself immediately to specifying topological relations based on the components topological relations. Relations other than *disjoint* and *contains*, such as *meet* or *covers*, may participate in the construction of compound spatial objects as well, but if they yield interior or exterior separations, the methods of partitioning (Section 5) may be necessary.

4   Boundary Contacts

To fully capture the topology of a spatial scene, one needs to consider the impact of the boundary intersections that the topological relations *meet, overlap, covers,* and *coveredBy* possess. These boundary intersections are a key ingredient for the reconstruction of a spatial scene from its scene description.

The method chosen to construct this scenario is based on following the boundary components of each object in the scene on a course through a Venn diagram view of the space, in a clockwise direction, combined with the semantic differences of two line-line relations—*touches* and *crosses* [6,21,28]—that are congruent under the 9-intersection.

Definition 1. Let A and B be two lines. A and B *touch* if the local intersection configuration is such that a consistent clockwise traversal visits the same line twice in a row (Fig. 6a).

Definition 2. Let A and B be two lines. A and B *cross* if the local intersection configuration is such that a consistent clockwise traversal visits the lines in an alternating sequence (Fig. 6b).

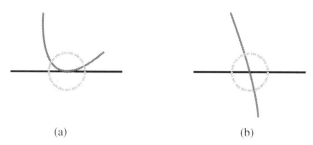

Fig. 6. Line intersections: (a) two lines in a touch configuration, where the order of contact in any cyclic path surrounding the intersection is of the form (black, red, red, black), and (b) two lines in a cross configuration where the order of contact in any cyclic path surrounding the intersection is of the form (black, red, black, red)

Theorem 1. Let A and B be two lines such that A *touches* B. This configuration cannot be transformed into a single A *crosses* B configuration without first going through another line-line relation with a different 9-intersection matrix.

Proof: Both *touches* and *crosses* have the same 9-intersection matrix, but their local intersection structure is different in $\mathbb{R}^2$ because of co-dimension constraints. To get from a single *touch* to a single *cross* without passing through another relation requires fixing the intersection of the two lines topologically, that is, to maintain that, the *intersection zone*—the part of B originally in common with A°—moves only along the interior of the line. If it moves to the boundary of either line, the 9-intersection matrix of the configuration changes. If it moves off of either line in any direction, one of two things must occur: (1) the intersection zone will move to the exterior and cease to be a *touches* or *crosses* configuration, thus going through at least another 9-intersection symbol, or (2) the intersection zone will be moved to the other side of the line, fixing the 9-intersection symbol, but producing two *crosses* configurations in the process. To remove either of these instances would require the boundary of A to enter the interior of B or the boundary of B to enter the interior of A, both of which producing a separate 9-intersection matrix. ∎

Theorem 1 shows that the line-line relations *touch* and a *cross* (though congruent under the 9-intersection) are different. Herring [28] also demonstrates the differences between these two configurations. To deal with boundaries of regions (bounded by Jordan Curves), the process differentiates occurrences of *touches* from *crosses*. Both *touch* and *cross* can exist in either 0-dimensional or 1-dimensional forms, leaving four possible types of *interactions* between any two boundaries: 0-*touch* (denoted as t_0), 0-*cross* (denoted as c_0), 1-*touch* (denoted as t_1), and 1-*cross* (denoted as c_1). Any region's boundary can be partitioned with respect to the other objects in a space based on these four units and a Venn diagram. The development of this method provides a shorthand notation for the process, which we refer to as the *o notation*.

$$\partial A_{comp} : o_S(dimension, T, C) \tag{1}$$

Symbol A_{comp} is a boundary component of a region A, S is the collection of regions of which that boundary component is currently *outside* (symbolized by the letter o), *dimension* is the qualitative length of the interaction (either 0 or 1), T is the collection of region boundaries that are undergoing a *touch* interaction, and C is the collection of regions undergoing a *cross* interaction. This notation accounts for all but the boundedness of the planar-disk classifying invariant [21], which is addressed in Section 5.

Theorem 2. Let A_1 and A_2 be two regions such that their boundaries *cross* (Fig. 7). Traverse ∂A_1 as it approaches A_2 from its exterior. The set C must be a subset of S.

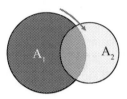

Fig. 7. Two regions whose boundaries cross as a c_0 intersection, as the cross goes from outside of A_2 to inside (symbolized by the arrow) and is of 0-dimension, yielding the o notation $\partial A_1 : o_{\{A_2\}}(0, \emptyset, A_2)$

Proof: Since ∂A_1 is in the exterior of A_2 as it approaches the intersection, $A_2 \in S$. Since ∂A_1 impacts A_2, $A_2 \in C$.   ∎

The same observation in Theorem 2 would also apply to a *touch* configuration upon the sets T and S. In the case of *crosses*, the next o notation (Eqn. 1) has its set S become the set $S\backslash C$, as the boundary crossed from the exterior to the interior of the regions in the set C. This observation is referred to as the *set difference rule*. In the case of *touches*, there are no corresponding changes in S, because the boundary line in the exterior never entered into the interior.

Theorem 3. Let A_1 and A_2 be two regions such that their boundaries *cross* (Fig. 8). Traverse ∂A_1 as it approaches A_2 from its interior. $C \cap S = \emptyset$.

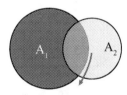

Fig. 8. Two regions whose boundaries cross in a c_0 intersection, as the cross goes from inside of A_2 to outside (symbolized by the arrow), is of 0-dimension, and the position is outside of nothing in the scene yielding the o notation $\partial A_1 : o_{\{\emptyset\}}(0, \emptyset, A_2)$

Proof: Since ∂A_1 is in the interior of A_2 as it approaches the intersection, $A_2 \notin S$. Since ∂A_1 impacts A_2, $A_2 \in C$.   ∎

The same observation in Theorem 3 also applies to a *touch* configuration upon the sets T and S. In the case of *crosses*, the next o notation (Eqn. 1) has its set S become the set $S \cup C$, as the boundary entered the exterior of those regions in the set C. This observation is referred to as the *union rule*. In the case of *touches*, there are no corresponding changes in S, because the boundary line in the interior never entered the exterior.

Theorem 4. Given the o notation (Eqn. 1) for an arbitrary spatial scene. $T \cap C = \emptyset$.

Proof: Since *touch* and *cross* have been shown to be two different configurations (Theorem 1), T and C can have no common elements. ∎

There are scenarios in which a spatial scene would produce sets T and/or C such that either or both of these sets would not be subsets of, or pairwise disjoint from, the corresponding set S. Let a group of objects constitute a spatial scene such that the set C from the o notation (Eqn. 1) is neither a subset of, nor pairwise disjoint from, S. This spatial scene contains an intersection of boundaries such that some go from interior to exterior, while others go from exterior to interior at the same location. The elements that are in common follow the set difference rule, while the elements not in common follow the union rule.

To completely traverse a boundary, the o notation is appended to itself until the entire boundary is traversed. Since the representation starts and ends with the same o notation, the redundant ending is dropped (Fig. 9).

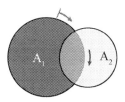

Fig. 9. Two regions whose boundaries cross in two c_0 intersections, yielding the o notation $\partial A_1 \colon o_{\{A_2\}}(0, \emptyset, A_2)o_{\{\emptyset\}}(0, \emptyset, A_2)$. In this convention, the last encountered state always cycles back to the initial outside condition, with the next impact being the first cross/touch

We now consider two scenarios where multiple boundaries come together at a single point (Fig. 10a) or with interrupting intersection (Fig. 10b).

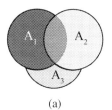 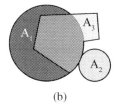

         (a)   (b)

Fig. 10. Two scenarios where multiple boundaries come together: (1) Three objects— A_1 *overlap* A_2, A_1 *meet* A_3, *and* A_2 *meet* A_3—to exhibit the boundary notation, and (b) The sequence for A_3 is $\partial A_3 \colon o_{\{A_1,A_2\}}(1, \emptyset, A_1)o_{\{A_1,A_2\}}(0, A_2, A_1) \ o_{\{A_1,A_2\}}(1, \emptyset, A_1)o_{\{A_2\}}(0, \emptyset, A_1)$. In this configuration, the change to the subscript based on the cross does not happen until the 1-dimensional cross is finished in its entirety

For Fig. 10a, we get three sequences (Eqs. 2-4), starting each boundary traversal in the common exterior in a clockwise direction. The union rule and set difference rule apply for the *crosses* configurations, while both rules do not apply for the *touches* configurations.

$$\partial A_1 : o_{\{A_2, A_3\}}(0, \emptyset, A_2) o_{\{A_3\}}(0, A_3, A_2) o_{\{A_2, A_3\}}(1, A_3, \emptyset) \tag{2}$$

$$\partial A_2 : o_{\{A_1, A_3\}}(1, A_3, \emptyset) o_{\{A_1, A_3\}}(0, A_3, A_1) o_{\{A_3\}}(0, \emptyset, A_1) \tag{3}$$

$$\partial A_3 : o_{\{A_1, A_2\}}(1, A_1, \emptyset) o_{\{A_1, A_2\}}(0, \{A_1, A_2\}, \emptyset) o_{\{A_1, A_2\}}(1, A_2, \emptyset) \tag{4}$$

For Fig. 10b, we get three sequences (Eqs. 5-7), starting each boundary traversal in the common exterior in a clockwise direction. This particular instance demonstrates how a shorter intersection that interrupts a longer intersection is handled for the case of *crosses*. The boundary traversal of A_1 (Eqn. 5) reveals that the union rule ($S \rightarrow S \cup C$) is not followed until the completion of the *cross*, despite the interruption of the *touch*. Similarly, the boundary traverse of A_3 (Eqn. 7) reveals that the set difference rule ($S \rightarrow S \backslash C$) is not followed until the completion of the *cross*, despite the interruption of the *touch*. The notation thus ensures that the *cross* has not been fully completed, distinguishing this scenario from other scenarios where a 0-*touch* and 0-*cross* would coincide between two 1-*cross* impacts. For the scenario of *touches* being interrupted, this cannot be maintained by the mere knowledge of boundary interactions, as it requires the consideration of the partitioning structure of the objects upon the space (Section 5).

$$\partial A_1 : o_{\{A_2, A_3\}}(0, \emptyset, A_3) o_{\{A_2\}}(1, \emptyset, A_3) o_{\{A_2\}}(0, A_2, A_3) o_{\{A_2\}}(1, \emptyset, A_3) \tag{5}$$

$$\partial A_2 : o_{\{A_1, A_3\}}(0, \{A_1, A_3\}, \emptyset) \tag{6}$$

$$\partial A_3 : o_{\{A_1, A_2\}}(1, \emptyset, A_1) o_{\{A_1, A_2\}}(0, A_2, A_1) o_{\{A_1, A_2\}}(1, \emptyset, A_1) o_{\{A_2\}}(0, \emptyset, A_1) \tag{7}$$

5   Exterior Partitions

The model presented thus far for describing a topological scene captures the relations between explicitly defined objects. However, a complete representation requires identifying relationships between objects that lie within separated components of the common exterior. These separated components are classified as *exterior partitions*, and a mechanism is provided for identifying them within a scene. This identification is necessary since existing methods, such as the 9-intersection, capture object-to-object relations, but not the relations between an object and a group of objects (Fig. 11).

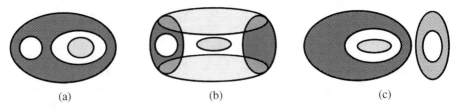

(a)   (b)   (c)

Fig. 11. Examples of topologically distinct configurations: (a) a holed region, (b) a union of objects forms a hole that contains other objects, and (c) a holed region with a separation

Intuitively, the exterior of an object is pictured as resting entirely outside of that object. It is commonly understood that the exterior of a set A within an embedding space X is A's complement; that is, $X \backslash \bar{A}$, or succinctly, A^-. In the simplest cases, exteriors are connected, but it need not be so.

An exterior partition is synonymous with a holed region within an object (Fig. 11a) or a hole formed from the union of multiple objects (Fig. 11b). In these instances the common exterior is not *path-connected*. This statement is equivalent to saying that the exterior is separated, or partitioned. Similarly, it can be seen that an object with a satellite (Fig. 11c) also lacks this property of path connectedness. In order to be self-contained we review the basic definitions of continuous, path, and path-connected [1].

Definition 3. Let X and Y be topological spaces. A function $f: X \to Y$ is *continuous* if $f^{-1}(V)$ is open in X for every open set $V \subset Y$.

Definition 4. Let Y be a topological space, and let $[0,1] \subset R$ have the standard topology. A *path* in Y is a continuous function $p: [0,1] \to Y$. The path *begins at* $p(0)$ and *ends at* $p(1)$.

Definition 5. Let A be a set in a topological space X. A is *path-connected* if for every $a_i, a_j \in A$, there exists a path $p \subset A$ that begins at a_i and ends at a_j.

If an object's interior is separated (Fig. 11c), it can safely be assumed that the separated area is surrounded by something else—generally the common exterior, but possibly other objects. In the same sense an exterior partition is subject to a similar surrounded relation—a singular object, or the union of multiple objects, can partition it (Fig. 11b). In the latter instance, the objects collectively surrounding a partition are also path-connected. A multigraph highlights such a collective surrounding (Fig. 12).

Each object depicted in the scene is represented as a vertex. Edges connect objects that intersect through the *meet* or *overlap* relations, with one edge representing each distinct intersection between the objects. Additional relations are not germane to the construction of edges and can be safely disregarded.

In such a graph, objects with explicitly defined holes are represented as self-loops. To generalize, however, unique loops of arbitrary length can possibly define a surrounds relation—encapsulating an exterior partition or another object. By identifying all such loops the search for exterior partitions can begin. In order to define the object topology for any partitioned segment of the exterior—and any object possibly residing within—the exterior partition must be represented explicitly as a set itself.

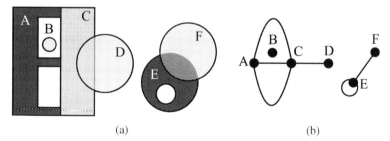

(a)	(b)

Fig. 12. A spatial scene represented (a) graphically and (b) by its corresponding graph

The first step in this process is to represent a holed object A without the exterior partition; in essence an operation is provided to fill the hole, creating a new object B. The concept of a region with its holes filled (Fig. 13) was introduced under the term of a *generalized region* [19], and has been redefined with the term *topological hull* [2] for a 3-dimensional digital embedding (i.e., $\mathbb{Z}^3$). To account for $\mathbb{R}^2$, we give a new definition of the topological hull.

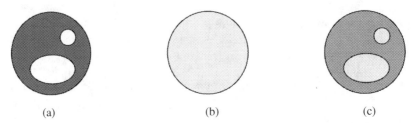

(a)   (b)   (c)

Fig. 13. (a) An object in a scene, (b) the object's topological hull, and (c) the object overlaid by its topological hull

Definition 6. Let A be a closed, path-connected set in $\mathbb{R}^2$ with co-dimension 0 within the standard topology, and let B be the smallest closed set homeomorphic to an n-disk such that $A \subseteq B$. B is called the *topological hull of A*, denoted as $[A]$.

The interior of the originating object, A, is a subset of the interior of the topological hull B. Importantly, $B°$ covers the region coincident to the interior of the partition if it exists, in addition to $A°$.

Theorem 5. If B is the topological hull of A, then $A° \subseteq B°$.

Proof: Since $A \subseteq B$ and B is closed, $A° \subseteq \bar{B}$. Since B is homeomorphic to an n-disk and the standard topology is enforced, $A° \subseteq B°$.   ∎

In a similar fashion, the boundary of the topological hull is a subset of the originating object's boundary as the originating object also bounds any exterior partitions within itself.

Theorem 6. If B is the topological hull of A, then $\partial A \supseteq \partial B$.

Proof: Since A is closed, and B is the smallest closed n-disk such that $A \subseteq B$, the boundary of B is part of the boundary of A. Furthermore, no part of the boundary of B can be outside of the boundary of A, however parts of the boundary of A may actually sit within $B°$, therefore the improper subset is allowed.   ∎

Using such a construction all of the topological hulls of an object A are represented, even if the object has separations, by taking the union of all topological hulls of A. The new object is the *aggregate topological hull* of A.

Definition 7: Let A be a closed set in $\mathbb{R}^2$ with co-dimension 0 within the standard topology. Consider the collection of path-connected subsets P such that $\bigcup_{p \in P} p = A°$ and $|P|$ is minimized. The *aggregate topological hull* $[A]$ is the set $B = \bigcup_{p \in P} [p]$.

Since Definition 6 is a special case of Definition 7 (namely when A has a single path-connected p), the same notation is used for the topological hull $[A]$ and the aggregate topological hull $[A]$.

The simplest scenario consists of an object A with no separations. If A is path-connected then the topological hull of A equals the aggregate topological hull of A (Fig. 14a).

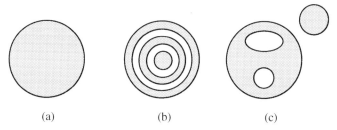

Fig. 14. (a) The topological hull of A equals the aggregate topological hull of A, (b) all separations in A are nested, and (c) A has unnested separations; the aggregate topological hull is not path-connected

Theorem 7. If B is the topological hull of A and C is the aggregate topological hull of A, then $B = C$.

Proof: Since B is the topological hull of A, A is path-connected. The collection of sets P that minimize the cardinality of the set P has exactly one member, A. By definition 5, the aggregate topological hull is constructed by considering the topological hull of each member set. Since A is the only member set, $C = B$, a symmetric relation. Since B is path-connected, C must similarly be path-connected. ∎

Exterior partitions and additional components of A can also be nested within A (Fig. 14b), similar to the construction of a Russian *matryoshka* doll. In this analogy, the outermost shell contains a number of smaller shells within itself, each transitively contained within the first. Unlike a *matryoshka* doll, however, multiple divisions can exist at the same level.

Theorem 8. If A is a set and C its aggregate topological hull such that all but one of A's separations are nested within another (Fig. 14b), then C is path-connected.

Proof: Consider the separation j that is not nested within any other separation. Since j cannot be nested within any other separation and is the only one that cannot be nested, j must surround all of the other separations. Now consider C_j and all immediately nested C_k. $C_k \subset C_j$. Any other C_* nested in a C_k, follows this relationship by the transitivity of subsets. This procedure exhausts all of separations. ∎

The next case is that of an object A with multiple unnested separations (Fig. 14c). In such a case the aggregate topological hull of A is not path-connected.

Theorem 9. If A has more than one separation that cannot be nested, C is not path-connected.

Proof: By Theorems 7 and 8, every set has at least one part that cannot be nested. Theorem 3 covers a path-connected A. Theorem 4 covers a completely nested A. This theorem is the final case: multiple unnested. Consider one of the unnested sets j and all other sets nested within it. By Theorem 8, the C_{j*} for this set is path-connected. Consider another one of the unnested sets k and all other sets nested within it. By Theorem 8, the C_{k*} for this set is path-connected. For C to be path-connected would imply that $\exists C_{j*}, C_{k*} \ni C_{j*}^\circ \cap C_{k*}^\circ = \neg \emptyset$. But C_j and C_k share nothing under definition 5. Therefore, C cannot be path-connected. ∎

By utilizing the properties of an aggregate topological hull, exterior separations within an object A can be identified. If the set difference between the aggregate topological hull C and the originating object A is nonempty, the resulting set coincides with the existence of any exterior partitions contained within.

Theorem 10. If C is the aggregate topological hull of A, C is path-connected, and $C \backslash A = \neg \emptyset$, then A separates A^- into separated components.

Proof: Assume not. There is thus a member of C that is in the aggregate topological hull of A and not in A that is path-connected to every other member of A^-. By Theorems 7–9, A is either a nested construction or itself path-connected. Consider the surrounding set j in the nested construction or the only set A in the path-connected construction. The boundary of C_j is a Jordan Curve, separating its exterior from its interior. The non-empty intersection suggests that at least one element is not in common with A, yet the boundary of C must be contained within the boundary of A (by Theorem 6), and the interior of A must be contained within the interior of C (by Theorem 5). There is thus a point in the interior of C that is not in the closure of A, but still bounded by the Jordan Curve. This point is excludable from the exterior, a contradiction of the assumption.   ■

Conversely there is an additional case, where A and the aggregate topological hull of A completely coincide. In this instance A does not contain an exterior partition.

Theorem 11. If C is the aggregate topological hull of A, and $A=C$, then A^- is path-connected.

Proof: Since C and A have the same membership, A is either an n-disk, or by Theorem 9, would be a combination of n-disks such that none were nested. In this case, the exterior remains path-connected.   ■

After proving the existence of one or more partitioned regions of the exterior, it is necessary to represent them explicitly. By Theorem 10, if $C \backslash A = \neg \emptyset$, A partitions some region of the exterior. Closure performed on this set produces an object H, which coincides with the partitioned regions of the exterior within A (Fig. 15), and can similarly contain nested separations, that is, satellites that are not path-connected, or have no separations.

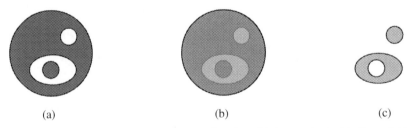

| (a) | (b) | (c) |

Fig. 15. (a) An object with holes and a separated region, (b) the topological hull overlaid over the object, and (c) the exterior partitions within the object

It is, therefore, necessary to take the topological hull of H, and recursively any resulting objects, until the scene has been exhaustively explored. In this manner the topology between objects, the exterior partitions that they form, and any objects contained within can be sufficiently represented.

Bounded and unbounded boundary components [21], which are not captured explicitly in the *o* notation, are distinguished with the aggregate topological hull. The two scenes in Fig. 16a and 16c have the same *o* notation, but they differ in the number of boundary intersections that are bounded and unbounded. Scene A has one bounded boundary intersection, while scene B has two. Unbounded intersections are in the boundary of the aggregate topological hull, while bounded intersections are in its interior. For bounded intersections, this difference is captured by the intersection of the aggregate topological hull's interior with the intersection of the two regions' boundaries, resulting in this example in sets of cardinality one (Fig. 16b) and two (Fig. 16d).

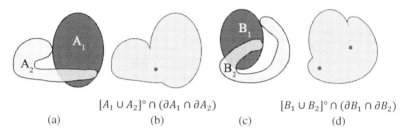

$$[A_1 \cup A_2]^\circ \cap (\partial A_1 \cap \partial A_2) \qquad\qquad [B_1 \cup B_2]^\circ \cap (\partial B_1 \cap \partial B_2)$$

(a)   (b)   (c)   (d)

Fig. 16. Two spatial scenes with different bounded and unbounded boundary intersections, which are captured through the aggregate topological hull

6   Integration of *o* Notation and Topological Hull

This section addresses how the combined use of the *o notation* and the *topological hull* enables the modeling of a complex spatial scene. While so far only scenes were considered that require only one of the two in order to determine completely the topology of a spatial scene, we consider here cases that need both simultaneously.

We start with the two scenes, A and B, in Fig. 17 that have the same *o* notation (Eqs. 8a-c for 16a, and replacing B for A to obtain the *o* notation for 16b). The two scenes are topologically different, however, as scene A has one path-connected exterior separation formed between regions A_1 and A_2, while scene B has two (between B_1 and B_2).

(a)   (b)

Fig. 17. Two topologically different spatial scenes with the same boundary intersections

$$\partial A_1: o_{\{A_2 A_3\}}(0, \emptyset, A_3)o_{\{A_2\}}(1, A_2, \emptyset)o_{\{A_2\}}(0, A_2, A_3)o_{\{A_2 A_3\}}(1, A_2, \emptyset) \tag{8a}$$

$$\partial A_2: o_{\{A_1 A_3\}}(0, A_1, A_3)o_{\{A_1\}}(1, \emptyset, A_1)o_{\{A_1\}}(0, \emptyset, A_3)o_{\{A_1 A_3\}}(1, A_1, \emptyset) \tag{8b}$$

$$\partial A_3: o_{\{A_1 A_2\}}(0, \emptyset, A_2)o_{\{A_1\}}(1, \emptyset, \{A_1, A_2\})o_{\{A_2\}}(0, \emptyset, A_1) \tag{8c}$$

The o notation alone cannot capture the difference for this construction, but the topological hull enables a distinction. Fig. 18 shows for both scenes the step-wise construction of the topological hulls (Fig. 18a and 18d), the unions of the regions (Fig. 18b and 18e), and their holes as the differences (Fig. 18c and 18f).

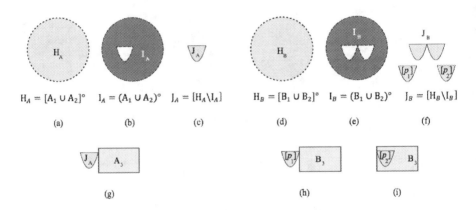

$H_A = [A_1 \cup A_2]^\circ$ $I_A = (A_1 \cup A_2)^\circ$ $J_A = [H_A \backslash I_A]$ $H_B = [B_1 \cup B_2]^\circ$ $I_B = (B_1 \cup B_2)^\circ$ $J_B = [H_B \backslash I_B]$

(a)   (b)   (c)   (d)   (e)   (f)

(g)   (h)   (i)

Fig. 18. Two complex spatial scenes that share the same boundary intersections, but are topologically different captured by using the aggregate topological hull

Since scene A has a single, bounded, path-connected exterior component, a single relation captures the exterior partition (Eqn. 9a). On the other hand, scene B has two bounded, path-connected exterior components. Therefore, an exterior partition assessment needs to be applied to both (Eqs. 9b and 9c), revealing the difference between scenes A and B.

$$J_A \ meet \ A_3 \tag{9a}$$

$$[p_1] \ meet \ B_3 \tag{9b}$$

$$[p_2] \ coveredBy \ B_3 \tag{9c}$$

The second scenario that considers o notation *and* topological hulls features two scenes (Fig. 19) with the same aggregate topological hulls, yet different o notations. In this case, the o notation *per se* is insufficient to capture the scenes' topological differences; therefore, both are needed.

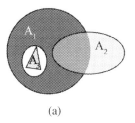   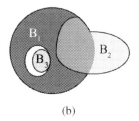

(a)   (b)

Fig. 19. Two spatial scenes, each consisting of three objects, for which the o notation *and* the topological hull are needed to completely distinguish them from each other

While the aggregate topological hulls $H_A = [A_1 \cup A_2]$ and $H_B = [B_1 \cup B_2]$ *contain*, respectively, A_3 and B_3, they cannot capture the differences in which A_1 and B_1 *surround* A_3 and B_3, respectively—namely in a 0-dimensional or a 1-dimensional boundary intersection. Likewise, the different types of *overlaps* of A_1 and A_2 and B_1 and B_2 are also only evident from their o notations. To differentiate these two properties, the boundary components need to be analyzed in each scene over the three objects involved, plus their respective holes $A_1H = H_A{}^\circ \backslash (A_1 \cup A_2)^\circ$ and $B_1H = H_B{}^\circ \backslash (B_1 \cup B)^\circ$. The differences appear then, for instance, the o notations of A_1 vs. B_1 (Eqs. 10a and 10c) and the o notations of A_3 and B_3 (Eqs. 10b and 10d).

$$\partial A_1: o_{\{A_1 H, A_2, A_3\}}(0, \emptyset, A_2) o_{\{A_1 H, A_3\}}(0, \emptyset, A_2) \tag{10a}$$

$$\partial A_3: o_{\{A_2\}}(0, A_1 H, \emptyset) \tag{10b}$$

$$\partial B_1: o_{\{B_1 H, B_2, B_3\}}(1, \emptyset, B_2) o_{\{B_1 H, B_3\}}(0, \emptyset, B_2) \tag{10c}$$

$$\partial B_3: o_{\{B_2\}}(1, B_1 H, \emptyset) \tag{10d}$$

7   Conclusions and Future Work

The topology of a scene as an ensemble of regions has been formalized. To this end, the boundary intersections between objects with the relations *meet, overlap, covers, coveredBy,* and *entwined* were further explored. These boundary intersections take the form *crosses* or *touches*, in either a 0-dimensional or 1-dimensional flavor. By recording these boundary impacts about each object in clockwise orientation the objects of the spatial scene can be related, even where there exist separations or more than two objects. The result of this approach is a symbolic representation of the topological scene, comprised of these intersection strings, called the o notation.

Furthermore, an operation is presented for creating the smallest, path-connected 2-disk of an object, in essence filling any holes within the *topological hull*. This enables the capture of relations between an object and a group of objects and also allows for the differentiation of separated regions of the common exterior. Such separations can be holes within individual objects, or holes formed by the union of multiple objects. Similarly, the *aggregate topological hull* enables nested or separated groups of objects to be considered. The set difference between a given topological hull and its originating object(s) captures surrounding or nested objects.

The methods to record boundary contacts and exterior partitions extend the established methods for capturing the topological relation between *two* simple regions [21] to an ensemble of spatial regions, where regions may have holes or separations, and the ensemble may create exterior separations (essentially holes in unions of regions).

These methods for representing a spatial scene can be expanded upon in the future through new lines of inquiry. First, the topological hull operation should be extended to allow for the inclusion of lines, the torus in $\mathbb{R}^3$, and any objects embedded in $\mathbb{R}^n$ generally, possibly relating to the topological hull definition in $\mathbb{Z}^3$ [2]. Neither lines in $\mathbb{R}^2$ nor regions and volumes in $\mathbb{R}^3$ are captured in the current model. The generalization of boundary contacts and exterior partitions for such settings is a challenge. Second, the description of a spatial scene involving objects such as handled or otherwise-spiked regions should be considered. A handled or spiked region fits within the compound-spatial object model by considering not only regions but also lines (e.g., for a region R5 and a line L1, their union R5 ∪ L1—when combined with the constraint R5 *doubleMeet* L1 [15]—yields a simple region with a handle, which creates an exterior partition.

Finally, the concepts of qualitative direction and distance could be added to the description of the spatial scene in order to further tighten the descriptive power of a scene by accounting for the continuous transformations objects represented within may undergo while preserving their relatedness to their neighbors, as well as accounting for their relative direction to one another using any number of qualitative methods for capturing such information.

Some may want to consider using first-order logic, in combination with either the 9-intersection relations or the RCC language an alternative approach. But one could also just use first-order predicate calculus altogether to capture the semantics of spatial scenes. Such comparative assessments may or may not have value in the future.

The method *per se*, even only for regions in $\mathbb{R}^2$, features two theoretical and two applied challenges: (1) a proof that the combination of boundary contacts and exterior partitions is fully sufficient to capture completely the topology of an ensemble of regions; (2) a normalization of the boundary components, as the *o* notation repeats for the same boundary interaction the same information along each participating region boundary; (3) the automatic generation of a graphical depiction of the topology of a spatial scene from its scene topology description; and (4) the automatic generation of a verbal description of the topology of a spatial scene from its scene topology description, at various levels of detail.

Acknowledgments. This work was partially supported by NSF Grants IIS-1016740 (PI: Max Egenhofer) and DGE-0504494 (PI: Kate Beard). COSIT reviewers provided valuable feedback.

References

1. Adams, C., Franzosa, R.: Introduction to Topology: Pure and Applied. Pearson Prentice Hall (2008)
2. Aktouf, Z., Bertrand, G., Perroton, L.: A Three-Dimensional Holes Closing Algorithm. Pattern Recognition Letters 23(5), 523–531 (2002)
3. Bruns, T., Egenhofer, M.: Similarity of Spatial Scenes. In: Kraak, M., Molenaar, M. (eds.) Seventh International Symposium on Spatial Data Handling, pp. 31–42 (1996)
4. Cassati, R., Varzi, A.: Holes and Other Superficialities. MIT Press (1994)
5. Chang, S.K., Shi, Q., Yan, C.: Iconic Indexing by 2-D Strings. IEEE Transactions on Pattern Analysis and Machine Intelligence 9(3), 413–428 (1987)
6. Clementini, E., Di Felice, P.: Topological Invariants for Lines. IEEE Transaction on Knowledge and Data Engineering 10(1), 38–54 (1998)
7. Clementini, E., Di Felice, P.: A Model for Representing Topological Relationships between Complex Geometric Features in Spatial Databases. Information Sciences 90(1-4), 121–136 (1996)
8. Clementini, E., Di Felice, P., Califano, G.: Composite Regions in Topological Queries. Information Systems 20(7), 579–594 (1995)
9. Cohn, A., Bennett, B., Gooday, J., Gotts, N.: Qualitative Spatial Representation and Reasoning with the Region Connection Calculus. GeoInformatica 1(3), 275–316 (1997)
10. Cohn, A., Hazarika, S.: Qualitative Spatial Representation and Reasoning: An Overview. Fundamenta Informaticae 46(1-2), 2–32 (2001)
11. Cohn, A., Renz, J.: Qualitative Spatial Representation and Reasoning. In: van Hermelen, F., Lifschitz, V., Porter, B. (eds.) Handbook of Knowledge Representation, pp. 551–596 (2008)
12. Cohn, A., Varzi, A.: Mereotopological Connection. Journal of Philosophical Logic 32(4), 357–390 (2003)
13. Egenhofer, M.: A Model for Detailed Binary Topological Relationships. Geomatica 47(3), 261–273 (1993)
14. Egenhofer, M., Franzosa, R.: Point-Set Topological Spatial Relations. International Journal of Geographical Information Systems 5(2), 161–174 (1991)
15. Egenhofer, M., Herring, J.: Categorizing Binary Topological Relationships Between Regions, Lines, and Points in Geographic Databases. Department of Surveying Engineering. University of Maine, Orono (1991)
16. Egenhofer, M.J.: A Reference System for Topological Relations between Compound Spatial Objects. In: Heuser, C.A., Pernul, G. (eds.) ER 2009. LNCS, vol. 5833, pp. 307–316. Springer, Heidelberg (2009)
17. Egenhofer, M.: Query Processing in Spatial-Query-by-Sketch. Journal of Visual Languages and Computing 8(4), 403–424 (1997)
18. Egenhofer, M.: Deriving the Composition of Binary Topological Relations. Journal of Visual Languages and Computing 5(2), 133–149 (1994)
19. Egenhofer, M., Clementini, E., Di Felice, P.: Topological Relations between Regions with Holes. International Journal of Geographical Information Systems 8(2), 129–142 (1994)
20. Egenhofer, M., Dube, M.: Topological Relations from Metric Refinements. In: Wolfson, O., Agrawal, D., Lu, C.-T. (eds.) 17th ACM SIGSPATIAL International Symposium on Advances in Geographic Information Systems, ACM-GIS 2009, pp. 158–167 (2009)

21. Egenhofer, M., Franzosa, R.: On the Equivalence of Topological Relations. International Journal of Geographical Information Systems 9(2), 133–152 (1995)

22. Egenhofer, M., Herring, J.: A Mathematical Framework for the Definition of Topological Relationships. In: Brassel, K., Kishimoto, H. (eds.) Fourth International Symposium on Spatial Data Handling, pp. 803–813 (1990)

23. Egenhofer, M., Vasardani, M.: Spatial Reasoning with a Hole. In: Winter, S., Duckham, M., Kulik, L., Kuipers, B. (eds.) COSIT 2007. LNCS, vol. 4736, pp. 303–320. Springer, Heidelberg (2007)

24. Freeman, J.: The Modeling of Spatial Relations. Computer Graphics and Image Processing 4(2), 156–171 (1975)

25. Galton, A.: Spatial and Temporal Knowledge Representation. Earth Science Informatics 2(3), 169–187 (2009)

26. Galton, A.: Modes of Overlap. Journal of Visual Languages and Computing 9(1), 61–79 (1998)

27. Guesgen, H.: Spatial Reasoning Based on Allen's Temporal Logic. Technical Report, International Computer Science Institute, Berkeley, CA (1989)

28. Herring, J.: The Mathematical Modeling of Spatial and Non-Spatial Information in Geographic Information Systems. In: Mark, D., Frank, A. (eds.) Cognitive and Linguistic Aspects of Geographic Space, pp. 313–350. Kluwer Academic (1991)

29. Li, S., Cohn, A.: Reasoning with Topological and Directional Spatial Information. Computational Intelligence 28(4), 579–616 (2009)

30. Nabil, M., Shephard, J., Ngu, A.: 2D Projection Interval Relationships: A Symbolic Representation of Spatial Relationships. In: Egenhofer, M., Herring, J.R. (eds.) SSD 1995. LNCS, vol. 951, pp. 292–309. Springer, Heidelberg (1995)

31. Nedas, K., Egenhofer, M.: Spatial-Scene Similarity Queries. Transactions in GIS 12(6), 661–681 (2008)

32. OGC: OpenGIS Geography Markup Language (GML) Encoding Standard (2005), http://www.opengeospatial.org/standards/gml

33. OGC: OGC Abstract Specifications OpenGIS Consortium (OGC) (1999), http://www.opengis.org/techno/specs.htm

34. Randell, D., Cui, Z., Cohn, A.: A Spatial Logic Based on Regions and Connection. In: Third International Conference on Knowledge Representation and Reasoning, pp. 165–176 (1992)

35. Randell, D., Cohn, A., Cui, Z.: Computing Transitivity Tables: A Challenge for Automated Theorem Provers. In: Kapur, D. (ed.) CADE 1992. LNCS, vol. 607, Springer, Heidelberg (1992)

36. Rodríguez, M.A., Egenhofer, M., Blaser, A.: Query Pre-processing of Topological Constraints: Comparing a Composition-Based with Neighborhood-Based Approach. In: Hadzilacos, T., Manolopoulos, Y., Roddick, J., Theodoridis, Y. (eds.) SSTD 2003. LNCS, vol. 2750, pp. 362–379. Springer, Heidelberg (2003)

37. Schneider, M., Behr, T.: Topological Relationships between Complex Spatial Objects. ACM Transactions on Database Systems 31(1), 39–81 (2006)

38. Sharma, J.: Integrated Topology- and Direction-Reasoning in GIS. In: Craglia, M., Onsrud, H. (eds.) Second ESF-NSF Summer Institute in Geographic Information, pp. 435–447. Taylor & Francis (1999)

39. Sistla, A., Yu, C.: Reasoning about Qualitative Spatial Relationships. Journal of Automated Reasoning 25(4), 291–328 (2000)

40. Tryfona, N., Egenhofer, M.: Consistency Among Parts and Aggregates: A Computational Model. Transactions in GIS 1(3), 189–206 (1996)
41. Varzi, A.: Parts, Wholes, and Part-Whole Relations: The Prospects of Mereotopology. Data & Knowledge Engineering 20(3), 259–286 (1996)
42. Varzi, A.: Spatial Reasoning and Ontology: Parts, Wholes, and Locations. In: Aiello, M., Pratt-Hartmann, I., van Benthem, J. (eds.) Handbook of Spatial Logics, pp. 945–1038 (2007)
43. Vasardani, M., Egenhofer, M.J.: Comparing Relations with a Multi-holed Region. In: Hornsby, K.S., Claramunt, C., Denis, M., Ligozat, G. (eds.) COSIT 2009. LNCS, vol. 5756, pp. 159–176. Springer, Heidelberg (2009)
44. Worboys, M., Bofakos, P.: A Canonical Model for a Class of Areal Spatial Objects. In: Abel, D.J., Ooi, B.-C. (eds.) SSD 1993. LNCS, vol. 692, pp. 36–52. Springer, Heidelberg (1993)

Algebraic Properties of Qualitative Spatio-temporal Calculi

Frank Dylla[1], Till Mossakowski[1,2], Thomas Schneider[3], and Diedrich Wolter[1]

[1] Collaborative Research Center on Spatial Cognition (SFB/TR 8), Univ. of Bremen
{dylla,till,dwolter}@informatik.uni-bremen.de
[2] DFKI GmbH, Bremen
Till.Mossakowski@dfki.de
[3] Department of Mathematics and Computer Science, University of Bremen
tschneider@informatik.uni-bremen.de

Abstract. Qualitative spatial and temporal reasoning is based on so-called qualitative calculi. Algebraic properties of these calculi have several implications on reasoning algorithms. But what exactly is a qualitative calculus? And to which extent do the qualitative calculi proposed meet these demands? The literature provides various answers to the first question but only few facts about the second. In this paper we identify the minimal requirements to binary spatio-temporal calculi and we discuss the relevance of the according axioms for representation and reasoning. We also analyze existing qualitative calculi and provide a classification involving different notions of relation algebra.

1 Introduction

Qualitative spatial and temporal reasoning is a sub-field of knowledge representation involved with representations of spatial and temporal domains that are based on finite sets of so-called qualitative relations. Qualitative relations serve to explicate knowledge relevant for a task at hand while at the same time they abstract from irrelevant knowledge. Often, these relations aim to relate to cognitive concepts. Qualitative spatial and temporal reasoning thus link cognitive approaches to knowledge representation with formal methods. Computationally, qualitative spatial and temporal reasoning is largely involved with constraint satisfaction problems over infinite domains where qualitative relations serve as constraints. Typical domains include, on the temporal side, points and intervals and, on the spatial side, regions or oriented points in the Euclidean plane. In the past decades, a vast number of qualitative representations have been developed that are commonly referred to as qualitative calculi (see [19] for a recent overview). Yet the literature provides us with several definitions of what a qualitative calculus exactly is. Nebel and Scivos [31] have introduced the rather general and weak notion of a *constraint algebra*, which is a set of relations closed under converse, finite intersection, and composition. Ligozat and Renz [20] focus on so-called non-associative algebras, which are relation algebras without associativity axioms, and which have a much richer structure. Both approaches assume

T. Tenbrink et al. (Eds.): COSIT 2013, LNCS 8116, pp. 516–536, 2013.

that the converse operation is *strong*, which is not the case for calculi like the Cardinal Direction (Relations) Calculus (CDR) [39] or its recently introduced rectangular variant (RDR) [30].

The goal of this paper is to relate the existing definitions and to identify the essential representation-theoretic properties that characterize a qualitative calculus. It is achieved by the following contributions:

- We propose a definition of a qualitative calculus that includes existing spatio-temporal calculi by weakening the conditions usually imposed on the converse and composition relation (Section 2).
- We generalize the notions of constraint algebra and non-associative algebra to cover calculi with weak converse (Section 3.1).
- We discuss the role of algebraic properties of calculi for spatial reasoning, especially in connection with general-purpose reasoning tools like SparQ [44,43] and GQR [11] (Section 3.2).
- We experimentally evaluate the algebraic properties of calculi and derive a reasoning procedure that is sensitive to these properties (Section 4).
- We examine information preservation properties of calculi during reasoning, i.e., how general relations evolve after several compositions (Section 5).

2   Qualitative Representations

In this section, we formulate minimal requirements to a qualitative calculus, discuss their relevance to spatio-temporal representation and reasoning, and list existing calculi. We restrict ourselves to calculi with binary relations because we want to examine their algebraic properties using the notion of a relation algebra, which is best understood for binary relations.

2.1   Requirements to Qualitative Representations

We start with minimal requirements used in the literature. Let us first fix some notation. Let r, s, t range over binary relations over a non-empty universe $\mathcal{U}$, i.e., $r \subseteq \mathcal{U} \times \mathcal{U}$. We use $\cup, \cap, \bar{\ }, \smile$ and $\circ$ to denote the union, intersection, complement, converse, and composition of relations, as well as the identity and universal relations $\mathrm{id} = \{(u, u) \mid u \in \mathcal{U}\}$ and $\mathsf{u} = \mathcal{U} \times \mathcal{U}$. A relation $r \subseteq \mathcal{U} \times \mathcal{U}$ is called *serial* if, for every $u \in \mathcal{U}$, there is some $v \in \mathcal{U}$ such that $(u, v) \in r$.

Ligozat and Renz [20] note that most spatial and temporal calculi are based on a set of JEPD (jointly exhaustive and pairwise disjoint) relations. The following definition is standard in the QSR literature [20,4].

Definition 1. Let $\mathcal{U}$ be a non-empty universe and $\mathcal{R}$ a set of non-empty binary relations over $\mathcal{U}$. $\mathcal{R}$ is called a set of *JEPD relations* over $\mathcal{U}$ if the relations in $\mathcal{R}$ are pairwise disjoint and $\mathcal{U} \times \mathcal{U} = \bigcup_{r \in \mathcal{R}} r$.

An *abstract partition scheme* is a pair $(\mathcal{U}, \mathcal{R})$ where $\mathcal{R}$ is a set of *JEPD relations* over $\mathcal{U}$. $(\mathcal{U}, \mathcal{R})$ is called a *partition scheme* [20] if $\mathcal{R}$ contains the identity relation id and, for every $r \in \mathcal{R}$, there is some $s \in \mathcal{R}$ such that $r^{\smile} = s$.

The universe $\mathcal{U}$ represents the set of all spatial or temporal entities, and $\mathcal{R}$ being a set of JEPD relations ensures that each two entities are in exactly one relation from $\mathcal{R}$. Incomplete information about two entities is modeled by taking the union of base relations, with the universal relation (the union of all base relations) representing that no information is available. Disjointness of the base relations ensures that there is a unique way to represent an arbitrary relation, and exhaustiveness ensures that the empty relation can never occur in a consistent set of constraints (which are defined in Section 2.2).

Ligozat and Renz [20] base their definition of a qualitative calculus on the notion of a partition scheme. This excludes calculi like CDR and RDR which do not have strong converses. Hence, we take a more general approach based on the notion of an abstract partition scheme. This accommodates existing calculi with these weaker properties: some existing spatio-temporal representations do not require an identity relation, and some representations are deliberately kept coarse and thus do not guarantee that the converse of a base relation is again a (base) relation. Furthermore, the computation of the converse operation may be easier when weaker properties are postulated. The same rationale applies to the composition operation. Thus, the following definition of a spatial calculus, based on abstract partition schemes, contains minimal requirements.

Definition 2. A *qualitative calculus* with binary relations is a tuple $(\mathsf{Rel}, \mathsf{Int}, \breve{\ }, \diamond)$ with the following properties.

- Rel is a finite, non-empty set of *base relations*. The subsets of Rel are called *relations*. We use r, s, t to denote base relations and R, S, T to denote relations.
- Int $= (\mathcal{U}, \varphi)$ is an *interpretation* with a non-empty universe $\mathcal{U}$ and a map $\varphi : \mathsf{Rel} \to 2^{\mathcal{U} \times \mathcal{U}}$ with $(\mathcal{U}, \{\varphi(r) \mid r \in \mathsf{Rel}\})$ being a weak partition scheme. The map φ is extended to arbitrary relations by setting $\varphi(R) = \bigcup_{r \in R} \varphi(r)$ for every $R \subseteq \mathsf{Rel}$.
- The *converse operation* $\breve{\ }$ is a map $\breve{\ } : \mathsf{Rel} \to 2^{\mathsf{Rel}}$ that satisfies

$$\varphi(r^{\smallsmile}) \supseteq \varphi(r)^{\smallsmile} \tag{1}$$

for every $r \in \mathsf{Rel}$. The operation $\breve{\ }$ is extended to arbitrary relations by setting $R^{\smallsmile} = \bigcup_{r \in R} r^{\smallsmile}$ for every $R \subseteq \mathsf{Rel}$.
- The *composition operation* $\diamond$ is a map $\diamond : \mathsf{Rel} \times \mathsf{Rel} \to 2^{\mathsf{Rel}}$ that satisfies

$$\varphi(r \diamond s) \supseteq \varphi(r) \circ \varphi(s) \tag{2}$$

for all $r, s \in \mathsf{Rel}$. The operation $\diamond$ is extended to arbitrary relations by setting $R \diamond S = \bigcup_{r \in R} \bigcup_{s \in S} r \diamond s$ for every $R, S \subseteq \mathsf{Rel}$.

We call Properties (1) and (2) *abstract converse* and *abstract composition*, following Ligozat's naming [18]. Our notion of a qualitative calculus makes weaker requirements on the converse operation than Ligozat and Renz's notions of a weak representation [18,20]. We have already discussed a rationale behind choosing these "weaker than weak" variants and will name another one in Section 2.2.

On the other hand, our notion makes stronger requirements on the converse than Nebel and Scivos's notion of a constraint algebra [31]. The following definition gives the stronger variants of converse and composition existing in the literature.

Definition 3. Let $C = (\mathsf{Rel}, \mathsf{Int}, \breve{\ }, \diamond)$ be a qualitative calculus.
C has *weak converse* if, for all $r \in \mathsf{Rel}$:

$$r^{\breve{\ }} = \bigcap \{S \subseteq \mathsf{Rel} \mid \varphi(S) \supseteq \varphi(r)^{\breve{\ }}\} \tag{3}$$

C has *strong converse* if, for all $r \in \mathsf{Rel}$:

$$\varphi(r^{\breve{\ }}) = \varphi(r)^{\breve{\ }} \tag{4}$$

C has *weak composition* if, for all $r, s \in \mathsf{Rel}$:

$$r \diamond s = \bigcap \{T \subseteq \mathsf{Rel} \mid \varphi(T) \supseteq \varphi(r) \circ \varphi(s)\} \tag{5}$$

C has *strong composition* if, for all $r, s \in \mathsf{Rel}$:

$$\varphi(r \diamond s) = \varphi(r) \circ \varphi(s) \tag{6}$$

The following fact captures that Properties (1)–(6) immediately carry over to arbitrary relations; the straightforward proof is given in [8]. It has consequences for efficient spatio-temporal reasoning, which are explained in Section 2.2.

Fact 4. *Given a qualitative calculus* $(\mathsf{Rel}, \mathsf{Int}, \breve{\ }, \diamond)$ *and relations* $R, S \subseteq \mathsf{Rel}$, *the following hold:*

$$\varphi(R^{\breve{\ }}) \supseteq \varphi(R)^{\breve{\ }} \tag{7}$$
$$\varphi(R \diamond S) \supseteq \varphi(R) \diamond \varphi(S) \tag{8}$$

If C *has weak converse, then, for all* $R \subseteq \mathsf{Rel}$:

$$R^{\breve{\ }} = \bigcap \{S \subseteq \mathsf{Rel} \mid \varphi(S) \supseteq \varphi(R)^{\breve{\ }}\} \tag{9}$$

If C *has strong converse, then, for all* $R \subseteq \mathsf{Rel}$:

$$\varphi(R^{\breve{\ }}) = \varphi(R)^{\breve{\ }} \tag{10}$$

If C *has weak composition, then, for all* $R, S \subseteq \mathsf{Rel}$:

$$R \diamond S = \bigcap \{T \subseteq \mathsf{Rel} \mid \varphi(T) \supseteq \varphi(R) \circ \varphi(S)\} \tag{11}$$

If C *has strong composition, then, for all* $R, S \subseteq \mathsf{Rel}$:

$$\varphi(R \diamond S) = \varphi(R) \circ \varphi(S) \tag{12}$$

Since base relations are non-empty and JEPD, we have

Fact 5. *For any qualitative calculus, φ is injective.*

Comparing Definitions 1–3 with the basic notions of a qualitative calculus in [20], a *weak representation* is a calculus with identity relation, strong converse and abstract composition. Our basic notion of a qualitative calculus is more general: it does not require an identity relation, and it only requires abstract converse and composition. Conversely, [20] are slightly more general than we are, because the map φ need not be injective. However, this extra generality is not very meaningful: if base relations are JEPD, φ could only be non-injective in giving multiple names to the empty relation. Furthermore, in [20], a *representation* is a weak representation with strong composition and an injective map φ.

2.2   Spatio-temporal Reasoning

The most important flavor of spatio-temporal reasoning is constraint-based reasoning. Like with a classical constraint satisfaction problem (CSP), we are given a set of variables and constraints. The task of constraint satisfaction is to decide whether there exists a valuation of all variables that satisfies the constraints. In calculi for spatio-temporal reasoning, variables range over the specific spatial or temporal domain of a qualitative representation. The relations defined by the calculus serve as constraint relations. More formally, we have:

Definition 6 (QCSP). Let $(\mathsf{Rel}, \mathsf{Int}, \breve{\,}, \diamond)$ be a binary qualitative calculus with $\mathsf{Int} = (\mathcal{U}, \varphi)$ and let X be a set of variables ranging over $\mathcal{U}$. A *qualitative constraint* is a formula $x_i\, R_j\, x_k$ with variables $x_i, x_k \in X$ and relation $R_j \in \mathsf{Rel}$. We say that a valuation $\psi : X \to \mathcal{U}$ satisfies $x_i\, R_j\, x_k$ if $(\psi(x_i), \psi(x_k)) \in \phi(R_j)$ holds.

A *qualitative constraint satisfaction problem* (QCSP) is the task to decide whether there is a valuation ψ for a set of variables satisfying a set of constraints.

In the following we use X to refer to the set of variables and $r_{x,y}$ stands for the constraint relation between variables x and y. For simplicity and wlog. it is assumed that for every pair of variables exactly one constraint relation is given.

Several techniques originally developed for finite domain CSP can be adapted to spatio-temporal QCSPs. Since deciding CSP instances is already NP-complete for search problems with finite domains, heuristics are important. One particular valuable technique is constraint propagation which aims at making implicit constraints explicit in order to identify variable assignments that would violate some constraint. By pruning away these variable assignments, a consistent valuation can be searched more efficiently. A common approach is to enforce k-consistency.

Definition 7. A CSP is k-consistent if for all subsets of variables $X' \subset X$ with $|X'| = k - 1$ we can extend any valuation of X' that satisfies the constraints to a valuation of $X' \cup \{x\}$ also satisfying the constraints, where $x \in X \setminus X'$ is any additional variable.

QCSPs are naturally 1-consistent as the domains are infinite and there are no unary constraints. A QCSP is 2-consistent if $r_{x,y} = r_{y,x}\breve{\,}$ and $r_{x,y} \neq \emptyset$ as relations are typically serial. A 3-consistent QCSP is also called *path-consistent* and Definition 7 can be rewritten using compositions as

$$r_{x,y} \subseteq \bigcap_{z \in X} r_{x,z} \circ r_{z,y} \tag{13}$$

and we can enforce the 3-consistency by iterating the refinement operation

$$r_{x,y} \leftarrow r_{x,y} \cap r_{x,z} \circ r_{z,y} \tag{14}$$

for all variables $x, y, z \in X$ until a fix point is reached. This procedure is known as the path-consistency algorithm [5]. For finite constraint networks the algorithm always terminates since the refinement operation is monotone and there are only finitely many relations.

If a qualitative calculus does not provide strong composition, iterating Equation (14) is not possible as it would lead to relations not contained in Rel. It is however straightforward to weaken Equation (14) using weak composition.

$$r_{x,y} \leftarrow r_{x,y} \cap r_{x,z} \diamond r_{z,y} \tag{15}$$

This procedure is called enforcing *algebraic closure* or *a-closure* for short. The reason why, in Definition 2, we require composition to be at least abstract is that the underlying inequality guarantees that reasoning via a-closure is sound.

Enforcing k-consistency or algebraic closure does not change the solutions of a CSP, as only impossible valuations are removed. If during application of Equation (15) an empty relation occurs, the QCSP is thus known to be inconsistent. By contrast, an algebraically closed QCSP may not be consistent though. However, for several qualitative calculi (or at least sub-algebras thereof) algebraic closure and consistency coincide.

Though we speak about composition in the following two paragraphs, the same statements hold for converse.

Fact 4 has the consequence that the composition operation of a calculus is uniquely determined if the composition of each pair of base relations is given. This information is usually stored in a table, the *composition table*. Then, computing the composition of two arbitrary relations is just a matter of table look-ups which allows algebraic closure to be enforced efficiently. Speaking in terms of composition tables, abstract composition implies that each cell corresponding to $r \diamond s$ contains *at least* those base relations t whose interpretation intersects with $\varphi(r) \circ \varphi(s)$. In addition, weak composition implies that each cell contains *exactly* those t. If composition is strong, then Rel and φ even have to ensure that whenever $\varphi(t)$ intersects with $\varphi(r) \circ \varphi(s)$, it is contained in $\varphi(r) \circ \varphi(s)$ – i.e., the composition of the interpretation of any two base relations has to be the union of interpretations of certain base relations.

2.3   Existing Qualitative Spatio-temporal Representations

This paper is concerned with properties of binary spatio-temporal calculi that are described in the literature and implemented in the spatial representation and reasoning tool SparQ [44,43]. Table 1 lists these calculi.

Table 1. Overview of the binary calculi tested

Name	Ref.	Domain	#BR	RM
9-Intersection	[9]	simple 2D regions	8	I [12,16]
Allen's interval relations	[1]	intervals (order)	13	A [42]
Block Algebra	[2]	n-dimensional blocks	13^n	A [2]
Cardinal Dir. Calculus CDC	[10,17]	directions (point abstr.)	9	A [17]
Cardinal Dir. Relations CDR	[38]	regions	218	P
CycOrd, binary CYC_b	[14]	oriented lines	4	U
Dependency Calculus	[33]	points (partial order)	5	A [33]
Dipole Calculus[a] DRA_f	[25,24]	directions from line segm.	72	I [46]
$\quad DRA_{fp}$	[24]	directions from line segm.	80	I
$\quad$ DRA-connectivity	[45]	connectivity of line segm.	7	U
Geometric Orientation	[7]	relative orientation	4	U
INDU	[32]	intervals (order, rel. dur.n)	25	P
$OPRA_m$, $m = 1, \ldots, 8$	[23,28]	oriented points	$4m \cdot (4m + 1)$	
(Oriented Point Rel. Algebra)				I [46]
Point Calculus	[42]	points (total order)	3	A [42]
Qualitat. Traject. Calc. QTC_{B11}	[40,41]	moving point obj.s in 1D	9	U
$\quad QTC_{B12}$	"	"	17	U
$\quad QTC_{B21}$	"	moving point obj.s in 2D	9	U
$\quad QTC_{B22}$	"	"	27	U
$\quad QTC_{C12}$	"	"	81	U
$\quad QTC_{C22}$	"	"	305	U
Region Connection Calc. RCC-5	[34]	regions	5	A [15]
$\quad$ RCC-8	[34]	regions	8	A [35]
Rectangular Cardinal Rel.s RDR	[30]	regions	36	A [30]
Star Algebra $STAR_4$	[36]	directions from a point	9	P

[a]Variant DRA_c is not based on a weak partition scheme – JEPD is violated [24].

#BR: number of base relations
RM: reasoning method used to decide consistency of CSPs with base relns only:
      A-closure; **P**olynomial: reducible to linear programming;
      Intractable (assuming P $\neq$ NP); **U**nknown

3   Relation Algebras

3.1   Definition

If we focus our attention on spatio-temporal calculi with binary relations, it is reasonable to ask whether they are relation algebras (RAs). If a calculus is a RA, it is guaranteed to have properties that allow several optimizations in constraint reasoners. For example, associativity of the composition operation $\diamond$ ensures that, if the reasoner encounters a path $ArBsCtD$ of length 3, then the relation between A and D can be computed "from left to right". Without associativity, $(r \diamond s) \diamond t$ as well as $r \diamond (s \diamond t)$ would have to be computed. RAs have been considered in the literature for spatio-temporal calculi [20,6,26].

An (abstract) RA is defined in [22]; here we use the symbols $\cup$, $\diamond$, and id instead of $+$, $;$, and $1'$. Let A be a set containing id and 1, and let $\cup$, $\diamond$ be binary

Table 2. Axioms for relation algebras and weaker variants [22]

R_1	$r \cup s = s \cup r$	$\cup$-commutativity
R_2	$r \cup (s \cup t) = (r \cup s) \cup t$	$\cup$-associativity
R_3	$\overline{\overline{r} \cup \overline{s}} \cup \overline{\overline{r} \cup s} = r$	Huntington's axiom
R_4	$r \diamond (s \diamond t) = (r \diamond s) \diamond t$	$\diamond$-associativity
R_5	$(r \cup s) \diamond t = (r \diamond t) \cup (s \diamond t)$	$\diamond$-distributivity
R_6	$r \diamond \mathrm{id} = r$	identity law
R_7	$(r^\smile)^\smile = r$	$\smile$-involution
R_8	$(r \cup s)^\smile = r^\smile \cup s^\smile$	$\smile$-distributivity
R_9	$(r \diamond s)^\smile = s^\smile \diamond r^\smile$	$\smile$-involutive distributivity
R_{10}	$r^\smile \diamond \overline{r \diamond s} \cup \overline{s} = \overline{s}$	Tarski/de Morgan axiom
WA	$((r \cap \mathrm{id}) \diamond 1) \diamond 1 = (r \cap \mathrm{id}) \diamond 1$	weak $\diamond$-associativity
SA	$(r \diamond 1) \diamond 1 = r \diamond 1$	$\diamond$ semi-associativity
R_{6l}	$\mathrm{id} \diamond r = r$	left-identity law
PL	$(r \diamond s) \cap t^\smile = \emptyset \Leftrightarrow (s \diamond t) \cap r^\smile = \emptyset$	Peircean law

and $\overline{}$, $\smile$ unary operations on A. The relevant axioms (R_1–R_{10}, WA, SA, and PL) are given in Table 2. All axioms except PL can be weakened to only one of two inclusions, which we denote by a superscript $\supseteq$ or $\subseteq$. For example, $R_7^{\supseteq}$ denotes $(r^\smile)^\smile \supseteq r$. Likewise, we use $PL^{\Rightarrow}$ and $PL^{\Leftarrow}$. Then, $\mathfrak{A} = (A, \cup, \overline{}, \diamond, \smile, \mathrm{id})$ is a

- *non-associative relation algebra (NA)* if it satisfies Axioms R_1–R_3, R_5–R_{10};
- *semi-associative relation algebra (SA)* if it is an NA and satisfies Axiom SA,
- *weakly associative relation algebra (WA)* if it is an NA and satisfies WA,
- *relation algebra (RA)* if it satisfies R_1–R_{10},

for all $r, s, t \in A$. Every RA is a WA; every WA is an SA; every SA is an NA.

In the literature, a different axiomatization is sometimes used, for example in [20]. The most prominent difference is that R_{10} is replaced by PL, "a more intuitive and useful form, known as the Peircean law or De Morgan's Theorem K" [13]. It is shown in [13, Section 3.3.2] that, given R_1–R_3, R_5, R_7–R_9, the axioms R_{10} and PL are equivalent. The implication $PL \Rightarrow R_{10}$ does not need R_5 and R_8.

Furthermore, Table 2 contains the redundant axiom R_{6l} because it may be satisfied when some of the other axioms are violated. It is straightforward to establish that R_6 and R_{6l} are equivalent given R_7 and R_9, see [8].

Due to our minimal requirements to a qualitative calculus given in Def. 2, certain axioms are always satisfied; see [8] for a proof of the following

Fact 8. *Every qualitative calculus satisfies* R_1–R_3, R_5, $R_7^{\supseteq}$, R_8, $WA^{\supseteq}$, $SA^{\supseteq}$ *for all (base and complex) relations. This axiom set is maximal: each of the remaining axioms in Table 2 is not satisfied by some qualitative calculus.*

3.2   Discussion of the Axioms

We will now discuss the relevance of the above axioms for spatio-temporal representation and reasoning. Due to Fact 8, we only need to consider axioms R_4, R_6, R_7, R_9, R_{10} (or PL) and their weakenings R_{6l}, SA, WA.

R$_4$ (and SA, WA). Axiom R$_4$ is helpful for modeling. It allows for writing chains of compositions without parentheses, which have an unambiguous meaning. For example, consider the following statement in natural language about the relative length and location of two intervals A and D. *Interval A is before some equally long interval that is contained in some longer interval that meets the shorter D.* This statement is just a conjunction of relations between A, the unnamed intermediary intervals B, C, and D. When we evaluate it, it intuitively does not matter whether we give priority to the composition of the relations between A, B and B, C or to the composition of the relations between B, C and C, D.

However, INDU does not satisfy Axiom R$_4$ and, therefore, here the two ways of parenthesizing the above statement lead to different relations between A and D. This behavior is sometimes attributed to the absence of strong composition, which we will refute in Section 4. Conversely, strong composition implies R$_4$ since composition of binary relations over $\mathcal{U}$ is associative:

Fact 9. *Let $C = (\mathsf{Rel}, \mathsf{Int}, \check{\ }, \diamond)$ be a qualitative calculus with strong composition. Then C satisfies R$_4$.*

Note that INDU still satisfies the weakenings SA and WA of R$_4$, and we already know from Fact 8 that the inequalities SA$^{\supseteq}$ and WA$^{\supseteq}$ are always satisfied.

Furthermore, Axiom R$_4$ is useful for optimizing reasoning algorithms: suppose a scenario that contains the constraints $\{W r X, X s Y, Y t Z, W r' Z\}$ with variables W, X, Y, Z needs to be checked for consistency. If *one* of the inclusions R$_4^{\supseteq}$ and R$_4^{\subseteq}$ is satisfied – say, $r \diamond (s \diamond t) \subseteq (r \diamond s) \diamond t$ – then it suffices to compute the "finer" composition result $r \diamond (s \diamond t)$ and check whether it contains r'. Otherwise, both results have to be computed and checked for containment of r'.

R$_6$ and R$_{6I}$. Axioms R$_6$ and R$_{6I}$ do not seem to play a significant role in (optimizing) satisfiability checking, but the presence of an id relation is needed for the standard reduction from the correspondence problem to satisfiability: to test whether a constraint system admits the equality of two variables x, y, one can add an id-constraint between x, y and test the extended system for satisfiability.

Furthermore, the absence of an id relation may lead to an earlier loss of precision. For example, assume two variants of the 1D Point Calculus [42]: PC$_=$ with the relations *less than* ($<$), *equal* ($=$), and *greater than* ($>$), interpreted as the natural relations $<, =, >$ over the domain of the reals, and its approximation PC$_\approx$ with the relations *less than* ($<$), *approximately equal* ($\approx$), and *greater than* ($>$), where $\approx$ is interpreted as the set of pairs of points whose distance is below a certain threshold. Then, $=$ is the id-relation of PC$_=$ and $= \diamond =$ results in $\{=\}$, whereas PC$_\approx$ has no id-relation and $\approx \diamond \approx$ results in the universal relation.

R$_7$ and R$_9$. These axioms allow for certain optimizations in decision procedures for satisfiability based on algebraic operations like algebraic closure. If R$_7$ holds, the reasoning system does not need to store both constraints $A r B$ and $B r' A$, since r' can be reconstructed as $r\check{\ }$ if needed. Similarly, R$_9$ grants that, when enforcing algebraic closure by using Equation (15) to refine constraints between variable A and B, it is sufficient to compute composition once and, after applying converse, reuse it to refine the constraint between B and A too.

Current reasoning algorithms and their implementations use the described optimizations; they produce incorrect results for calculi violating R_7 or R_9.

R_{10} and PL. These axioms reflect that the relation symbols of a calculus indeed represent binary relations, i.e., pairs of elements of a universe. This can be explained from two different points of view.

1. If binary relations are considered as sets, R_{10} is equivalent to $r^\smile \diamond \overline{r \diamond s} \subseteq \bar{s}$. If we further assume the usual set-theoretic interpretation of the composition of two relations, the above inclusion reads as: *For any X, Y, if $Z r X$ for some Z and, $Z r U$ implies not $U s Y$ for any U, then not $X s Y$.* This is certainly true because X is one such U.
2. Under the same assumptions, each side of PL says (in a different order) that there can be no triangle $X r Y, Y s Z, Z t X$. The equality then means that the "reading direction" does not matter, see also [6]. This allows for reducing nondeterminism in the a-closure procedure, as well as for efficient refinement and enumeration of consistent scenarios.

3.3   Prerequisites for Being a Relation Algebra

The following correspondence between properties of a calculus and notions of a relation algebra is due to Ligozat and Renz [20].

Proposition 10. *Every calculus C based on a partition scheme is an NA. If, in addition, the interpretations of the base relations are serial, then C is an SA.*

Furthermore, R_7 is equivalent to the requirement that a calculus has strong converse. This is captured by the following lemma.

Lemma 11. *Let $C = (\mathsf{Rel}, \mathsf{Int}, \smile, \diamond)$ be a qualitative calculus. Then the following properties are equivalent.*

1. *C has strong converse.*
2. *Axiom R_7 is satisfied for all base relations $r \in \mathsf{Rel}$.*
3. *Axiom R_7 is satisfied for all relations $R \subseteq \mathsf{Rel}$.*

Proof. Items (2) and (3) are equivalent due to distributivity of $\smile$ over $\cup$, which is introduced with the cases for non-base relations in Definition 2.

For "(1) $\Rightarrow$ (2)", the following chain of equalities, for any $r \in \mathsf{Rel}$, is due to C having strong converse: $\varphi(r^{\smile\smile}) = \varphi(r^\smile)^\smile = \varphi(r)^{\smile\smile} = \varphi(r)$. Since Rel is based on JEPD relations and φ is injective, this implies that $r^{\smile\smile} = r$.

For "(2) $\Rightarrow$ (1)", we show the contrapositive. Assume that C does not have strong converse. Then $\varphi(r^\smile) \supsetneq \varphi(r)^\smile$, for some $r \in \mathsf{Rel}$; hence $\varphi(r^\smile)^\smile \supsetneq \varphi(r)^{\smile\smile}$. We can now modify the above chain of equalities replacing the first two equalities with inequalities, the first of which is due to Requirement (1) in the definition of the converse (Def. 2): $\varphi(r^{\smile\smile}) \supseteq \varphi(r^\smile)^\smile \supsetneq \varphi(r)^{\smile\smile} = \varphi(r)$. Since $\varphi(r^{\smile\smile}) \neq \varphi(r)$, we have that $r^{\smile\smile} \neq r$.   □

4 Algebraic Properties of Existing Calculi

In this section, we report on tests for algebraic properties we have performed on spatio-temporal calculi. We want to answer the following questions. *(1) Which existing calculi correspond to relation algebras? (2) Which weaker notions of relation algebras correspond to calculi that do not fall under (1)?*

We examined the corpus of the 31 calculi[1] listed in Table 1. This selection is restricted to calculi with (a) binary relations – because the notion of a relation algebra is best understood for binary relations – and (b) an existing implementation in SparQ.

To answer Questions (1) and (2), we use the axioms for relation algebras listed in Table 2 using both the heterogeneous tool set HETS [27] and SparQ. Due to Fact 8, it suffices to test Axioms R_4, R_6, R_7, R_9, R_{10} (or PL) and, if necessary, the weakenings SA, WA, and R_{6I}. The weakenings are relevant to capture weaker notions such as semi-associative or weakly associative algebras, or algebras that violate either R_6 or some of the axioms that imply the equivalence of R_6 and R_{6I}. Because all axioms except R_{10} contain only operations that distribute over the union $\cup$, it suffices to test them for base relations only. Therefore, we have written a CASL specification of R_4, R_6, R_7, R_9, PL, SA, WA, and R_{6I}, and used a HETS parser that reads the definitions of the above listed calculi in SparQ to test them against our CASL specification. In addition, we have tested all definitions against R_4, R_6, R_7, R_9, PL, and R_{6I} using SparQ's built-in function `analyze-calculus`.

A part of the calculi have already been tested by Florian Mossakowski [26], using a different CASL specification based on an equivalent axiomatization from [20]. He comprehensively reports on the outcome of these tests, and on repairs made to the composition table where possible.

The results of our and Mossakowski's tests are summarized in Table 3; details are listed in [8]. With the exceptions of QTC, Cardinal Direction Relations (CDR) and Rectangular Direction Relations (RDR), all tested calculi are at least semi-associative relation algebras; most of them are even relation algebras. Hence, these calculi enjoy the advantages for representation and reasoning optimizations discussed in Section 3.2. In particular, current reasoning procedures, which already implement the optimizations described for R_7 and R_9, yield correct results for these calculi, and they could be optimized further by implementing the optimizations described for R_4, R_{10}, and PL.

The three groups of calculi that are SAs but not RAs are the Dipole Calculus variants DRA_f (variants DRA_{fp} and DRA-connectivity are even RAs!), as well as INDU and $OPRA_m$ for $m = 1, \ldots, 8$. These calculi do not even satisfy one of the inclusions $R_4^{\supseteq}$ and $R_4^{\subseteq}$, which implies that the reasoning optimizations described in Section 3.2 for Axiom R_4 cannot be applied, but this is the only disadvantage of these calculi over the others. Our observations suggest that the meaning of the letter combination "RA" in the abbreviations "DRA" and "OPRA" should stand for "Reasoning Algebra", not for "Relation Algebra".

---

[1] For the parametrized calculi DRA, OPRA, QTC, we count every variant separately.

Table 3. Overview of calculi tested and their properties. The symbol "✓" means that the axiom is satisfied; otherwise the percentage of counterexamples (relations, pairs or triples violating the axiom) is given

Calculus	Tests[a]	R_4	SA	WA	R_6	R_{6l}	R_7	R_9	PL	R_{10}
Allen	MHS	✓	✓	✓	✓	✓	✓	✓	✓	✓
Block Algebra	HS	✓	✓	✓	✓	✓	✓	✓	✓	✓
Cardinal Direction *Calculus*	MHS	✓	✓	✓	✓	✓	✓	✓	✓	✓
CYC_b, Geometric Orientation	HS	✓	✓	✓	✓	✓	✓	✓	✓	✓
DRA_{fp}, DRA-conn.	HS	✓	✓	✓	✓	✓	✓	✓	✓	✓
Point Calculus	HS	✓	✓	✓	✓	✓	✓	✓	✓	✓
RCC-5, Dependency Calc.	MHS	✓	✓	✓	✓	✓	✓	✓	✓	✓
RCC-8, 9-Intersection	MHS	✓	✓	✓	✓	✓	✓	✓	✓	✓
$STAR_4$	HS	✓	✓	✓	✓	✓	✓	✓	✓	✓
DRA_f	MHS	19	✓	✓	✓	✓	✓	✓	✓	✓
INDU	MHS	12	✓	✓	✓	✓	✓	✓	✓	✓
$OPRA_n$, $n \leqslant 8$	MHS	21–91[b]	✓	✓	✓	✓	✓	✓	✓	✓
QTC_{Bxx}	MHS	✓	✓	✓	89–100	✓	✓	✓	✓	✓
QTC_{C21}	HS	55	✓	✓	99	99	✓	2	<1	1
QTC_{C22}	HS	79	✓	✓	99	99	✓	3	<1	1
Rectang. Direction Relations	HS	✓	✓	✓	97	92	89	66	7	52
Cardinal Direction *Relations*	HS	28	17	✓	99	99	98	12	<1	88

[a]calculus was tested by: M = [26], H = HETS, S = SparQ
[b]21%, 69%, 78%, 83%, 86%, 88%, 90%, 91% for $OPRA_n$, $n = 1, \ldots, 8$

In principle, it cannot be completely ruled out that associativity is reported to be violated due to errors in either the implementation of the respective calculus or the experimental setup. This even applies to non-violations, although it is much more likely that errors cause sporadic violations than systematic non-violations. In the case of DRA_f, INDU and $OPRA_m$, $m = 1, \ldots, 8$, the relatively high percentage of violations make implementation errors seem unlikely to be the cause. However, to obtain certainty that these calculi indeed violate R_4, one has to find concrete counterexamples and verify them using the original definition of the respective calculus. For DRA_f and INDU, this has been done in the literature [24,3]. Interestingly, the violation of associativity has been attributed to the absence of strong converse and strong composition, respectively. We remark, however, that the latter cannot be responsible because, for example, DRA_{fp} has an associative, but only weak, composition operation. While DRA_{fp} has been proven to be associative due to strong composition in [24], for $OPRA_m$, it can be shown that *none* of the variants for any m are associative (see [29]).

The B-variants of QTC violate only the identity law R_6 and R_{6l}. As observed in [26], it is possible to equip them with a new id relation, modify the interpretation of the other relations such that they become JEPD, and adapt the converse and composition table accordingly. The thus modified calculi are then relation algebras.

The C-variants of QTC additionally violate R_4, R_9, R_{10}, and PL. We call the corresponding notion of algebra semi-associative Boolean algebra with converse-involution. As a consequence, most of the reasoning optimizations described in

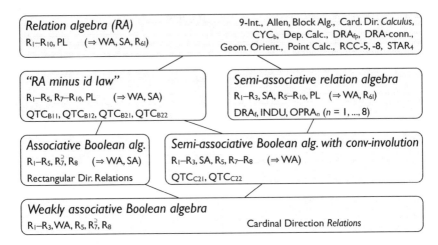

Fig. 1. Overview of algebra notions and calculi tested

Section 3.2 cannot be applied to the C-variants of QTC; hence, reasoning with these calculi is expected to be less efficient than with the calculi described so far. It is possible that the noticeably few violations of R_9, R_{10}, and PL are due to errors in the composition table; the non-trivial verification is part of future work.

Cardinal Direction Relations and Rectangular Direction Relations are the only calculi with weak converse that we have tested. The former satisfies only WA in addition to the axioms that are always satisfied by a Boolean algebra with distributivity. We call the corresponding notion of algebra weakly associative Boolean algebra. Hence, this calculus enjoys none of the advantages for representation and reasoning discussed in Section 3.2. Similarly to the C-variants of QTC, the relatively small number of violations of PL may be due to errors in the implementation. Rectangular Direction Relations additionally satisfies R_4 and therefore corresponds to what we call an associative Boolean algebra. Since both calculi satisfy neither R_7 nor R_9, current reasoning algorithms and their implementations yield incorrect results for them, as seen in Section 3.2.

An overview of the algebra notions identified is given in Figure 1.

When making use of algebraic closure as inference mechanism it is essential to acknowledge that some axiom violations require special procedures in order to compute algebraic closure. Our analysis reveals that there indeed exist calculi that do not meet axioms that have been taken for granted. For example, the current version of GQR can fail to compute algebraic closure correctly for calculi that violate R_9. In Algorithm 1 we present a universal algorithm to compute algebraic closure. For clarity and brevity of the presentation we stick to the well-known but simple control structure of PC-1. A real implementation would use an advanced control structure to avoid unnecessary invocations of the RE-VISE function, i.e., to use at least PC-2 [21]. Conformance with R_7 allows CSP storage to be restricted (flag s in the algorithm), while violation of R_9 requires two computations for the refinement operation Eq. 15, namely $C_{i,j} \diamond C_{j,k}$ and

Algorithm 1. Universal algebraic closure algorithm

```
 1: function LOOKUP((C, i, j, s))
 2:     if s ∨ (i < j) then
 3:         return C_{i,j}
 4:     else
 5:         return C_{j,i}ˇ
 6: end function

 7: function REVISE((C, i, j, k, s))
 8:     u ← false                          ▷ update flag to signal whether relation was updated
 9:     r ← C_{i,j} ∩ LOOKUP(C, i, k, s) ◇ LOOKUP(C, k, j, s)
10:     if s ∨ R₉ does not hold then
11:         r′ ← LOOKUP(C, j, i, s) ∩ (LOOKUP(C, j, k, s) ◇ LOOKUP(C, k, i, s))
12:         r ← r ∩ r′ˇ
13:         r′ ← r′ ∩ rˇ
14:         if r′ ≠ C_{j,i} then
15:             assert r′ ≠ ∅                              ▷ stop if inconsistency is detected
16:             u ← true
17:             C_{j,i} ← r′
18:     if r ≠ C_{i,j} then
19:         assert r ≠ ∅                                  ▷ stop if inconsistency is detected
20:         u ← true
21:         C_{i,j} ← r
22:     return (C, u)
23: end function

24: function A-CLOSURE((n, {x₁ r₁ y₁, . . . , x_m r_m y_m}))
25:     if R₇ does not hold then
26:         s ← true                       ▷ without R₇ we must store converse relations
27:         C_{i,j} ← U, i = 1, . . . , n, j = 1, . . . , n
28:     else
29:         s ← false  ▷ for small calculi/CSPs storing converses may be more efficient
30:         C_{i,j} ← U, i = 1, . . . , n, j = i + 1, . . . , n      ▷ use triangular matrix storage
31:     C_{i,i} ← id, i = 1, . . . , n
32:     for i = 1, . . . , m do
33:         x ← x_i, r ← r_i, y ← y_i                          ▷ process constraint x_i r_i y_i
34:         if ¬s ∧ (x > y) then
35:             (x, y) ← (y, x), r ← rˇ               ▷ only write into upper half of matrix
36:         C_{x,y} ← C_{x,y} ∩ r
37:         assert (x = y) → (id ∈ C_{x,y})
38:     end for
39:     update ← true
40:     while update do
41:         update ← false
42:         for i = 1, . . . , n, j = i+1, . . . , n, k = 1, . . . n, k ≠ i, k ≠ j do
43:             (u, C) ← REVISE(C, i, j, k, s)
44:             update ← update ∨ u
45:         end for
46:     return C                                           ▷ fix point reached
47: end function
```

$(C_{j,k}{}^\vee \diamond C_{i,j}{}^\vee)^\vee$ (lines 10–17). R_4 and R_{10} are not used by the algorithm, since this would complicate the algorithm unduly.

5   A Quantitative Account of Qualitative Calculi

In this section, we report on computational properties of specific calculi which are beyond the computational complexity of constraint-based reasoning. For example, one might be interested to know how many relations are typically sufficient to describe a scene of n objects unequivocally or with a specific residual uncertainty. To this end, we developed two empirical measures that characterize certain aspects of qualitative calculi that are arguably relevant to applications. We want to answer two questions: (1) How well do calculi with many relations make use of the usually higher information content? (2) Does information content differ significantly between the six classes of calculi established in Section 4?

The first measure we consider is *information content* of the composition operation. Our motivation is to estimate how much additional information can be gained by applying a composition operation. This allows us to estimate whether, for example, having observed relations $r(A, B)$ and $r'(B, C)$ in a scene, it is worthwhile to observe $r''(A, C)$ too, as it may be improbable to derive r'' by composition $(r \circ r')$. To obtain more general results we consider sequences of compositions $r \circ r' \circ r'' \circ \ldots$ for several lengths. We define the information content I of a relation $R \subseteq \mathsf{Rel}$ to be

$$I(R) = 1 - \frac{|R|}{|\mathsf{Rel}|} \tag{16}$$

where $|\mathsf{Rel}|$ denotes the number of base relations of the calculus, and $|R|$ the number of base relations R consists of. In case of the universal relation this results in $I(U) = 0$ as nothing is known, $I(r) = 1 - \frac{1}{|\mathsf{Rel}|}$ for all base relations $r \in \mathsf{Rel}$, and $I(\emptyset) = 1$. Obviously, the more base relations a calculus involves, the higher the information content *can be* for base relations. Therefore, we define

$$I_C^k = \frac{\sum_{R \in r^k} I(R)}{|\mathsf{Rel}|^{k+1}} \tag{17}$$

for a calculus C with $r^k = \{r^0 \circ \ldots \circ r^k | r^0, \ldots, r^k \in \mathsf{Rel}\}$ to be the average information content after k composition operations, i.e., how restrictive relations are on average after information propagation with composition. In particular, I_C^k is 1 minus the average proportion of base relations in any cell in the composition table. For example, for QTC_{C22} ($|\mathsf{Rel}| = 209$) or the Cardinal Direction Relations (CDR) ($|\mathsf{Rel}| = 218$) $I^0 \approx 1$, whereas for the Point Calculus with three base relations $I_{PC}^0 \approx 0.67$. We apply an iterative method to derive the values of I_C^k that constructs r^k for $k = 0, 1, \ldots$ rather than looping across combinations of base relations. Despite the potentially exponential size of r^k, the calculation remains feasible in many cases. Only for OPRA_m with $m \geq 3$ and some QTC variants we were not able to derive values for higher k in reasonable time.

For the other calculi, computation was terminated after 14 compositions or if I_C^k drops below 0.5.

As a second measure we determine the average degree of overlap that occurs after k steps of composition for selected calculi. The degree of overlapping $O(R_i, R_j)$ is determined by counting the number of atomic relations shared by two relations, normalized by the total number of base relations:

$$O(R_i, R_j) = \frac{|R_i \cap R_j|}{|\mathsf{Rel}|} \tag{18}$$

For example, if two relations in a calculus with eight base relations share four base relations, the overlap is 0.5. This value indicates how the information content differs between dealing with base relations only versus dealing with arbitrary relations (and thus how the results on information content generalize to arbitrary relations). Similar to $I(R)$ and I_C^k, we define O_C^k to be the average overlap over all composition chains of length k.

$$O_C^k = \frac{\sum_{R_i, R_j \in r^k} O(R_i, R_j)}{|\mathsf{Rel}|^{k+1}} \tag{19}$$

The results of the two measures are summarized in Figure 2 and Table 4, showing information content versus length k of composition chains.

Figure 2 shows that the average information content for the Point Calculus after 1 step is ≈ 0.52 and additionally, the overlap of ≈ 0.33 is already quite high after a single composition. Therefore, in order to obtain detailed information it is reasonable to also observe r_{AC} between objects A and C even if r_{AB} and r_{BC} are already known. By contrast, the INDU calculus has a very high information

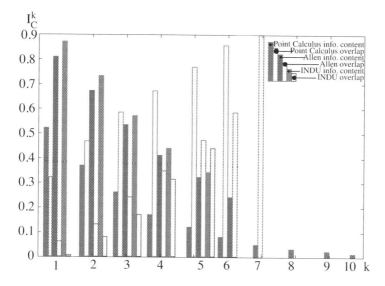

Fig. 2. Information content and overlap after k compositions for selected calculi

Table 4. Information content I_C^k for calculi in %

Calculus	0	1	2	3	4	5	6	7	8	9	10	11	12	13	14
Allen	92.3	81.4	66.8	52.8	41.1	31.8	24.5	18.9	14.5	11.2	8.6	6.6	5.1	3.9	3.0
Block Algebra	99.4	96.5	89.0	77.7	65.3	53.4	43.0	34.1	27.0	21.1	16.4	12.8	9.9	7.7	5.9
CDC	88.9	76.8	60.4	44.5	31.6	21.9	14.9	10.1	6.8	4.6	3.1	2.0	1.4	0.9	0.6
CYC_b	75.0	62.5	46.9	32.8	21.9	14.1	8.8	5.4	3.2	1.9	1.1	0.6	0.4		
DRA_{fp}	98.8	89.9	69.0	45.0	25.8	13.4	6.5	3.0	1.3	0.6	0.2				
DRA-con	85.7	74.6	59.0	43.4	30.4	20.5	13.5	8.7	5.6	3.5	2.2	1.3	0.8	0.5	0.3
Point Calculus	66.7	51.9	37.0	25.5	17.3	11.6	7.8	5.2	3.5	2.3	1.5	1.0	0.7	0.5	
RCC-5	80.0	56.8	34.9	19.7	10.6	5.5	2.7	1.3	0.6	0.3					
RCC-8	87.5	62.3	38.0	21.1	11.0	5.5	2.6	1.2	0.6	0.3					
$STAR_4$	88.9	66.9	45.0	28.5	17.4	10.3	6.0	3.5	2.0	1.1	0.6	0.4			
DRA_f	98.6	90.6	70.4	46.3	26.7	13.9	6.7	3.0	1.3	0.6	0.2				
INDU	96.0	86.9	72.5	57.5	44.1	33.2	24.7	18.2	13.4	9.9	7.2	5.3	4.0	2.9	2.1
$OPRA_1$	95.0	82.0	55.8	30.8	14.5	6.2	2.4	0.9	0.3						
$OPRA_2$	98.6	90.3	64.1	32.9	13.0	4.3	1.3	0.3							
$OPRA_3$	99.4	93.1	71.4	40.2	16.7	5.6									
$OPRA_4$	99.6	94.6	76.7	48.0											
QTC_{B11}	88.9	90.0	93.2	95.8	97.5	98.6	99.1	99.5	99.7						
QTC_{B12}	94.1	91.2	90.5	91.3	92.8	94.2	95.6								
QTC_{B21}	88.9	0.0	0.0	0.0	0.0	0.0	0.0	0.0	0.0	0.0	0.0	0.0	0.0	0.0	0.0
QTC_{B22}	96.3	51.9	37.0	25.5	17.3	11.6	7.8	5.2	3.5	2.3	1.5	1.0	0.7	0.5	0.3
QTC_{C21}	98.8	92.5	76.6	68.6	69.5	73.0	76.5	79.4	81.8	83.7	85.2	86.4	87.4		
QTC_{C22}	99.5	95.1	78.0	69.3	51.2										
RDR	97.2	82.6	63.2	45.7	32.0	22.0	15.0	10.1	6.8	4.6	3.1	2.0	1.4	0.9	0.6
CDR	99.5	78.8	60.9	48.9	39.6	32.1	26.1	21.2	17.2	14.0	11.4	9.3	7.6	6.2	5.1

content (≈ 0.87) and a much smaller overlap. Therefore, it is not so informative to observe r_{AC} as a lot of information is preserved after a composition. It is clear that the O_C^k grows for increasing k as composition results become coarser step by step. Nevertheless, information loss for PC is much higher than compared to Allen and INDU calculus: I_{INDU}^5 and I_{Allen}^5 are close to I_{PC}^2 (O_C^k respectively).

Our results show that there is no evidence for a relation between the information content of a calculus and its classification as per Figure 1. The only exceptions are some of the QTC calculi as I_C^k starts to increase after some k with increasing k.

Although the calculi start with quite different values for I^0, most calculi have an information content less than 0.1 after six steps. The most notable exception is the Block Algebra where $I_{BA}^6 \approx 0.43$ and even after ten compositions it remains above 16%. Only Allen, INDU and CDR are somehow comparable. Concerning the classes we derived in Section 4 no uniform behavior can be observed. Thus, from a perspective of expressive power of calculi, there is no argument against working with calculi that are not relation algebras. We have to note that the comparison of the values for calculi where it is known that a-closure decides consistency and those where it does not (or is unknown) is problematic. The latter ones may contain relations which are not physically realizable and thus reduce the value of information content.

There are some interesting observations wrt. the various QTC variants. The QTC_{B1x}, QTC_{B21} and QTC_{C21} calculi behave differently from other calculi, whereas QTC_{B22} behaves 'normally', i.e., I^k increases, although it is very closely related to the other QTC variants. Interestingly, QTC_{B1x} and QTC_{C21} are the only calculi where I^k increases with growing k. From our perspective, the reason lies in the multimodal structure of the calculus. As it combines points with line segments, the composition table (CT) contains empty relations, since an object cannot be interpreted as a point and a line segment at the same time. Additionally, the CT contains only fairly small relations, i.e., with small $|R|$. For example, the CT of QTC_{B12} contains 29% empty relations, 29% atomic relations, and 42% other relations which have a maximal size of $|R| \leq 3$. The results for QTC_{B21} are not surprising as the composition table only contains the universal relation and thus, for all k, $I^k = 0.0$ and $O^k = 100.0$. For QTC_{C21} we observe that I^k decreases to 69.5% at step 4, but starts to increase for $k \geq 5$. We assume that this is also the case for QTC_{C22}, as it is a refinement of QTC_{C21}, but we were not able to calculate necessary values due to the high complexity. So far, we have no explanation for this decrease.

An additional observation is that PC and QTC_{B22} are similar with respect to information content, i.e., $I^k_{PC} \approx I^k_{QTC_{B22}}$ for $k \leq 14$. This congruence is interesting as the overlap values vary, the underlying partition scheme is different and the difference in base relations is significant (three for PC vs. 27 for QTC_{B22}). We leave the question of connections between these two calculi for future research.

6   Conclusion

We have looked at spatio-temporal representation and reasoning from an algebraic perspective, examining the implications of algebraic properties on modeling and reasoning algorithms, and testing these properties for a representative corpus of existing calculi. The resulting classification shows that calculi which have been described early in the literature tend to reside in the upper part of Figure 1; that is, they tend to have a rich algebraic structure. Few more recently developed calculi are based on generalizations and have a weaker structure. We have been able to conclude that common reasoning procedures are incorrect for the latter class of calculi, and have proposed a corrected universal a-closure algorithm that makes use of reasoning optimizations where they are allowed. Furthermore, we found that algebraic properties do not necessarily relate to how much information is preserved in successive reasoning steps.

An interesting and significant line of future work is to extend this study to ternary calculi, which requires an extension of binary relation algebras to ternary relations, see also [37].

Acknowledgments. We would like to thank Immo Colonius, Arne Kreutzmann, Jae Hee Lee, André Scholz and Jasper van de Ven for inspiring discussions during the "spatial reasoning tea time". This work has been supported by the DFG-funded SFB/TR 8 "Spatial Cognition", projects R3-[QShape] and R4-[LogoSpace].

References

1. Allen, J.F.: Maintaining knowledge about temporal intervals. Communications of the ACM 26(11), 832–843 (1983)
2. Balbiani, P., Condotta, J., Fariñas del Cerro, L.: Tractability results in the block algebra. J. Log. Comput. 12(5), 885–909 (2002)
3. Balbiani, P., Condotta, J., Ligozat, G.: On the consistency problem for the INDU calculus. J. Applied Logic 4(2), 119–140 (2006)
4. Cohn, A., Renz, J.: Qualitative spatial representation and reasoning. In: van Harmelen, F., Lifschitz, V., Porter, B. (eds.) Handbook of Knowledge Representation, ch. 13, pp. 551–596. Elsevier (2008)
5. Dechter, R.: Constraint processing. Elsevier Morgan Kaufmann (2003)
6. Düntsch, I.: Relation algebras and their application in temporal and spatial reasoning. Artif. Intell. Rev. 23(4), 315–357 (2005)
7. Dylla, F., Lee, J.H.: A combined calculus on orientation with composition based on geometric properties. In: ECAI 2010. pp. 1087–1088 (2010)
8. Dylla, F., Mossakowski, T., Schneider, T., Wolter, D.: Algebraic properties of qualitative spatio-temporal calculi. Tech. rep., University of Bremen, Cognitive Systems (2013), http://arxiv.org/abs/1305.7345
9. Egenhofer, M.: Reasoning about binary topological relations. In: Günther, O., Schek, H.-J. (eds.) SSD 1991. LNCS, vol. 525, pp. 143–160. Springer, Heidelberg (1991)
10. Frank, A.: Qualitative spatial reasoning with cardinal directions. In: Proc. of ÖGAI 1991. Informatik-Fachberichte, vol. 287, pp. 157–167. Springer (1991)
11. Gantner, Z., Westphal, M., Wölfl, S.: GQR - A Fast Reasoner for Binary Qualitative Constraint Calculi. In: Proc. of the AAAI 2008 Workshop on Spatial and Temporal Reasoning (2008)
12. Grigni, M., Papadias, D., Papadimitriou, C.H.: Topological inference. In: Proc. of IJCAI 1995 (1), pp. 901–907. Morgan Kaufmann (1995)
13. Hirsch, R., Hodkinson, I.: Relation algebras by games, Studies in logic and the foundations of mathematics, vol. 147. Elsevier (2002)
14. Isli, A., Cohn, A.: A new approach to cyclic ordering of 2D orientations using ternary relation algebras. Artif. Intell. 122(1-2), 137–187 (2000)
15. Jonsson, P., Drakengren, T.: A complete classification of tractability in RCC-5. J. Artif. Intell. Res (JAIR) 6, 211–221 (1997)
16. Kontchakov, R., Pratt-Hartmann, I., Wolter, F., Zakharyaschev, M.: Spatial logics with connectedness predicates. Log. Meth. Comp. Sci. 6(3) (2010)
17. Ligozat, G.: Reasoning about cardinal directions. J. Vis. Lang. Comput. 9(1), 23–44 (1998)
18. Ligozat, G.: Categorical methods in qualitative reasoning: The case for weak representations. In: Cohn, A.G., Mark, D.M. (eds.) COSIT 2005. LNCS, vol. 3693, pp. 265–282. Springer, Heidelberg (2005)
19. Ligozat, G.: Qualitative Spatial and Temporal Reasoning. Wiley (2011)

20. Ligozat, G., Renz, J.: What is a qualitative calculus? A general framework. In: Zhang, C., Guesgen, H.W., Yeap, W.-K. (eds.) PRICAI 2004. LNCS (LNAI), vol. 3157, pp. 53–64. Springer, Heidelberg (2004)
21. Mackworth, A.K.: Consistency in networks of relations. Artif. Intell. 8, 99–118 (1977)
22. Maddux, R.: Relation algebras, Studies in logic and the foundations of mathematics, vol. 150. Elsevier (2006)
23. Moratz, R.: Representing Relative Direction as a Binary Relation of Oriented Points. In: Proc. of ECAI 2006. pp. 407–411. IOS Press (2006)
24. Moratz, R., Lücke, D., Mossakowski, T.: A condensed semantics for qualitative spatial reasoning about oriented straight line segments. Artif. Intell. 175, 2099–2127 (2011), http://dx.doi.org/10.1016/j.artint.2011.07.004
25. Moratz, R., Renz, J., Wolter, D.: Qualitative spatial reasoning about line segments. In: Proc. of ECAI 2000. pp. 234–238. IOS Press (2000)
26. Mossakowski, F.: Algebraische Eigenschaften qualitativer Constraint-Kalküle. Diplom thesis, Dept. of Comput. Science, University of Bremen (2007) (in German)
27. Mossakowski, T., Maeder, C., Lüttich, K.: The heterogeneous tool set, HETS. In: Grumberg, O., Huth, M. (eds.) TACAS 2007. LNCS, vol. 4424, pp. 519–522. Springer, Heidelberg (2007)
28. Mossakowski, T., Moratz, R.: Qualitative reasoning about relative direction of oriented points. Artif. Intell. 180-181, 34–45 (2012), http://dx.doi.org/10.1016/j.artint.2011.10.003
29. Mossakowski, T., Lücke, D., Moratz, R.: Relations between spatial calculi about directions and orientations. Technical report, University of Bremen
30. Navarrete, I., Morales, A., Sciavicco, G., Cárdenas-Viedma, M.: Spatial reasoning with rectangular cardinal relations – the convex tractable subalgebra. Ann. Math. Artif. Intell. 67(1), 31–70 (2013)
31. Nebel, B., Scivos, A.: Formal properties of constraint calculi for qualitative spatial reasoning. KI 16(4), 14–18 (2002)
32. Pujari, A.K., Sattar, A.: A new framework for reasoning about points, intervals and durations. In: Proc. of IJCAI 1999. pp. 1259–1267 (1999)
33. Ragni, M., Scivos, A.: Dependency calculus: Reasoning in a general point relation algebra. In: Furbach, U. (ed.) KI 2005. LNCS (LNAI), vol. 3698, pp. 49–63. Springer, Heidelberg (2005)
34. Randell, D.A., Cui, Z., Cohn, A.G.: A spatial logic based on regions and "Connection". In: Proc. of KR 1992, pp. 165–176 (1992)
35. Renz, J.: Qualitative Spatial Reasoning with Topological Information. LNCS (LNAI), vol. 2293. Springer, Heidelberg (2002)
36. Renz, J., Mitra, D.: Qualitative direction calculi with arbitrary granularity. In: Zhang, C., Guesgen, H.W., Yeap, W.-K. (eds.) PRICAI 2004. LNCS (LNAI), vol. 3157, pp. 65–74. Springer, Heidelberg (2004)
37. Scivos, A.: Einführung in eine Theorie der ternären RST-Kalküle für qualitatives räumliches Schließen. Diplom thesis, University of Freiburg (2000) (in German)
38. Skiadopoulos, S., Koubarakis, M.: Composing cardinal direction relations. Artif. Intell. 152(2), 143–171 (2004)
39. Skiadopoulos, S., Koubarakis, M.: On the consistency of cardinal direction constraints. Artif. Intell. 163(1), 91–135 (2005)
40. Van de Weghe, N.: Representing and Reasoning about Moving Objects: A Qualitative Approach. Ph.D. thesis, Ghent University (2004)

41. Van de Weghe, N., Kuijpers, B., Bogaert, P., De Maeyer, P.: A qualitative trajectory calculus and the composition of its relations. In: Rodríguez, M.A., Cruz, I., Levashkin, S., Egenhofer, M. (eds.) GeoS 2005. LNCS, vol. 3799, pp. 60–76. Springer, Heidelberg (2005)

42. Vilain, M., Kautz, H., van Beek, P.: Constraint propagation algorithms for temporal reasoning: a revised report. In: Readings in Qualitative Reasoning about Physical Systems, pp. 373–381. Morgan Kaufmann (1989)

43. Wallgrün, J.O., Frommberger, L., Wolter, D., Dylla, F., Freksa, C.: Qualitative spatial representation and reasoning in the sparQ-toolbox. In: Barkowsky, T., Knauff, M., Ligozat, G., Montello, D.R. (eds.) Spatial Cognition 2007. LNCS (LNAI), vol. 4387, pp. 39–58. Springer, Heidelberg (2007)

44. Wallgrün, J.O., Frommberger, L., Dylla, F., Wolter, D.: SparQ User Manual V0.7. User manual, University of Bremen (January 2009)

45. Wallgrün, J.O., Wolter, D., Richter, K.F.: Qualitative matching of spatial information. In: Proceedings of ACM GIS (2010)

46. Wolter, D., Lee, J.H.: Qualitative reasoning with directional relations. Artificial Intelligence 174(18), 1498–1507 (2010)

Author Index